Simulation and Optimization of Digital Circuits

Vazgen Melikyan

Simulation and Optimization of Digital Circuits

Considering and Mitigating Destabilizing Factors

Springer

Vazgen Melikyan
Director of Educational Department
Synopsys Armenia CJSC
Yerevan, Armenia

ISBN 978-3-319-71636-7 ISBN 978-3-319-71637-4 (eBook)
https://doi.org/10.1007/978-3-319-71637-4

Library of Congress Control Number: 2018932032

Printed on acid-free paper

This Springer imprint is published by the registered company Springer International Publishing AG part of Springer Nature.
The registered company address is: Gewerbestrasse 11, 6330 Cham, Switzerland

Dedicated to the bright memory of my parents.
Shavarsh Melikyan and Siranush Hakobyan.

Preface

The book systematically expounds the main results obtained by the author in the field of gate-level simulation and optimization of digital ICs, with consideration of destabilizing factors (radiation exposure, ambient temperature, nonideality of the power source, input signals, interconnects, power rails, etc.).

The basis of most methods, models, and algorithms of gate-level simulation of digital ICs proposed in the monograph is to provide a practical compromise between the accuracy of the results obtained and the costs of the required computer resources.

The book is anticipated for scientists and engineers specializing in the field of IC simulation, as well as for students and postgraduate students studying the disciplines related to IC design.

Yerevan, Armenia Melikyan V. Sh.

Introduction

Designing modern digital integrated circuits (IC) is impossible without the use of effective electronic design automation (EDA) tools. High rates of the development of IC fabrication technologies constantly lead to the necessity to update EDA tools, including systems of gate-level simulation and optimization of digital ICs.

Owing to the success in microelectronics, leading companies-manufacturers such as Intel, Taiwan Semiconductor Manufacturing Company (TSMC), Samsung, and Global Foundries (GF) have already developed 16-, 14-, and even 7-nanometer technologies of IC fabrication. At present, already produced ICs contain tens of billions of active devices on a single die with an area up to hundreds of mm^2. Taking into account the planned transition to 5-nanometer technology in the near future and increasing the crystal size to a thousand square millimeters, the number of circuit elements will soon increase another several times. Traditional systems of circuit-level simulation based on models in the form of systems of ordinary differential equations are not suitable for an entire simulation of modern ICs due to unacceptably high costs of machine resources (computing time and memory size). The way out is the use of either pure gate-level simulation in case of digital ICs or a multilevel mixed-mode simulation that uses different models and algorithms of fundamentally different nature (circuit, logic, etc.) for different IC parts in case of digital and mixed-signal ICs. In case of mixed-mode analysis, the role of gate-level simulation is also extremely important, because it is just due to its application that the dimensionality of the problem is reduced.

Thus, gate-level simulation has become the main tool for the analysis of modern digital ICs and digital fragments of ICs. However, the tasks assigned to gate-level simulation tools have now changed radically. Traditional gate-level simulation was focused on the qualitative reflection of the processes occurring in digital circuits and was based on the apparatus of Boolean algebra, and high speed of simulation was achieved precisely due to the simplified representation of models of logic gates. For detailed simulation, circuit-level simulation tools were used. As a result of fundamental change of gate-level simulation role, a rather detailed imitation of signals just at the logical level of abstraction is required, and not at the expense of the main advantage—the speed of simulation. Otherwise, the dimensionality issue of simulating circuits will be not solved.

Thus, an extreme need arises for the development of digital circuit simulation systems that simultaneously possess the speed, inherent in gate-level simulation and accuracy, close to circuit-level simulation.

Existing gate-level simulation tools are not able to solve this problem, since at the time of their creation the ratios of various physical phenomena were fundamentally different. Because of the unjustifiably low accuracy of existing gate-level models, in most cases they do not even provide qualitatively accurate results of simulation of modern digital ICs.

A series of reasons exist that make it difficult to apply the available methods of gate-level simulation in the real IC design practice. For example, with the transition to 90-nanometer and newer technologies, the leakage currents of transistors, ignored by existing simulation programs, already have a dominant role. In the development of modern high-speed ICs, operating in the range of tens of gigahertz, the role of parasitic inductances changes radically. There are many other examples of this kind. However, among all reasons, nowadays the dominant is the absence of the possibility of considering the influence of various external (ambient temperature, radiation, etc.) and internal (delay and distortion of the waveform when passing through the interconnect, crosstalk, IR drop, etc.) destabilizing factors (DF) on IC operation in existing gate-level simulation tools. The fact is that due to the success in the development of microelectronics, the ratio of the parameters of useful and parasitic signals has changed qualitatively. If, for example, the amplitude of logic gate signals in recent years has decreased from a few to a part of volts, the magnitude of the changes in the output signals of logic gates due to fluctuations in ambient temperature or radiation exposure has not only not decreased but also increased. The latter is due to the gradual strengthening of requirements in relation to the ranges of external DF, under which the designed ICs should remain operational. As a result, the influence of external DF became commensurable, and sometimes exceeded their nominal values. If earlier it was possible to ignore the changes in the levels of logical signals due to the mentioned external DF, at present this in most cases simply leads to qualitatively inaccurate simulation results. A similar situation occurs with internal DFs. For example, the ratio of the delay on the logic gate and on the interconnect has changed from 100:1 sometimes to 1:100 over the last decade, and it's impossible to ignore this factor in simulating circuits. Approximately the same situation is with other external and internal DFs. Therefore, the existing means of gate-level simulation of digital circuits, created at times when the role of DF was insignificant, are not suitable for analysis of modern ICs and need either fundamental restructuring or replacement with tools built by new principles. This also refers to gate-level simulation subsystems, embedded in systems of mixed-mode multilevel simulation. Therefore, it is necessary to develop gate-level simulation means, used autonomously or as part of mixed-mode simulators, since they are not adapted to consider DF effects.

The development of gate-level simulation and optimization tools with consideration of DFs has now become a decisive part in the progress of design automation of electronic circuits. Therefore, intensive work is underway to create fundamentally new gate-level simulation approaches and optimize digital ICs, allowing with the inherent speed of gate-level simulation and accuracy close to circuit-level simulation to reproduce the operation of digital ICs taking DFs into account.

However, despite the small progress, the successes in this direction are more than modest. There is no general theoretical approach to solve different aspects of the problem, which significantly reduces the effectiveness of the developed methods. Attempts are being made to solve the problem by partially using simpler but having nonlogical (often circuit-level) nature of models, which leads to disadvantages that are unacceptable from the viewpoint of design practice. There is no clear classification of DFs and their parameters. The main causes (e.g., the discrete nature of variables) of the inapplicability of existing gate-level simulation systems to consider DF effects are not eliminated.

There are a number of unresolved problems in both parametric and structural optimization of DCs, and there are simply no means of considering DFs in them.

Gate-level simulation and optimization of DCs with consideration of DFs requires a fundamentally new approach to develop such class of EDA tools.

This monograph is devoted to the description of the developed new principles for constructing such systems.

The author expresses deep gratitude to his teachers in science:

- Archangelsky Alexei Yakovlevich,
- Agakhanyan Tatevos Mamikonovich,
- Norenkov Igor Petrovich,
- Rusakov Sergey Grigorievich.

The author would also like to thank his colleague Ruzanna Goroyan who assisted in the preparation of this monograph.

Contents

The original version of the book backmatter was revised: References has been added.

Chapter 1
General Issues of Gate-Level Simulation and Optimization of Digital Circuits with Consideration of Destabilizing Factors

1.1 Significance of DF Influence on Functioning of Modern Digital Circuits

Digital circuits (digital ICs, digital fragments of mixed-signal ICs and systems based on them) [1, 2] operate in the environment of different destabilizing factors (DFs) [3–15], which significantly influence their behavior and often disrupt the normal operation [16–19].

According to the place of application, DFs can be classified into external and internal. Modern digital circuits function in the environment shown in Fig. 1.1 where $V_i(t)$, $_{i\,=\,1,2,\ldots,\,k}$-external, and $W_j(t)$, $_{j\,=\,1,\,2,\ldots,\,l}$-internal DFs, and k and l are the numbers of different types of DFs.

The degree of influence of different DFs on the functioning of digital circuits is different. Studies [2, 4, 20–40] show that at the present stage of the development of circuitry and the technology of fabricating digital circuits, the most significant effect is the influence of DF, given in Table 1.1.

Features of the presented DF types from the point of view of their significance on the distortion of circuit signals are shown below.

1.1.1 External DF

In this section the significance, mechanisms of the influence of radiation (V_1) on the operation of digital circuits, and the methods used to reduce the effect of DF of this type are relatively presented in detail. For the other types of external DFs, only the issue of their significance on the operation of digital ICs is considered, since the mechanisms and methods of combating them are well known.

The issue of the influence of radiation on the behavior of digital circuits has been studied by many groups of authors, among which the works of [3, 4, 16, 17, 20, 21,

V. Melikyan, *Simulation and Optimization of Digital Circuits*,
https://doi.org/10.1007/978-3-319-71637-4_1

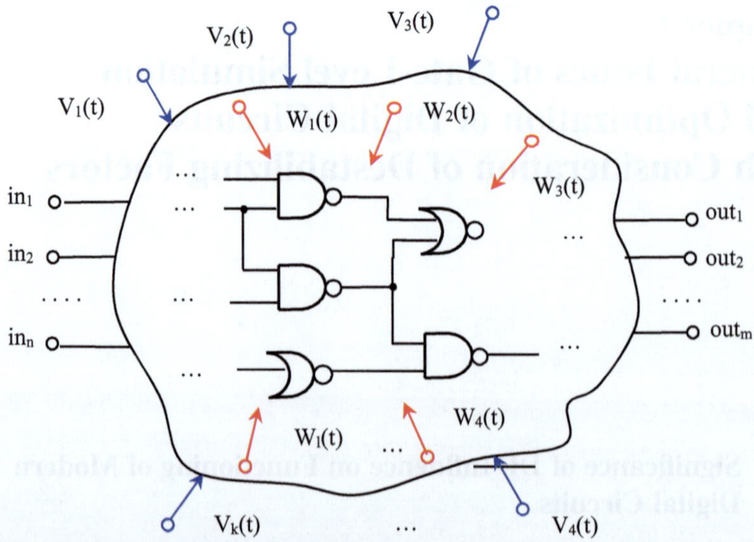

Fig. 1.1 Environment of functioning of a digital circuit

Table 1.1 Classification of DF

According to the place of application of DF	According to the nature of DF	Symbol
External	Radiation exposure	V_1
	Ambient temperature	V_2
	Non-ideality of power source	V_3
	Non-ideality of input signal source	V_4
	Non-ideality of load	V_5
	Electromagnetic field	V_6
	Mechanical impacts (vibration, strike, acceleration, etc.)	V_7
	Climatic conditions (humidity, atmospheric pressure, etc.)	V_8
Internal	Non-ideality of interconnects	W_1
	Non-ideality of power rails	W_2
	Features of the internal structure of the circuit (number of loads, etc.)	W_3
	Distorted input signal form of the element	W_4

41–58] are particularly distinguished by the depth of the study of the significance of the effect of this type of DF on the operation of digital ICs.

Digital circuits function in the radiation environment and are subject to its influence [3, 17, 20, 21, 41, 42, 46, 48, 49, 59, 60], which occurs as follows: the charged particle passes through the substance of digital IC, loses energy in the interaction with an atomic lattice, the energy lost by the charged particle is

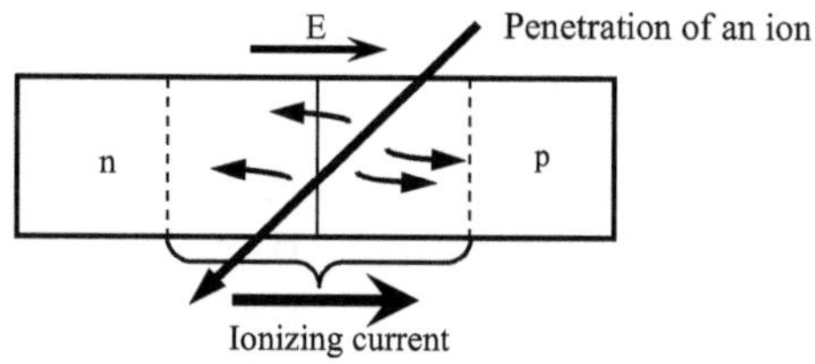

Fig. 1.2 The radiation effect on the p-n junction

transferred to the bound electrons, as a result of which ionization occurs, new electron-hole pairs are formed [61, 62].

Thus, particles of ionizing radiation passing through the components of digital IC interact with the atoms of IC substance and transmit their energy to them. The further propagation of this energy in the volume of IC leads to different radiation effects.

These effects are manifested [41, 43, 48] in the form of changes in the parameters and characteristics of separate semiconductor devices of digital circuits (transistors, diodes, passive components, etc.) (Fig. 1.2) [61]. As a result, the integrated parameters and characteristics of entire digital circuits change. The behavior of digital circuits in processing the same input signals depends on the parameters of surrounding radiation environment. With the development of IC manufacturing technology, the role of the influence of radiation on the behavior of digital circuits is continuously increasing [3, 20, 41, 47]. This is due to the fact that over time, the geometric sizes of semiconductor devices of digital ICs and, as a result, capacities decrease [1, 2, 31, 63–66]. At the same time, the values of the supply voltage of digital circuits and the power consumption are also reduced [67–70]. Due to the noted reasons for storing useful information, less charge or current is required. However, in this case, the digital circuit becomes more sensitive to radiation, since the ratio of the induced charge of the ionizing particle to the charge necessary for storing useful information increases [21, 41, 48]. Thus, ionizing particles with lower energy, which could previously be ignored by digital circuits, can cause radiation damage and failures in modern digital ICs.

Table 1.2 shows the classification of the most significant radiative effects for digital circuits, based on generalization of research results [3, 16, 20, 41, 42, 48–50, 52, 55, 58, 61].

According to [41, 47, 48, 62], during radiation of digital IC, the characteristics of their active and passive devices change, which, in turn, leads to failures of ICs of two types:

- Long-term (residual)
- Short-term (transitional)

These two groups differ from each other in the nature of the processes occurring in the circuit, in terms of their duration, as well as in parameters characterizing the size of their influence on digital circuits.

Long-term radiation damage in digital ICs is caused by displacement and ionization effects [3, 41, 48, 52]. Due to displacement effects, the lifetime, mobility and carrier concentrations change, leading to a change in the parameters of electronic

Table 1.2 Classification of radiation effects, important for digital circuits

Classification 1	Kind of radiation effect	Classification 2	Kind of radiation effect
By the nature of physical processes	Displacement (displacement of atoms in the crystal lattice, leading to radiation defects)		
	Ionization (ionization of substrate by radiation, i.e. formation of free charge carriers)		
By the nature of power flux	Uniform		
	Heterogeneous		
	Equilibrium		
	Nonequilibrium		
Due to occurrence	Primary, caused directly by radiation energy that is absorbed by a digital circuit (displacement defects, conductivity modulation, etc.)		
	Secondary, caused by the auxiliary radiation, redistribution of energy (radiative latching, secondary photocurrent, breakdown, etc.)		
By the ratio between the duration of the exposure to radiation T_i and the relaxation time T_p	Residual, long-term		
	Transitional, short-term	On the consequences of transient ionization currents	Recoverable
			Non-recoverable
			Destructive

components of digital ICs. Because of ionization, charges in dielectric layers accumulate (Fig. 1.3) [61], which also lead to a change in the parameters of components of digital circuits.

For example, the accumulation of a positive charge in a dielectric can lead to the formation of inverse channels in p regions of bipolar transistors that cause the occurrence of uncontrolled current components [3, 41, 48], and the change in the parameters of CMOS transistors is mainly associated with the formation of a positive charge in the layer of the gate dielectric (Fig. 1.4) [62], modeling the channel conductivity of a given type of transistor [20].

This, in turn, causes a change in the threshold voltage of the transistor, occurrence of parasitic channels, changes in carrier mobility, etc. These phenomena lead to a change in the parameters of the device, which, in turn, can cause its functional damage. As seen from Fig. 1.5 [56], which shows the dependence of the drain current of the n-MOS transistor on the gate-source voltage in the presence and absence of radiation, the increase of the pre-threshold currents is obvious. As a result of irradiation, the leakage currents of the transistor can become commensurate

Fig. 1.3 An example of distribution of charges in a dielectric layer

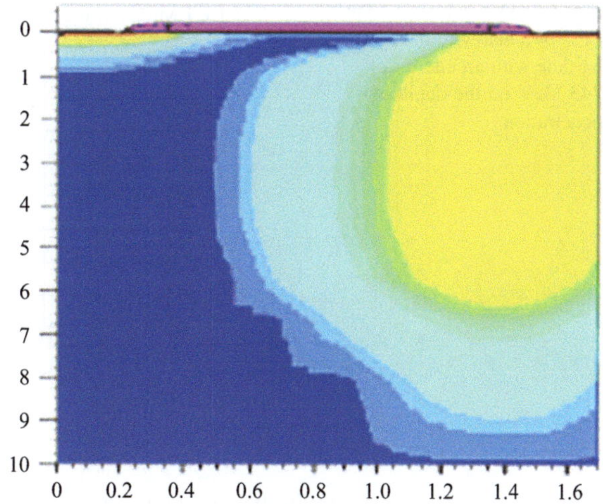

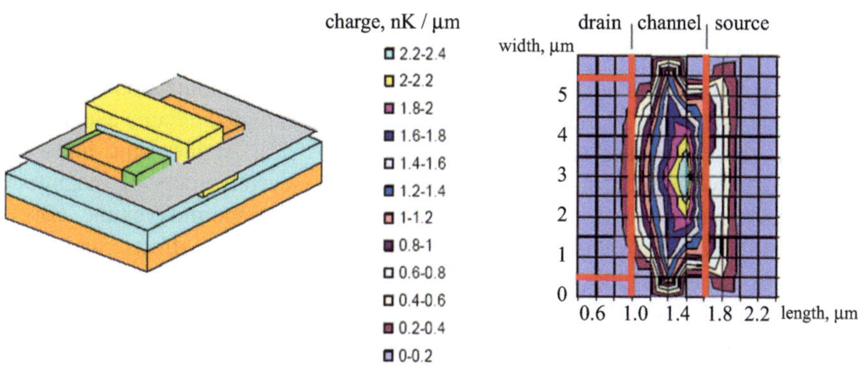

Fig. 1.4 An example of the distribution of charges in a layer of a gate dielectric of an MOS transistor

Fig. 1.5 Dependence of the drain current of the n-MOS transistor on the gate-source voltage in the presence and absence of radiation (technology—0.7 μm, *W*/*L* = 2000/1.5)

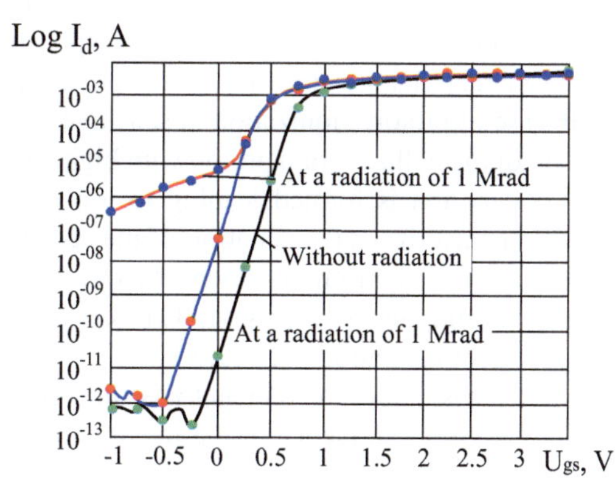

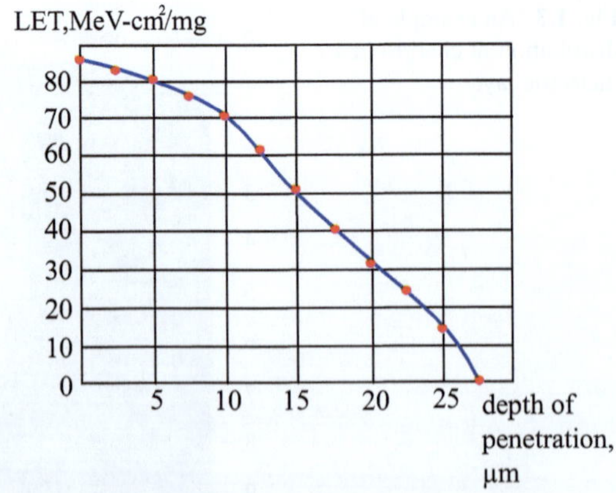

Fig. 1.6 Dependence of the LET of a heavy ionizing particle with an energy of 345 MeV on the depth of penetration

with the operating currents. In the study and calculation of residual radiation effects, all components of the radiation environment (heavy ions, protons, etc.) are represented by unit dose (rad), and the summary dose is used as the key calculation parameter [41].

Short-term radiation effects are caused by the appearance of ionization currents in the volume of digital ICs, causing a change in the state of digital circuits with a short duration, and sometimes causing damage to it [47, 53]. At the same time, the minimum power of the absorbed radiation dose is several orders of magnitude lower than that required for digital IC failures due to thermodynamic effects [41, 47, 48]. The environment that generates transient radiative effects can consist of particles of various types (protons, heavy ions, etc.). Since the forms of interaction of these particles with the IC substance have different character, the models for describing the transient radiative effects for them are also different. Heavy ions, penetrating the sensitive areas of digital ICs, lead to direct ionization.

Generation of one electron-hole pair in a semiconductor requires about 3.6 eV of energy, while an ionizing particle can have energy up to 4 ... 9 MeV [61].

For this reason, the linear energy transfer (LET), measured in MeV-cm^2/mg [41, 62] and decreasing as a function of penetration depth (Fig. 1.6) [61], is used as the characterizing parameter of the radiation environment with heavy ions.

In case of protons, ionization in sensitive regions is mainly formed indirectly. As a result of the collision of a proton with high energy, with the substance of digital IC, a secondary ion is formed leading to ionization. In this case, the proton energy, which characterizes the excitation of the secondary ion, is used as the main parameter of the radiation environment. In rare cases, when the LET threshold for the electronic component of a digital IC is too low (LET <1 MeV-cm^2/mg), protons can form a direct ionization [41, 42, 61].

In such cases, it is convenient to use the proton LET as the determining parameter of the size of the influence of the radiation environment. Depending on the type of particle, its energy, and the type of semiconductor device exposed, transient radiation effects can lead to various disruptions in the operation of the digital IC. These violations by nature are divided into three groups [4, 41, 48, 62]:

1. Recoverable (light) failures. Radiation induces a fault in one bit of a digital circuit. Failures of this type can be fixed by resetting the system or by overwriting the data in digital circuits. This category includes:

   – Single switching, when particle with high energy, passing through the sensitive area of the cell (for example, memory cells), forms accumulation of charges, leading to switching of the state of the cell (memory cell). In the static CMOS memory cells, which are two inverters covered by feedbacks (Fig. 1.7a), the sensing nodes are the drains of transistors (Fig. 1.7b). If the LET of the penetrating particle exceeds the critical value for this cell, this results in a single switching in the device (Fig. 1.7c).

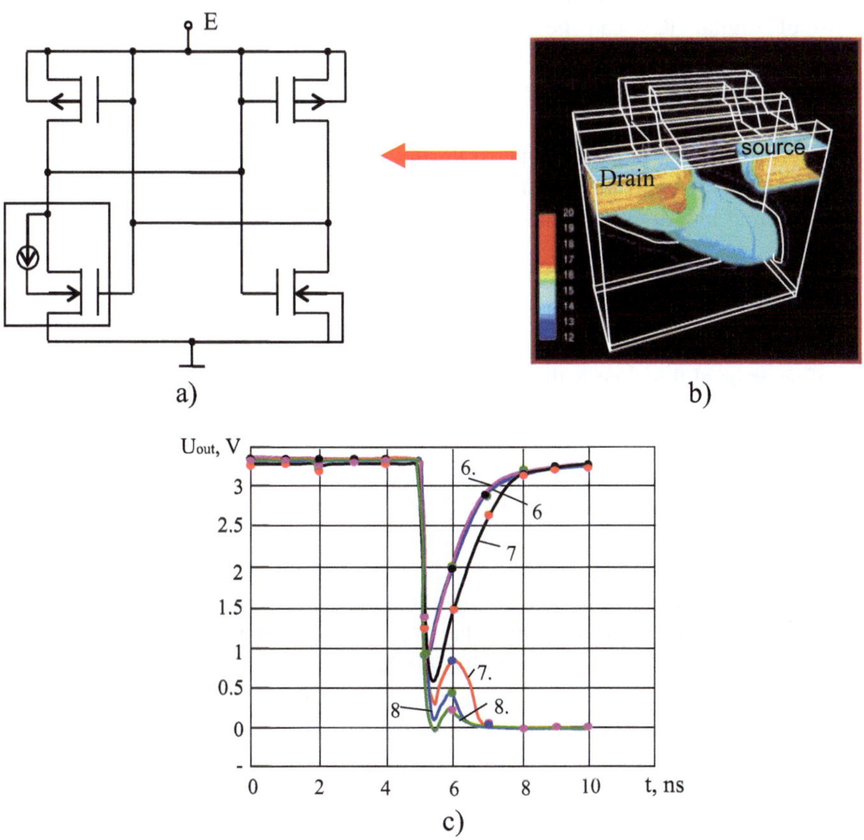

Fig. 1.7 Single switching of the static CMOS cell (**a**) the scheme of the affected static CMOS cell; (**b**) accumulated charge in the drain area of CMOS transistor; (**c**) the output voltage of the static CMOS cell at different energy values of ionizing particle (6 … 8.5 MeV)

Table 1.3 Classification of single recoverable errors by device types and their sensitive areas

Device type	Sensitive area	Type of recoverable failure
Memory	Memory cells	A bit switch
	Control device	Switching the bit into sequential and false short-term pulses in combinational parts
Combinational logic	Combinational logic	False short-time pulses
Sequential logic	Sequential logic	A bit switch
Microprocessors	Registers, cache memory, sequential logic, control device	A bit switch
	Combinational logic	False short-time pulses

- Multi-bit switching, having a similar mechanism, but leading to simultaneous switching of several bits of the device. Charge accumulation in various nodes of digital ICs can be formed by means of one or several particles simultaneously.
- A single functional failure, which is a special case of a single switching. In complex memory devices, memory cells and external circuits are connected to other devices, designed to perform additional functions, for example, to detect and fix errors. If an ionizing particle affects the circuit of this type, then the normal functioning of the entire cell is disrupted.

Table 1.3 shows the classification of single recoverable errors by device types and their sensitive areas.

2. Non-recoverable failures. This category includes permanent errors. An example of an error of this type is the non-recoverable failure of one digital circuits bit.

   It should be noted that for the same radiation exposure of the same digital circuit, both recoverable (Fig. 1.8a) and non-recoverable (Fig. 1.8b) failures can occur [62]. The reason is that the accumulation of charges as a result of transferring the ionizing particle generates a transient noise impulse, competing with normal synchronized signals in digital circuits.

   The resulting effect depends on: the vulnerability of the affected node of a digital circuit, the activity of combinational logical paths of signal propagation in a digital circuit, the propagation delay in the signal path, the dynamic parameters of latches and flip-flops in a digital circuit, and many other circuit-level and technological features of implementing a digital circuit.

   If a false signal disappears with time, then a recoverable error is obtained; if a single switch results in a permanent error at the output of the circuit, a non-recoverable error is obtained.

3. Destructive failures (DR). This category includes:
   - Single "latching." This phenomenon is specific for CMOS circuits. It is associated with the opening of a parasitic pnpn thyristor under certain

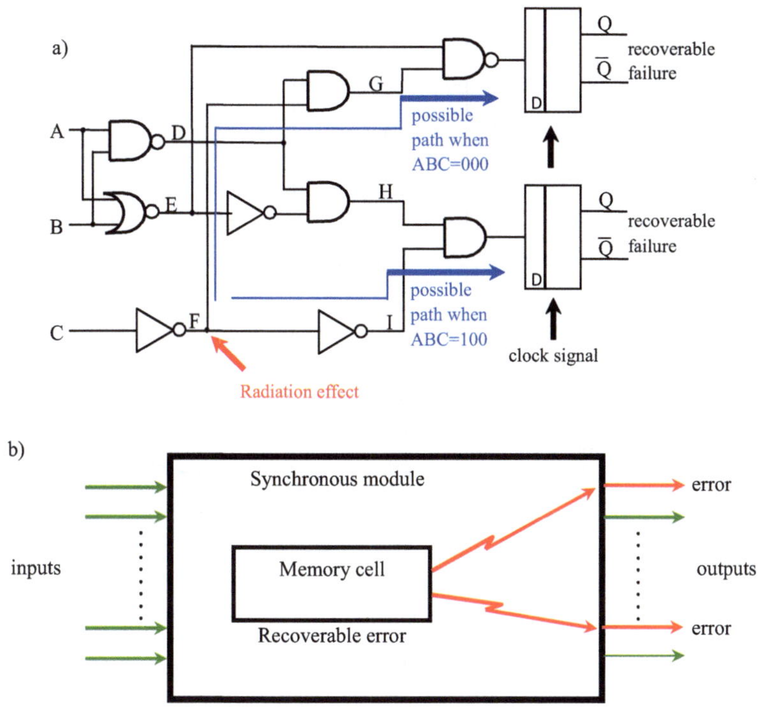

Fig. 1.8 Recoverable (**a**) and non-recoverable (**b**) failures of digital circuits

conditions. This leads to a short circuit of the power source, which generates large currents, capable of disabling digital circuits. The cause of this phenomenon, in addition to electrical overloads [71, 72], can also be the penetration of an ionizing particle.

- Single "unwinding." This phenomenon is also based on parasitic feedback. However, in this case, the presence of a parasitic pnpn thyristor is not mandatory. It occurs when a parasitic npn transistor is opened in an n-MOS transistor with a large drain current, and a feedback is formed under the influence of the penetrating particle. This phenomenon occurs mainly at high values of the supply voltage.
- Single breakdown of the gate. In this case, the penetrating ionizing particle leads to a breakdown of the gate oxide of a CMOS transistor. This phenomenon is typical for electronic components in which a large electric field is formed in the oxide layer.
- A single heavy error. In this case, the high-energy particle passes through the gate oxide and leads to displacements of the threshold voltage.
- Single burning. This phenomenon occurs in power MOS transistors, where parasitic bipolar transistors are present. Under the influence of the ionizing particle, the parasitic transistor opens. The resulting high currents and power can lead to the melting of the transistor.

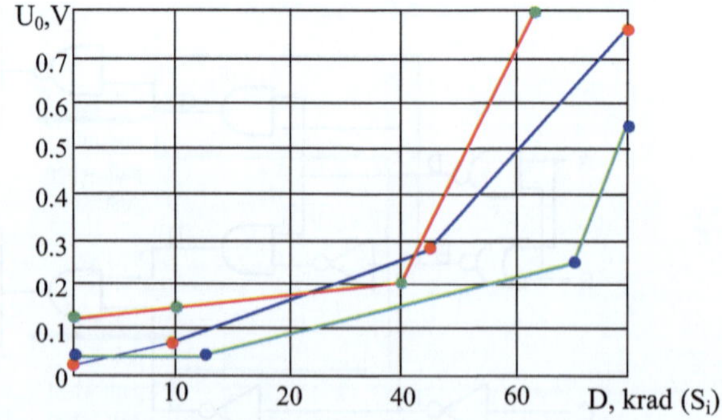

Fig. 1.9 Dependence of the logical "0" voltage on the absorbed radiation dose for various instances of the *S/*390 microprocessor

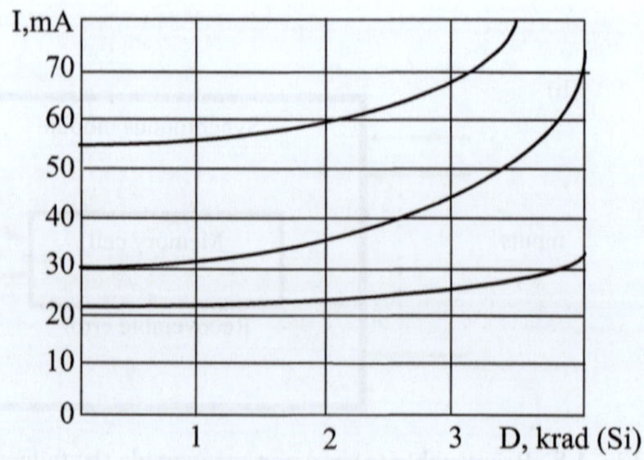

Fig. 1.10 Dependence of consumption current on the absorbed radiation dose for various instances of 8080A microprocessors

According to [48], radiation affects the static, dynamic, and functional parameters of digital ICs. The static parameters of digital circuits are mainly influenced by the particle fluence φ or the absorbed radiation dose D, and the dynamic parameters—power of absorbed dose P. The functional parameters of the digital IC depend on all of the above written characteristics of irradiation. The most sensitive to radiation exposure, the static parameters of digital circuits are the levels of logical signals "0" and "1." For example, according to [48], the logical "0" level at the internal nodes of the *S*/390 microprocessor, depending on the absorbed dose of radiation, changes as shown in Fig. 1.9.

It is not difficult to see that even for small values of radiation, the level of U_o can vary by more than an order of magnitude. The level of radiation, according to [41, 48], greatly affects the amount of current consumed from the power source, which, in turn, determines all modes of operation of the digital IC. As an example, Fig. 1.10 shows the dependencies of the current consumed by the microprocessor 8080A (manufactured by n-MOS technology by different companies) on the absorbed radiation dose [44]. As seen from the charts, depending on the absorbed dose of radiation, the power consumption varies several times, which, undoubtedly, leads to various failures in the circuit operation.

The effect of radiation exposure on the inertia (in particular, on signal propagation delay) of the logic gates of digital ICs is due to a change in the parameters of active

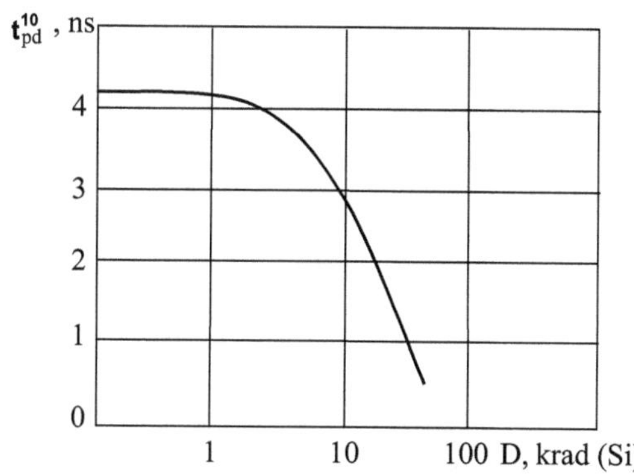

Fig. 1.11 Dependence of propagation delay $\left(t_{pd}^{10}\right)$ of PowerPC/740 microprocessor gates on the absorbed dose of radiation

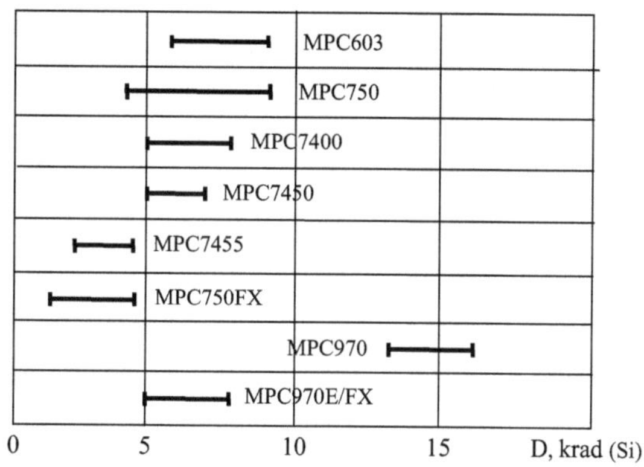

Fig. 1.12 Areas of functional failures of PowerPC microprocessors

elements and passive regions of semiconductor structures, as well as an increase in leakage current. According to the data of [3], changes in timing parameters can sometimes reach up to several orders, and, depending on IC fabrication technology, the gate delay can both increase and vice versa. For example, Fig. 1.11 shows the time dependence of propagation delay $\left(t_{pd}^{10}\right)$ of the PowerPC/740 microprocessor gates from the absorbed radiation dose (*D*). As seen from the chart, the dependence is significant.

To represent the significance of radiation effect on functional parameters of digital ICs, as an example, a diagram of functional failures of some PowerPC microprocessors [73] is illustrated on Fig. 1.12.

It can be seen from the diagrams that functional failures, which are a consequence of changes in the static and dynamic parameters of IC gates, appear even in case of insignificant values of radiation exposure (Figs. 1.13, 1.14, and 1.15).

Figures 1.16, 1.17, and 1.18 show different dependencies for the standard TTL cell (Fig. 1.13), and Figs. 1.19, 1.20, and 1.21—for the ECL cell (Fig. 1.14) obtained by simulating these circuits using the SPICE circuit simulator [74, 75]. These charts also confirm the strong dependence of digital circuit parameters on radiation.

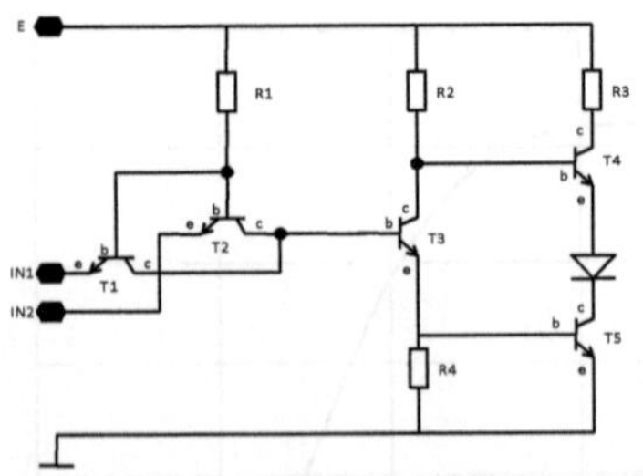

a)

```
R1  E     N113  R = 4K
R2  E     N35   R = 1.6K
R3  E     N93   R = 130
R4  N100  GND   R = 1K
T1  N108  N113  IN1   npn
T2  N108  N113  IN2   npn
T3  N35   N108  N100  npn
T4  N93   N35   OUT   npn
T5  N116  N100  GND   npn
D1  OUT   N116  d
E   E     GND   5
```

b)

```
.MODEL NPN NPN (LEVEL=1 BF=42 NF=0.995 IS=2.2E-17 ISC=2.2E-17 ISE=2.2E-17 +NE=1.45
RB=151.4797694 IRB=1.791514E-4 RBM=1.3776E-3 RE=1.16 IKF=1.257515E-3 +NKF=0.59
VAF=210.8833929 BR=0.0108 NR=1 NC=1.0645678 RC=20.2 IKR=8.978E-3 +VAR=9.95 TBF1=5.5E-3 TBF2=-
2E-6 TNE1=-2.88029E-4 TNE2=8.875512E-5
+TNF1=-1.534813E-4 TNF2=1.449935E-6 TRB1=1E-4 TIRB1=4.485779E-7
+TRM1=9.44068E-6 TRE1=2.130384E-5 TIKF1=-5.73196E-3 TIKF2=2.523403E-5
+TVAF1=-6.950157E-5 TBR1=1E-4 TBR2=5E-6 TNR1=1E-5 TNR2=-1E-6 TNC1=1E-4 +TNC2=-1.1E-6 XTB=0
XTI=3 TRC1=1E-3 TIKR1=-8.9251E-3 EG=1.18
+ CJE=1.353304E-13 VJE=0.941157 MJE=0.4592962 FC=0 CJC=9.430542E-14 +VJC=0.5445709
MJC=0.2932769 TLEVC=1 CTE=8.658836E-4 CTC=1.571976E-3 +TVJE=1.189865E-3 TVJC=3.303052E-3
TREF=25 SUBS=1 TLEV=0) *
.ENDL BIP
```

c)

The frequency of input pulses is 100 MHz, the duration of rise and fall times of input signal is 0.5 ns, the ambient temperature is 25° C, the number of loads is 4.

d)

Fig. 1.13 Standard TTL cell. (**a**) circuit, (**b**) SPICE description, (**c**) parameters of the transistor model, (**d**) nominal operating conditions

Due to the importance of the effect of radiation on the functioning of modern digital IC, various methods are used to increase their resistance to this type of DF. The classification of digital circuits on radiation hardness is also known [46] (Table 1.4).

There are three ways to increase the hardness of CMOS digital circuits to radiation:

1. Improvement of technological process. By changing some technological parameters (for example, by improving the quality of the gate oxide of a CMOS transistor) and steps, it is possible to increase the radiation hardness of digital ICs [54].
2. The use of special topological solutions that allow to some extent solve the problems of increasing static currents and eliminating a single "latching," and also reducing the sensitivity to single switching. Figure 1.22 [56] shows various topological implementations of an n-MOS transistor, which to some extent prevent the increase of parasitic leakage currents. From this point of view, the most acceptable variant is *D*, known as the "Transistor with Protected Topology" (TPT).

However, in this case, there are some difficulties associated with large transistor sizes, with large values of drain and source capacitance, and also with the choice of the *W*/*L* ratio. Despite the mentioned shortcomings, TPT has wide practical

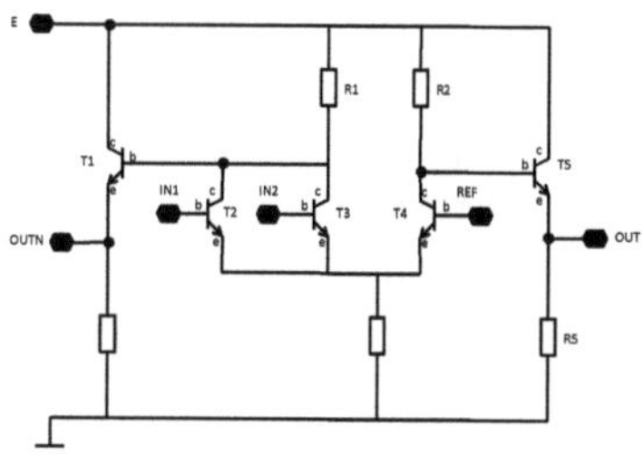

a)

T1	E	N32	OUTN	npn
T2	N32	IN1	N13	npn
T3	N32	IN2	N13	npn
T4	N10	REF	N13	npn
T5	E	N10	OUT	npn
R1	E	N32	R = 300	
R2	E	N10	R = 300	
R3	OUTN	GND	R = 2K	
R4	N13	GND	R = 1.24K	
R5	OUT	GND	R = 2K	
E	E	GND	5	

b)

```
.MODEL NPN NPN (LEVEL=1 BF=42 NF=0.995 IS=2.2E17 ISC=2.2E-17 ISE=2.2E-17
+NE=1.45 RB=151.4797694 IRB=1.791514E-4 RBM=1.3776E-3 RE=1.16 IKF=1.257515E-3 +NKF=0.59
VAF=210.8833929  BR=0.0108 NR=1 NC=1.0645678 RC=20.2 IKR=8.978E -3 +VAR=9.95 TBF1=5.5E -3
TBF2=-2E-6  TNE1=-2.88029E-4 TNE2=8.875512E-5
+TNF1=-1.534813E-4 TNF2=1.449935E-6 TRB1=1E-4 TIRB1=4.485779E-7
+TRM1=9.44068E-6 TRE1=2.130384E-5 TIKF1=-5.73196E-3 TIKF2=2.523403E-5 +TVAF1= -6.950157E-5
TBR1=1E-4 TBR2=5E-6 TNR1=1E-5 TNR2=-1E-6 TNC1=1E-4 +TNC2=-1.1E-6 XTB=0 XTI=3 TRC1=1E-3
TIKR1=-8.9251E-3 EG=1.18
+CJE=1.353304E-13 VJE=0.941157 MJE=0.4592962 FC=0 CJC=9.430542E -14 +VJC=0.5445709
MJC=0.2932769 TLEVC=1 CTE=8.658836E -4 CTC=1.571976E-3 +TVJE=1.189865E-3 TVJC=3.303052E-3
TREF=25 SUBS=1 TLEV=0) *
.ENDL BIP
```

c)

The frequency of the input pulses is 100 MHz, the duration of rise and fall times of input signal is 0.3 ns, the ambient temperature is 25° C, the number of loads is 3.

d)

Fig. 1.14 Standard ESL cell. (**a**) circuit, (**b**) SPICE description, (**c**) parameters of the transistor model, (**d**) nominal operating conditions

application. Figure 1.5 shows the advantage of TPT in relation to the usual topology of an MOS transistor. With the help of topological solutions, it is also possible to reduce leakage currents between separate semiconductor devices (Fig. 1.23 [56]). Figure 1.24 [56] illustrates an example of the topology of a rad-hard CMOS inverter, which allows reducing the mentioned currents.

3. Selection of special circuit architectures, less sensitive to changes in the parameters of circuit components as a result of radiation exposure. For a digital circuit to become more resistant to long-term radiation effects, various means can be used. One of such means is simulation of changes in some parameters (threshold voltage, conductivity, etc.) of transistors and obtaining their dependencies on the total radiation dose. Such simulation allows predicting the impact of general dose on the behavior of digital circuits and, based on this, design a digital circuit. There are several ways to increase the hardness to short-term radiation effects [17, 45–48, 53, 54, 56]. For example, a single switching can be prevented by changing the critical value of the charge or by filtering the transition signals resulting from exposure to a high-energy particle. Such an example is shown in Fig. 1.25 [56]. Figures 1.26 and 1.27, respectively, illustrate examples of rad-hard memory cells [56], and also circuit-level solutions for synchronizing circuits to increase radiation hardness [46].

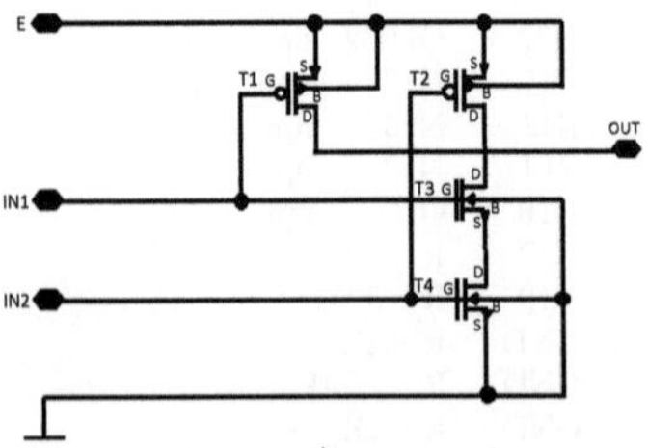

a)

```
T1    OUT   IN1   E     E
      +pmos l = 0.1u  w = 2u
T2    OUT   IN2   E     E
      +pmos l = 0.1u  w = 2u
T3    OUT   IN1   N160  GND
      +nmos l = 0.1u  w = 1u
T4    N160  IN2   GND   GND
      +nmos l = 0.1u  w = 1u
      E     E     GND   1
```

b)

.MODEL nmos NMOS (LMIN=1.0E-07 LMAX='2.4000E-07-dxl' WMIN='1.0000E-06-dxw' +WMAX='1.0000E-05-dxw' LEVEL=54 VERSION=4.2 BINUNIT=2 PARAMCHK=1 +MOBMOD=0 CAPMOD=2 IGCMOD=1 IGBMOD=1 DIOMOD=1 RDSMOD=0 +RBODYMOD=0 RGATEMOD=0 PERMOD=1 ACNQSMOD=0 TRNQSMOD=0 +TNOM=25 TOXE='toxn' TOXM='toxn' DTOX=3e -010 EPSROX=3.9 WINT=6e-009 +LINT=2.5e-009 LL=0 WL=0 LLN=1 WLN=1 LW=0 WW=0 LWN=1 WWN=1 LWL=0 +WWL=0 LLC=0 WLC=0+LWC=0 WWC=0 LWLC=0 WWLC=0 DLC=1.172e -008 DWC=0 +XPART=1 TOXREF=3e -009 DLCIG=2.5e-009 XL='-3.5E-08+dxl' XW='1.5e-8+dxw' +VTH0='0.22744278+dvthn' WVTH0='5.5738939e -009+dwvthn' LVTH0 = '0+dlvthn' +PVTH0='0+dpvthn' K1=0.32433703 LK1=-2.4740667e-010 WK1=-1.2374044e-008 +PK1=2.4748089e-015 K2=-0.011014366 LK2=1.2028732e-009 WK2 = 3.4532661e-009 +PK2=-6.9065323e-016 K3=-2.9571429 LK3=1.5428571e-008 K3B=1.2 W0=-5e-008 +DVT0=13.9 DVT1=1.25 DVT2=0 DVT0W=0 DVT1W=0 DVT2W=0 DSUB=0.5 MINV=0 +VOFFL=0 DVTP0=4e-007 DVTP1=1.5714286 LDVTP1=8.5714286e-008 LPE0=4.45e-008 +LPEB=0 XJ=1.5e-007 NGATE=4e020 NDEP=8.3687e017 NSD=1e020 PHIN=0 CDSC=0 +CDSCB=0 CDSCD=0 CIT=-0.00012285714 LCIT=2.2457143e-010 VOFF=-0.11586017 +LVOFF= -2.4718229e-009 WVOFF=-1.6403173e-008 PVOFF=1.318624e-015 +NFACTOR=0.5 ETA0=0.22133733 WETA0= -1.3377345e-008ETAB=-0.1 +U0=0.045463996 LU0=-5.1444e-010 WU0=-1.2300788e-008 PU0=5.1598332e-016 +UA=-6.284e-010 UB=2.2915951e-018 LUB=-9.1553714e-027 WUB=-1.0877056e-024 +PUB=4.185198e-032 UC=2.4114338e-010 LUC=-4.1075238e-019 WUC=-1.3447417e-016 +PUC=4.1087561e-024 VSAT=114000 A0=2.375 AGS=0.95 A1=0 A2=1 B0=0 B1=0 +KETA=-0.044142857 LKETA=3.4285714e-009 DWG=0 DWB=0 PCLM=0.84 PDIBLC1=0 +PDIBLC2=1.4285714e-005 LPDIBLC2=1.7142857e-011 PDIBLCB=0 DROUT=0 +PVAG=1.5 DELTA=0.0075 PSCBE1=9.264e008 PSCBE2=1e-020 FPROUT=200 PDITS=0 +PDITSD=0 PDITSL=0 RSH=10.5 RDSW=140 PRWG=0 PRWB =0 WR=1 ALPHA0=0 +ALPHA1=0.03 BETA0=8.7 AGIDL=9e -008 BGIDL=2.3e009 CGIDL=0.5 EGIDL=0.53 +AIGBACC=0.01134 BIGBACC=0.003249 CIGBACC=0.1416 NIGBACC=4.05 +AIGBINV=0.35 BIGBINV=0.03 CIGBINV=0.006 EIGBINV=1.1 NIGBINV=1 +AIGC=0.010347143 LAIGC= -9.4285714e-012 +BIGC=0.001525 CIGC=0 +AIGSD=0.0083547333 LAIGSD=5.716e-012 WAIGSD=9.5552467e-011 +PAIGSD=-5.733148e-018 BIGSD=0.0004021 CIGSD=0.001463 NIGC=1 POXEDGE=1 +PIGCD=2.3 NTOX='ntoxn' XRCRG1=12 XRCRG2=1 CGSO='cgon' CGDO='cgon' +CGBO=0 CGDL='cgln' CGSL='cgln' CLC=1e-007 CLE=0.6 CF='cfn' CKAPPAS=0.6 +ACDE=1.2 +MOIN=12.6 NOFF=1.686 VOFFCV= -0.05472 KT1=-0.2624 LKT1=-1.62e-009 +KT1L=0 KT2=-0.04464 UTE=-1.5502921 LUTE=1.1143371e-008 WUTE=7.437167e-008 +PUTE=-1.7199444e-015 UA1=2.684e-009 UB1=-3.673e-018 UC1=-6.421e-011 PRT=0 +AT=35357.143 LAT=0.0019285714 JSS=1.77E -07 JSWS=4.23E-13 JSWGS=9E-12 NJS=1 +IJTHSFWD=0.01 IJTHSREV=0.01 BVS=10 XJBVS=1 PBS=0.5 CJS='cjn' MJS=0.25 +PBSWS=0.8 CJSWS='cjswn' MJSWS=0.01 PBSWGS=0.78 CJSWGS='cjswgn' +MJSWGS=0.52 +TPB=1.00E -03 TCJ=7.32E-04 TPBSW=1.90E-03 TCJSW=3.57E-04 +TPBSWG=1.90E-03 +TCJSWG=3.57E-04 XTIS=3 DMCG=9.25e-008 DMCI=9.25e-008 +DMDG=0 DMCGT=0 DWJ=0 XGW=0 XGL=0 RSHG=0.1 GBMIN=1e -012 RBPB=50 +RBPD=50 RBPS=50 RBDB=50 RBSB=50 NGCON=1 STIMOD=2 SA0=0.63E -6 +SB0=0.63E-6 KVSAT=0.20 WLOD =0.8E-6 TKU0=0.01 KU0=-7.36E-8 LLODKU0=1 +WLODKU0=1 LKU0=4.027E-7 WKU0=8E-7 PKU0=-2.0E-13 KVTH0=3.5E-9 +STK2=1.35E-9 STETA0=0 LKVTH0=-2.68E-11 WKVTH0=1.5E-6 PKVTH0=0 +LLODVTH=1 WLODVTH=1 LODK2=0.5 LODETA0=1 FNOIMOD=1 NOIA=4.2926E+41 +NOIB=1.612E+24 NOIC=8. 75 EM=1.192E+07 EF=0.8841 TNOIMOD=0)

.MODEL pmos PMOS (LMIN='1.0000E -05-dxl' LMAX=2.001E-05 WMIN='1.0000E-06-+dxw' +WMAX='1.0000E-05-dxw' LEVEL=54 VERSION=4.2 BINUNIT=2 PARAMCHK=1 +MOBMOD=0 CAPMOD=2 IGCMOD=1 IGBMOD=1 DIOMOD=1 RDSMOD=0 +RBODYMOD=0 RGATEMOD =0 PERMOD=1 ACNQSMOD=0 TRNQSMOD=0 +TNOM=25 TOXE='toxp' TOXM='toxp' DTOX=5.25e -010 EPSROX=3.9 WINT=5e-009 +LINT=2.5e-009 LL=0 WL=0 LLN=1 WLN=1 LW=0 WW=0 LWN=1 WWN=1 LWL=0 +WWL=0 LLC=0 WLC=0 LWC=0 WWC=0 LWLC=0 WWLC=0 DLC=1.525e -008 DWC=0 +XPART=1 TOXREF=3e-009 DLCIG=2.5e-009 XL='-3.5E-8+dxl' XW='1.5E-8+dxw' +VTH0='-0.16574717+dvthp' WVTH0='5.4744025e -009+dwvthp' LVTH0='0+dlvthp' +PVTH0='0+dpvthp'

Fig. 1.15 Standard CMOS cell. (**a**) circuit, (**b**) SPICE description, (**c**) parameters of the transistor model, (**d**) nominal operating conditions

K1=0.1623 K2=0.012789333 WK2=-8.9378e-010 K3=-0.58 K3B=0.8 +W0=0 DVT0=10 DVT1=1.71
DVT2=0.0928 DVTOW=0 DVT1W=0 DVT2W=0 DSUB=0.5 +MINV=0 VOFFL=0 DVTP0=0 DVTP1=0
LPE0=2.805e-007 LPEB=1e-007
+XJ=1.7000001e-007 NGATE=8e019 NDEP=1.7200001e017 NSD=1e020 PHIN=0 CDSC=0 +CDSCB=0
CDSCD=0 LCDSCD='0+dlcdscdp' CIT=0.00011166667 WCIT= -1.117225e-009 +VOFF=-0.09669 NFACTOR=0.5
ETA0=0.085 ETAB=-0.02 U0=0.0085405017
+WU0=-7.8540918e-010 UA=-2.77941e-010 WUA=1.5417705e-017 UB=2.0172933e-018 +WUB=-1.430048e-025 UC=8.136e-011 VSAT=120000 A0=3.62 AGS=1.096 A1=0 A2=1 +B0=0 B1=0 KETA= -0.04058 DWG=0
DWB=0 PCLM=0.2 PDIBLC1=0 PDIBLC2=0.005 +P DIBLCB=0 DROUT=0 PVAG=0 DELTA=0.02
PSCBE1=9.264e008 PSCBE2=1e-020 +FPROUT=1000 PDITS=0 PDITSD=0 PDITSL=0 RSH=9.68
RDSW='210+drdswp' +PRWG=0 PRWB=0 WR=1 ALPHA0=0 ALPHA1=0.06 BETA0=12 AGIDL=9e -008
+BGIDL=2.3e009 CGIDL=0.5 EGIDL=0.53 AIGBACC=0.01069 BIGBACC=0 .003132 +CIGBACC=0.1498
NIGBACC=4.049 AIGBINV=0.35 BIGBINV=0.03 CIGBINV=0.006 +EIGBINV=1.1 NIGBINV=1
AIGC=0.00606 BIGC=0.0007625 CIGC=1e-010 +AIGSD=0.0045821833 WAIGSD=7.820575e -011 BIGSD=0
CIGSD=0.002267 NIGC=1 +POXEDGE=1 PIGCD=2.3 NTOX='ntoxp' XRCRG1=12 XRCRG2=1 CGSO='cgop'
+CGDO='cgop' CGBO=0 CGDL='cglp' CGSL='cglp' CLC=1e-007 CLE=0.6 CF='cfp' +CKAPPAS=0.6 ACDE=1.2
MOIN=11.09 NOFF=2.175 VOFFCV=-0.03885 KT1=-0.29497017 +WKT1=5.1727518e-008 KT1L=0 KT2=-0.027639233 WKT2=-1.9417371e-008
+UTE=-1.193365 WUTE=6.3681825e-008 UA1=4.1822917e-009 WUA1=-1.0334331e-015 +UB1=-7.7915117e-018 WUB1=1.4758542e-024 UC1=-6.421e-011 PRT=0 AT=280000 +JSS=1.245E -07 JSWS=3.77E-13
JSWGS=3.00E-12 NJS=1 IJTHSFWD=0.01
+IJTHSREV=0.01 BVS=10 XJBVS=1 PBS=0.8 CJS='cjp' MJS=0.36 PBSWS=0 .8 +CJSWS='cjswp' MJSWS=0.01
PBSWGS=1 CJSWGS='cjswgp' MJSWGS=0.99 TPB=1.9E-03
+TCJ=8.54E-04 TPBSW=1.9E-03 TCJSW=3.57E-04 TPBSWG=1.9E-03 TCJSWG=3.12E-03 +XTIS=3
DMCG=9.25e-008 DMCI=9.25e-008 DMDG=0 DMCGT=0 DWJ=0 XGW=0 +XGL=0 RSHG=0.1 GBMIN=1e -012 RBPB=50 RBPD=50 RBPS=50 RBDB=50 RBSB=50 +NGCON=1 STIMOD=2 SA0=0.63E -6 SB0=0.63E-6
KVSAT=-0.9 WLOD=0.8E-6 +TKU0=0.1 KU0=1.3249E-7 LLODKU0=1 WLODKU0=1 LKU0=1.33E -7
WKU0=6.5E-7 +PKU0=0 KVTH0=-1.5E-9 STK2=4E-10 STETA0=0 LKVTH0=-1.68E-15 WKVTH0=0.5E-6
+PKVTH0=0 LLODVTH=2 WLODVTH=1 LODK2=2 LODETA0=1 FNOIMOD=1 +NOIA=1.635E+42
+NOIB=7.241E+26 NOIC=8.75 EM=1.50E+08 EF=1.19 TNOIMOD=0)

c)

The frequency of the input pulses is 400 MHz, the duration of rise and fall times of input signal is 0.1 ns, the ambient temperature is 25° C, and the capacitive load is 20φF.

d)

Fig. 1.15 (continued)

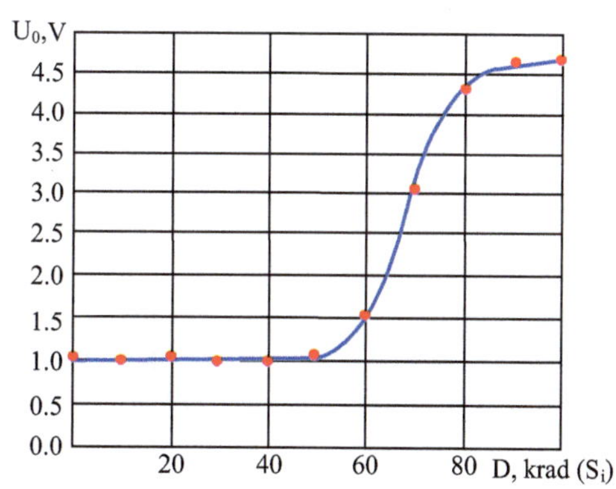

Fig. 1.16 Dependence of the TTL cell output voltage on the absorbed dose of radiation in “0”

4. Use of special packages (Fig. 1.28). Note that methods to increase the radiation hardness of digital ICs only slightly reduce the effect of radiation on the behavior of digital circuits.

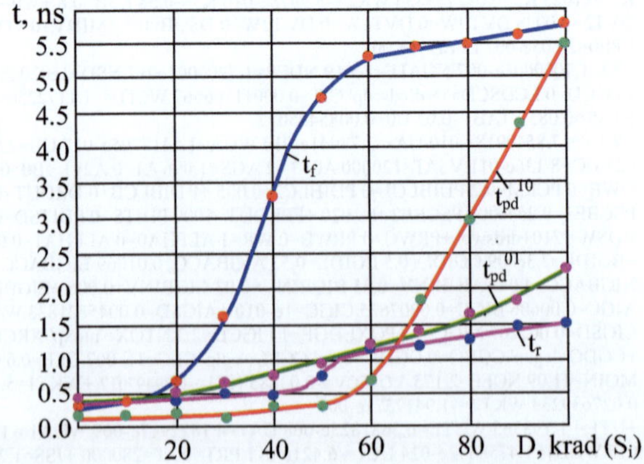

Fig. 1.17 Dependences of timing parameters of TTL cells from absorbed dose of radiation

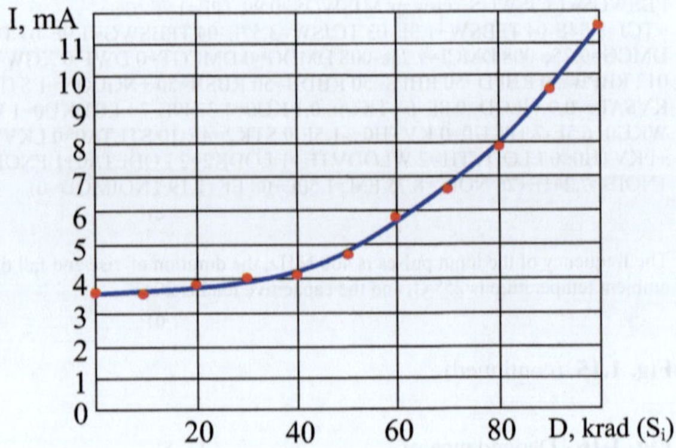

Fig. 1.18 Dependence of current consumed by TTL cell from the absorbed dose of radiation

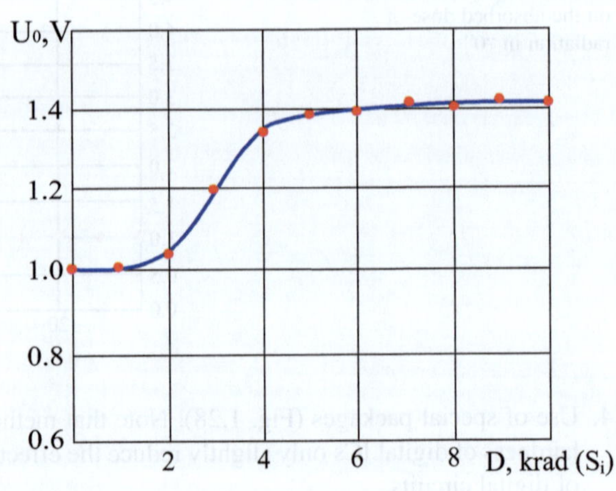

Fig. 1.19 Dependence of the output voltage of ESL cell on the absorbed radiation dose in the state "0"

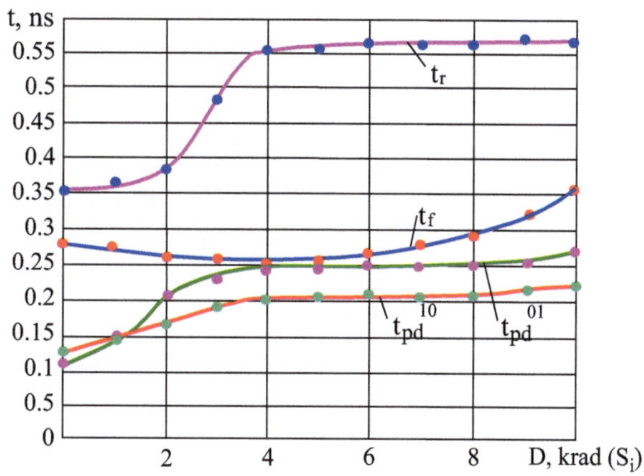

Fig. 1.20 Dependences of timing parameters of the ECL cell on the absorbed dose of radiation

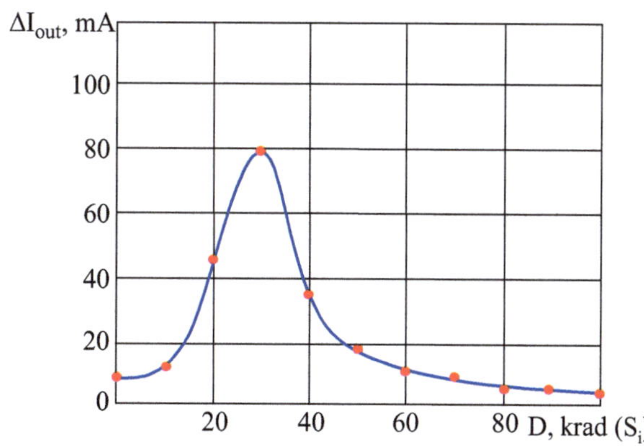

Fig. 1.21 Dependence of the difference in the output currents of the ECL cell on the absorbed dose of radiation

Table 1.4 Classification of digital ICs as per radiation hardness

	Commercial	Rad-hard	Radiation-Protected
Long-term (krad) (S_i)	<20	20…100	100…1000
Short-term (rad) (S_i)/s	$<10^7$	$10^7\ldots10^9$	$>10^9$
Non-recoverable failures (MeV-cm^2/mg)	<20	20…80	>80

Summarizing the above, it can be said that radiation has a significant and decisive influence on the behavior of digital circuits: reduces the switching speed of digital cells, increases the static and dynamic power consumption of digital circuits, leads to various failures in digital circuits (Fig. 1.29).

Thus, the parameters of digital circuits, functioning in the environment of radiation exposure are strongly dependent on the level of exposure. The dimensions of the changes in the parameters of digital IC cells due to radiation exposure can reach from one to several orders and are capable of qualitatively changing the behavior of the circuit in case of the same input signals. For confirmation, Fig. 1.30 shows the

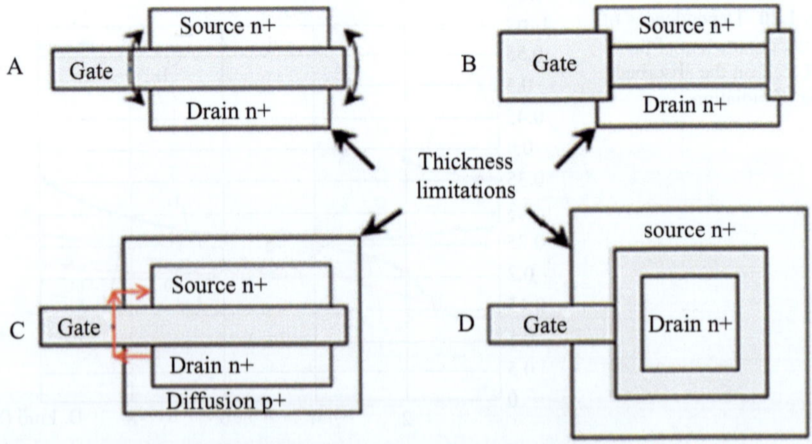

Fig. 1.22 Various topological implementations of an n-MOS transistor

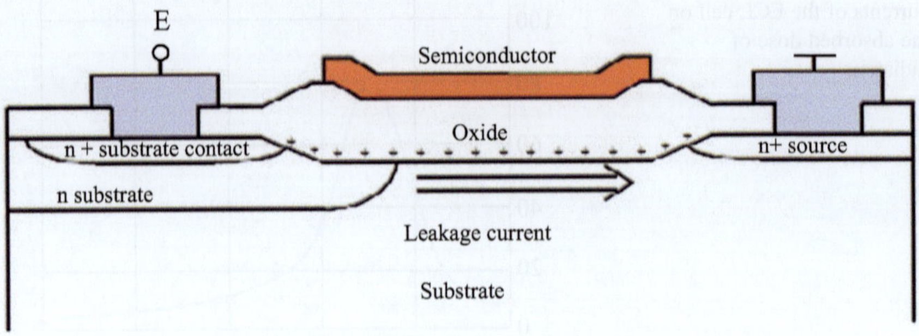

Fig. 1.23 Leakage current between separate transistors, induced by radiation exposure

results of circuit-level simulation [76] of a D flip-flop (Fig. 1.29), with (Fig. 1.30a) and without consideration (Fig. 1.30b) of radiation. As seen from the diagrams, the difference is qualitative.

An important circumstance is the significant dependence of the values of changes of parameters of the same circuit, noted by many researchers, in case of the same radiation exposure from the processed signal transformations [3, 20, 41, 48, 53, 57, 59].

The situation is similar with respect to the significance of the influence of other types of external DFs (Table 1.1) (Fig. 1.31).

It is known that the parameters of both passive and active components of digital ICs are strongly dependent on the ambient temperature (V_2) [1, 2, 6, 32–35, 38, 67–70, 77–81]. Therefore, the integrated parameters of logic gates are also significantly changed in the operating range of the ambient temperature, which is confirmed by the dependencies, shown in Fig. 1.32 for the PLL (Fig. 1.31) [82], as well as in Figs. 1.33, 1.34, and 1.35—for the standard TTL cell (Fig. 1.13), in Figs. 1.36, 1.37,

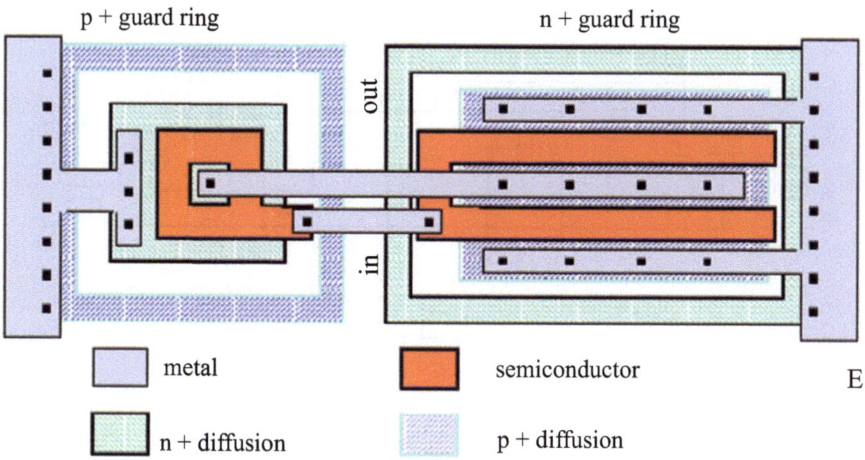

Fig. 1.24 An example of topology of a rad-hard CMOS inverter

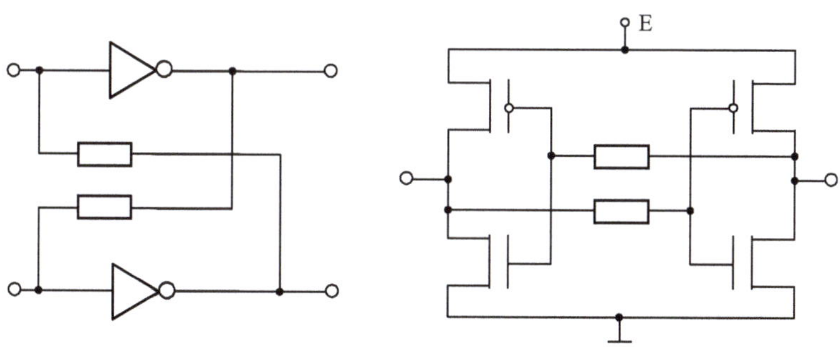

Fig. 1.25 Increase of radiation hardness of static CMOS memory cell by adding resistances in feedback circuits

and 1.38 for the standard ECL cell (Fig. 1.14) and in Figs. 1.39 and 1.40—for the standard CMOS cell (Fig. 1.15), obtained by simulating with SPICE [59]. The situation is similar with a change in the supply voltage (V_3).

The dependence, illustrated in Fig. 1.41 [79], indicates significant displacements of circuit parameters even in insignificant limits of the change in the supply voltage (E_n). Several times, and often even by orders of magnitude, the values of other parameters of digital circuits change even with small shifts of E_n from the nominal value.

The same is confirmed by the dependencies in Fig. 1.42 for the standard TTL cell (Fig. 1.13), in Fig. 1.43 for the standard ECL cell (Fig. 1.14), obtained by simulating with SPICE [59].

The non-ideality of the sources of input signals (V_4, Table 1.1) are not only the real parameters (output impedance, capacitance, etc.) of the pulse generators of the entire circuit, but also formation of signals with distorted shapes (amplitude,

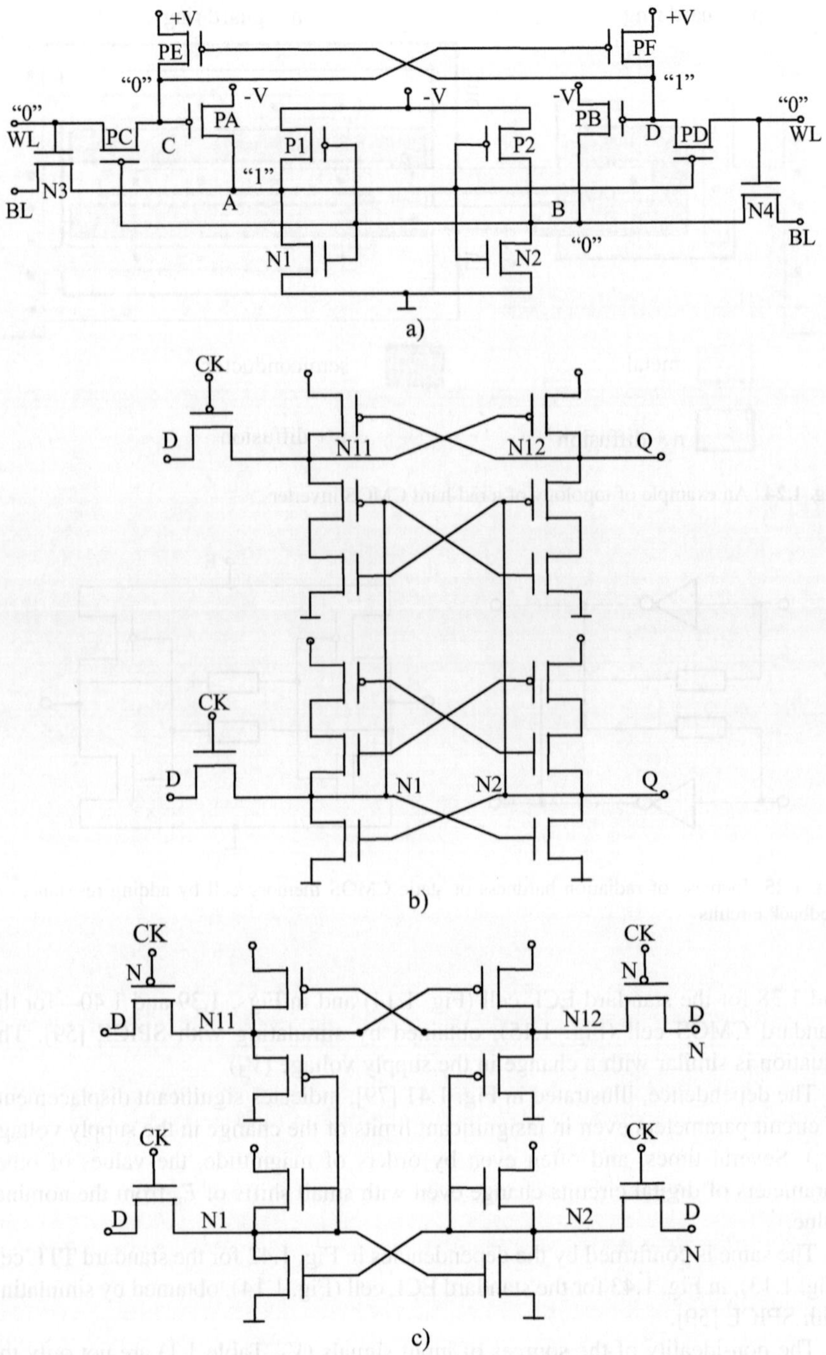

Fig. 1.26 Examples of rad-hard memory cells

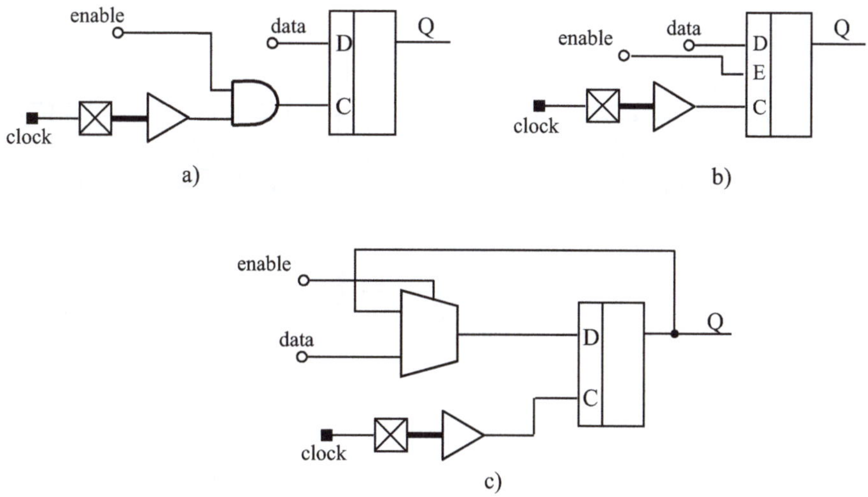

Fig. 1.27 Examples of circuit-level solutions of circuit synchronization to increase radiation hardness (**a**-standard circuit, **b**-rad-hard solutions of the company Actel)

Fig. 1.28 Special package of the company Atmel

duration, etc.) in the internal nodes of the circuit. The latter is more often found in mixed-signal ICs, when signals are received on logic gates from analog blocks. In both cases, the operating modes of the loading cells are violated, since experimental studies show the existing strong dependence of digital cell parameters on the shape of input signals. As an example, Figs. 1.44, 1.45, 1.46, 1.47, 1.48, 1.49, and 1.50 shows the dependences of timing parameters of standard TTLs (Fig. 1.13), ECL (Fig. 1.14), and CMOS (Fig. 1.15) cells obtained by simulation using SPICE simulator [59].

The curves indicate the existing strong dependence of the shape of the output pulse of digital gates from the parameters of input signals, and this phenomenon cannot be ignored.

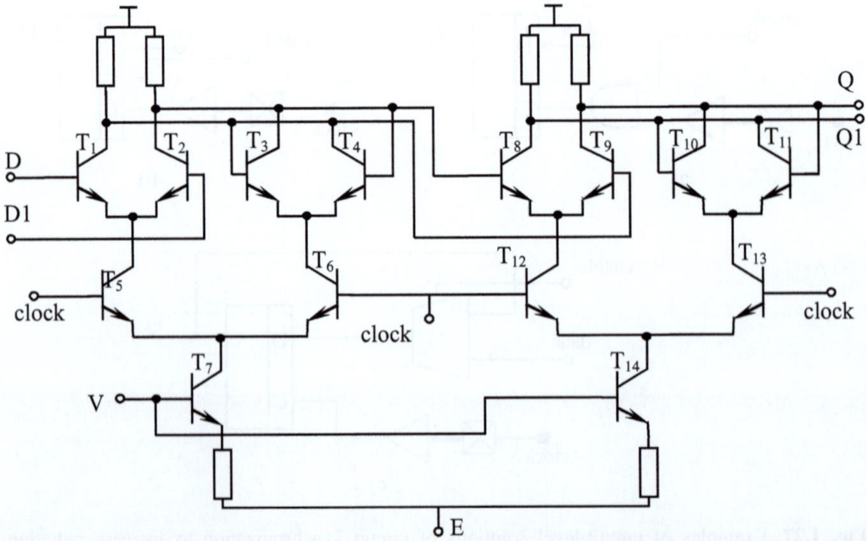

Fig. 1.29 D flip-flop

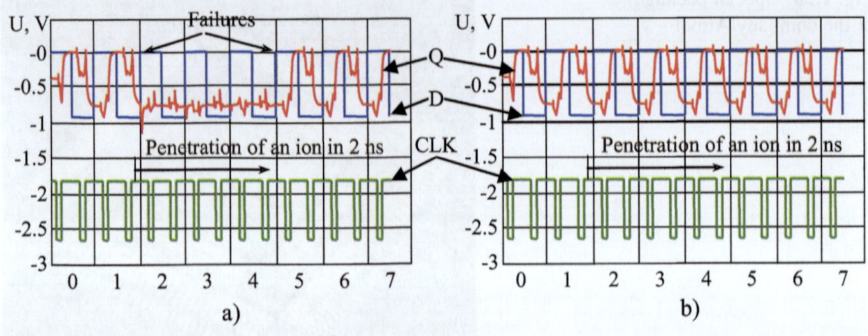

Fig. 1.30 Results of circuit-level simulation of a D flip-flop

Fig. 1.31 Phase-locked loop (PLL)

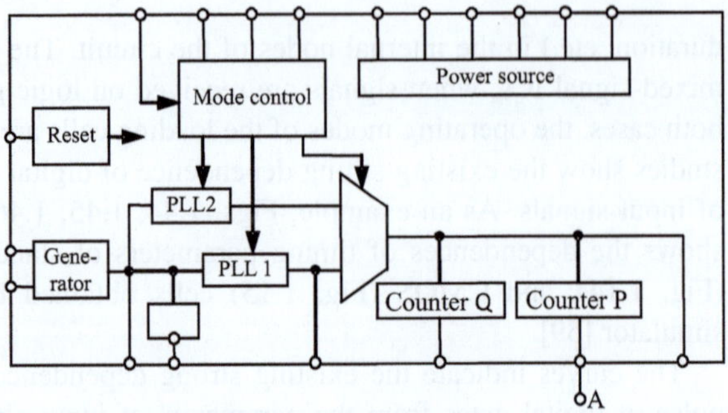

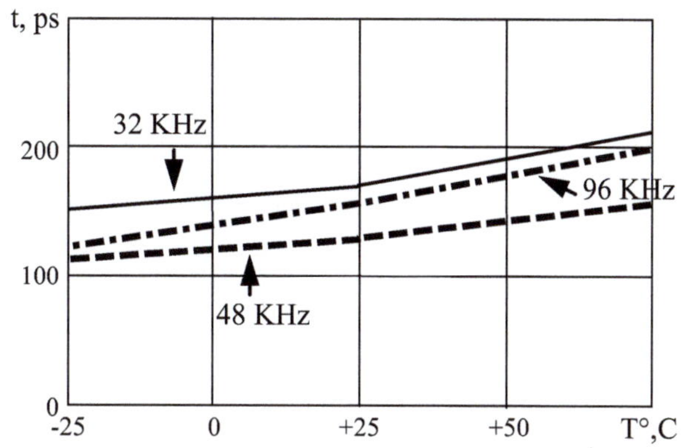

Fig. 1.32 The bounce of the signal in node A of the PLL (power supply 5 V, load capacity 20 pF), depending on the temperature

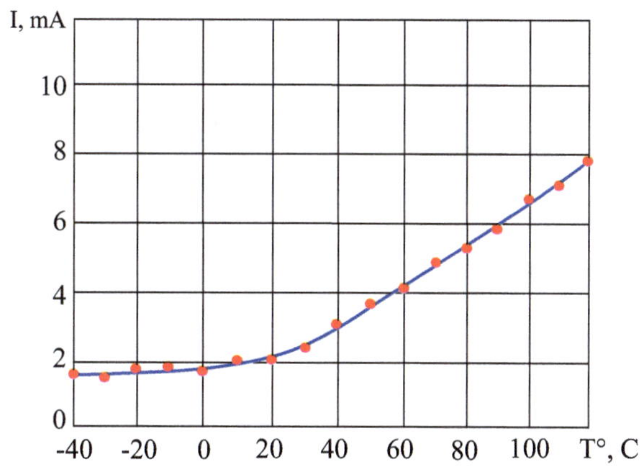

Fig. 1.33 Dependence of current consumed by TTL cell on the ambient temperature

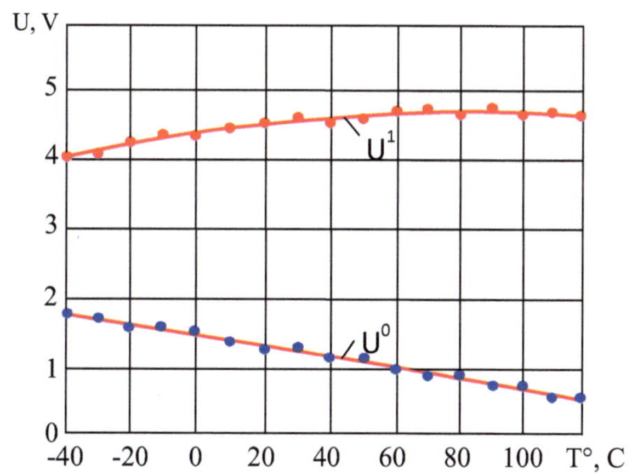

Fig. 1.34 Dependences of the output voltages of TTL cell in the states "0" and "1" on the ambient temperature

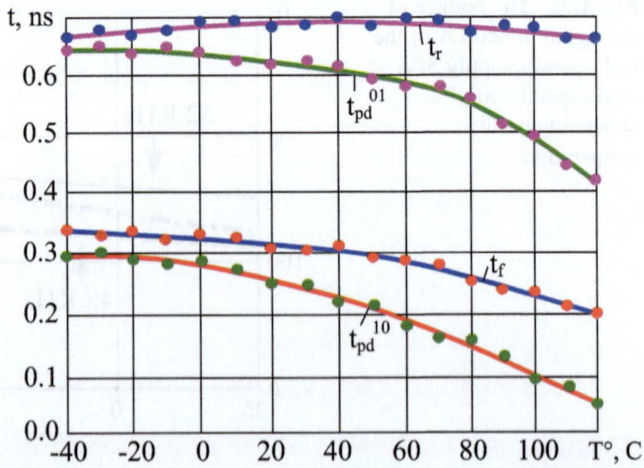

Fig. 1.35 Dependences of timing parameters of the TTL cell on the ambient temperature

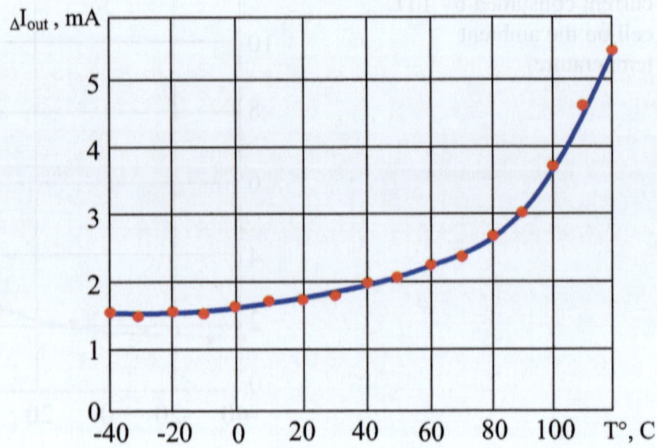

Fig. 1.36 Dependence of the difference of output currents of ECL cell on the ambient temperature

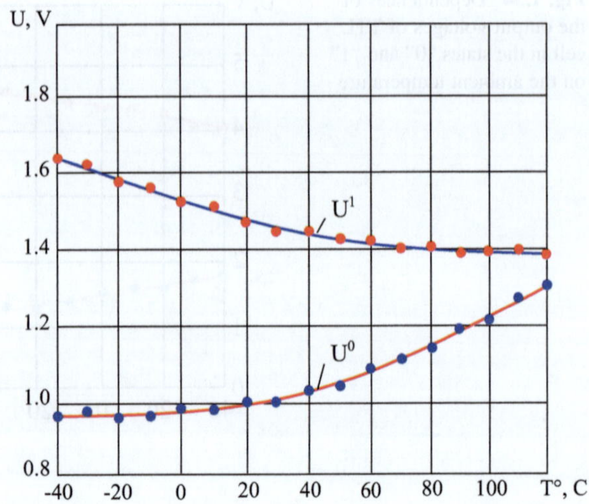

Fig. 1.37 Dependences of the output voltages of the ECL cell in the states "0" and "1" on the ambient temperature

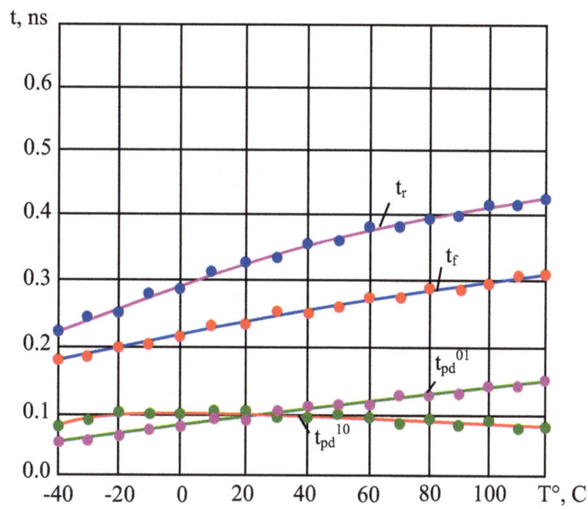

Fig. 1.38 Dependences of timing parameters of the ECL cell on the ambient temperature

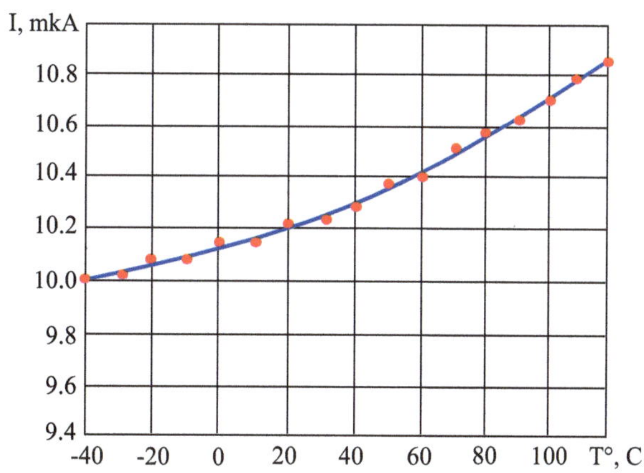

Fig. 1.39 Dependence of the current consumed by CMOS cell on the ambient temperature

The non-ideality of the loads of digital circuits (V_5, Table 1.1) is also considerable, especially in case of simulation of mixed-signal ICs, since in the internal nodes of the circuit the load is often an analog element whose input *I–V* characteristic can significantly differ from the same curves of the standard logic cells (TTL, ECL, I^2L, CMOS, etc.). For this reason, the characteristics of the gate itself are shifted.

Similar experiments using SPICE make it easy to witness that the nonstandard *I–V* characteristic of the inputs of the gate load significantly affects the parameters of the gate source (this is also confirmed by the dependences in Figures 1.51, 1.52, 1.53, 1.54, 1.55, 1.56, 1.57, 1.58, and 1.59, obtained by simulating with SPICE [59]), and this factor should also be taken into account while simulating digital circuits.

As electromagnetic noise (V_6) is widely covered in the literature [15, 40, 83], and the DF influence V_7 and V_8 is less significant, they are not discussed here.

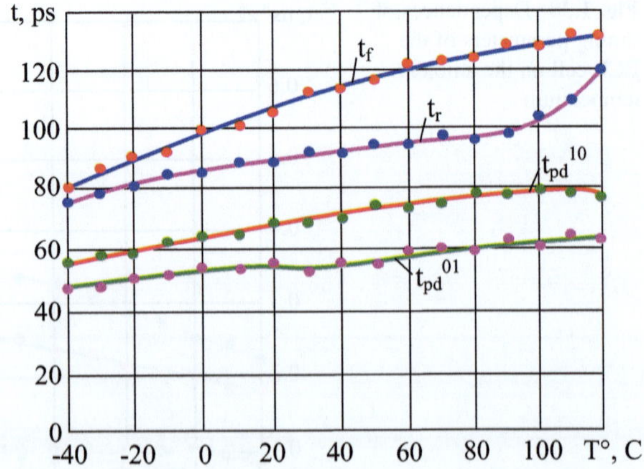

Fig. 1.40 Dependences of timing parameters of CMOS cell on the ambient temperature

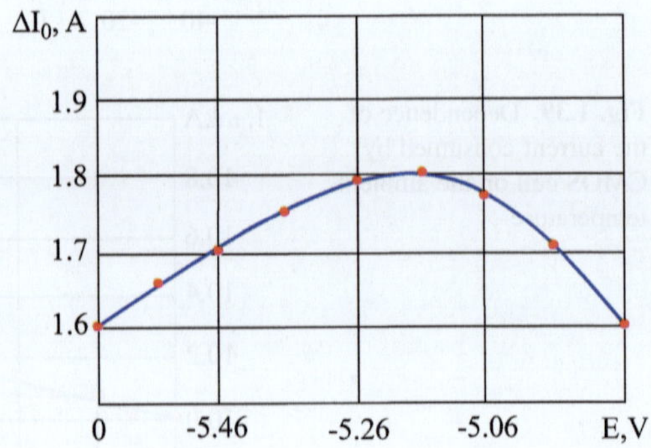

Fig. 1.41 Dependence of the difference of output currents of the ECL cell on the supply voltage

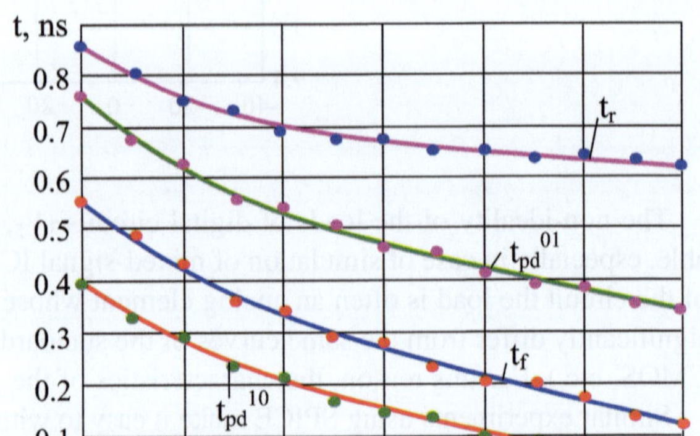

Fig. 1.42 Dependences of timing parameters of a TTL cell on the supply voltage

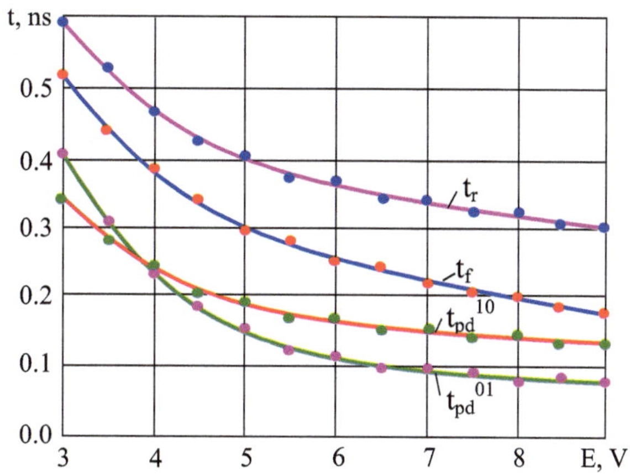

Fig. 1.43 Dependences of timing parameters of an ECL cell on the supply voltage

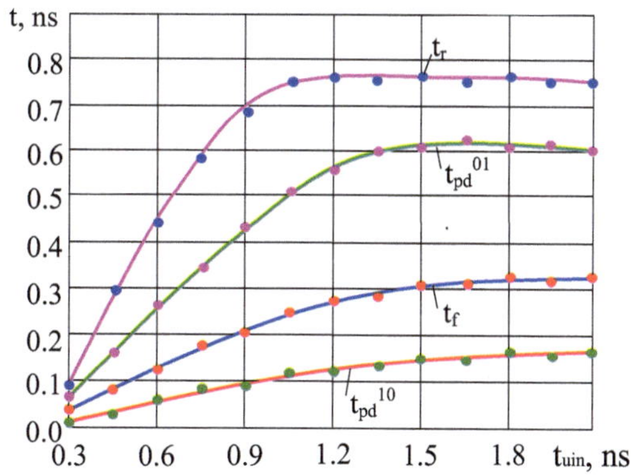

Fig. 1.44 Dependences of timing parameters of a TTL cell on the duration of the input pulse

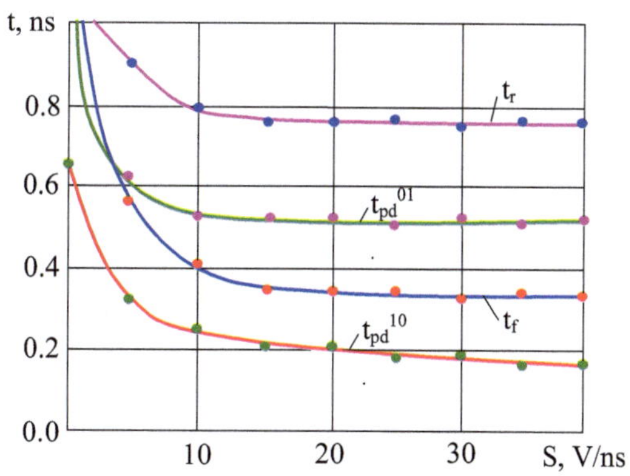

Fig. 1.45 Dependences of timing parameters of a TTL cell on the slope of the input signal

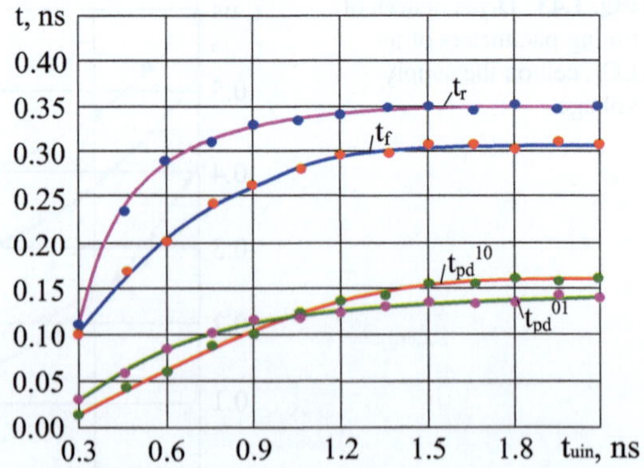

Fig. 1.46 Dependences of timing parameters of an ECL cell on the duration of the input pulse

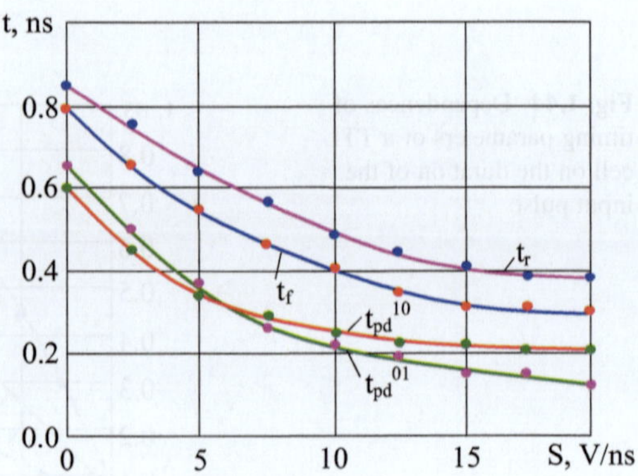

Fig. 1.47 Dependences of timing parameters of an ECL cell on the slope of the input signal

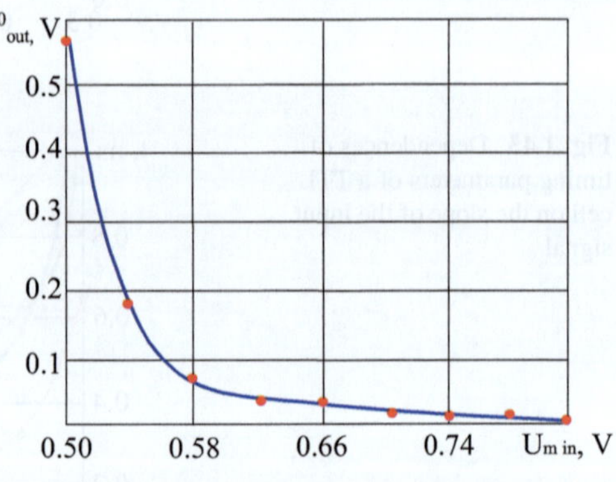

Fig. 1.48 Dependence of the output voltage of a CMOS cell on the amplitude of the input signal in the state "0"

The dependencies presented in this section from literature sources, as well as the results of circuit simulation, indicate the considerable significance of the influence of external DFs (Table 1.1) on the behavior of digital circuits. The neglect of the noted DFs during their simulation can lead to significant, and often even to qualitative, errors of simulation of digital circuits.

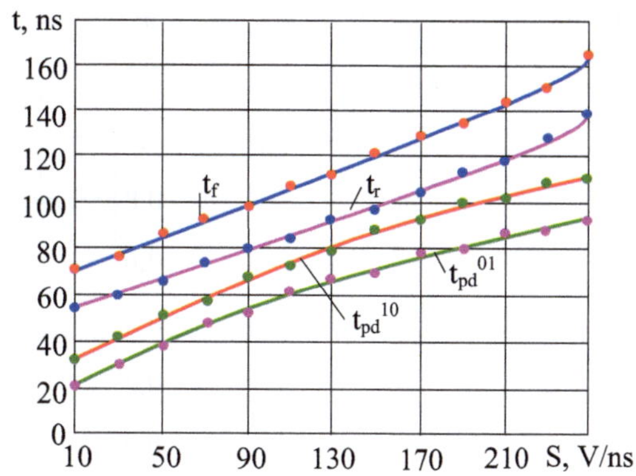

Fig. 1.49 Dependences of timing parameters of a CMOS cell on the slope of the input signal

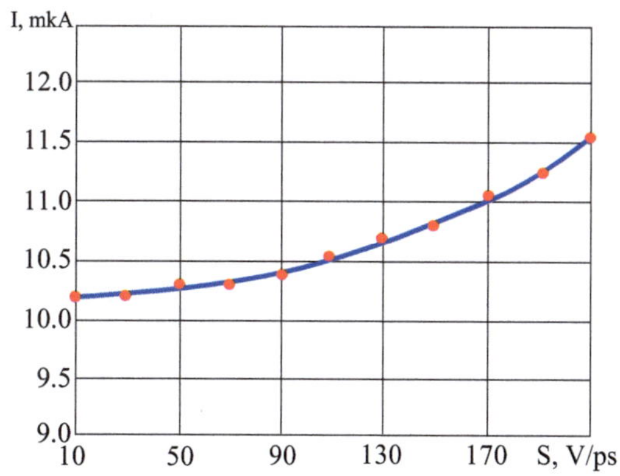

Fig. 1.50 Dependence of the current consumed by a CMOS cell on the slope of the input signal

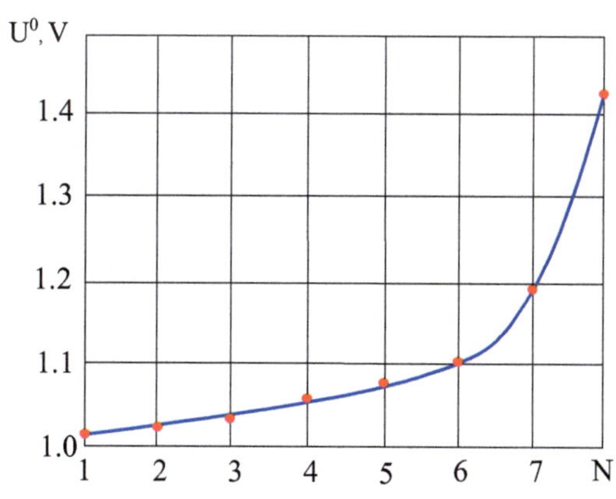

Fig. 1.51 Dependence of the output voltage of a TTL cell on the number of loads in the state "0"

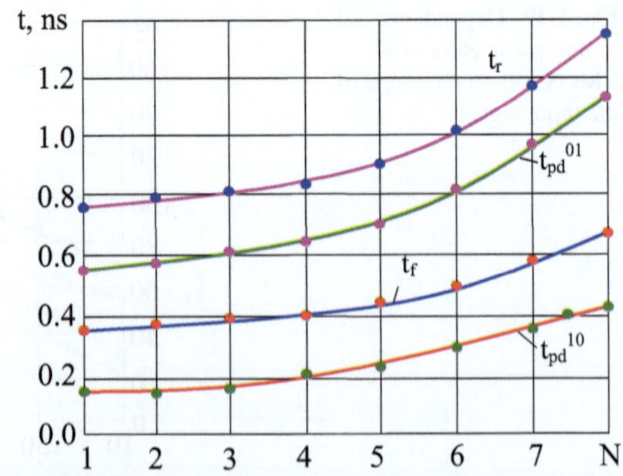

Fig. 1.52 Dependences of timing parameters of a TTL cell on the number of loads

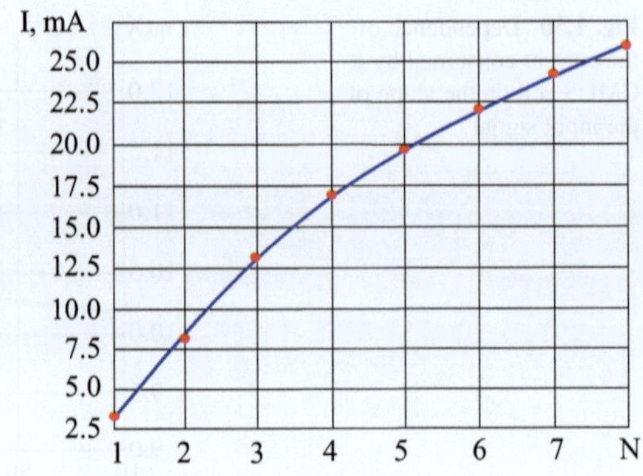

Fig. 1.53 The dependence of the current consumed by a TTL cell on the number of loads

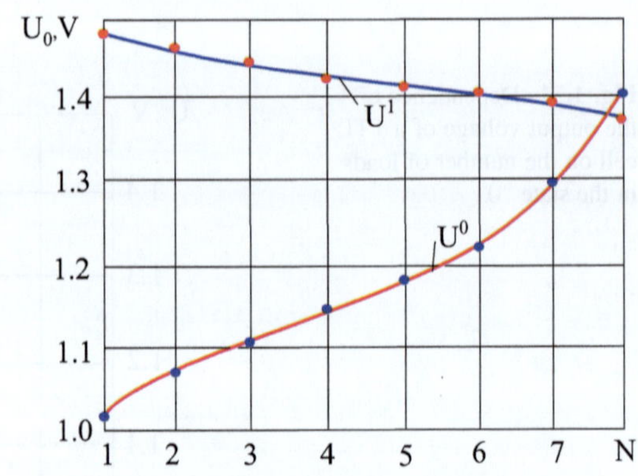

Fig. 1.54 Dependences of the output voltage of an ECL cell on the number of loads

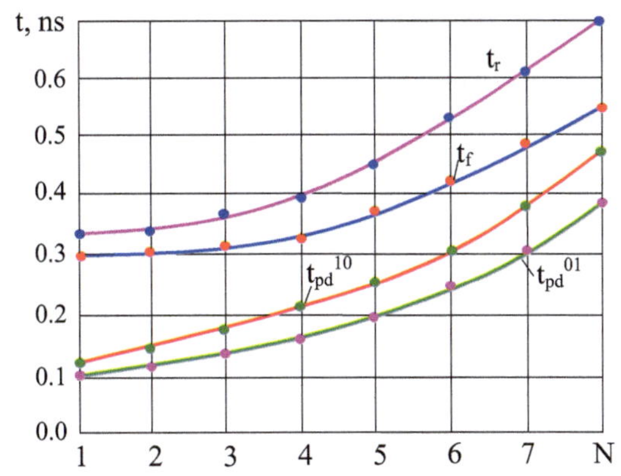

Fig. 1.55 Dependences of timing parameters of an ECL cell on the number of loads

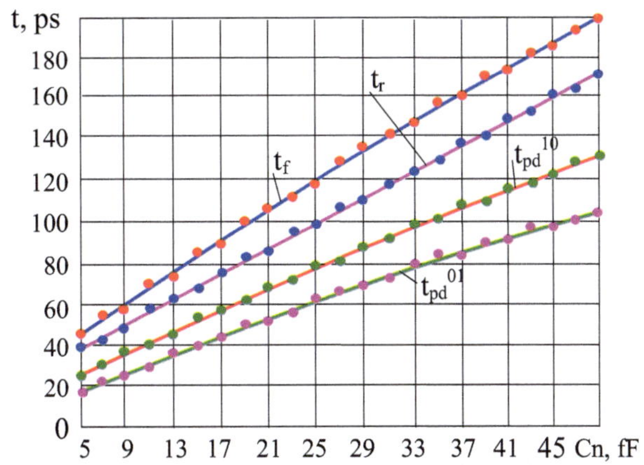

Fig. 1.56 Dependences of timing parameters of a CMOS cell on the load capacitance

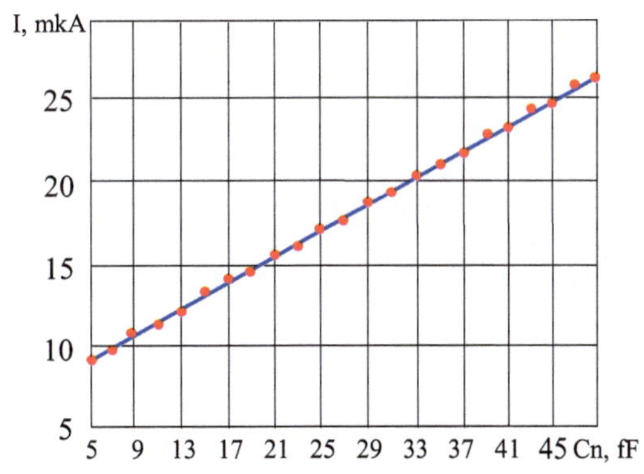

Fig. 1.57 Dependence of the current consumed a CMOS cell on the load capacitance

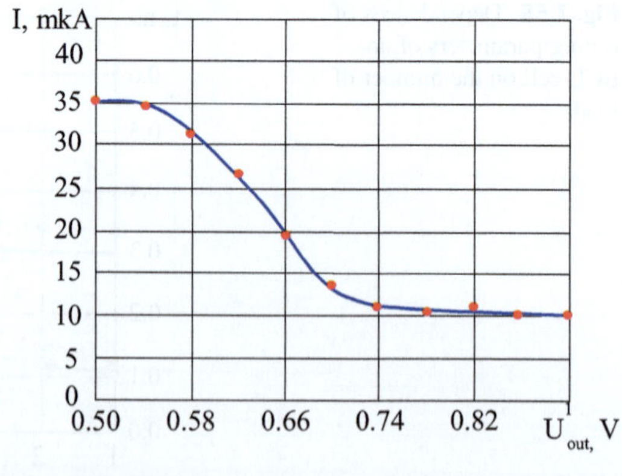

Fig. 1.58 Dependence of the cell current consumed by a CMOS cell on the level of the output voltage in the state "1"

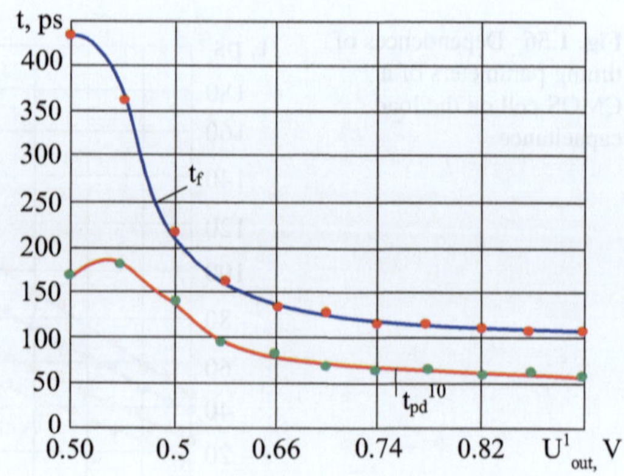

Fig. 1.59 Dependences of timing parameters of a CMOS cell on the level of the output voltage in the state "1"

1.1.2 Internal DF

Non-ideality of interconnects (W_1) has a dominant role in the characteristics of modern digital circuits [22, 25, 84, 85]. The influence of interconnects on the behavior of digital circuits has been studied by many groups of authors, among which the works of [7–9, 19, 22–27, 86–117] are particularly notable for the depth of the study of the significance of the effect of this type of DF on the operation of digital ICs.

The behavior of a digital circuit is affected both by internal interconnects of digital ICs (Fig. 1.60) [117], which are divided into global and local (Fig. 1.61) [22], and interconnects on the PCB (Fig. 1.62) [118] and within the package of a digital circuit (Fig. 1.63) [119].

The behavior of a digital circuit significantly depends on:

1. The intrinsic parasitic parameters R, C, L (1.1–1.3) [22] of a separate interconnect (Fig. 1.64) [25, 99], leading to distortion of the information signals passing through them, i.e. $Q_1(t) \neq Q_2(t)$ (Fig. 1.65);

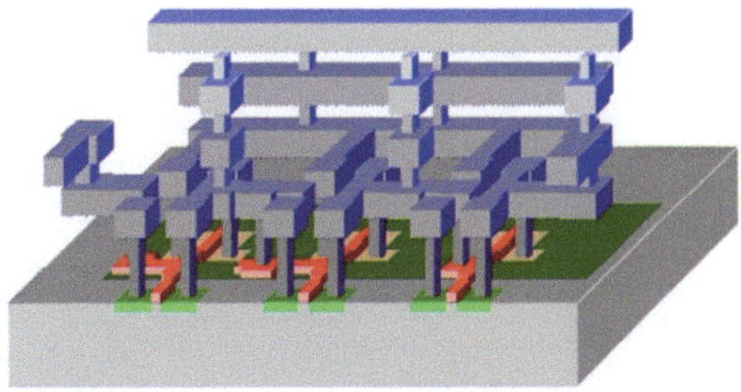

Fig. 1.60 Internal interconnects of digital ICs

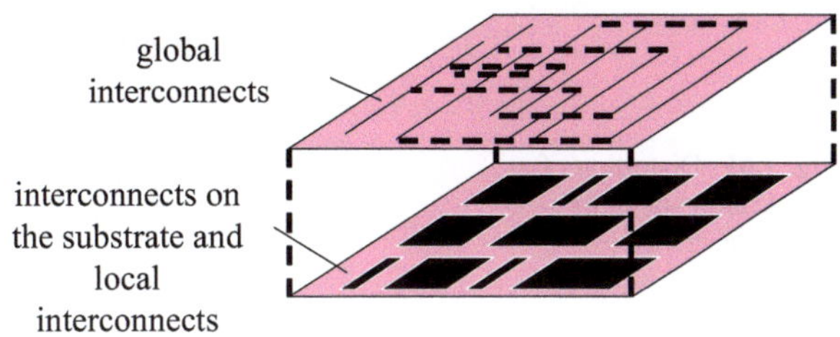

Fig. 1.61 Global and local interconnects of digital ICs

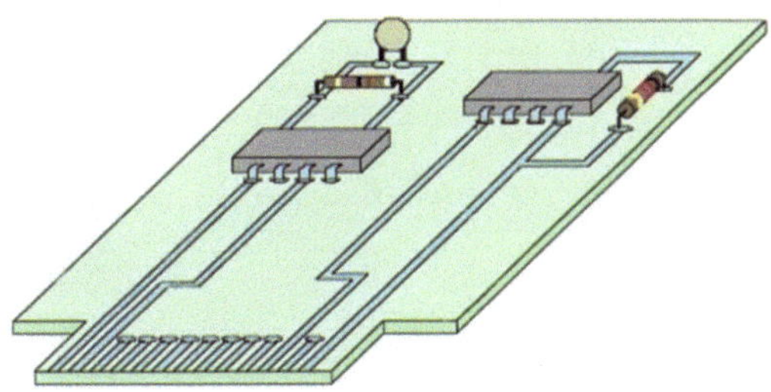

Fig. 1.62 Interconnects on PCB

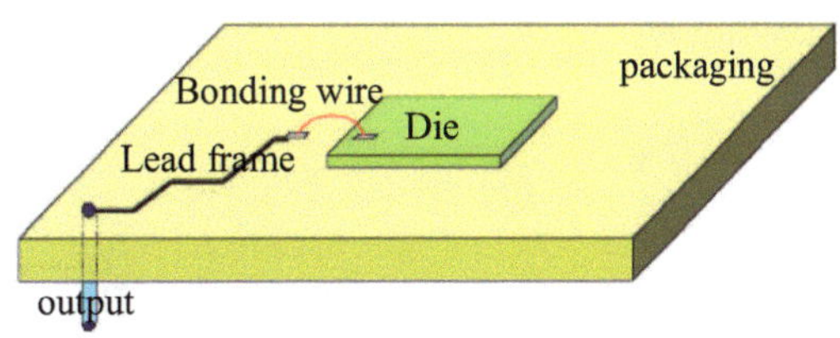

Fig. 1.63 Interconnects within a package

$$R = \frac{\rho \cdot l}{t \cdot w}, \tag{1.1}$$

$$C = \frac{\varepsilon \cdot w \cdot l}{t_{0X}}, \tag{1.2}$$

$$L = \frac{\mu_0 \cdot l}{2 \cdot \pi} \cdot \left[\ln \frac{2 \cdot l}{w + l} + \frac{1}{2} + \frac{0.447 \cdot (w + l)}{2 \cdot l} \right] \tag{1.3}$$

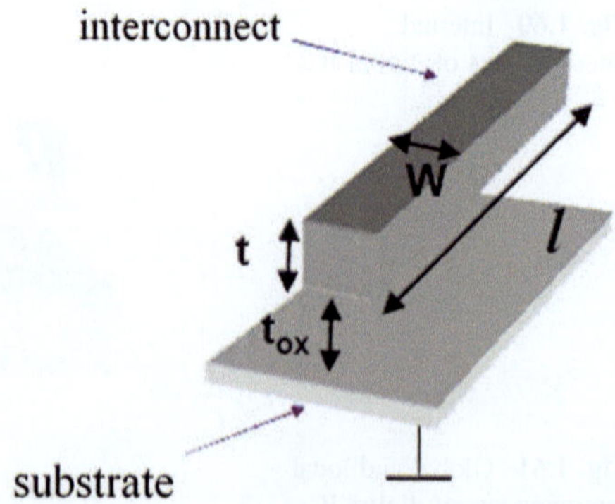

Fig. 1.64 A separate interconnect

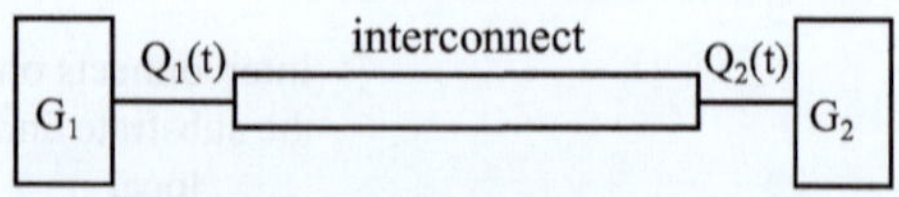

Fig. 1.65 Distortion of information signal by interconnect

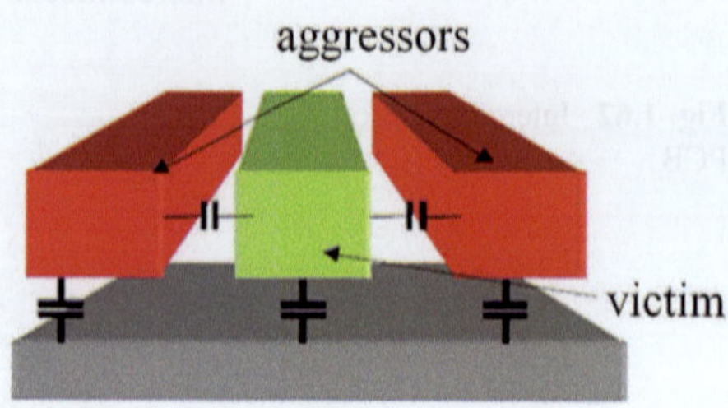

Fig. 1.66 Mutual parasitic interconnect parameters

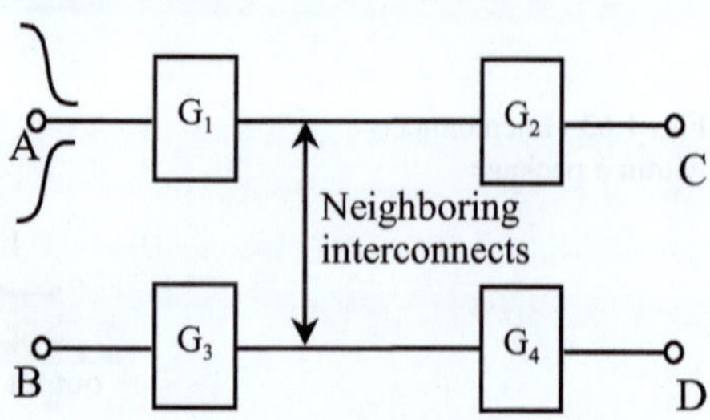

Fig. 1.67 Crosstalk of neighboring intradie interconnects

2. Mutual parasitic parameters between interconnects, as well as between interconnects and the IC substrate (Fig. 1.66) [110], leading to mutual influences of nearby intradie connections (Fig. 1.67).

The influence of the intrinsic parasitic parameters of a single interconnect has a greater effect on timing parameters of the information signal [5, 7, 26, 100–103]. Figure 1.68 illustrates graphs of the change in timing parameters at the output of gate (G_1) (Fig. 1.65), depending on the length of the interconnect, obtained using

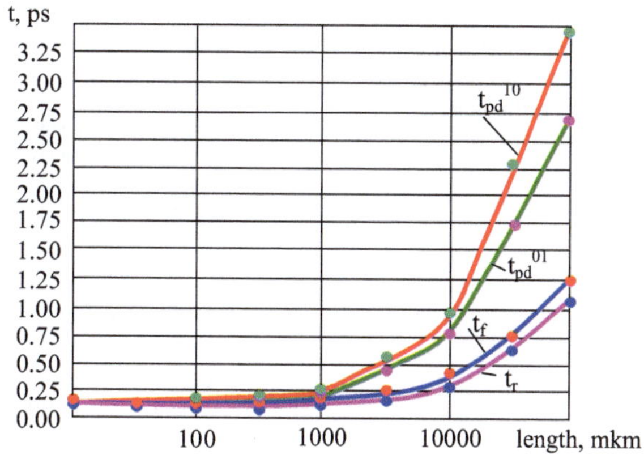

Fig. 1.68 Dependences of timing parameters of G_1 on the interconnect length

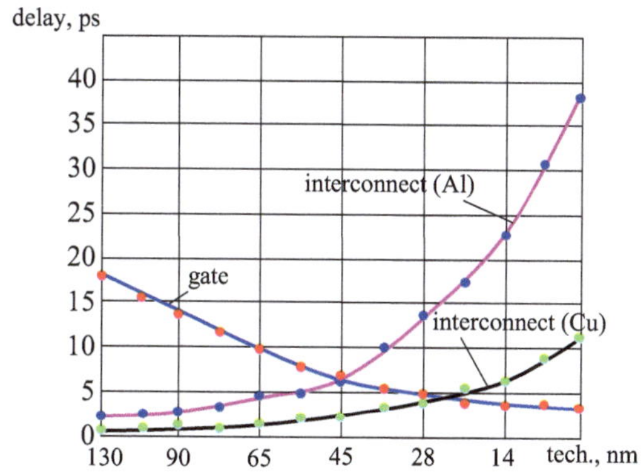

Fig. 1.69 Dependences of timing parameters of gates and interconnects from various materials on technology

SPICE. A standard CMOS cell was used as G_1 and G_2 (Fig. 1.15), the technology-45 nm, the interconnect parameters-$w = 0.14$ μm, $t = 0.31$ μm.

With the advance of the technology of manufacturing integrated circuits [64, 120, 121], the role of intradie interconnects in the operation of digital ICs became more and more tangible, and now the interconnects are already crucial in the formation of useful signals [8, 22, 121–124]. This is witnessed by the dependences (Fig. 1.69 [125] and Fig. 1.70 [126]) of delays of gate and interconnects from technology. As seen from the above graphs, the role of interconnect delays has now changed radically and often exceeds the intrinsic gate delay by several orders.

The explanation of this fact is as follows. The successes of microelectronics allow constantly the increase in the performance of logic gates [2, 35]. Ways to reduce the specific interconnect delays have almost exhausted themselves [92, 126]. The die area due to the success of the technology of their fabrication with time increases [126, 127]. Because of this, inter-cell distances increase (Fig. 1.71) [106], which leads to an increase in interconnect delays. Since a digital circuit is too sensitive to signal delays on its paths [15, 31, 128], there is a situation when the logic of the functioning of a digital circuit is largely determined by interconnect delays.

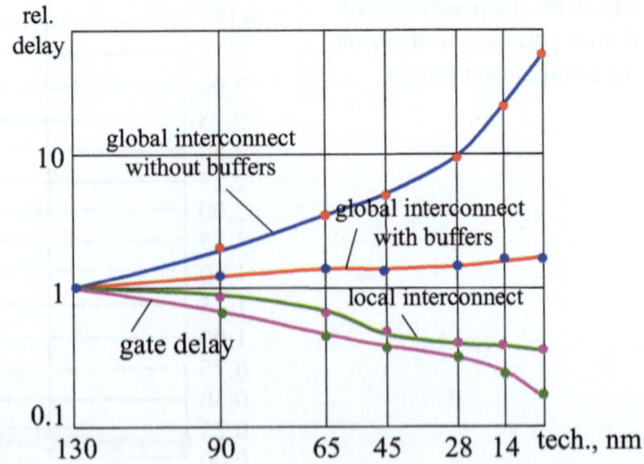

Fig. 1.70 Dependences of timing parameters of gates and interconnects of various types on technology

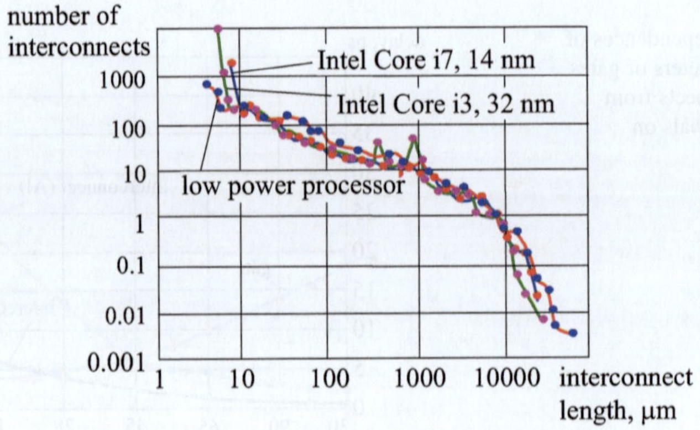

Fig. 1.71 Distribution of interconnect lengths in various microprocessor ICs

This circumstance forces digital circuit designers to look for radically new principles for their synthesis [31, 63, 128–132].

However, the intrinsic parasitic parameters of a single interconnect for nonstandard signal amplitudes (which, as already noted, often occurs during simulation of digital parts of mixed-signal ICs) may lead not only to signal delay during transition through the interconnects (Fig. 1.65) but also to the occurrence of such phenomena as gradual loss of the signal amplitude when passing through the interconnect (Fig. 1.72 shows signals in equidistant interconnection points with the parameters: l = 125 mm, $w = 0.14$ μm, $t = 0.31$ μm, the length of each segment-25 mm, the technology-28 nm) or the propagation of the "series" of pulses (Fig. 1.73 shows the signals at the inputs of G_1 and G_2 with the following interconnect parameters-$l = 30$ mm, $w = 0.14$ μm, $t = 0.31$ μm, technology-28 nm). The diagrams in Figs. 1.72 and 1.73 were obtained by SPICE simulation of the circuit in Fig. 1.65 [100].

As the technology of manufacturing digital ICIs [64, 120, 121] has advanced, the roles of various parasitic parameters of interconnects have also changed (Fig. 1.74).

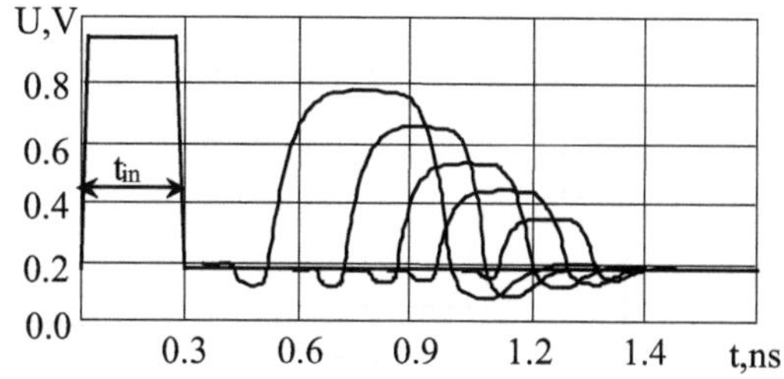

Fig. 1.72 Signal attenuation while passing through interconnects

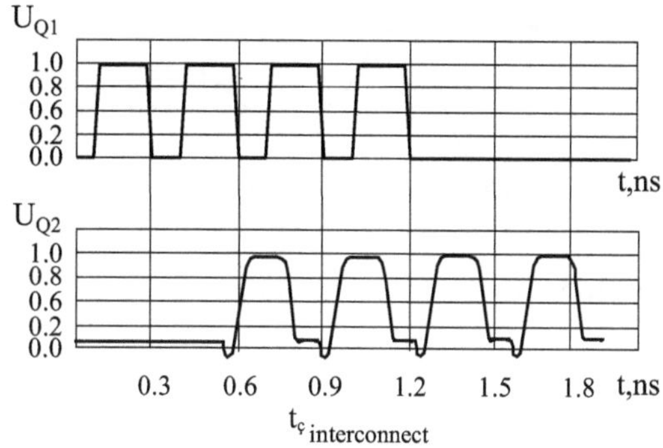

Fig. 1.73 Propagation of the "series" of pulses through interconnects

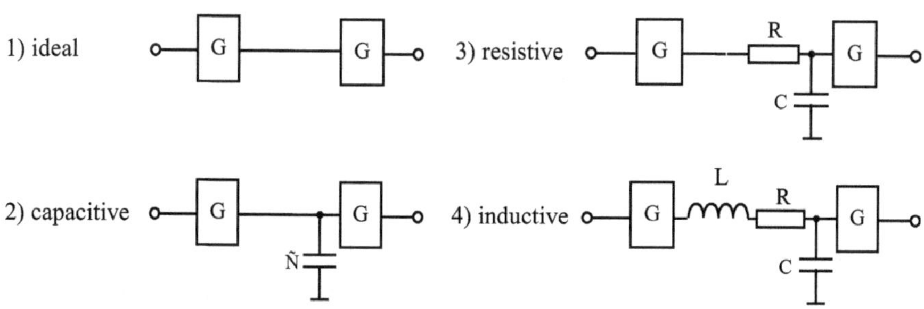

Fig. 1.74 Changing the roles of parasitic interconnect parameters

In particular, recently the role of the parasitic inductance of interconnects has increased dramatically [7, 91, 95, 96, 104, 107, 108]. It is known that the complex resistance of interconnects is defined in this form:

$$Z = R + j \cdot \omega \cdot L + \frac{1}{j \cdot \omega \cdot C}. \tag{1.4}$$

Increasing the frequency leads to an increase in the role of the inductive component. This, in turn, depending on the ratio between the parameters of gates and

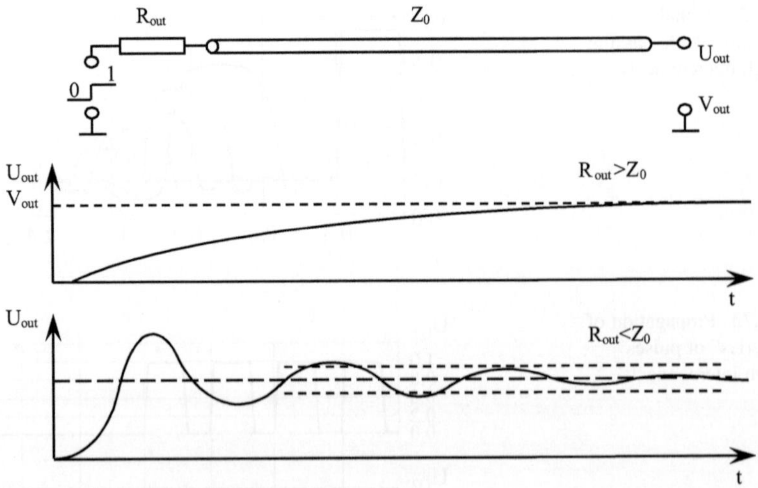

Fig. 1.75 Signal source with interconnects

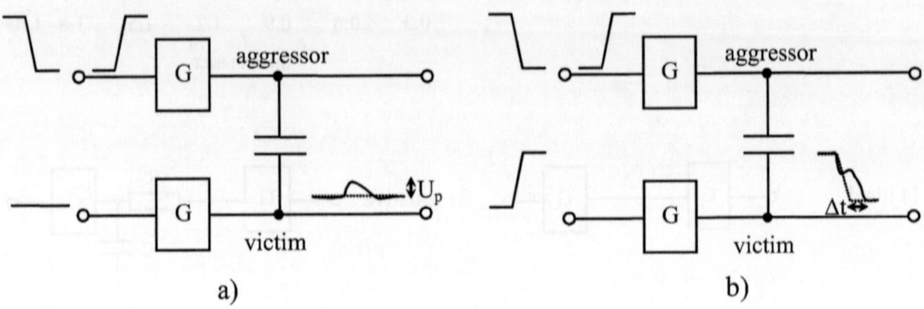

Fig. 1.76 Critical consequences of interference noise: (**a**) glitch; (**b**) delay change

interconnects, can lead to new phenomena, in particular, to oscillation of the signal (Fig. 1.75 [95]).

As seen from Fig. 1.75, the output resistance of most logic gates is loaded into an interconnect with a complex resistance Z_0. Depending on the ratio R_{out} and Z_0, different waveforms are possible. In particular, when R_{out} is smaller than Z_0, signal oscillations appear. As for the mutual parasitic parameters between interconnects (Fig. 1.66), their significance for the functioning of modern digital circuits is too large [114, 115]. They lead to crosstalk, which has two critical consequences for the victim when switching aggressors: glitches (U_P, Fig. 1.76a) in the static state of the victim and delay changes (Δt, Fig. 1.76b) in the state of switching the victim. Figure 1.77 shows U_P and Δt (Fig. 1.77b) for fragment of a digital circuit, the equivalent circuit of which is shown in Fig. 1.77a [22].

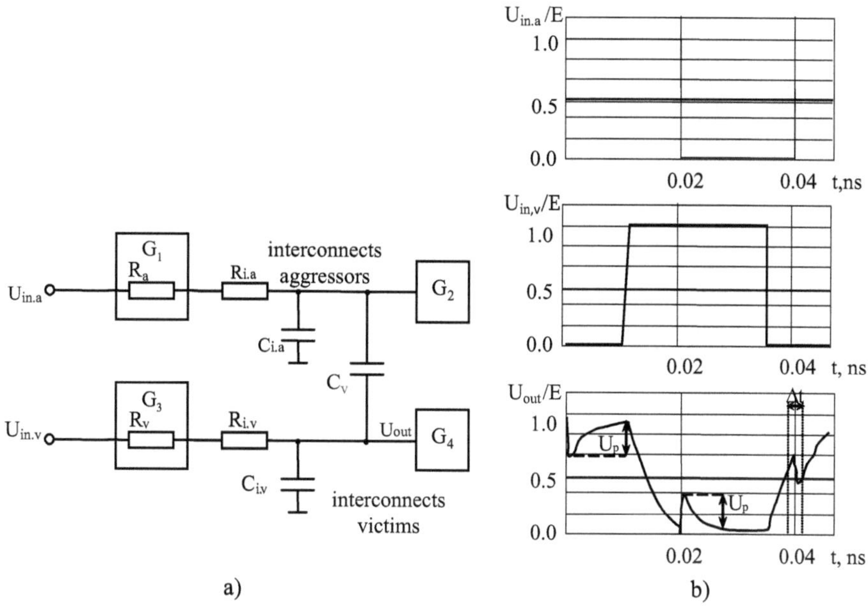

Fig. 1.77 Glitches and delay changes: (**a**) an equivalent circuit; (**b**) timing diagrams

Fig. 1.78 Crosstalk signal parameters

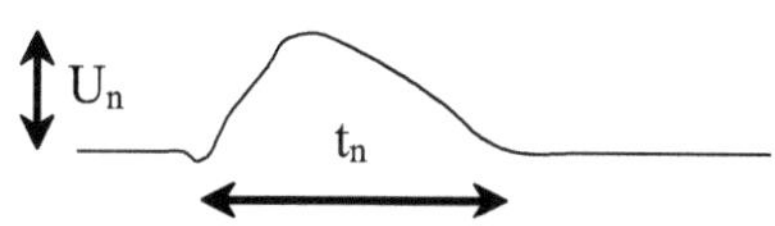

The effect of the crosstalk signal on the functioning of a digital circuit depends on:

- The ratios of parameters of noise signal (Fig. 1.78) and the gate victims. Figure 1.79 shows an example where, depending on the U_n ratio and the threshold voltage of the victim, the output of a digital circuit can switch to an unintended error state (Fig. 1.79a) or remain in the same state (Fig. 1.79b).
- From the location of noise signal. If the noise with a small U_n can be extinguished during propagation in the combinational part of a digital circuit (Fig. 1.79b), then the noise signal with the same U_n on arrival directly to the sequence fragments can switch the digital circuit to the wrong state (Fig. 1.79c) (Fig. 1.80).

Figures 1.81 and 1.82 show the dependences of U_n and t_n on the interconnect length for the circuit in Fig. 1.80, where a standard CMOS cell was used as a gate (Fig. 1.15).

As seen from the graphs given by SPICE, the crosstalk is commensurate with useful signals even with small interconnect length. If the presence of a large number of interconnects with a much longer length (Fig. 1.71), as well as the summation factor of crosstalk (1.5 [110], where X_i is the capacity of interference between the victim and ith aggressor) in the presence of several aggressors (Fig. 1.83) is taken into account in digital circuits,

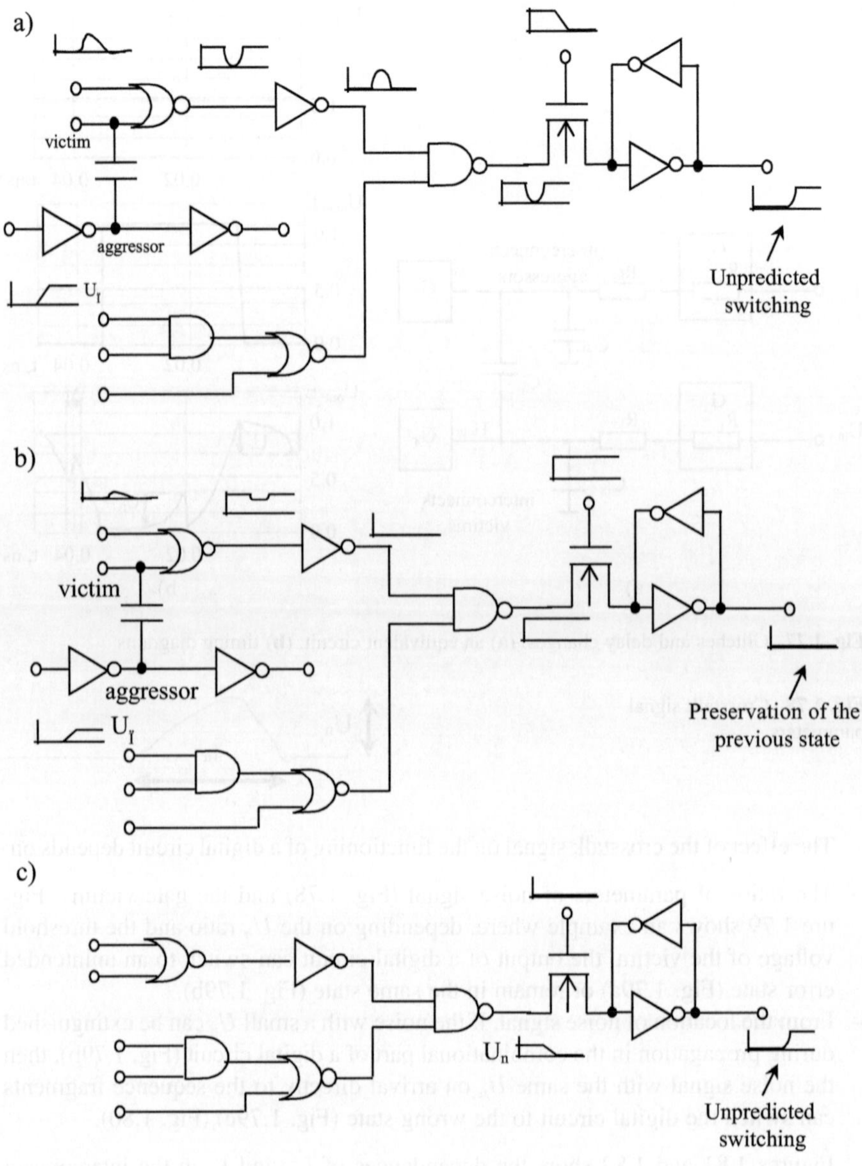

Fig. 1.79 Example of the effect of crosstalk on a digital circuit: (**a**) noise with a large U_n; (**b**) noise with a small U_n; (**c**) arrival of noise directly to sequential fragments

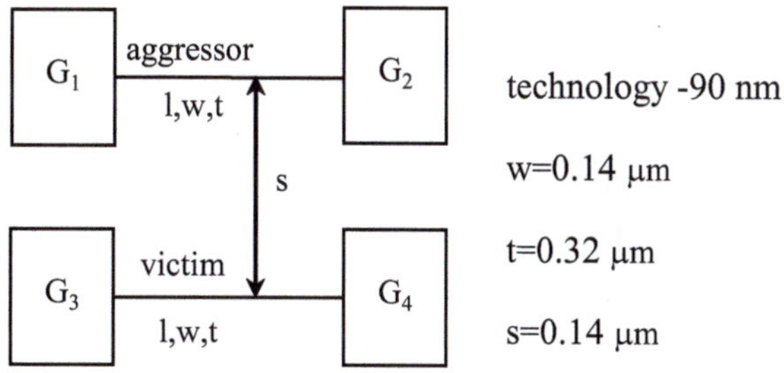

Fig. 1.80 The scheme for obtaining dependencies of the parameters of crosstalk signal

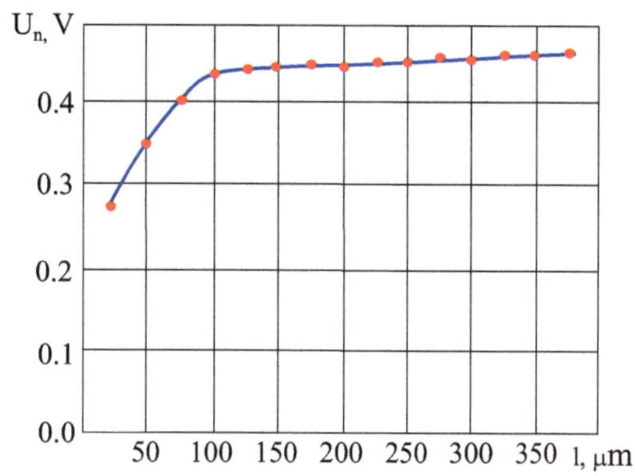

Fig. 1.81 Dependence of the amplitude of crosstalk signal on interconnect length

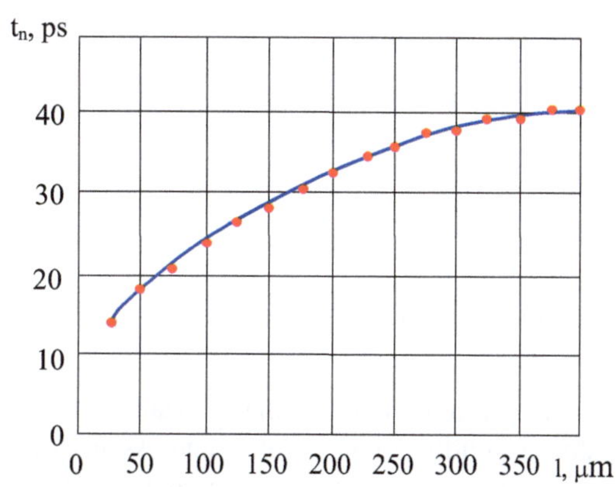

Fig. 1.82 Dependence of duration of crosstalk signal on interconnect length

$$U_{\mathrm{p}} = \frac{\sum\limits_{R_i \in P(0)} X_i R_i}{\sum\limits_{C_i \in C} C_i R_{ii}} \tag{1.5}$$

then it is not difficult to see that crosstalk plays a huge role in the functioning of digital circuits, and their consideration in the simulation of digital ICs is mandatory. Otherwise, this can lead to both quantitative (Fig. 1.84 [19]) and qualitative errors (Fig. 1.85 [133])).

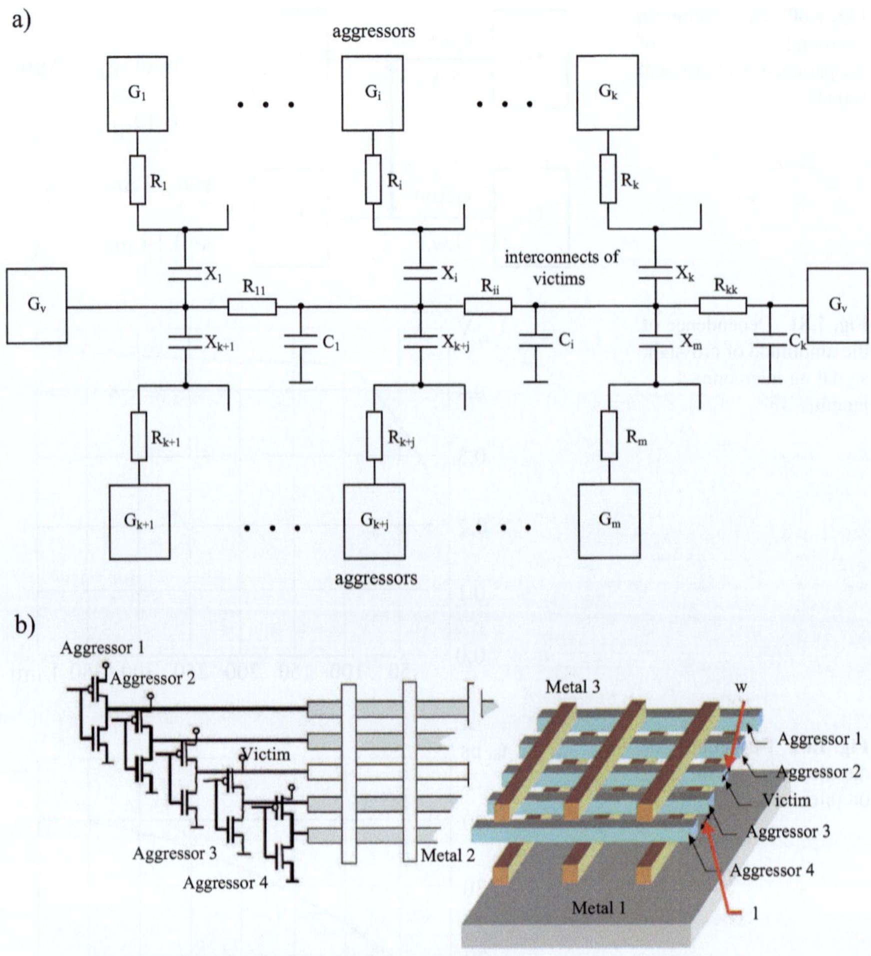

Fig. 1.83 Summation of noise from several aggressors

It is also important that the crosstalk can propagate in digital circuits on different paths, depending on its current state. As an example, Fig. 1.86b illustrates a graph of propagation of the interference in the circuit of Fig. 1.86a. This circumstance complicates the predictability of the behavior of a digital circuit in the conditions of crosstalk.

The role of the influence of crosstalk due to parasitic interconnection parameters on the functioning of digital circuits is so important that it is considered one of the main indicators of the quality of the design of modern digital circuits (Fig. 1.87) [111]. It becomes even more because of the increasing role of the inductive component [11, 124] ($L \cdot di/dt$) of crosstalk (Figs. 1.88 [9], 1.89, 1.90), although it is important mainly for global interconnects (Fig. 1.91 [26]). Figure 1.92 shows the results of the simulation of the circuit of Fig. 1.80 using SPICE. A standard CMOS

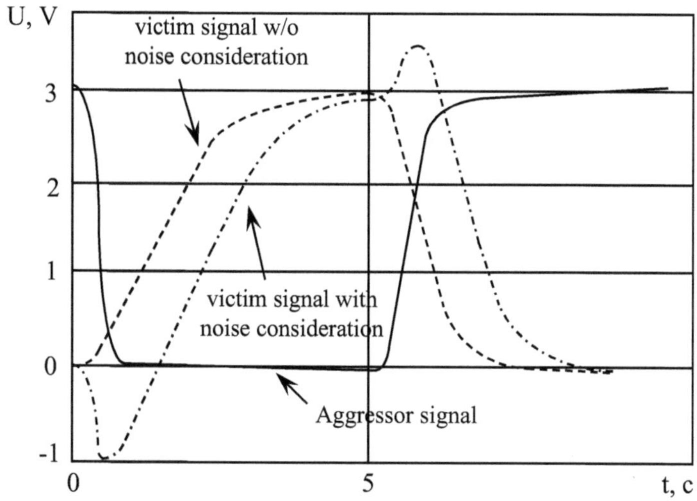

Fig. 1.84 Example of a quantitative error in simulation results with and without crosstalk

cell (Fig. 1.15) was used as a gate, taking into account the parasitic inductances ($L_c = 0.5$ pG/μm, $L_B = 0.3$ pH/μm), capacitances ($C_c = 36$ aF/μm, $C_s = 62$ aF/Mm) and resistance ($R = 0.4$ mΩ/μm) of interconnects. The results clearly demonstrate the importance of crosstalk on the behavior of digital circuits, including the inductive component, which generates signal oscillations.

Because of the significance of the effect of crosstalk due to the parasitic parameters of interconnects on the functioning of modern digital circuits, various methods are used to increase the resistance to this type of DF [22, 24, 25, 89, 111, 132, 134, 135]:

1. Logical methods (coding of polarity of signals, etc.)
2. Shielding of interconnects and increasing distances between interconnects
3. Rerouting of critical paths
4. Selection of transistor sizes in gates which serve as a source of the signal
5. Addition of buffers on long interconnects (Fig. 1.93)
6. Other strategies for setting up interconnects (Fig. 1.94)
7. Some new approaches to the synthesis of digital circuits

One way to reduce crosstalk is to improve technology. In particular, Cu ($\rho_{Cu} \approx 1/3\rho_{Al}$) is used as an interconnect material instead of Al (Fig. 1.95 [24]), the metal layer and contacts are made simultaneously, reduce the k factor from 3.9 to 2, etc.

However, all the mentioned methods of increasing the noise immunity of a digital circuit only slightly reduce the influence of parasitic parameters of interconnects on the behavior of IC, and they must be taken into account while simulating modern digital circuits.

The non-ideality of power supply chains (W_2), which are also distributed chains of parasitic R, L, C parameters (Fig. 1.96), causes mutual interference of gates,

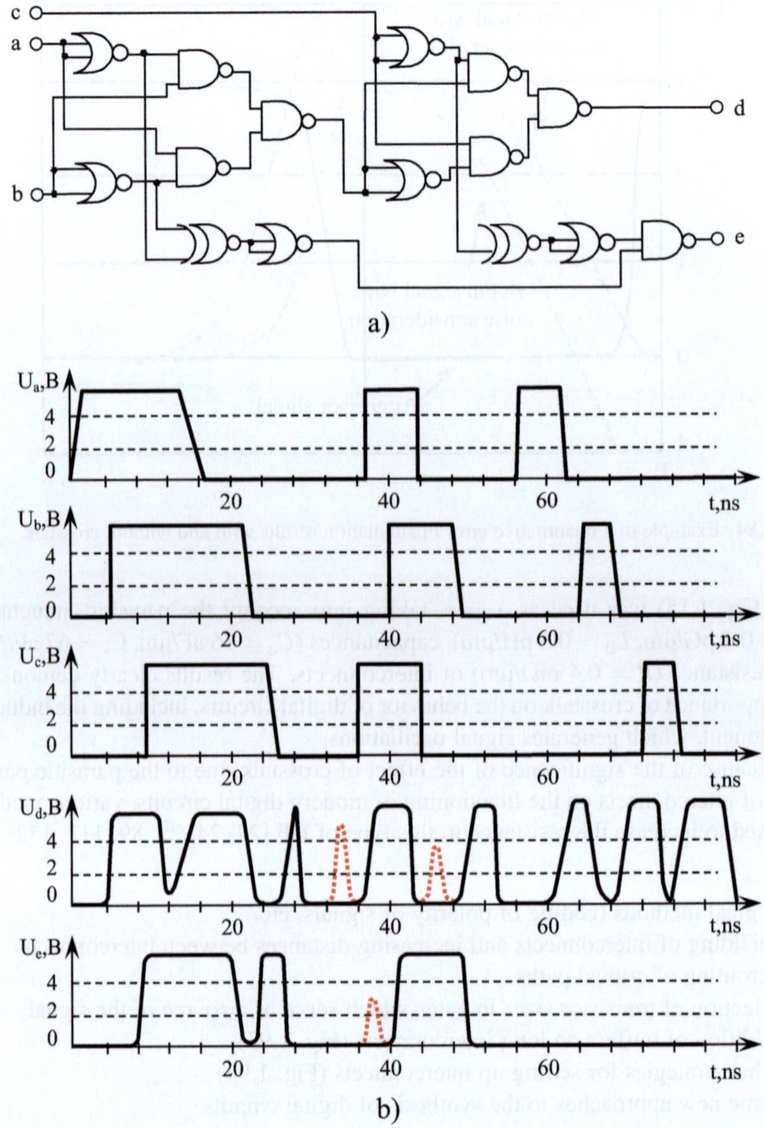

Fig. 1.85 An example of a qualitative error in simulation results with and without consideration of crosstalk

functionally unrelated with each other, through the power rail (Fig. 1.97). The situation with the common bus is similar.

The issue of influence of parasitic parameters of power rails on the behavior of digital circuits has been studied by many groups of authors, among which the works of [12, 14, 19, 30, 124, 136–146] are particularly distinguished by the depth of the

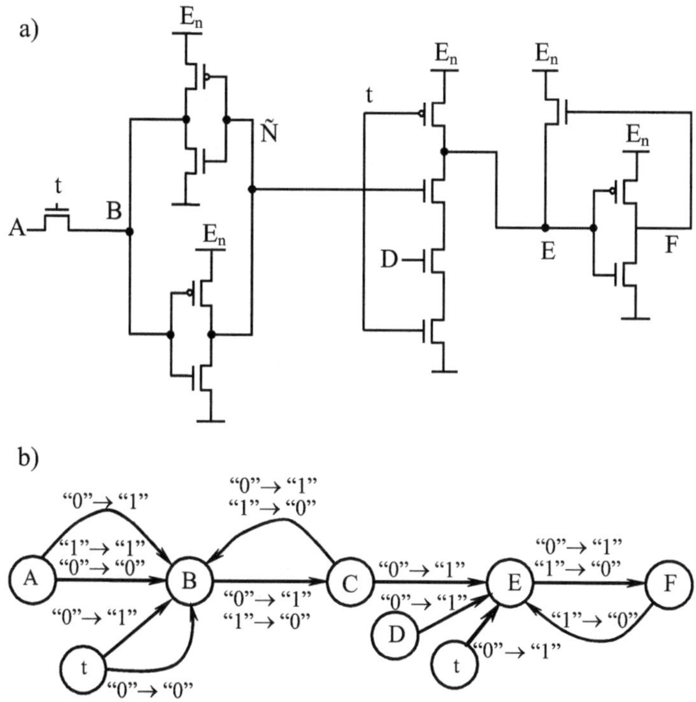

Fig. 1.86 Noise propagation graph (**a**) an example of a scheme; (**b**) graph of propagation

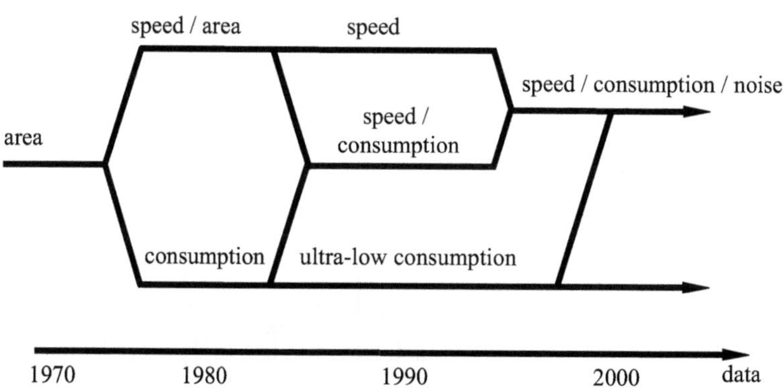

Fig. 1.87 Dynamics of the importance of the design parameters of a digital circuit

study of the significance of the effect of this type of DF on the functioning of digital circuits.

It is known that, depending on the logical state of the G_1, especially during its switching, the value of the current I^1_{cons} consumed by this power supply from the power circuit varies. Therefore, the supply voltage of G_2 changes:

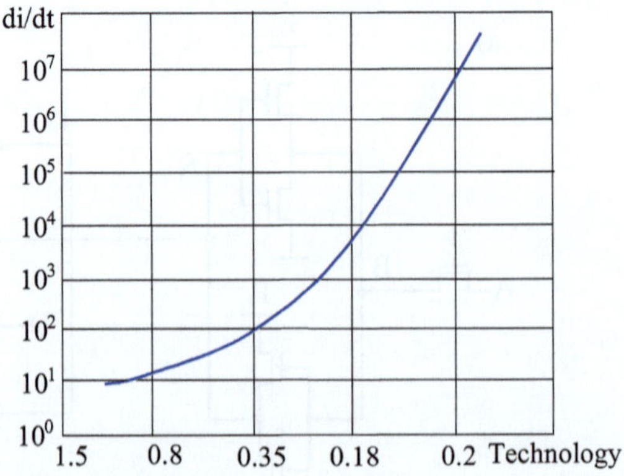

Fig. 1.88 Change of *di*/*dt* depending on technology

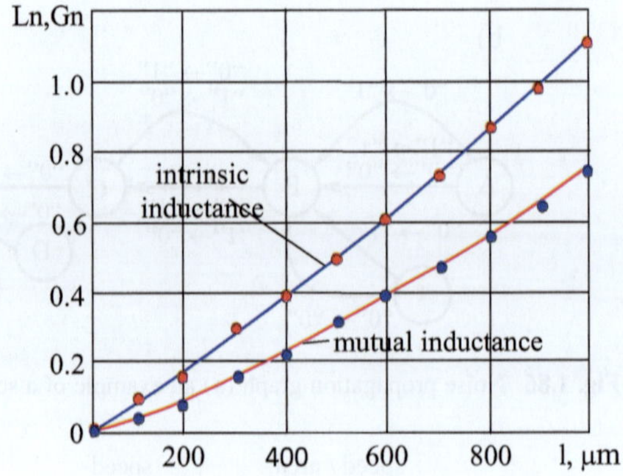

Fig. 1.89 Dependence of parasitic inductances of interconnect on its length ($w_1 = w_2 = s = 10$ µm, $t = 0.31$ µm)

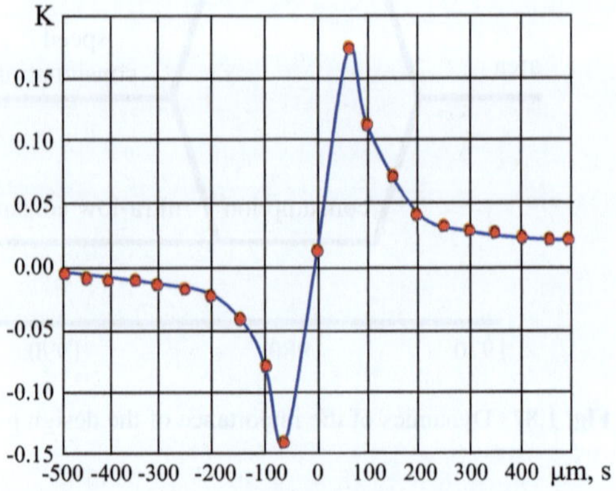

Fig. 1.90 Dependence of normalized mutual inductance (*k*-factor) on the distance between interconnects

$$E_{G_2} = E_{ideal} + L \cdot \frac{dI^1}{dt} + R \cdot dI^1, \tag{1.6}$$

i.e., the noise through the power circuit has two main components:

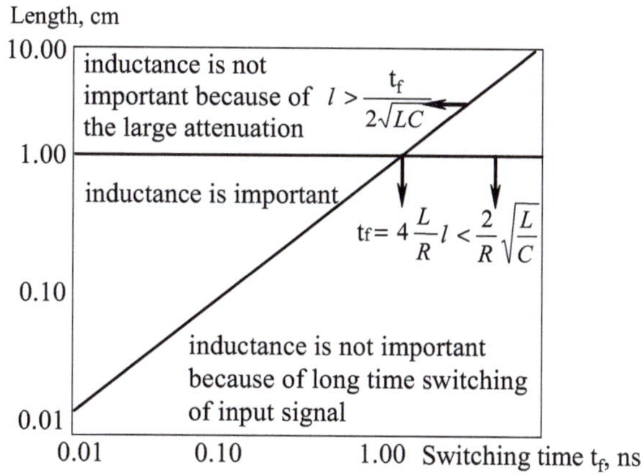

Fig. 1.91 The area where the role of parasitic inductance of interconnects is important

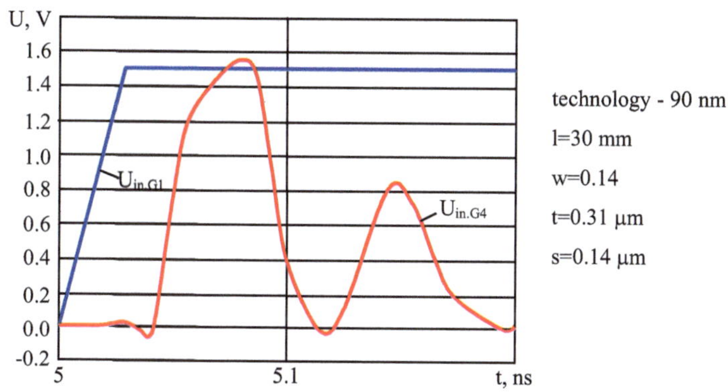

Fig. 1.92 Results of simulating the circuit in Fig. 1.80 with parasitic parameters of interconnects

1. Noise due to switching of signal $L \cdot di/dt$ [11]
2. Noise due to voltage drop in the power circuit *IR* [14]

As a result, the I^2_{cons} changes. This, in turn, due to parasitic parameters of G_2 (Fig. 1.98) affects all parameters of the signal Q_{out2} and can lead to various kinds of errors and failures.

Thus, the function $Q_{\text{out2}} = f_1(Q_{\text{in2}})$ is actually replaced by $Q_{\text{out2}} = f_2(Q_{\text{in2}}, Q_{\text{in1}})$. If one takes into account that on G_2, except G_1, all other gates of the circuit can affect, which appear through the power supply circuit of noise, even with consideration of difference in the signs of separate components, can become significant.

Noise in the power supply can also lead to indirect effects:

1. Crosstalk between the power supply and the signal nets of digital circuits (Fig. 1.99 [19])
2. Interference through the substrate (Fig. 1.100 [145]).

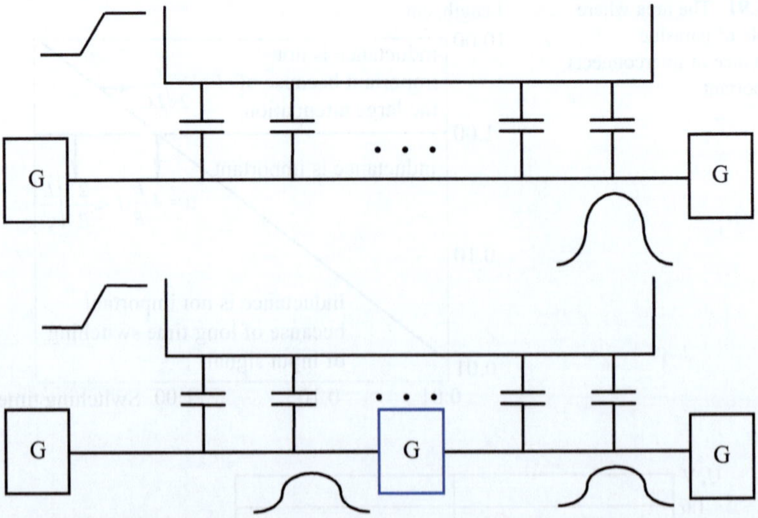

Fig. 1.93 Adding buffers on long interconnects

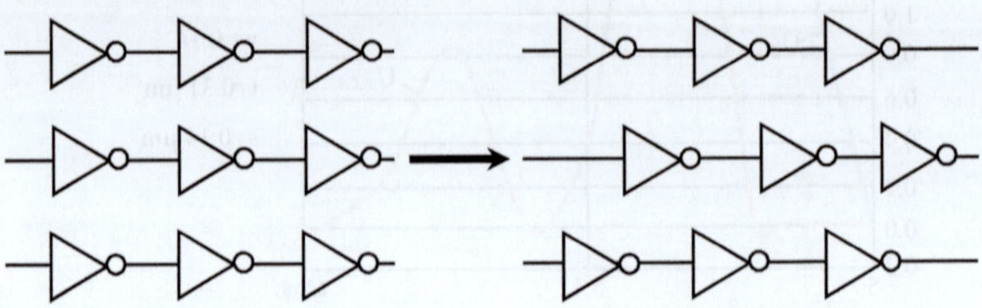

Fig. 1.94 Example of setting up interconnects

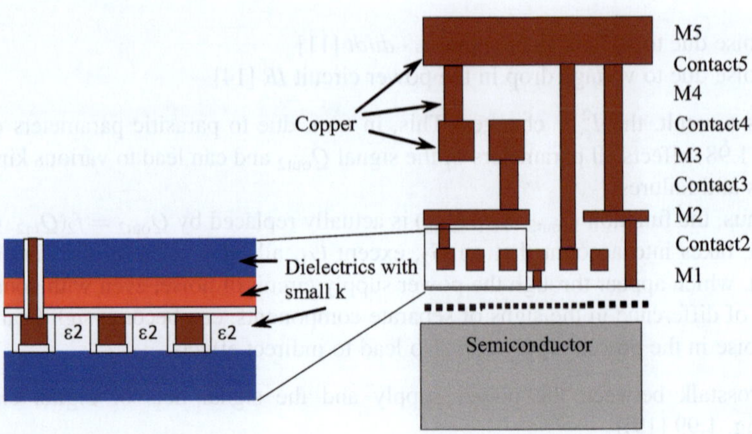

Fig. 1.95 Technological ways to reduce crosstalk

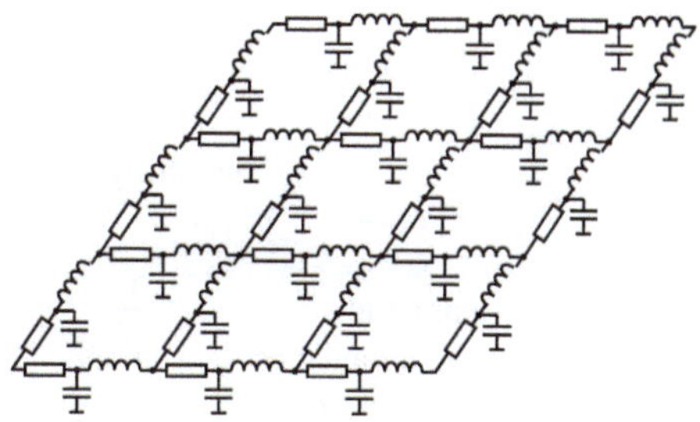

Fig. 1.96 Distributed network of parasitic parameters of power nets of digital ICs

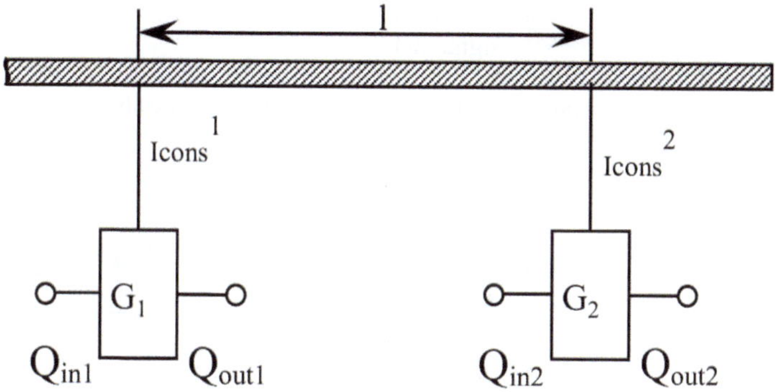

Fig. 1.97 Interaction of functionally unrelated gates through the power rail

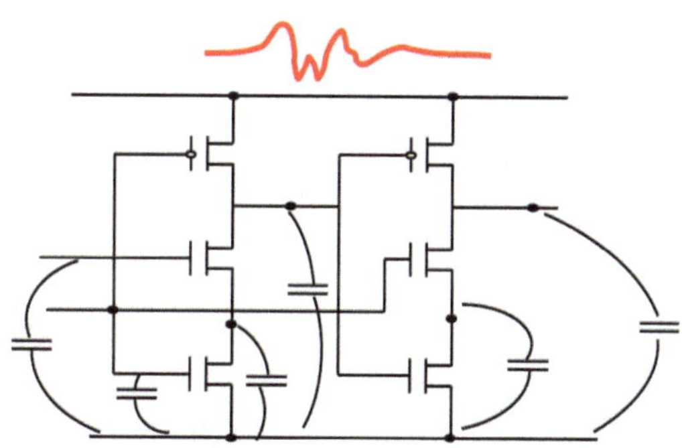

Fig. 1.98 The mechanism of failures due to crosstalk in the power rail

The level of crosstalk through the power supply circuit depends on the number of simultaneously switching gates (Fig. 1.101 [138]), as well as the distance l between the *G* aggressor and the *G* victim (Fig. 1.97). Figure 1.102 shows the dependence of the amplitude of the noise on l for the circuit in Fig. 1.97, obtained with the help of SPICE, when a standard CMOS cell was used as *G* (Fig. 1.15).

In modern digital circuits, the number of simultaneously switching gates and their distances to the *G*-victims are such [115] that noise of this type is significant and can influence the behavior of digital circuits. Figure 1.103 illustrates examples of measuring crosstalk through a substrate in a real digital IC [145].

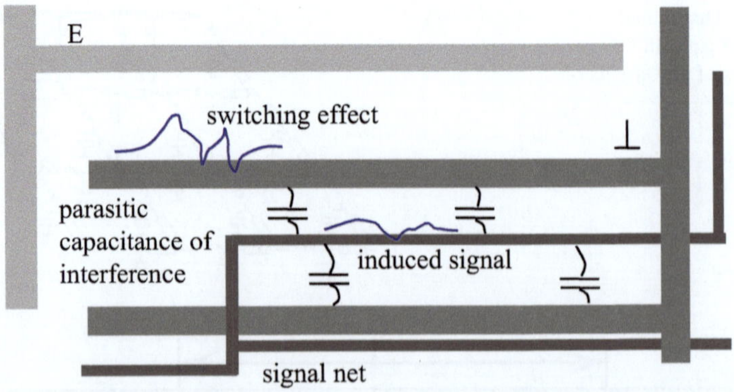

Fig. 1.99 Crosstalk between power supply and signal nets of digital circuits

Fig. 1.100 Interference through substrate

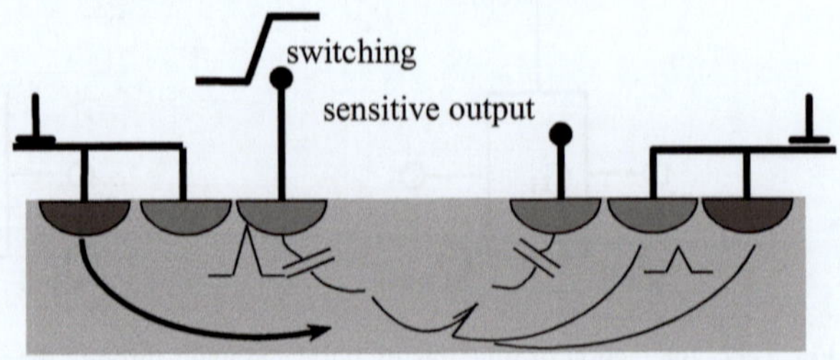

Fig. 1.101 The form of the crosstalk signal in the power supply net of a digital IC at different numbers (1869 and 1239) of simultaneously switching gates at a time of 200 ps

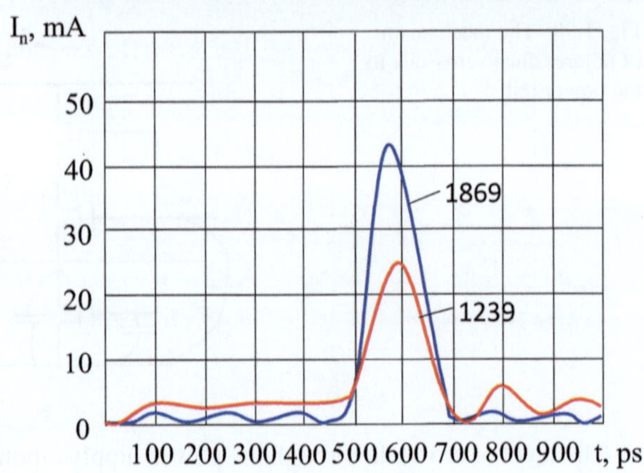

To avoid noise through power circuits, various solutions are also used [110, 112, 136, 138]: technological (use of copper, low *k*-factor, etc.), circuit level (use of shunting capacitors, current limitation, etc.), physical design (use of special placement and routing strategies). However, all the mentioned methods of increasing the noise immunity of digital circuits only slightly reduce the influence of parasitic parameters of the power rails on the behavior of IC, and they must be taken into account in the simulation of modern digital circuits.

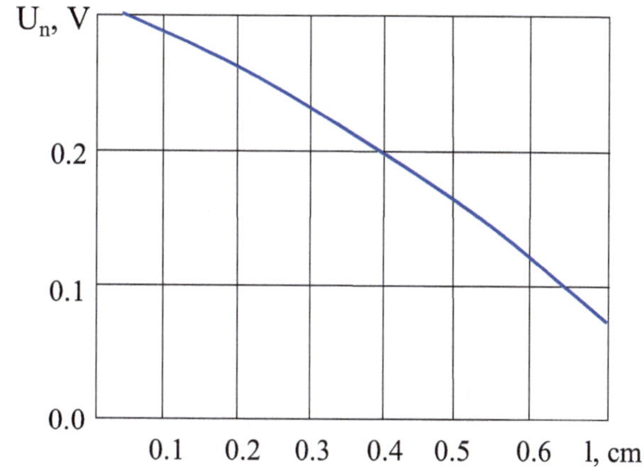

Fig. 1.102 Dependence of noise in the supply net on the distance between the gate aggressor and the *G*-victim (technology-28 nm, $w = 3$ μm, $t = 0.31$ μm, $s = 0.18$ μm)

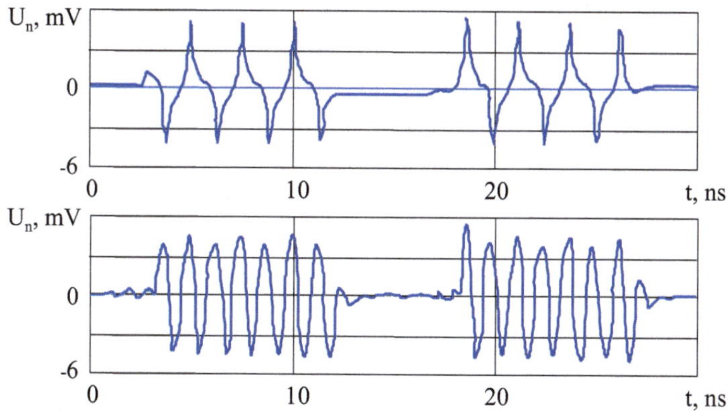

Fig. 1.103 Results of measuring crosstalk through a substrate in a real digital IC

Thus, the noise that occurs in the power supply nets of digital circuits, when switching gates, is one of the most significant internal DFs, the consideration of which in the simulation programs is mandatory.

As for the influence of the internal structure of digital ICs (W_3) on its behavior, here, first of all, the variability of the number of load cells for different gates inside the IC is meant. Thus, different gates operate in different modes, which, in turn, significantly affects the shape of their output signals. This is also confirmed by the dependences in Figs. 1.51–1.57.

Other structural features include the use of a different number of parallel outputs of gates (non-inverse and inverse), implementation of logic functions by means of interconnects with different number of inputs, etc. These circumstances can also significantly affect the shape of the output signal of the gate (Fig. 1.104).

The significance of non-ideality of the input signal sources of the entire circuit (V_4) was shown above. However, the distortions of the input signal form for internal gates of ICs (W_4) are no less significant.

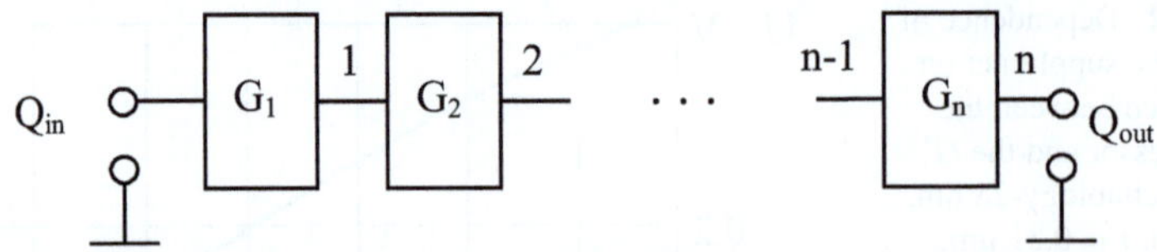

Fig. 1.104 A chain of sequentially connected TTL gates

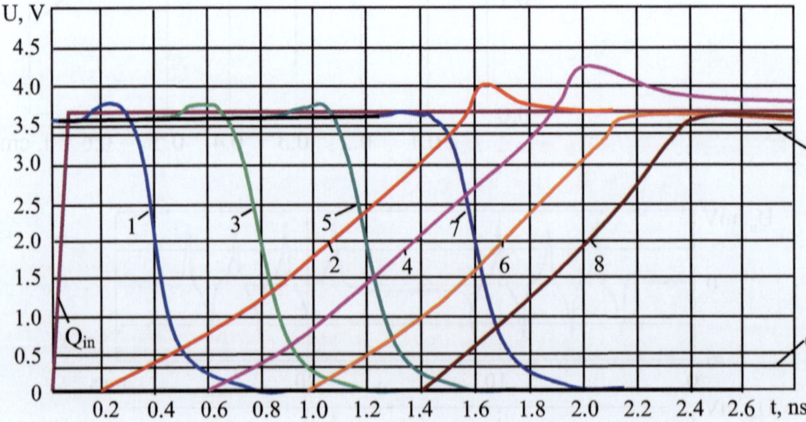

Fig. 1.105 Transient processes in a chain of sequentially connected TTL gates

At first glance, it seems that even if inputs of digital part of the circuit receive signals with nonstandard parameters, then when the signal passes through a certain number of elements, the signal characteristics are restored. This is really so. For example, Fig. 1.105 shows diagrams of voltage changes in a chain of sequentially connected TTL cells (Fig. 1.104), when a rectangular signal switching is applied to its input (the circuit of Fig. 1.13 is used as a TTL cell).

It can be seen from the graphs that if you do not take into account partial parameter shifts before the signal is established, this in itself can become a source of incorrect simulation results. In addition, signals with arbitrary amplitudes and durations can appear in any node of a digital circuit (for example, due to signal race). At the same time, nonstandard signals can always arrive at the input of gates in digital parts of mixed-mode ICs, connected to the analog part. And this, as can be seen from Figs. 1.47 to 1.50, significantly affects the parameters of gates.

From Figs. 1.33 to 1.40, a significant dependence of the behavior of a digital circuit on the ambient temperature (V_2) is seen. However, due to the difference in the switching activity of a digital cell in the IC, the local temperature (W_5) in different regions of the die can be different (Fig. 1.106 [4]), which can lead to a change in the gate parameters just as in Figs. 1.33–1.40. And this means that the behavior of a digital circuit is strongly dependent on the local temperature.

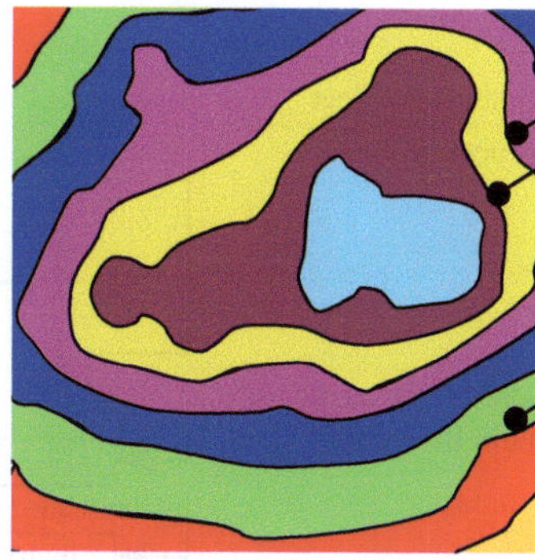

Fig. 1.106 Example of local temperature distribution in real digital ICs

Summarizing the analysis of external and internal DFs (Table 1.1), this can be concluded:

1. The significance of the influence of external and internal DF on the behavior of modern digital circuits is decisive.
2. With the advance of IC fabrication technologies [64, 121], the role of DF in the functioning of a digital circuit is steadily increasing [1, 31, 67, 81, 147–149].
3. All DFs listed in Table 1.1, play a significant role in the functioning of a digital circuit, and ignoring them with a high probability will lead not only to quantitative, but also to qualitatively inaccurate simulation results.

1.2 Analysis of the Current State of Simulation and Optimization of Digital Circuits in Terms of DF Consideration

As follows from Sect. 1.1, DFs play a decisive role in the operation of modern digital circuits, and if they are ignored, the simulation results can often prove to be even qualitatively incorrect. Therefore, the EDA tools for simulating and optimizing digital circuits must necessarily take into account the influence of DF.

Let's consider the possibilities of existing systems of simulating digital circuits in this aspect.

The available tools of analyzing digital circuits are classified according to simulation levels [1–3, 81, 150–155], given in Table 1.5. This classification is based on the degree of detail of the simulated digital circuit.

At the system level of simulation, a device consisting of large blocks such as a processor, RAM, data transmission channel, etc., is analyzed. The characteristic tasks solved at this level are information matching of separate blocks, determination of the capacity of channels, etc. The mathematical apparatus used is, as a rule, the theory of queuing systems. The most well-known systems of this level are: GPSS,

Table 1.5 Simulation levels of a digital circuit

N	Level	Simulated object	Example of a simulated object	Mathematical apparatus
1	System	Structural scheme	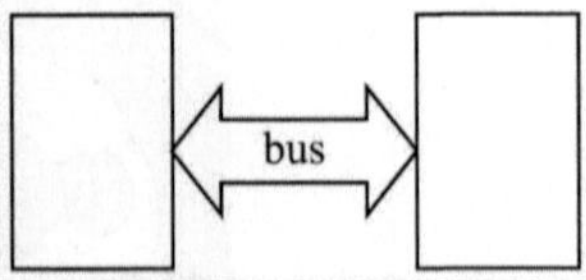	Queuing theory
2	Register-transfer	Scheme, consisting of multi-bit circuits	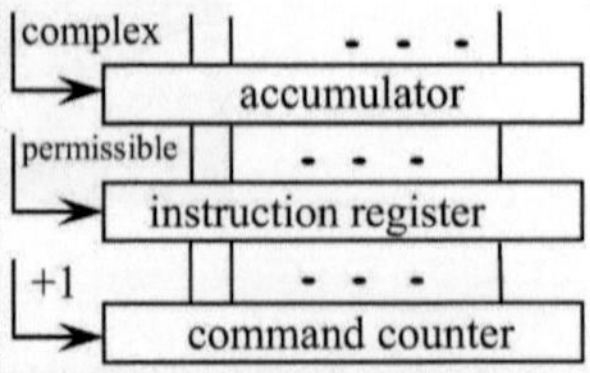	Boolean algebra
	Gate	The circuit at the level of gates and flip-flops	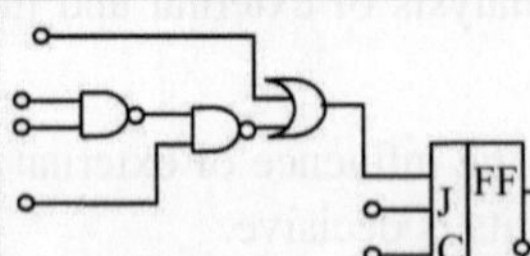	
3	Circuit	Circuits at the level of semiconductor devices	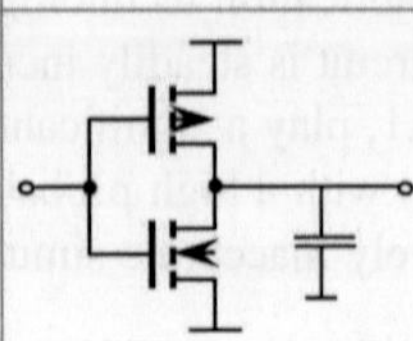	Systems of differential equations
4	Device	Internal processes of semiconductor device	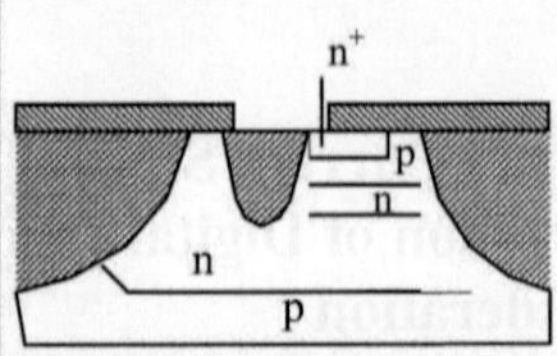	Systems of partial differential equations

DISS, CSS, OASIS, COCENTRIC SYSTEM STUDIO, VCC, SABER, etc. [150, 151, 156–159].

At the register-transfer level (Fig. 1.107a), the simulated digital circuit is detailed to multi-bit elements such as registers, counters, and decoders. At the gate level, the simulated digital circuit is represented by separate gates and flip-flops. The simulation tasks are: verification of the logic of functioning of microprograms, timing matching of the operation of separate cells, calculation of the maximum speed of a digital circuit, detection of various failures in the functioning of the circuit, etc. The basis for constructing gate-level simulation systems is Boolean algebra. The most popular programs for digital IC designers at different times were: F/LOGIC, TEGAS, LAMP, MOSSIM, SLS, LOSTIN, MIMIC, LMOSII, LECSIM, HILO, ADM, SCIROCCO, LEAPFROG, SPEEDSIM, COBRA, CYCLONE, VCS, MODELSIM, etc. [150, 152, 155, 156, 160–200].

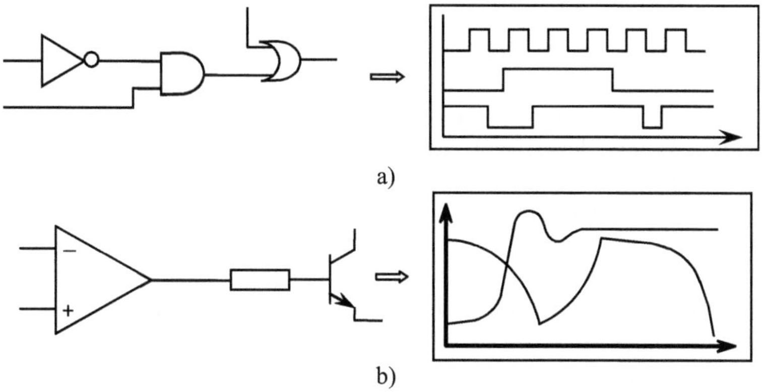

Fig. 1.107 Gate-level (**a**) and circuit-level (**b**) simulation

At the circuit-level simulation (Fig. 1.107b), the functioning of small fragments of circuits, detailed to separate semiconductor devices (transistors, diodes, resistors, etc.) is checked. In this case, detailed forms of voltage and current variations in the circuit nodes are obtained, on the basis of which it is possible to solve such problems as electrical matching of separate devices, sensitivity analysis with respect to changes in the values of a particular circuit device, frequency analysis, etc. The most well-known systems of circuit-level simulation were: MASCOT, MOT IS, CADEC, EMOTA, ASTAP, RELAX, MCAP, VIRTUOSO ULTRASIM, NANOSIM, MAXWELL, SPECTRE, ARTIST, WORKBENCH, MDS, ACCUSIM, METACIRCUIT, ECA, DYMOLA, APLAC, SPICE, HSPICE, SMARTSPICE, PSPICE, etc. [74, 75, 150–155, 157, 203–207]. For the implementation of the circuit-level simulation, these systems automatically build a model of the circuit, which is represented by a system of ordinary differential equations and is solved by one or another numerical method.

At the device level, separate fragments of semiconductor structure (often a separate transistor) are simulated, represented as physical-topological objects. The electrical parameters of the devices are calculated on the basis of physical processes occurring in them. Popular systems for simulation in the device level are: MINIMOS, SAP, SIMON, SPIN, PROMIS, LIBRA, HARMONICA, TECHIS, TRANS, etc. [33, 65, 72, 97, 150, 152, 155, 208–211]. The implementation of models of the device level is carried out in the form of systems of partial differential equations.

Traditionally, the main and most popular simulation levels of digital IC designers are the gate-level and circuit-level simulations (Fig. 1.107). The reasons are as follows.

Existing system-level simulation tools, as a rule, require manual building of the circuit model. Often this is even impossible. In addition, the degree of detalization does not allow simulation of processes occurring within digital fragments of large blocks. As for the device level, the situation here is vice versa. The degree of detailing of the simulated object makes it possible to obtain so detailed information about the phenomena inside the semiconductor device that they often do not even interest the developers of digital circuits. In addition, the digital circuit in its simulation is inconceivable to detail up to the level of semiconductor structures in terms of the dimensionality of the problem being solved, since even small digital fragments at this level will require unrealistic expenditure of machine resources [150].

Gate-level and circuit-level simulation designs (Fig. 1.107) began to develop independently from each other since the late 50s. During this time, a huge number of models, methods, algorithms, and tools were created, which at one time allowed solving specific tasks of digital IC designers. It is accepted [152–154, 212] that the developed IC simulation methods of these two levels are divided into generations. An analysis of the development process of IC simulation tools shows that the main driving factor was the contradiction between the requirements for the accuracy of calculation results and the necessary costs of machine resources. At each stage of development, there was a situation where the means used to simulate a digital circuit only partially met the requirements of developers of digital circuits. Therefore, a search was made for new ways to improve existing simulation systems. The reason for this situation was the faster pace of development of the technology for IC fabrication [127, 148, 149] in relation to the progress in the development of simulation tools. The noted trend became more noticeable in connection with the transition to integrated technology and was especially aggravated with the appearance of circuits with a high degree of integration [149, 150].

The transition to each new IC fabrication technology (65 nm, 45 nm, 32 nm, 28 nm, 14 nm, 7 nm) led to two major problems for developers of circuit-level or gate-level simulators of digital circuits:

1. The dimensionality (the number of semiconductor devices) of devices being designed is sharp, even often increasing by several orders of magnitude, and the simulation tools available at the current time proved incapable of simulating circuits with so many elements.
2. There was a qualitative reconsideration of the significance of various physical phenomena affecting the functioning of a digital circuit. The phenomena that could be ignored before the development of the new technology became decisive for the functioning of the new type of a digital circuit (many such examples are given in Sect. 1.1), and the existing simulation systems turned out to be not anticipated for this.

At the gate-level simulation, especially after the idea of event-driven simulation [160, 181, 182, 194], the dimension problem has always not been as acute as in circuit-level simulation. The main reason is the use of simple gate-level models of digital cells, based on Boolean algebra, which do not require complex calculations for their solution. For electrical simulation, the most important problem has always been the dimension of the solved systems of differential equations. Because of the nonlinear nature of the dependence of the number of calculations on the number of elements in the circuit [150, 212], the computer time spent in the circuit-level simulation of modern ICs is unacceptably large in terms of the requirements for

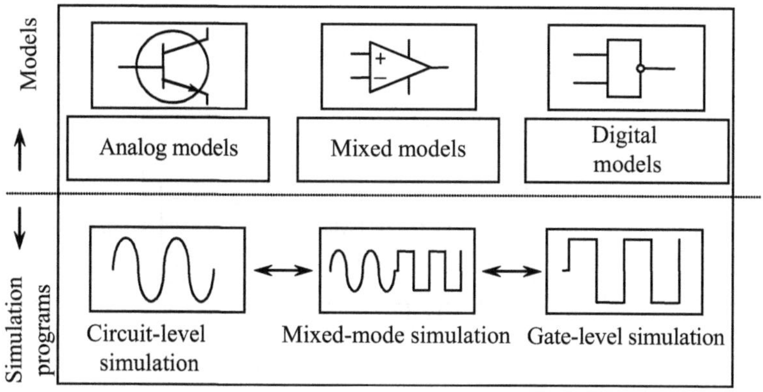

Fig. 1.108 Mixed logic-electric simulation

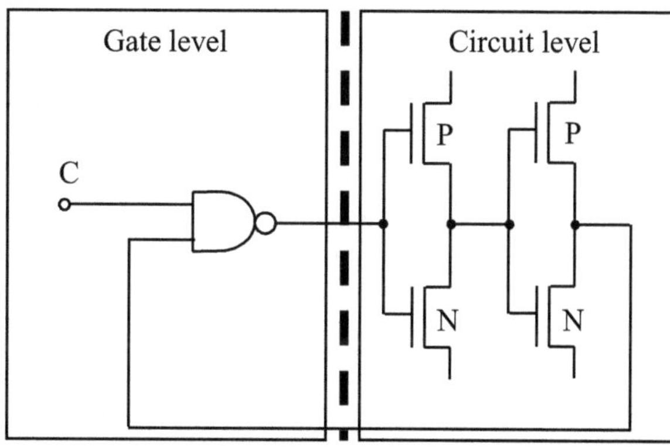

Fig. 1.109 Detailing the various parts of the circuit to a logic or gate level

the scheduling of IC design process. For admissible times of calculations with the help of the best modern tools of circuit-level simulation it is possible to really simulate only the circuits containing no more than 200–300 transistors [150, 152, 212]. And this is in case of using all the advanced ideas that emerged during the development of circuit simulation: new, faster numerical methods for solving systems of differential equations [150, 154, 204, 206, 212], macromodeling [213–227], eventual and structural decomposition of simulated circuits [152, 228–230], timing [150, 152], symbol [218] and parallel simulation [154, 204], etc.

Among the number of ways mentioned to reduce the dimensionality of simulation problem, the idea of mixed-mode gate-level–circuit-level simulation (Fig. 1.108), first proposed in [231, 232] and later developed in [228, 233–250], was the most tangible. It lies in the fact that, depending on the degree of detail required by the user for reflecting the processes occurring in the circuit, various fragments are detailed to different levels (Fig. 1.109): circuit- (CM) or gate level (GM) (Fig. 1.110) [239].

AM represents the fragments of the circuit, which require a very careful calculation of the waveform, is described by a system of differential equations and solved by conventional methods of circuit simulation. GM includes the fragments of the

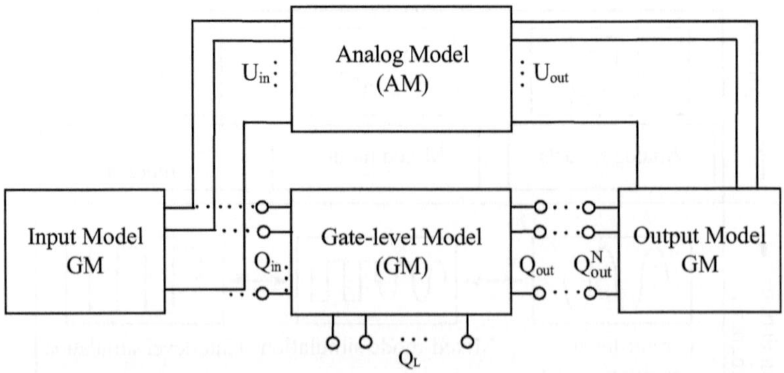

Fig. 1.110 Block diagram of mixed-mode simulation

circuit in which it is necessary to accurately simulate timing relationships, and the waveform is of secondary importance. GM is defined by a system of Boolean equations or in terms of register transfer level (RTL) and is calculated by traditional gate-level simulation methods. Thus, algorithms of fundamentally different nature interact in one tool: the numerical solution of systems of differential equations describing AM, and the solution of logical equations for GM. Mutual transformations of signals of different nature are carried out by models of inputs and outputs. In essence, a compromise is obtained between the advantages and disadvantages of the two levels of simulation: circuit and gate. The use of event-driven gate-level simulation for GM in mixed-mode simulation program provides a significant reduction in the costs of machine resources (time and memory), reaching from 1 to 3 orders of magnitude in comparison with circuit-level simulation [228, 233, 239, 246]. However, this does not satisfy modern requirements.

The mixed-mode simulation has practically exhausted itself from the point of view of connection of new levels of simulation, as various researches confirm [233, 239, 246], the greatest gain in time of simulation is obtained at combining the mentioned levels. Attempts to further spread the idea of mixed-mode simulation to other levels (both higher-RTL and system, and lower-device) increase the efficiency of calculations slightly.

From the above written it follows that for the simulation of modern complex digital circuits, the role of gate-level simulation, from the point of view of ensuring acceptable costs of machine resources, becomes the main one. Only gate-level simulation can ensure the calculation of modern digital circuits for acceptable time. There is a large class of simulation tasks built on digital ICs, which require circuit-level simulation of an entire circuit: determination of performance, debugging of microprograms, testbench generation, etc. In this series, a special place is occupied, of course, by calculations of the stability of the circuit with respect to various types of DF. With the help of the above and other well-known systems of circuit-level simulation, in which practically all types of models of external and internal DFs of the circuit level are developed, only certain fragments of the circuits

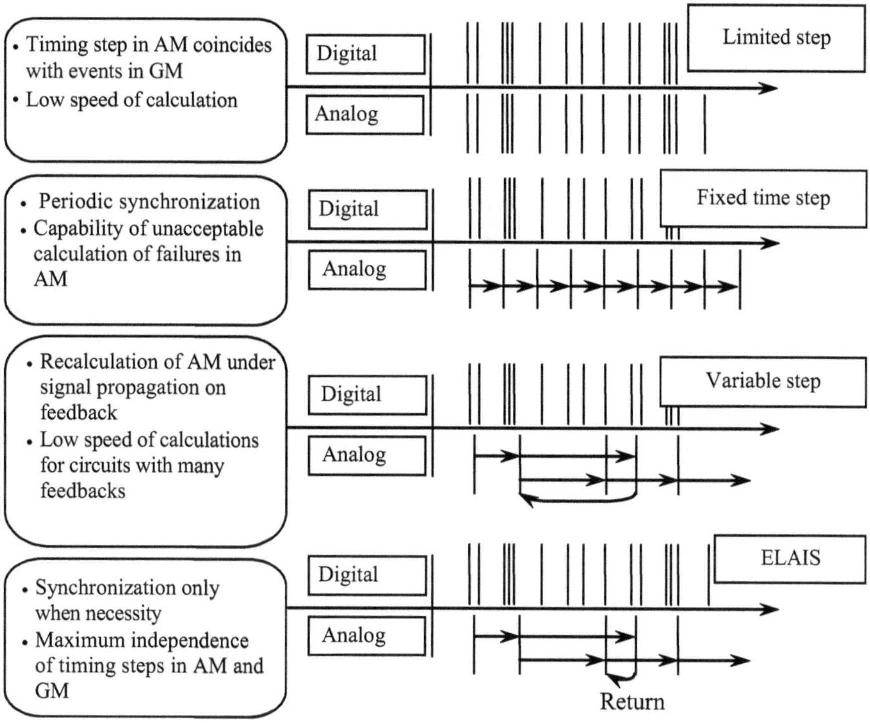

Fig. 1.111 Methods of timing matching of AM and GM

can be simulated. Therefore, for a holistic simulation of ICs or circuits based on them, only the application of either purely gate or mixed-mode gate–circuit-level simulation is possible. In the latter case, most of the elements of the circuit are presented at a gate level.

Thus, gate-level simulation is now the only possible level of simulation of modern digital ICs and circuits based on them, since only it allows them to be calculated for acceptable computer time costs.

However, the existing gate-level simulation methods were not designed for that reassessment of the significance of physical phenomena, which is described in Sect. 1.1. And it has natural roots. Gate-level simulation developed at a time when the ratio of useful signals to parasitic phenomena given in Sect. 1.1 had cardinally different values. This situation took place even during the development of mixed-mode simulation systems. Therefore, when developing tools of mixed-mode gate-circuit simulation, the main attention was paid to the development of circuit-based models from the point of view of their joint functioning with gate-level models. In addition, the issues of information and timing matching of models of various nature were intensively developed (Fig. 1.111 [228, 233, 234, 239]). In this case, as a rule, existing gate-level simulation subsystems were used as GM without their special modification (Fig. 1.112). Only in separate works [228, 238, 239] new gate-level

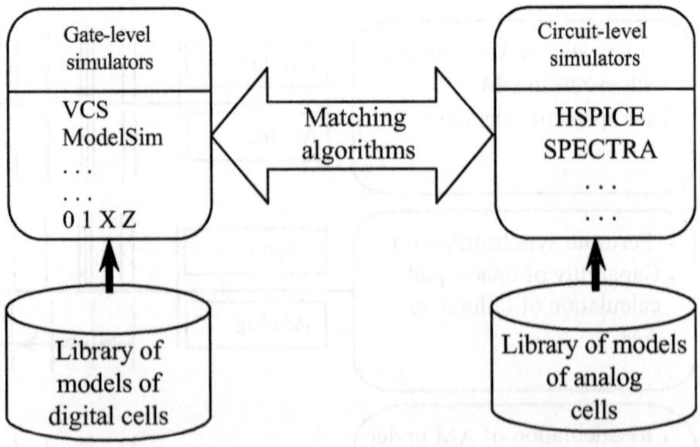

Fig. 1.112 Use of existing subsystems for building mixed-mode simulation

models were developed specifically for mixed-mode simulation. And the main goal of such developments was only to increase the effectiveness of mixed-mode simulation due to better matching of circuit- and gate-level models. In this period, the issues of taking DFs in GM into account due to their small significance were simply not considered.

Summarizing the above, the following can be stated:

1. Ignoring the effect of DF in the simulation of modern digital circuits is unacceptable.
2. It is necessary to take into account the effect of DF on digital circuits precisely during gate-level simulation as well as in the digital part of the circuit-during mixed-mode simulation.
3. The existing means of gate-level simulators are not able to take into account the DF effects, presented in Sect. 1.1 on the functioning of digital circuits.

It follows that there was an extreme need to develop models, methods, algorithms and EDA tool of gate-level simulation that can take into account the influence of various external and internal DFs on the functioning of modern digital circuits during simulation. Such tools can be used both for purely gate-level simulation and for the analysis of digital fragments of ICs in mixed-mode gate-circuit simulation. Therefore, at the present time, various authors and teams engaged in the creation of simulators of electronic circuits are actively working on the development of gate-level simulation tools and optimization of digital circuits, focused on considering DF. However, it should be noted that, in fact, such methods have not been developed yet. The first publications on these issues [87, 114, 139, 212, 250, 251] show:

1. In each specific program, means are allocated for consideration of only separate DFs (e.g. interconnects), but there is no comprehensive consideration of the issue.
2. Models of nonlogical nature continue to be developed (more often extremely simplified circuit-level models and macromodels), and an attempt is made to use them as part of gate-level simulation systems. Of course, in this case, due to simplifications, there is a significant (on average, several dozen times) gain in

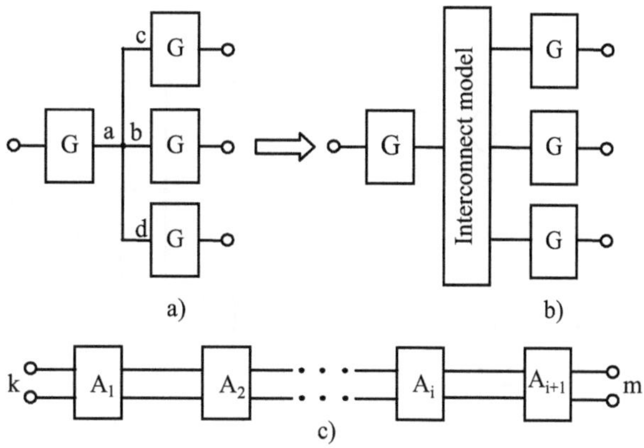

Fig. 1.113 VAMP model: (**a**) a simulated circuit; (**b**) a transformed circuit; (**c**) cascade connection of quadripoles

machine resources in comparison with circuit simulation. But, as it was said above, this too is not enough, and it is necessary to apply complete gate-level simulation.

For example, in [250, 251] only the influence of interconnects on the functioning of digital circuits is considered, using the so-called VAMP model that implements "accurate" simulation approach. Its essence can be shown by the following example: the simulated digital circuit (Fig. 1.113a) is converted to the view shown in Fig. 1.113b.

In this model, gates are presented at a logical level, and interconnect models are presented at a simplified circuit level. In particular, each interconnection segment between the *k*th and *m*th nodes of the circuit (for example, between a and b, b and c, etc.) is represented as a cascade connection of quadripoles A_i, implementing the transfer function of the elementary interconnection fragment in the complex plane:

$$A_i = \begin{bmatrix} \cos(\gamma \cdot l) & Z \cdot \sin(\gamma \cdot l) \\ \dfrac{\sin(\gamma \cdot l)}{Z} & \cos(\gamma \cdot l) \end{bmatrix}, \tag{1.7}$$

where

$$Z = \sqrt{\frac{R' + sL'}{sC'}}, \quad \gamma = \sqrt{(R' + sL')(sC')}, \quad s = j\omega; \tag{1.8}$$

R', L' and C'—specific values of parasitic resistance, capacitance and inductance of interconnects; *L*—the length of the fragment; $\grave{A}_{(k,m)}$-the transfer function between the nodes *k* and *m* and is calculated as

$$A_{(k,m)} = \prod_n A_n. \tag{1.9}$$

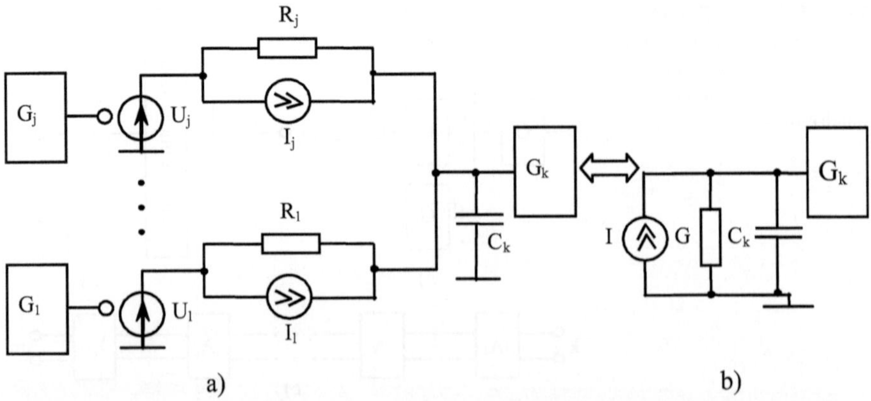

Fig. 1.114 SPECS model: (**a**) N-pole macromodel; (**b**) generalized macromodel

The input signal of the interconnect segment $u_i(t)$ is represented as the sum of components $u_{i,k}(t)$, each of which is defined as one of the components $u_{j,k}(t)$ of the output signal $u_j(t)$. Each component of the output signal $u_{j,k}(t)$ is computed by $u_{i,k}(t)$ using Laplace transforms, in which $A_{(k,m)}$ is used.

Although a number of simplifications have been implemented in this model (the assumption of the linearity of the dependences of interconnect delays on their length, rough approximations in the calculation of hyperbolic functions cosh, sinh, etc.), an unjustified cost of machine time is entirely required for IC simulation. The reason is the use of yet simplified, but still circuit-level models. Even the circuit level of interconnects, simplified to the described degree, transmits gate signals to the logical models with much more detailed form than necessary. The fact is that in the same program, the gate outputs yield rectangular or trapezoidal signals, which determine the overall accuracy of calculations. Other advantages of gate-level simulation are not used, either: calculation with a time step, often significantly exceeding the delays of separate gates, etc. All these factors ultimately affect the calculation time of the entire digital circuit.

In [133], only the influence of interconnect is considered and another (SPECS) model is proposed (Fig. 1.114).

The SPECS model is not much different from the VAMP model and has the same drawbacks. The fact is that SPECS is an N-pole circuit-level macromodel that operates with circuit-level categories. For example, the output current of the kth gate is determined by the following formula:

$$I_k = \frac{u_k}{z_{kk}} - \frac{z_{k1}}{z_{kk}} I_1 - \cdots - \frac{z_{k,k-1}}{z_{kk}} I_{k-1} - \frac{z_{k,k+1}}{z_{kk}} I_{k+1} - \cdots - \frac{z_{kN}}{z_{kk}} I_N, \qquad (1.10)$$

where voltages in the outputs of gates are figures (U_k), the currents through them (I_k), the conductance of the branches (z_k), mutual conductivities (for example, $z_{k,\,k+1}$), etc.

In [113] only crosstalk of capacitive nature is taken into account (C_n in Fig. 1.115a). Here the interference of signal lines is represented purely by a

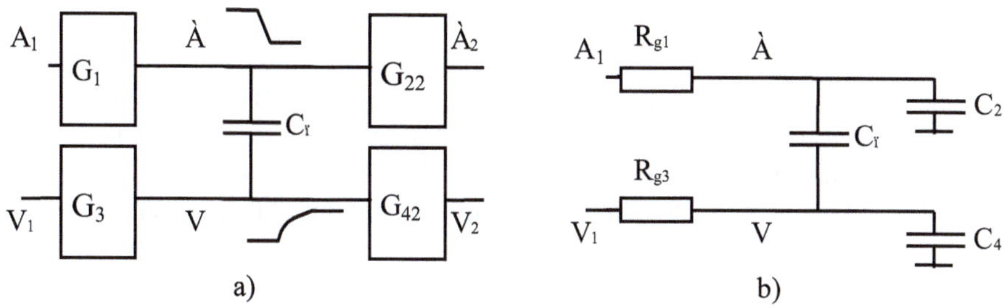

Fig. 1.115 Crosstalk noise of capacitive nature: (**a**) simulated circuit; (**b**) equivalent circuit

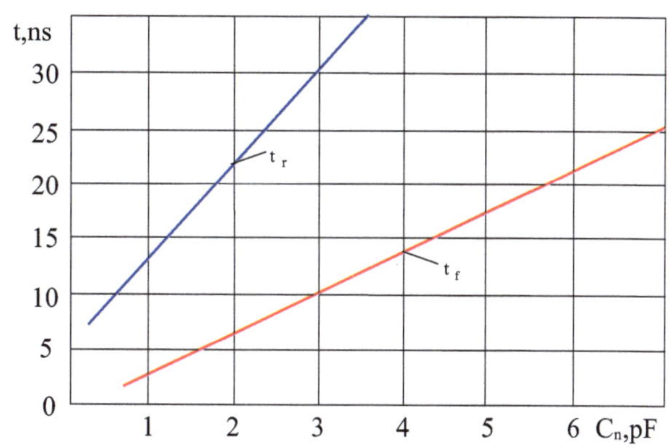

Fig. 1.116 Models of dependences of the edges of the output signal of the gate used in the LOSTIN program

circuit-level model, operating by output resistances of the gate when working on a separate, isolated signal line (R_g), by input capacitances C, etc. In [139], a circuit-level but already resistive model was also used to calculate the crosstalk in gate-level simulation.

The closest to DF consideration at a purely gate-level simulation came from the LOSTIN program [210]. However, as in the previous works, the effect of only one, a separate factor (in this case, the load capacitance of the G C_L) on the duration of rise and fall (t_r, t_f) of the output signal of the gate is considered (Fig. 1.116) in the form of dependences:

$$t_r = \frac{y_0 + z(t - t_0)}{1 + z(t - t_0)}, \quad t_f = \frac{y_0}{1 + z(t - t_0)}. \tag{1.11}$$

In addition, two-way restrictions on the values of t_r and t_f are imposed, since the traditional two- and three-valued simulation, which was not intersected in the mentioned system, does not allow simulating signals with nonstandard amplitudes.

Thus, the problems of considering DF effects, given in Sect. 1.1, at the gate-level simulation in the literature are practically not covered and need careful elaboration.

As seen from what follows, they require a fundamental change in the principles of gate-level simulation.

The issues of complex consideration of the influence of DF at the gate level are not considered in the works on optimization of digital circuits, either, in particular, on power consumption [131, 252–254] and timing parameters of critical paths [128, 129, 132, 134, 135, 255–269]. Limitations of systems of this type are discussed separately in Chap. 4.

1.3 Requirements for Gate-Level Simulation Tools and Optimization of Digital Circuits, Focused on DF Consideration

From the consideration of existing methods and programs of simulating digital circuits it follows that, from the point of view of computer resources, consideration of DF is possible to implement only in the subsystems of gate-level simulation. However, in existing programs at this level, there are no such capabilities. Moreover, study of traditional gate-level simulation methods shows: because they were not designed for the set goal, they need fundamental changes. But at the same time, their main advantages should be preserved-speed of calculations and universality. The fundamental limitations of the traditional gate-level simulation are analyzed from the observed point of view and, based on this, formulate the requirements for gate-level simulation tools that are oriented toward considering DF.

In traditional gate-level simulation systems, the gate model is represented as a finite state machine (FSM) (Fig. 1.117).

$$\mathrm{FSM} = \langle \vec{X}, \vec{Y}, \vec{Z}, \varphi, \psi \rangle, \tag{1.12}$$

where $\vec{X} = \{x_1, x_2, \ldots, x_n\}$, $\vec{Y} = \{y_1, y_2, \ldots, y_m\}$—the vectors of the input and output variables, respectively; $\vec{Z} = \{z_1, z_2, \ldots, z_k\}$—the internal state vector of the gate ($\vec{Z}'$—new value of the vector $\vec{Z}$); *n*, *m*, *k*—the number of inputs, outputs and internal states of the gate; φ—the generalized function of the outputs: $\vec{X} \times \vec{Z} \Rightarrow \vec{Y}$; ψ—the generalized transition function: $\vec{X} \times \vec{Z} \Rightarrow \vec{Y}'$, i.e.

$$\vec{Y}(t) = \varphi[\vec{X}(t), \vec{Z}(t)], \tag{1.13}$$

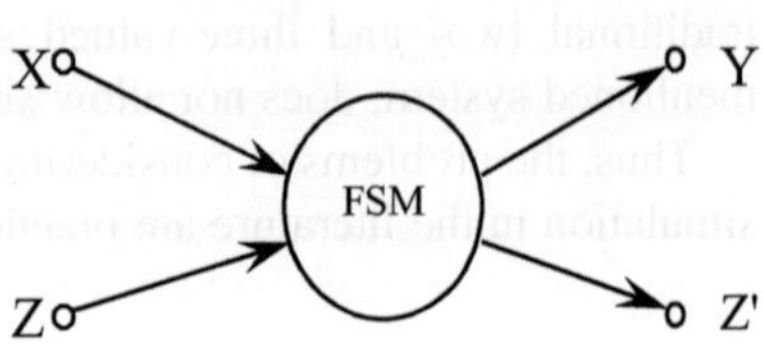

Fig. 1.117 Gate model in the form of an FSM

$$\vec{Z}'(t) = \psi\left[\vec{X}(t), \vec{Z}(t)\right]. \tag{1.14}$$

Variables of vectors $\vec{X}$, $\vec{Y}$, $\vec{Z}$ take values from some finite set: {0,1}-for two-valued, {0,1,x}-three-valued, {0,1,x,ε,h}-five-valued, etc. simulation.

The model of the entire digital circuit is represented as a network of FSM:

$$\text{MC} = \left\langle \vec{I}, \vec{O}, \vec{E}, \vec{S} \right\rangle, \tag{1.15}$$

where $\vec{I} = \{i_1, i_2, \ldots, i_p\}$, $\vec{O} = \{o_1, o_2, \ldots, o_q\}$—the vectors of the input and output variables of the general circuit respectively, p and q are their numbers; $\vec{E} = \{e_1, e_2, \ldots, e_g\}$—is the set of gates, each of which is a separate FSM, g is their number; $\vec{S} = \{s_1, s_2, \ldots, s_t\}$—is the set of connections between elements, t is their number, s_i is the ith connection.

The task of gate-level simulation of a digital circuit is concentrated to the calculation of the MC at discrete instants of time t_1, t_2, …, t_u in order to obtain the dependence $\vec{O}(t) = f\left[\vec{I}(t)\right]$, where $0 \leq t_1 \leq t_2 \leq \cdots \leq t_u \leq T_{\text{sim}}$, where T_{sim} is the total time of the transient processes of the circuit being simulated, u—the number of simulation time points.

According to [150, 151, 203, 204, 212, 270–274], mathematical models and other means of simulation require accuracy, universality and economy. It is also known that these requirements are contradictory. Study of existing tools of gate-level simulation shows that their main advantages, economy (speed of calculations) and universality (independence from the circuit realization), are achieved due to loss of the accuracy of the results. In the traditional gate-level models a number of simplifications have been introduced, the main ones of which are:

1. The identity of physical nature of all variables (x, y, z) of the model. At other levels of simulation in the same model, different physical values (for example, on circuit-level—voltages, currents, conductivities, etc.) are operated, and additional transformations of variables from one type to another are needed. Logical models use only dimensionless variables that reflect the logical state of the digital gate. In addition to gaining time, such a view also contributes to abstracting from a specific implementation of digital gate.
2. Finiteness of the state variables of the model (x, y, z). Although there are many-valued [174, 178, 182–184, 201] and even infinite-valued [173] logics, in most practical programs only two-valued and very rarely three, five, etc. valued models are used. The latter are applied only in case of solving special simulation tasks (determining the static and dynamic hazards, signal race, etc.). It is due to the discrete nature of the variables that it is possible to use the fast apparatus of Boolean algebra in gate-level simulation tools.
3. The discreteness of the simulation time (t). It is believed that the state variables of the model can change their value only at certain instants of time $t_{i,\ i=1,2,\ldots,u}$. This simplification makes it possible in gate-level simulation tools to calculate a circuit with a time step $\Delta t = t_i - t_{i-1}$, which is much higher than the analogous value of

other levels (for example, circuit-level simulation), and also apply the principle of event-driven simulation [160, 181, 182, 194]. The latter factor, in turn, is one of the main reasons of gate-level simulation high performance.

4. The invariance of the properties of gate in the course of simulation. It is due to the consistency of timing parameters of circuit elements that the need for a "fragmentation" of simulation step during the calculation of the transient processes of digital circuits is eliminated and the mechanism for planning future events is substantially simplified. Ultimately, the simulation time is significantly reduced.
5. Constancy of the structure of the MC during simulation. A unidirectional character of transmission of signals from the inputs to the outputs of the gate is adopted, i.e. dependencies $\vec{X}$ on $\vec{Y}$ available in some modern digital circuits (bidirectional cells [31]) are ignored, leading to the need to change the structure of the circuit during the simulation. It is clear that the latter can significantly delay the calculations.
6. Isolation of gate from other elements of the circuit that are not in the structural connection ($\vec{S}$) with it. This circumstance makes it possible to perform a structural decomposition of simulated device [152, 228–230] and, as a result, substantially reduce the costs of machine resources.
7. Uniqueness of the functions of outputs (φ) and transitions (ψ) of the gate. It is assumed that the supply of any admissible set of input signals to the inputs of the gate transforms it into some uniquely determined stable state $\vec{Z}$, which substantially simplifies simulation algorithms.
8. Ideality of signal transmission chains. It is believed that if there is a s_i connection between the ath output of the bth input of the gate, then $\vec{X}_b = \vec{Y}_a$ during the whole simulation time $0 \ldots T_{\text{sim}}$. Therefore, the connections s_i are considered only at the stage of automatic formation of MC, and when it is solved, they do not affect the computational process.

However, it must be assumed that the experimental data (Sect. 1.1) confirm the violation of the accuracy of the majority of the above assumptions (2, 3, 4, 6, 7, 8) for digital IC functioning in the environment of external ($V_{i,\ i=1,2,\ldots}$) and internal ($W_{j,\ j=1,2,\ldots}$) DF. Therefore, the traditional means of gate-level simulation, in order to take into account the influence of DF, must undergo fundamental changes in order to meet certain requirements. The main one is the combination of economy and universality of gate-level simulation with increased (close to the circuit) accuracy of simulation results. Such a combination can be formed as a result of the application of new principles for the construction of gate-level simulation tools.

From the foregoing it follows that the effectiveness of gate-level simulation systems, taking into account the influence of the DF, can be influenced by the correct solution of the following issues:

1. A methodology for consideration of DF effects on the operation of the circuit
2. Gate-level models
3. Models for determining the effect of DF on gate

4. Organization of a system for simulating and optimizing digital circuits with consideration of DF

The requirements for the gate-level simulation tools of digital circuits are formulated separately for each of the above questions.

Requirements for the methodology for consideration of DF effects:

1. Compromise satisfaction of the necessary accuracy, universality and economy, as well as orientation to modern practical limitations of the design process according to these indicators.
2. Complex imitation of the influence of all most tangible digital circuits, DF.
3. Complete invariance with respect to the features of the circuit being simulated: structure, types of elements used, modes of operation, manufacturing techniques, etc.
4. Information and timing compatibility of all models of DF and gate, i.e. absence of the need for various additional transformations in the transition from one model to another and the use of a single mechanism for the expiration of simulation time.
5. Good consistency with advanced known, as well as perspective methods for calculating electronic circuits: through multi-level mixed-mode simulation, macromodeling, the use of the principle of latency, symbol analysis, etc.

Requirements for the gate-level model:

1. Correct imitation of the behavior of a digital circuit both at standard and non-standard (with arbitrary amplitude, duration, abruptness, etc.) input signals. Nonstandard input signals of circuit elements (including internal ones) are characterized precisely by simulation with consideration of DF.
2. Differentiated accuracy of reflection of different static states and switching stages of the gate, considering their significance from the point of view of influence on other nodes of the circuit, i.e. adaptability of the gate-level model to the current computational situation.
3. Sufficient universality from the point of view of simulation of all or at least most types of digital circuits, regardless of their configuration, circuit-level basis, etc.
4. Compatibility not only with DF models, but also with models of other (for example, circuit) levels in order to provide, apart from purely logical, also a mixed-mode multilevel simulation, taking into account the influence of real conditions of the circuit operation.
5. The possibility of establishing a unique correspondence between the standard parameters of modern digital circuits and the values used in the gate-level model.
6. Smallest computer time and memory resources required when using the gate-level model. In other words, in the gate-level model, the amount of calculations and the number of parameters should not differ much from traditional analogues of the gate-level due to the addition of new features in them.

In relation to the models for determining the influence of DF on the operation of digital circuits, in addition to compatibility with the gate-level models, the following requirements are also imposed:

1. The total reflection of the effect of as many physical phenomena as possible and only on the integrated parameters of the gate-level model. Both the specification of DF models for separate physical processes and the increase in the details of the parameters of the gate-level model are fraught with a significant increase in the costs of machine resources.
2. Correspondence of the accuracy of simulation to the current nature (for example, the rate of change) of the dependence of the gate parameters on the DF value, since these functions often have areas of fundamentally different character.
3. The level of complexity identical with gate-level models.
4. A small number of parameters in order to simplify the process of identifying them and establishing compliance with physical quantities.

In the organization of digital circuit simulation system taking into account the DF, in addition to the traditional requirements (openness, compatibility of traditional and automated design, effective software implementation of algorithms and models, etc.), the issues of information consistency come to the fore, not only within a separate gate-level simulation subsystem (because its composition should include various programs that transmit the initial data to the models of determining the DF: topological design, calculation of the radiation level, etc.), but also existing in other EDA tools of digital circuits. To ensure the latter, the system must contain a set of translators that translate intermediate information from the formats of data representation of one program to another, also into standard formats (for example, VHDL, Verilog [162, 275, 276]).

Of no small importance is the speed of the development of the system due to the rapid pace of progress in microelectronics. From the point of view of this requirement, the most expedient is the use of ready-made subsystems that have proved themselves in practice in designing.

1.4 Principles of Gate-Level Simulation and Optimization of Digital Circuits with Consideration of DF

The proposed gate-level model with consideration of effects (MCE) is presented in the following form (Fig. 1.118):

$$\mathrm{MCE} = \langle \vec{X}, \vec{Y}, \vec{Z}, \vec{P}, \vec{F}, \varphi, \psi, \chi \rangle, \tag{1.16}$$

where $\vec{X} = \{x_1, x_2, \ldots, x_n\}$, $\vec{Y} = \{y_1, y_2, \ldots, y_m\}$, $\vec{Z} = \{z_1, z_2, \ldots, z_k\}$ have the same meaning as in the FSM model. But, unlike the FSM, the variables of the marked vectors take arbitrary values from the continuous range [*a*, *b*], i.e.

$$a \le x_i \le b, \quad a \le y_i \le b, \quad a \le z_i \le b. \tag{1.17}$$

It is often assumed that $a = 0$, $b = 1$. But in the general case, there is

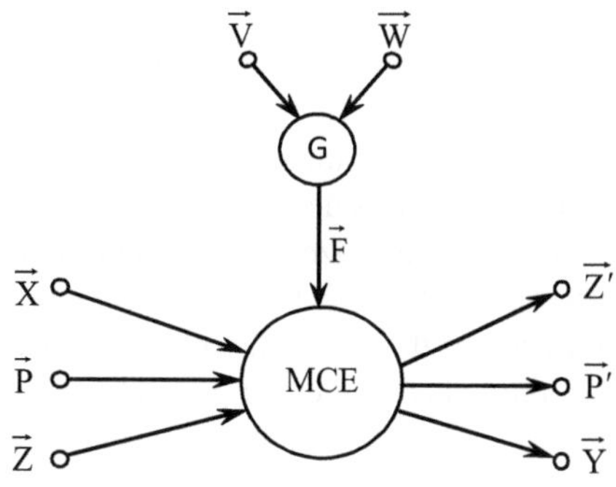

Fig. 1.118 Gate-level model with consideration of effect (MCE)

$$a < 0 < 1 < b. \tag{1.18}$$

For example, from Fig. 1.105 it follows that the output voltage of the gate during certain time intervals can be either less than the stationary value U_0 ("0") or more than U_1 ("1"). However, a and b usually do not differ much from 0 and 1, respectively (in practical circuits—no more than 20–30%), because otherwise the gate loses its efficiency.

$\vec{P} = \{p_1, p_2, \ldots, p_u\}$–vector of gate parameters (mainly timing), and u is their number; $\vec{P}'$—new value of the vector $\vec{P}$. The value of the ith parameter of the gate p_i varies in the continuous range $[p_{i\min}, p_{i\max}]$, i.e.

$$p_{i\min} \le p_i \le p_{i\max}. \tag{1.19}$$

Moreover, the values of the limits $p_{i\min}$ and $p_{i\max}$ depend on the physical nature of the ith parameter. Thus, unlike the FSM, it is assumed that the gate parameters can change during its operation.

$\vec{F} = \{f_1, f_2, \ldots, f_m, f_{m+1}, f_{m+2}, \ldots, f_{m+k}, f_{m+k+1}, f_{m+k+2}, \ldots, f_{m+k+u}\}$—vector of DF effect. Each of the first m components of the given vector ($f_{l,\ 1\le l\le m}$) reflects the generalized effect of all the considered DF ($V_{i,\ i=1,2,\ldots}$; $W_{j,\ j=1,2,\ldots}$) on the state of one of the outputs of the gate (y_l), the following k ($f_{m+l,\ 1\le l\le k}$)—to its internal states (z_l), and the last u ($f_{m+k+l,\ 1\le l\le u}$) to the parameters of the gate (p_l). Each ith element of the vector F, in turn, is a function of all considered external and internal DFs: $f_l = G(V_{i,\ i=1,2,\ldots}; W_{j,\ j=1,2,\ldots})$, and represents the value to which the corresponding model variable should be adjusted, intended without consideration of DF effect:

$$y_l' = \mathrm{RSE}_y(y_l, f_l), \quad z_l' = \mathrm{RSE}_z(z_l, f_{m+l}), \quad p_l' = \mathrm{RSE}_p(y_l, f_{m+k+l}), \tag{1.20}$$

where y_l, z_l, p_l are the values of the MCE variables without taking into account the influence of the DF, and y_l', z_l', p_l' are the same values after correcting their values. In this case, of course, the following conditions must be observed:

$$a \le y_l' \le b, \quad a \le z_l' \le b, \quad p_{l\min} \le p_1' \le p_{l\max}. \tag{1.21}$$

In (1.20) RSE_y, RSE_z, RSE_p—the rules for summing the effects of DF with the values of the outputs, internal states and parameters of the gate respectively. The RSE receives the value of the model parameter (for example, p_l), the effect of the DF (f_l) and transmits the corrected parameter value (for example, p_l'). Possible variants of RSE are considered in Chap. 2, but in the simplest case (when the physical meaning is the f_l-size of the deviation of the corresponding variable as a result of DF effect), the RSE can be a simple algebraic sum. In this case, the sign f_l indicates the direction of the effect (increasing or decreasing the variable of MCE). The function G represents the rules for generalizing the effects (RGE) of various DFs ($V_{i,\ i=1,2,\ldots}$; $W_{j,\ j=1,2,\ldots}$) on a separate MCE variable, i.e.

$$f_l = \sum\nolimits_i \mathrm{RGE} f_{li}(v_i) + \sum\nolimits_j \mathrm{RGE} f_{lj}(w_j). \tag{1.22}$$

In the particular case, the RGE can be represented as an algebraic sum with the corresponding weight coefficients (other variants are considered in Sect. 3.1):

$$f_l = \sum_i a_i \cdot f_{li}(v_i) + \sum_j a_j \cdot f_{lj}(w_j), \tag{1.23}$$

where $f_{li}(v_i)$ is the model of effects (ME) of the ith external, and $f_{lj}(w_j)$—of the jth internal DF on the lth MCE variable.

ME, RSE and RGE for different circuit-level basis have different character and can be obtained by approximating the corresponding experimental data obtained with the help of circuit-level simulation tools (such models were obtained in Chap. 3). The criterion for choosing these models is the compromise satisfaction of the requirements for the accuracy of calculations and the costs of machine resources.

φ—generalized function of outputs: $\vec{X} \times \vec{Z} \times \vec{P} \times \vec{F} \Rightarrow \vec{Y}$;
ψ—generalized function of transitions: $\vec{X} \times \vec{Z} \times \vec{P} \times \vec{F} \Rightarrow \vec{Y}$;
χ—generalized function of changing parameters: $\vec{X} \times \vec{Z} \times \vec{P} \times \vec{F} \Rightarrow \vec{P}'$, i.e.

$$\vec{Y}(t) = \varphi\left[\vec{X}(t), \vec{Z}(t), \vec{P}(t), \vec{F}(t)\right], \tag{1.24}$$

$$\vec{Z}'(t) = \psi\left[\vec{X}(t), \vec{Z}(t), \vec{P}(t), \vec{F}(t)\right], \tag{1.25}$$

$$\vec{P}'(t) = \chi\left[\vec{X}(t), \vec{Z}(t), \vec{P}(t), \vec{F}(t)\right]. \tag{1.26}$$

Since MCE variables have a continuous character, the rules of Boolean algebra cannot be applied for the calculation of functions φ, ψ, and χ. Therefore, other rules (given in Sect. 2.1) are applied, which ensure the work with continuous values and the speed of calculations close to Boolean algebra. Such are, for example, the rules of minimax logic [248]:

$$x_1 \wedge x_2 \wedge \ldots = \min(x_1, x_2, \ldots), \tag{1.27}$$

$$x_1 \vee x_2 \vee \ldots = \max(x_1, x_2, \ldots), \tag{1.28}$$

$$x = a + b - x, \tag{1.29}$$

where $x_1, x_2, \ldots \in [a, b]$.

The model of the entire digital circuit with consideration of effect (MEE) is presented in the form of an MCE chain:

$$\text{MEE} = \left\langle \vec{I}, \vec{O}, \vec{V}, \vec{W}, \vec{E}, \vec{S} \right\rangle, \tag{1.30}$$

where $\vec{I} = \{i_1, i_2, \ldots, i_p\}$, $\vec{O} = \{o_1, o_2, \ldots, o_q\}$, $\vec{E} = \{e_1, e_2, \ldots, e_g\}$ *and* $\vec{S} = \{s_1, s_2, \ldots, s_t\}$ have the same meaning as in MC; $\vec{V} = \{v_1, v_2, \ldots\}$ is a vector of external DFs; $\vec{W} = \{w_1, w_2, \ldots\}$ is the vector of internal DFs.

However, it is proposed in MEE to organize connections (s) between arbitrary *a*th and *b*th gate-level circuits and implement through the developed gate-level model of effects (GME, Fig. 1.119) which is as follows. Suppose that in the calculated circuit the output of the *a*th gate is directly connected with the input of the *b*th gate; $\vec{X}_b = \vec{Y}_a$. GME breaks this connection, replaces the value $\vec{Y}_a$ by some other values $\vec{X}'_b$, taking into account the influence of all DFs (the current state of the circuit), i.e. $\vec{X}'_b \neq \vec{Y}_a$, but $\vec{X}'_b = f(\vec{Y}, C)$, where C is the current state of the circuit. In the future this value $\vec{X}'_b$ is transferred to the input of the *b*th gate.

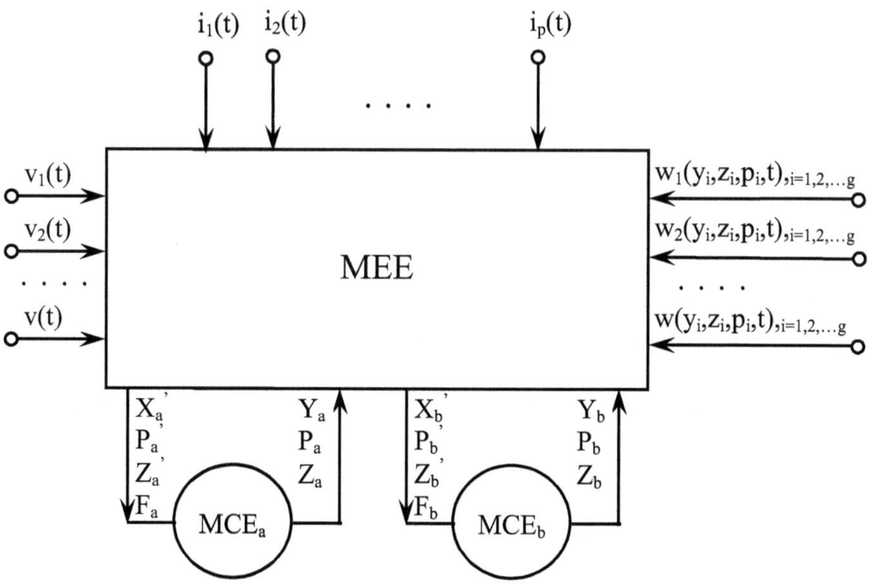

Fig. 1.119 Organization of connection between two MCEs

Thus, the connections between arbitrary elements of the circuit are carried out only through GME. Each time the GME receives the following information about the state (S) of the circuit:

1. The values of the input signals of the circuit: $\vec{I} = \{i_1(t), i_2(t), \ldots, i_p(t)\}$;
2. The values of internal effects: $\vec{V} = \{v_1(t), v_2(t), \ldots\}$;
3. The current state of all elements of the circuit, i.e. output signals (y_i), values of internal variables (z_i) and parameters (p_i) for $_{i=1,2,\ldots,g}$;
4. The dimensions of internal effects $\vec{W} = \{w_1(t), w_2(t), \ldots\}$, which in turn are the state functions of all elements of the circuit (i.e. y_i, z_i, p_i, $_{i=1,2,\ldots,g}$).

On the basis of this information, the GME calculates and transmits information about its input signals ($\vec{X}_i'$), parameters ($\vec{P}_i'$) and state ($\vec{Z}_i'$) to each gate of the circuit at the next instant of the model time. In addition, based on the information received, the vector value for each circuit element is calculated.

The main task of the solution of the MEE, as in case of MC (see Sect. 1.3), is to derive dependences of the values of the outputs of digital circuits from the input signals in the time interval $[0, T_{\text{sim}}]$, but already considering DF effects, i.e. $\vec{O}(t) = f\left[\vec{I}(t), \vec{V}(t), \vec{W}(t)\right]$. It is clear that, under the conditions, $\vec{V}(t) = 0$, $\vec{W}(t) = 0$, $\vec{P}_i = \text{const}(\text{for all } i = 1, 2, \ldots, p)$ the MEE automatically turns into MC. In this case, traditional types of gate-level simulation are possible:

1. Calculation of static mode when it is assumed that $\vec{I}(t) = \text{const}$.
2. Analysis of transient process, in which the input signals of the circuit are switched, i.e. $\vec{I}(t) \neq \text{const}$. It is known [156, 161] that the second mode, in turn, has two options: synchronous simulation, when the distances between two neighboring switching times of input signals significantly exceed the inertia time of gates, and asynchronous simulation—with commensurability of the noted values.

But the spectrum of problems of simulating digital circuits with the use of MEE is broader:

1. Calculation of static mode, when $\vec{I}(t) = \text{const}$, $\vec{V}(t) = \text{const}$, $\vec{W}(t) = \text{const}$.
2. Analysis of transient process without taking into account the influence of DF, when $\vec{I}(t) \neq \text{const}$, $\vec{V}(t) = \text{const}$, $\vec{W}(t) = \text{const}$.
3. Study of the influence of DF on the static state of the circuit. When $\vec{I}(t) = \text{const}$, $\vec{V}(t) \neq \text{const}$, $\vec{W}(t) \neq \text{const}$. It should be noted that this mode is fundamentally different from the first one, because states of separate nodes of the circuit as a result of the influence of DF can deviate so much from their stationary values that switching will begin, i.e. transient processes with constant input signals of the whole circuit. In particular, with the help of this mode it is possible to determine the stability of the circuit in relation to the values of different effects (radiation resistance, heat resistance, etc.) and to calculate the boundary values of these disturbances, under which the circuit still maintains its static state.

4. Analysis of transient process with consideration of DF effects, when $\vec{I}(t) \neq \text{const}$, $\vec{V}(t) \neq \text{const}$, $\vec{W}(t) \neq \text{const}$. This mode also differs significantly from the second one, as the picture of switching the gate-level circuit is dictated not only by the logic of the circuit, but also by changes in its properties as a result of the influence of the DF.

As in the case of MC, the MEE can also implement other traditional types of gate-level simulation: probabilistic calculation and simulation of the circuit with a dispersion of timing parameters of digital gates [156, 160, 161]. This practical necessity arises, in view of the fact that often not only the parameters of separate gates (delay, front, etc.), but also the features of the implementation of the circuit (parameters of interconnects, power circuits, etc.), as well as the characteristics of external DFs have probabilistic character. To ensure probabilistic calculation, as well as simulation with a dispersion of various parameters, the MEE undergoes changes of the same type as in case of traditional gate-level simulation:

1. Variables of the model do not take deterministic values from a certain range (1.12), (1.13), but are determined by the probability distribution of the parameters taking values from these ranges (in case of probabilistic MCE) or by the intervals of changes (in case of MCEs with the dispersion of parameters).
2. New calculation rules (RSE, RGE, ME, functions φ, ψ, χ) are used.

Comparison of the proposed models of the gate (MCE) and digital circuits (MEE), as well as simulation techniques with traditional gate-level simulation, in terms of their limitations and advantages, shows:

1. Traditional gate-level simulation automatically follows from the proposed means when taking a number of assumptions: $\vec{W}(t) = 0$; $\vec{V}(t) = 0$; $a = 0$; $b = 1$; $\vec{P}' = \vec{P}$; use of Boolean algebra rules for calculating functions φ and ψ, etc.
2. All the main advantages of gate-level simulation have been preserved:
   (a) Similarity of the physical nature of the variable models is achieved by using the ME ($f_{li}(v_i)$ and $f_{lj}(w_j)$), which establish connection between the values of other types (v_i and w_j) with the values of the dimensionless (f_l) model;
   (b) Although transition from discrete variables of the model to continuous ones is carried out, rules for calculating functions (for example, minimax logic) are used, which only slightly reduce the calculation speed with respect to Boolean algebra;
   (c) Use of continuous parameter values (in particular, timing) of the p_i model does not exclude the use of event-driven algorithms for simulating digital circuits, since, depending on the required accuracy of calculations, it is possible to make an appropriate approximation of the parameter function from the DF value.
3. The main limitations of gate-level simulation described in Sect. 1.3 have been eliminated: the invariability of the properties of the elements and the entire circuit during simulation, the isolation of separate fragments of the device from each

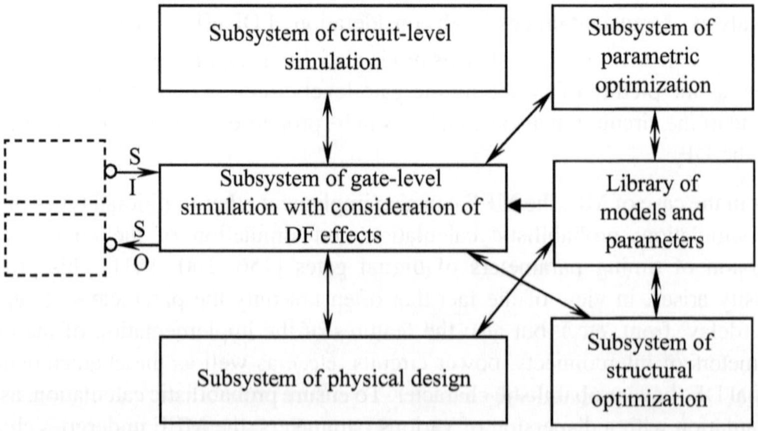

Fig. 1.120 Organization of the system of simulation and optimization of digital circuits with consideration of DF

other, the ideality of signal transmission chains, etc. All these issues are addressed in the MEE.

With regard to the principles of the organization of a software system for simulating and optimizing digital circuits with consideration of DF, from the previous presentation it follows that:

1. The gate-level simulation subsystem (Fig. 1.120) in the course of simulating digital circuits, taking into account the influence of the DF, functions not in isolation but interacts with other packages of computer-aided design programs:
   (a) Topological calculation—in order to extract the necessary information for the ME, RSE, RGE.
   (b) Parametric optimization—for calculating the coefficients of the approximation functions of ME, RSE, RGE as well as improving the indicators of the circuit being simulated.
   (c) Structural optimization—to improve the indicators of digital circuits on power consumption and timing parameters of critical paths of digital circuits.
   (d) Other levels (for example, circuit-level simulation)—for the purpose of organizing a multilevel mixed-mode simulation.

   Moreover, from the point of view of the requirements of efficiency (see Sect. 1.3), it seems expedient to use ready subsystems of the mentioned types, too.

2. The gate-level simulation program must have standard inputs (SI) and outputs (SO) (HDL and other means of specifying the descriptions of digital circuits, input signals, cell parameters, and presentation of calculation results) for convenient integration with existing advanced EDA tools. As noted in Sect. 1.3, they can be obtained by introducing a set of translators into the subsystem that convert input and output information into formats of other programs and vice versa, as

well as standard formats (for example, VHDL, Verilog) of representing simulated information.

3. The subsystem should contain a rich library of models of the above mentioned types (MCE, ME, RSE, RGE) and their parameters for all circuit-level bases. Even for the same basis and the same element, the library should have multiple models with different accuracy and complexity in order to best combine the requirements of the adequacy of calculations and the required computer time. For this purpose, the subsystem must also contain means of adaptation to a particular computing situation, as well as the possibility of self-learning.

Chapter 2
Models of Logical Elements for DF Consideration

2.1 Capabilities of MCE to Define Functional Failures

The MCE proposed in Sect. 1.4 is a new class of digital cell models for consideration of DF effect. In the MCE, the main advantages of traditional gate-level models are retained [164, 172, 174, 181, 182, 191, 196, 207]. At the same time, it is possible to explicitly determine the dependence of digital cells' behavior on different DFs (Table 1.1). This, in turn, allows identifying various kinds of functional failures in modern digital circuits.

From (1.16) it follows that a digital cell with n inputs, m outputs, k internal states, p parameters, and $q = m + k + u$ inputs of DF effect is $\mathrm{MCE}_{n,\ m,\ k,\ p,\ q}$, if it is represented in the form of the system (1.24)–(1.26). However, in accordance with Fig. 2.1, $\mathrm{MCE}_{n,\ m,\ k,\ p,\ q}$ can have specific varieties. For each of possible varieties $\mathrm{MCE}_{n,\ m,\ k,\ p,\ q}$ system of equations in the general form (1.24)–(1.26) is transformed into one of the specific types, presented in Table. 2.1.

Varieties of MCE differ from each other also by means of defining functional failures degree (FFD) in digital circuits as a result of influence of various DFs.

To determine the FFD in the static mode (for example, for $\mathrm{MCE}^{\mathrm{CWSt}}$), the values $\varphi(x_1, x_2, \ldots, x_n, t)$ for various combinations $x_1, x_2, \ldots, x_n$ in the normal disjunctive form of the logic function $\mathrm{MCE}^{\mathrm{CWSt}}$ are first computed:

$$\varphi_l(x_1, x_2, \ldots, x_n, t) = \theta_{l1}\lambda_1 \vee \theta_{l2}\lambda_2 \vee \ldots \vee \theta_{lj}\lambda_j \vee \ldots \vee \theta_{lr}\lambda_r, \tag{2.1}$$

$$P_l(x_1, x_2, \ldots, x_n, t) = \eta_{l1}\lambda_1 \vee \eta_{l2}\lambda_2 \vee \ldots \vee \eta_{lj}\lambda_j \vee \ldots \vee \eta_{lr}\lambda_r, \tag{2.2}$$

where $l = 1, 2, \ldots, m$ (m is the number of $\mathrm{MCE}^{\mathrm{CWSt}}$ outputs); t is model time; r—number of timing intervals of simulation:

$$\lambda_j = \begin{cases} 1, & \text{if } t = j, \\ 0, & \text{if } t \neq j; \end{cases}$$

V. Melikyan, *Simulation and Optimization of Digital Circuits*,
https://doi.org/10.1007/978-3-319-71637-4_2

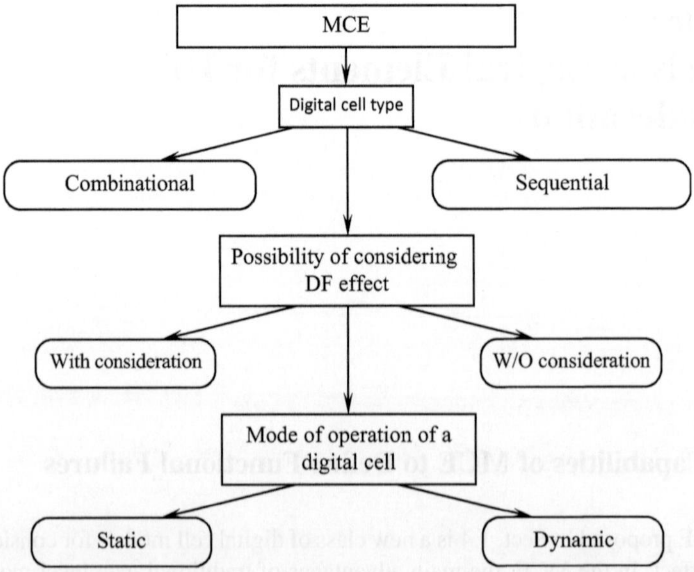

Fig. 2.1 Classification of varieties model of MCE

$$Q_{lj} = \underset{\beta_1^{lj} \ldots \beta_n^{lj}}{V} X_1\left(\beta_1^{lj}\right) \ldots X_n\left(\beta_n^{lj}\right) \ \mathrm{RSE}_y f\left(\beta_1^{lj}, \ldots, \beta_n^{lj}\right), \tag{2.3}$$

$$\eta_{lj} = \underset{\beta_1^{lj} \ldots \beta_n^{lj}}{V} X_1\left(\beta_1^{lj}\right) \ldots X_n\left(\beta_n^{lj}\right) \ \mathrm{RSE}_p f\left(\beta_1^{lj}, \ldots, \beta_n^{lj}\right); \quad \beta_p^{lj} \in [a, b], p = \{1, 2, \ldots, n\};$$

$$X_p\left(\beta_p^{lj}\right) = a + b - \beta_p^{lj}$$

In $\mathrm{MCE}^{\mathrm{CWSt}}$ $f_i \in \vec{F}$ are considered as additional inputs of digital cells.

In order to construct a model of an entire digital circuit of combinational type, with the aim of determining the FFD in the static mode, an inductive circuit [187] of connecting MCE is used (Fig. 2.2).

However, as it was said in Sect. 1.4, the value of the output signal of the previous element is transferred to the next input via MCE, and the inputs $f_{ij,\, j=1,\, 2,\, \ldots,\, qi}$ of each MCE_i are left independent. Then, using all axioms and rules for calculating logical functions, the model of the entire digital circuit is simplified for the input combinations $\vec{X}$. As a result, for each output of a digital circuit $y_{ij,\, j=1,\, 2,\, \ldots,\, mi}$, values from the range [a, b] corresponding to different combinations at the inputs of digital circuits are obtained. Based on the analysis of the values of digital circuit outputs, the FFD of this output is detected. In case of sequential digital cells (for example, $\mathrm{MCE}^{\mathrm{SWSt}}$), as in case of general gate-level models [173, 174], the feedback loops are broken (Fig. 2.3):

Table 2.1 Varieties of MCE

Type of a digital circuit (combinational-C, sequential-S)		Consideration of DF effect (with consideration-W, without consideration-W/O)		Mode of operation of a digital circuit (static-St, dynamic-D)	
$\vec{Y}(t) = \varphi[\vec{X}(t), \vec{P}(t), \vec{F}(t)]$ $\vec{P}'(t) = \chi[\vec{X}(t), \vec{P}(t), \vec{F}(t)]$	MCE^{C}	$\vec{Y}(t) = \varphi[\vec{X}(t)]$	$MCE^{CW/O}$	$\vec{Y} = \varphi[\vec{X}]$	$MCE^{CW/OSt}$
				$\vec{Y}(t) = \varphi[\vec{X}(t)]$	$MCE^{CW/OD}$
		$\vec{Y}(t) = \varphi[\vec{X}(t), \vec{P}(t), \vec{F}(t)]$ $\vec{P}'(t) = \chi[\vec{X}(t), \vec{P}(t), \vec{F}(t)]$	MCE^{CW}	$\vec{Y}(t) = \varphi[\vec{X}, \vec{P}(t), \vec{F}(t)]$ $\vec{P}'(t) = \chi[\vec{X}, \vec{P}(t), \vec{F}(t)]$	MCE^{CWSt}
				$\vec{Y}(t) = \varphi[\vec{X}(t), \vec{P}(t), \vec{F}(t)]$ $\vec{P}'(t) = \chi[\vec{X}(t), \vec{P}(t), \vec{F}(t)]$	MCE^{CWD}
$\vec{Y}(t) = \varphi[\vec{X}(t), \vec{Z}(t), \vec{P}(t), \vec{F}(t)]$ $\vec{Z}'(t) = \psi[\vec{X}(t), \vec{Z}(t), \vec{P}(t), \vec{F}(t)]$ $\vec{P}'(t) = \chi[\vec{X}(t), \vec{Z}(t), \vec{P}(t), \vec{F}(t)]$	MCE^{S}	$\vec{Y}(t) = \varphi[\vec{X}(t), \vec{Z}(t)]$ $\vec{Z}'(t) = \psi[\vec{X}(t), \vec{Z}(t)]$	$MCE^{SW/O}$	$\vec{Y} = \varphi[\vec{X}, \vec{Z}]$ $\vec{Z}' = \psi[\vec{X}, \vec{Z}]$	$MCE^{SW/OSt}$
				$\vec{Y}(t) = \varphi[\vec{X}(t), \vec{Z}(t)]$ $\vec{Z}'(t) = \psi[\vec{X}(t), \vec{Z}(t)]$	$MCE^{SW/OD}$
		$\vec{Y}(t) = \varphi[\vec{X}(t), \vec{Z}(t), \vec{P}(t), \vec{F}(t)]$ $\vec{Z}'(t) = \psi[\vec{X}(t), \vec{Z}(t), \vec{P}(t), \vec{F}(t)]$ $\vec{P}'(t) = \chi[\vec{X}(t), \vec{Z}(t), \vec{P}(t), \vec{F}(t)]$	MCE^{SW}	$\vec{Y}(t) = \varphi[\vec{X}, \vec{Z}(t), \vec{P}(t), \vec{F}(t)]$ $\vec{Z}'(t) = \psi[\vec{X}, \vec{Z}(t), \vec{P}(t), \vec{F}(t)]$ $\vec{P}'(t) = \chi[\vec{X}, \vec{Z}(t), \vec{P}(t), \vec{F}(t)]$	MCE^{SWSt}
				$\vec{Y}(t) = \varphi[\vec{X}(t), \vec{Z}(t), \vec{P}(t), \vec{F}(t)]$ $\vec{Z}'(t) = \psi[\vec{X}(t), \vec{Z}(t), \vec{P}(t), \vec{F}(t)]$ $\vec{P}'(t) = \chi[\vec{X}(t), \vec{Z}(t), \vec{P}(t), \vec{F}(t)]$	MCE^{SWD}

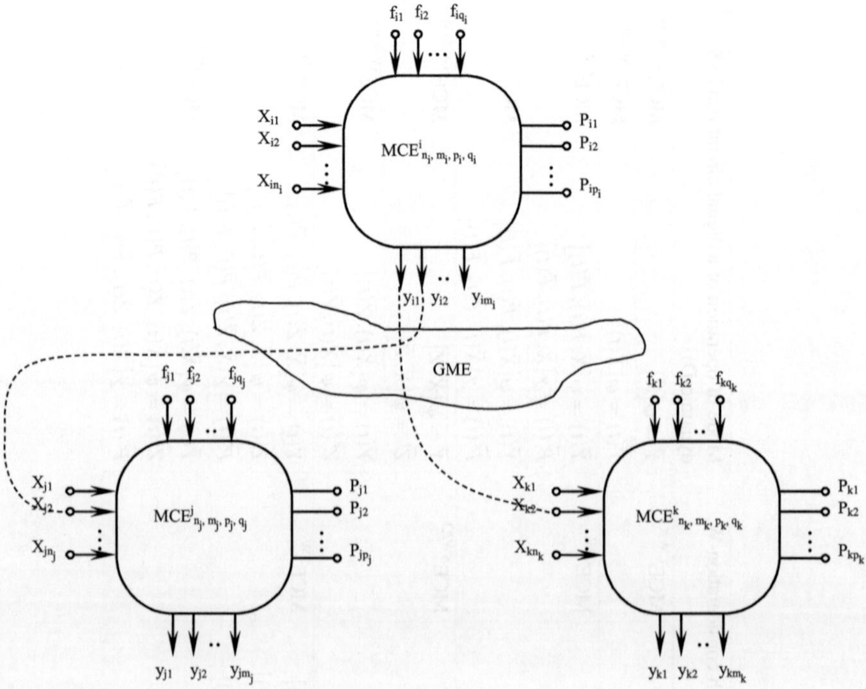

Fig. 2.2 Inductive circuit of connecting MCE

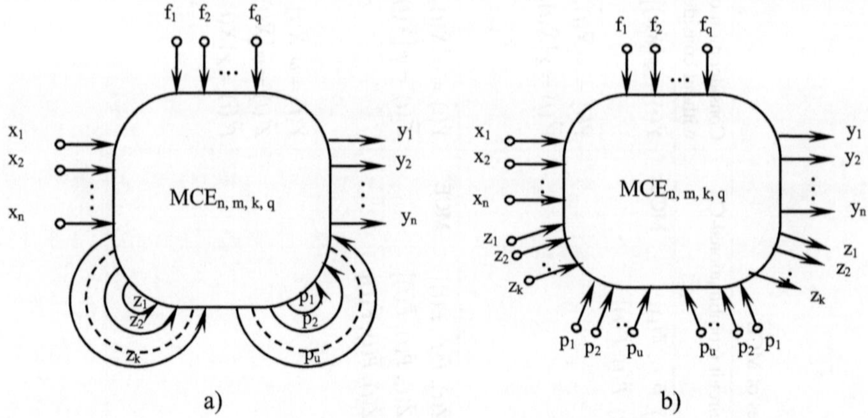

Fig. 2.3 MCE of sequential digital cell before (**a**) and after (**b**) cutout of feedbacks

In this case, the system of equations (1.24)–(1.26) is transformed into the following form:

$$\vec{Y}(t) = \varphi'\left[\vec{X}(t), \vec{Z}(t_0), \vec{P}(t_0), \vec{F}(t)\right], \tag{2.4}$$

$$\vec{Z}(t) = \psi'\left[\vec{X}(t), \vec{Z}(t_0), \vec{P}(t_0), \vec{F}(t)\right], \tag{2.5}$$

$$\vec{P}(t) = \chi'\left[\vec{X}(t), \vec{Z}(t_0), \vec{P}(t_0), \vec{F}(t)\right], \tag{2.6}$$

$$\vec{Z}(t) = \vec{Z}(t_0), \tag{2.7}$$

$$\vec{P}(t) = \vec{P}(t_0), \tag{2.8}$$

where $\vec{Z}(t) = (z_1(t), \ldots, z_k(t))$, $\vec{P}(t) = (p_1(t), \ldots, p_k(t))$, $t > t_0$, φ', ψ', χ'—new functions corresponding to MCE with broken feedbacks. The solution of system (2.4)–(2.8) is carried out by the iterative algorithm in Fig. 2.4.

To determine the FFD in the dynamic mode (for example, for $\mathrm{MCE}^{\mathrm{SWD}}$), one can use the relations (2.1)–(2.3). But in this case the function λ_j is replaced by another, taking into account the variability of the simulation time step. The latter is associated with changes in timing parameters of digital cells due to the effect of DF.

The use of the MCE allows entering indices that reflect the FFD in digital circuits and the relative contribution of each digital cell that is part of a digital circuit. Table 2.2 shows examples of indices of this type. It is accepted that $\vec{B} = \vec{V} \cup \vec{W}$. The operability index $\mathrm{FFD}_1\omega_j(\vec{B})_{,j=1,2,\ldots,m}$ is defined in such a way that the standard is the value of the function φ in (1.24) with $\vec{B} = 0$. In this case, $\omega_j(\vec{B}) = 0$, if the digital cell is fully operational, and 1 if it is completely inoperative. Intermediate values of FFD_1 characterize the degree of inoperability of a digital cell. Similarly, the remaining FFDs in Table 2.2 are defined. The calculation of the values of one or several FFD allows solving a number of useful practical problems:

1. Determining the operability of a particular implementation of digital circuits in a certain range of changes $\mathrm{DF}\left[\vec{\mathrm{B}}_{\min}, \vec{\mathrm{B}}_{\max}\right]$ when the condition is carried out:

$$\mathrm{FFD}_i \Big/ \vec{B}_{\min} \le B \le \vec{B}_{\max} < \varepsilon_{\mathrm{FFD}_i, i=1,2,3,4,5}, \tag{2.9}$$

where $\varepsilon_{\mathrm{SFN}_i}$—given permissible value of FFD_i.

2. Comparison of s different variants, implementation of digital circuits and selection of the best one by $\min\limits_k \left\{\mathrm{FFD}_{ij,1,2,\ldots,s}\right\}$. In fact, it becomes possible to carry out optimization (both structural and parametric).

   Determination for the given digital circuit of the allowed ranges of changes of values DF $\vec{B}_{\min}$, $\vec{B}_{\max}$, in case of which the circuit remains operational.

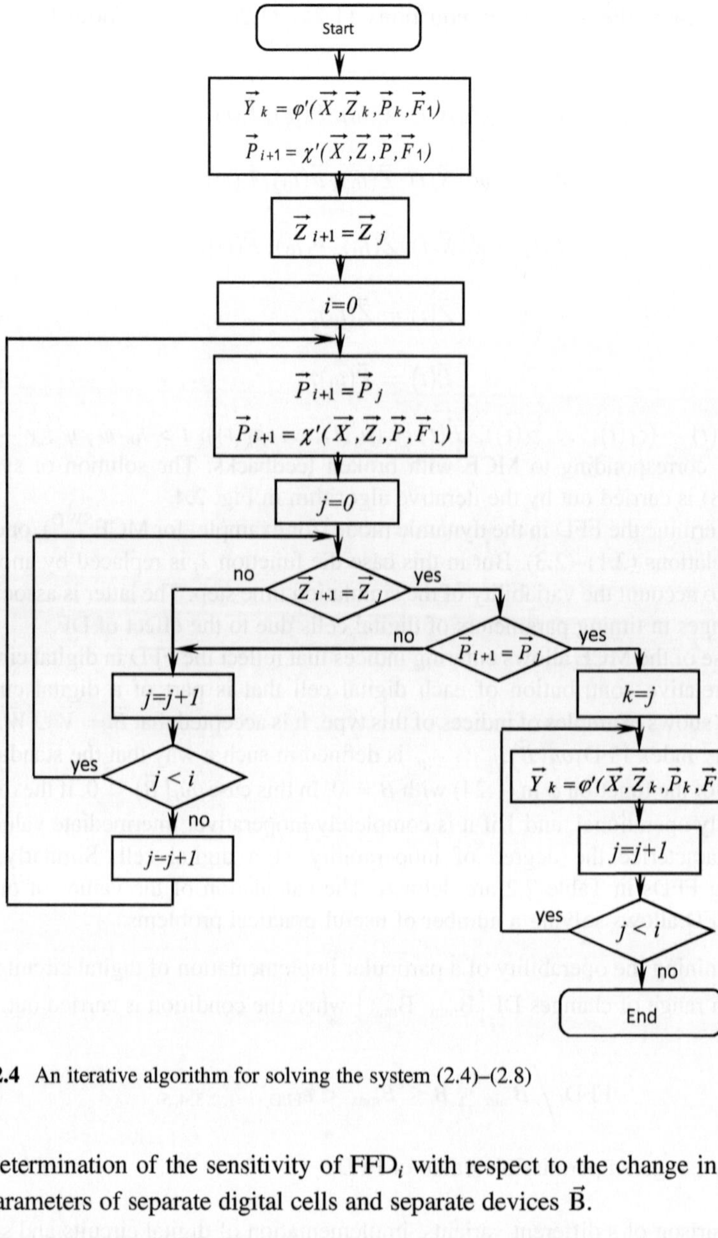

Fig. 2.4 An iterative algorithm for solving the system (2.4)–(2.8)

3. Determination of the sensitivity of FFD_i with respect to the change in both the parameters of separate digital cells and separate devices $\vec{\text{B}}$.

Thus, the proposed MCE has wide-ranging possibilities to define functional failures in digital circuits.

Table 2.2 FFD indices

Denotion	Name	Definition
FFD_1	Operability on *j*th output	$\omega_j(\vec{B}) = \max(\min(\varphi_j(\vec{B}), a + b - \varphi_j(0)), \min(a + b - \varphi_j(\vec{B}), \varphi_j(0)))$
FFD_2	Operability of *j*th output with *k*th input signal set	$\omega_j(\vec{X}_k, \vec{B}) = \max(\min(\varphi_j(\vec{X}_k, \vec{B}), a + b - \varphi_j(\vec{X}_k, 0)), \min(a + b - \varphi_j(\vec{X}_k, \vec{B}), \varphi_j(\vec{X}_k, 0)))$
FFD_3	Operability of *j*th output, regardless of the input sets	$\omega_j(\vec{B}) = \max_k \omega_j(\vec{X}_k, \vec{B})$
FFD_4	Operability of *k*th input set regardless of output sets	$\omega(\vec{X}_k, \vec{B}) = \max_j \omega_j(\vec{X}_k, \vec{B})$
FFD_5	Generalized operability	$\omega(\vec{B}) = \max_k \omega_j(\vec{B}) = \max \omega(\vec{X}_k, \vec{B})$

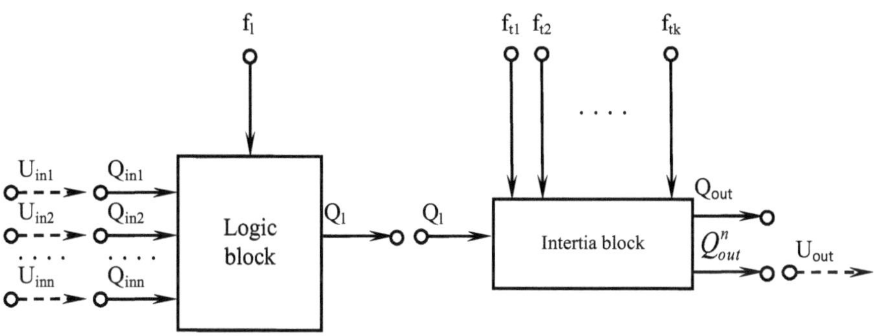

Fig. 2.5 Structure of the model of a gate-level element

2.2 Structure of the Model of Considering DF Effects (MCE)

For deterministic gate-level simulation of a digital circuit with consideration of DF, it is proposed to present the MCE (Sect. 1.4) in the form of an equivalent circuit consisting of models of gate-level elements (MGE) [59, 60, 167–170, 202, 277–282]. MGE is the basic component of MCE and has the structure shown in Fig. 2.5.

The first block implements the logic of switching the gates without taking its inertia into account. After calculating the new state of the gates (Q_l), the second block reproduces the switching inertia. With such a representation, the gates can be described both as a simple logic gate (AND, OR, NAND, etc.), and a more complex digital fragment if:

1. Its inertial properties with respect to all input signals can be considered identical
2. Has one output

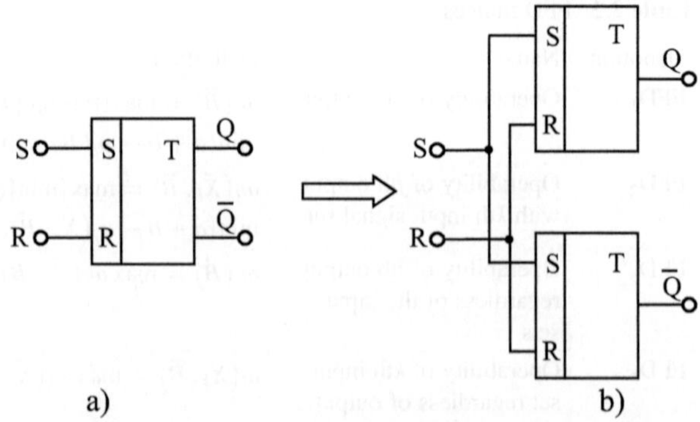

Fig. 2.6 Conversion to the equivalent MGE circuit: (**a**) R-S flip-flop; (**b**) equivalent circuit

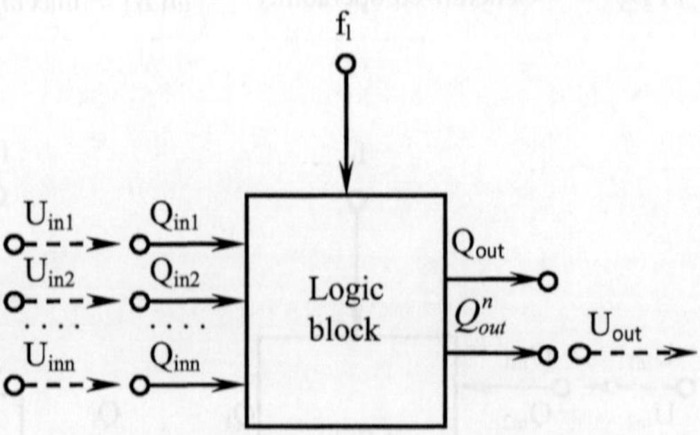

Fig. 2.7 Synchronous MGE

If these conditions are violated, i.e. delays at the inputs of the digital fragment are different, or there are several outputs, then the gate can always be represented as an equivalent circuit of several MGE. For example, if the digital fragment is a simple asynchronous RS flip-flop (Fig. 2.6a), which has two outputs, it can be converted to the equivalent circuit shown in Fig. 2.6b. The representation of the MGE in Fig. 2.5 is dictated by great effectiveness of the application of the principle of spatial event [181]. For example, if, in contrast to the above representation, it is assumed that the gate can have more than one output, then when the state of the gate changes only in one of its outputs, the whole fragment must be considered excited, despite the stability of the states of the other outputs.

The logic block [278] implements the synchronous part of the MGE, and together with the inertia block [281] turns into an asynchronous model. In other words, if the inertia of digital cells can be ignored (i.e., it is assumed that all timing parameters of the gates are zero: $t_1, t_2, \ldots, t_k = 0$), then the output signal of MGE Q_{out} is identified with the calculated logic block by the value Q_l, i.e. synchronous MGE is obtained (Fig. 2.7).

If the timing parameters of digital cells cannot be ignored because of their importance to the form of the output signal, then an asynchronous MGE, shown in Fig. 2.5, is obtained.

In the logic block, the input f_l reflects the generalized effect of all the considered DF($V_{i,\, i=1,\, 2,\, \ldots}$; $W_{j,\, j=1,\, 2,\, \ldots}$) on the value of Q_l, and in the inertia block f_{tl}—on the value of the ith timing parameter of the gate (the composition of timing parameters of the MGE will be discussed later). F_l and $f_{ti,\ i=1,\ 2,\ \ldots,\ k}$ are functions of all digital fragments (Sect. 1.4). The presence of two output signals of the model is due to the fact that, as will be seen from further consideration of the behavior of MGE in different situations, the value of Q_{out} that arrives at the input of the subsequent gate and the signal Q^n_{out}, given to the input of the analog element-load, have different forms. In particular, Q^n_{out} is adjusted for convenient conversion by output models (Fig. 1.110) into the form of output voltage U_{out} in case when the gate load is an analog cell. In the same way, if the source of the input signal $Q_{\text{in}i}$ is the output of the analog node, then the input model converts $U_{\text{in}i} \rightarrow Q_{\text{in}i}$. However, because of the continuous nature of the variables $Q_{\text{in}i,\ i=1,\ 2,\ \ldots,\ n,}$ and Q^n_{out} , the input and output models in this case, must differ fundamentally from the analogous threshold models proposed in [239]. The threshold nature of the functions of the transformation of logical states into voltages and vice versa was due to the discrete nature of signals of the model in the mentioned work. In the proposed MGE, logical signals take continuous values from the range $[a, b]$, therefore the corresponding transformation functions take the following form:

$$Q_{\text{in}i} = a + (b - a) \cdot \frac{U_{\text{in}i} - U_{\text{in}i\text{min}}}{U_{\text{in}i\text{max}} - U_{\text{in}i\text{min}}}, \tag{2.10}$$

$$U_{\text{out}} = U_{\text{outmin}} + (U_{\text{outmax}} - U_{\text{outmin}}) \cdot \frac{Q^i_{\text{out}} - a}{b - a}, \tag{2.11}$$

where $U_{\text{in}i\text{min}}$, $U_{\text{in}i\text{max}}$—minimum and maximum voltage values at the ith input of the gate; U_{outmin}, U_{outmax}—the corresponding values of the output signal of digital cells. It should be noted that $U_{\text{in}i\text{min}}$, $U_{\text{outmin}} \leq U_{\text{o}}$ and $U_{\text{in}i\text{min}}$, $U_{\text{outmax}} \geq U_1$ can differ significantly from voltage levels of logic "0" (U_0) and "1" (U_1) (Fig. 1.108).

2.2.1 Logic Block of MGE

The initial data for the logic block [278] of the proposed MGE at each current simulation time t_{sim} are:

1. Values of the variables of the vector of input signals:

$$\vec{Q}_{\text{in}} = \{Q_{\text{in}1}, Q_{\text{in}2}, \ldots, Q_{\text{in}i}, \ldots, Q_{\text{in}n}\}, Q_{\text{in}i} \in [a, b];$$

2. Quantity $f_l \in [a, b]$.

Based on this information, the logic block calculates the new value of Q_l. The calculation is carried out with the help of the function $Q_l = \varphi(Q_{\text{in}1}, Q_{\text{in}2}, \ldots, Q_{\text{in}n})$, then the calculated value of Q_l is corrected using the rules of summation of excitations (Sect. 1.4): $Q_n = \text{RSE}(Q_l, f_l)$.

The criterion for the adequacy of the logic block is the correspondence of the value of the U_{out} to the gates calculated by formula (2.11) (when the calculated value Q_l is used as Q^n_{out}), with the value of the same voltage obtained experimentally. If the first voltage is denoted by U^M_{out} and the second one is denoted by U^E_{out}, then when a specific input set $Q^l_{\text{in}1}, Q^l_{\text{in}2}, \ldots, Q^l_{\text{in}n}$ and the value f^l_l is given to the input of the model, the absolute error of the Δ_i model can be determined in the form of:

$$\Delta_i = \left|U^E_{\text{out}} - U^M_{\text{out}}\right|, \tag{2.12}$$

and a relative error ε_i

$$\varepsilon_i = \left|\frac{U^E_{\text{out}} - U^M_{\text{out}}}{U^E_{\text{out}}}\right|, \tag{2.13}$$

The integrated estimation of the absolute (Δ) and relative (ε) accuracy of the model can be performed by summing over all input sets:

$$\Delta = \sum_l \Delta_l, \tag{2.14}$$

$$\varepsilon = \frac{\sum_l \varepsilon_l}{l}. \tag{2.15}$$

In turn, the accuracy of calculation results of the logic block depends on:

1. Rules for calculating logical functions φ;
2. Physical interpretation of the variable f_l.

Moreover, the mentioned two factors are interrelated.

Several interpretations of the physical meaning of the variable f_l are possible.

Correction value of the output signal:

$$f_l = Q^{\text{DF}}_l - Q^{\varphi}_l, \tag{2.16}$$

where Q^{DF}_l-the value of the gate output as a result of effects of all DFs, and Q^{φ}_l is the value calculated from the functions φ. In this case, the RSE is simple:

$$Q_l = Q^{\varphi}_l + f_l. \tag{2.17}$$

As for the rules for calculating logical functions, the rules of minimax logic (1.27)–(1.29) are the most natural. However, the determination of f_l by the method described above makes it possible to use other rules for calculating the logical functions of fuzzy set theory [282–289], given in Table. 2.3.

For all these rules, the following are typical:

1. Performing axioms (commutativity, associativity, distributivity, etc.) of Boolean algebra.
2. A small (close to Boolean algebra) number of operations to calculate the result.

Table 2.3 Rules for calculating elementary logical functions

N	Title	AND	OR	NOT
1	Minimax fuzzy logic	$\min(x_1, x_2)$	$\max(x_1, x_2)$	$a+b-x$
2	Probable fuzzy logic	$x_1 \cdot x_2$	$x_1 + x_2 - x_1 \cdot x_2$	$a+b-x$
3	Boundary fuzzy logic	$\max(a, x_1 + x_2 - b - a)$	$\min(b - a, x_1 + x_2)$	$a+b-x$
4	The fuzzy logic of Lorentz	$\frac{x_1 \cdot x_2}{2 \cdot (b-a) - (x_1 + x_2 - x_1 \cdot x_2)}$	$\frac{x_1 + x_2}{b - a + x_1 \cdot x_2}$	$a+b-x$

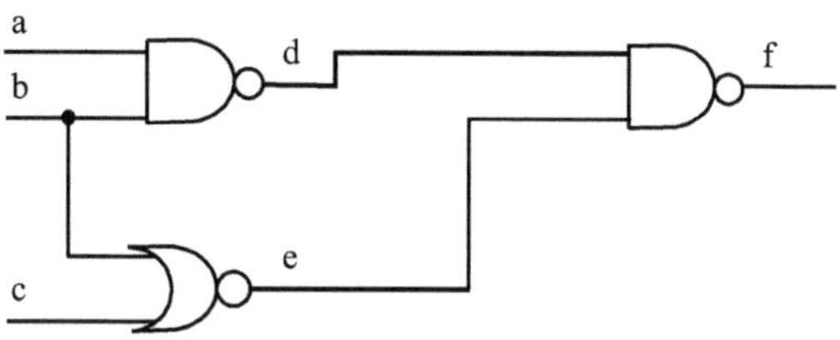

Fig. 2.8 Example of a simulated circuit

However, with the noted interpretation of the physical meaning of f_l, the choice of specific rules requires its own models for calculating f_l, since different values are obtained in each case Q_l^{φ}. As we will see after considering other interpretations of the physical meaning of f_l, there are also variants when the above rules are equivalent. When determining f_l as the correcting value (2.16), the models for calculating f_l, are obtained as follows:

1. The rules for calculating elementary logic functions are selected (Table 2.3).
2. Voltages $U_{\text{in}1}$, $U_{\text{in}2}$, …, U_{in} are given to the inputs of the basic logic cell (TTL (Fig. 1.13), ECL (Fig. 1.14), CMOS (Fig. 1.15), etc.).
3. By simulation at the circuit level, the output voltage of the basic logic cell is determined for a given DF value: U_{out}^E.
4. $U_{\text{in}i,\ i=1,\ 2,\ \ldots,\ n}$ is transformed into $Q_{\text{in}i,\ i=1,\ 2,\ \ldots,\ n}$ according to the formula (2.10).
5. Using the selected rules, the value Q_l^{φ} is calculated.
6. Q_l^{φ} it transformed into U_{out}^M according to formula (2.11).
7. Using the formulas (2.12)–(2.16) (the accuracy of calculations is given), the value f_l is determined for a given value of DF.

Then the experiments are repeated for other values of DF and the input voltages. To explain the operation of the logic block, consider the following example. Let the simple circuit shown in Fig. 2.8. be simulated.

Suppose that the influence of only one (external) DF is taken into account. It is the irradiation flux (V_1) that varies with time according to the function $D = f_1(t)$ (Fig. 2.9).

Suppose there are models of dependencies $f_l = f_2(D)$ of the NAND gate for different values of Q_l (Fig. 2.10). For simplicity, as an example, the models f_l obtained experimentally are given only for two values of Q_l: 0 and 0, 86.

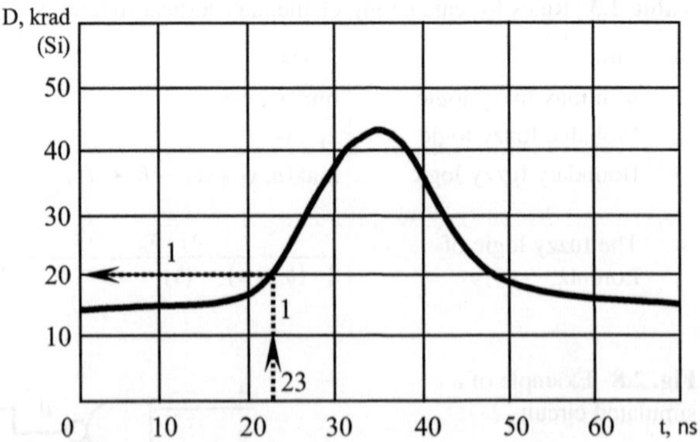

Fig. 2.9 The change in the absorbed dose of radiation in time: $D = f_1(t)$

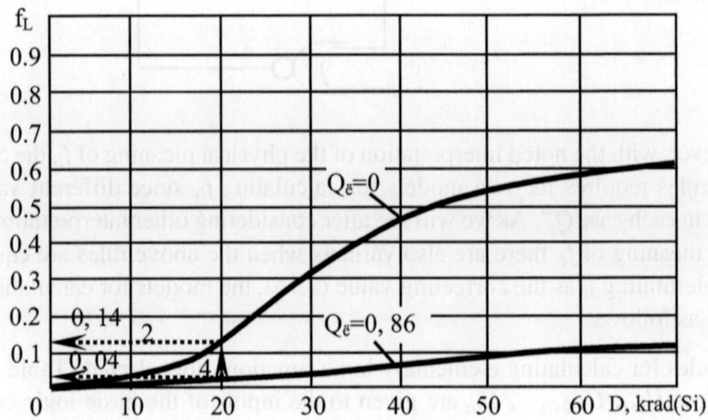

Fig. 2.10 Models of dependencies $f_l = f_2(D)$ of the NAND gate

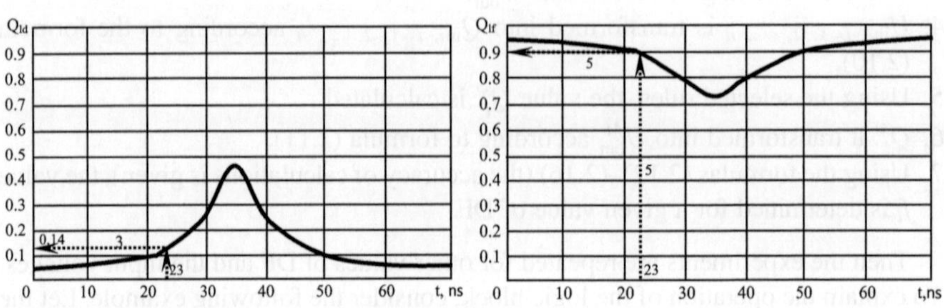

Fig. 2.11 Changes in the states of the nodes *d* and *f*

For simplicity, it is also assumed that for an OR cell, always $f_l = 0$ (which, as a rule, does not correspond to reality). Figure 2.11 shows the simulation results of the circuit in Fig. 2.8 with the change in the absorbed dose of radiation shown in Fig. 2.9 and when the input combination abc 110 is applied.

value of the *i*th timing parameter of the gate (the composition of timing parameters of the MGE will be discussed later). F_l and $f_{ti,\ i=1,\ 2,\ \ldots,\ k}$ are functions of all digital fragments (Sect. 1.4). The presence of two output signals of the model is due to the fact that, as will be seen from further consideration of the behavior of MGE in different situations, the value of Q_{out} that arrives at the input of the subsequent gate and the signal Q^n_{out}, given to the input of the analog element-load, have different forms. In particular, Q^n_{out} is adjusted for convenient conversion by output models (Fig. 1.110) into the form of output voltage U_{out} in case when the gate load is an analog cell. In the same way, if the source of the input signal $Q_{\mathrm{in}i}$ is the output of the analog node, then the input model converts $U_{\mathrm{in}i} \rightarrow Q_{\mathrm{in}i}$. However, because of the continuous nature of the variables $Q_{\mathrm{in}i,\ i=1,\ 2,\ \ldots,\ n,}$ and Q^n_{out} , the input and output models in this case, must differ fundamentally from the analogous threshold models proposed in [239]. The threshold nature of the functions of the transformation of logical states into voltages and vice versa was due to the discrete nature of signals of the model in the mentioned work. In the proposed MGE, logical signals take continuous values from the range [*a*, *b*], therefore the corresponding transformation functions take the following form:

$$Q_{\mathrm{in}i} = a + (b - a) \cdot \frac{U_{\mathrm{in}i} - U_{\mathrm{in}i\mathrm{min}}}{U_{\mathrm{in}i\mathrm{max}} - U_{\mathrm{in}i\mathrm{min}}}, \tag{2.10}$$

$$U_{\mathrm{out}} = U_{\mathrm{outmin}} + (U_{\mathrm{outmax}} - U_{\mathrm{outmin}}) \cdot \frac{Q^i_{\mathrm{out}} - a}{b - a}, \tag{2.11}$$

where $U_{\mathrm{in}i\mathrm{min}}$, $U_{\mathrm{in}i\mathrm{max}}$—minimum and maximum voltage values at the *i*th input of the gate; U_{outmin}, U_{outmax}—the corresponding values of the output signal of digital cells. It should be noted that $U_{\mathrm{in}i\mathrm{min}}$, $U_{\mathrm{outmin}} \leq U_{\mathrm{o}}$ and $U_{\mathrm{in}i\mathrm{min}}$, $U_{\mathrm{outmax}} \geq U_1$ can differ significantly from voltage levels of logic "0" (U_0) and "1" (U_1) (Fig. 1.108).

2.2.1 Logic Block of MGE

The initial data for the logic block [278] of the proposed MGE at each current simulation time t_{sim} are:

1. Values of the variables of the vector of input signals:

$$\vec{Q}_{\mathrm{in}} = \{Q_{\mathrm{in}1}, Q_{\mathrm{in}2}, \ldots, Q_{\mathrm{in}i}, \ldots, Q_{\mathrm{in}n}\}, Q_{\mathrm{in}i} \in [a, b];$$

2. Quantity $f_l \in [a, b]$.

Based on this information, the logic block calculates the new value of Q_l. The calculation is carried out with the help of the function $Q_l = \varphi(Q_{\mathrm{in}1}, Q_{\mathrm{in}2}, \ldots, Q_{\mathrm{in}n})$, then the calculated value of Q_l is corrected using the rules of summation of excitations (Sect. 1.4): $Q_n = \mathrm{RSE}(Q_l, f_l)$.

The criterion for the adequacy of the logic block is the correspondence of the value of the U_{out} to the gates calculated by formula (2.11) (when the calculated value Q_l is used as Q^n_{out}), with the value of the same voltage obtained experimentally. If the first voltage is denoted by U^M_{out} and the second one is denoted by U^E_{out}, then when a specific input set $Q^l_{\text{in}1}, Q^l_{\text{in}2}, \ldots, Q^l_{\text{in}n}$ and the value f^l_l is given to the input of the model, the absolute error of the Δ_i model can be determined in the form of:

$$\Delta_i = \left|U^E_{\text{out}} - U^M_{\text{out}}\right|, \tag{2.12}$$

and a relative error ε_i

$$\varepsilon_i = \left|\frac{U^E_{\text{out}} - U^M_{\text{out}}}{U^E_{\text{out}}}\right|, \tag{2.13}$$

The integrated estimation of the absolute (Δ) and relative (ε) accuracy of the model can be performed by summing over all input sets:

$$\Delta = \sum_l \Delta_l, \tag{2.14}$$

$$\varepsilon = \frac{\sum_l \varepsilon_l}{l}. \tag{2.15}$$

In turn, the accuracy of calculation results of the logic block depends on:

1. Rules for calculating logical functions φ;
2. Physical interpretation of the variable f_l.

Moreover, the mentioned two factors are interrelated.

Several interpretations of the physical meaning of the variable f_l are possible.

Correction value of the output signal:

$$f_l = Q^{\text{DF}}_l - Q^{\varphi}_l, \tag{2.16}$$

where Q^{DF}_l-the value of the gate output as a result of effects of all DFs, and Q^{φ}_l is the value calculated from the functions φ. In this case, the RSE is simple:

$$Q_l = Q^{\varphi}_l + f_l. \tag{2.17}$$

As for the rules for calculating logical functions, the rules of minimax logic (1.27)–(1.29) are the most natural. However, the determination of f_l by the method described above makes it possible to use other rules for calculating the logical functions of fuzzy set theory [282–289], given in Table. 2.3.

For all these rules, the following are typical:

1. Performing axioms (commutativity, associativity, distributivity, etc.) of Boolean algebra.
2. A small (close to Boolean algebra) number of operations to calculate the result.

Table 2.3 Rules for calculating elementary logical functions

N	Title	AND	OR	NOT
1	Minimax fuzzy logic	$\min(x_1, x_2)$	$\max(x_1, x_2)$	$a+b-x$
2	Probable fuzzy logic	$x_1 \cdot x_2$	$x_1 + x_2 - x_1 \cdot x_2$	$a+b-x$
3	Boundary fuzzy logic	$\max(a, x_1 + x_2 - b - a)$	$\min(b - a, x_1 + x_2)$	$a+b-x$
4	The fuzzy logic of Lorentz	$\frac{x_1 \cdot x_2}{2 \cdot (b-a) - (x_1 + x_2 - x_1 \cdot x_2)}$	$\frac{x_1 + x_2}{b - a + x_1 \cdot x_2}$	$a+b-x$

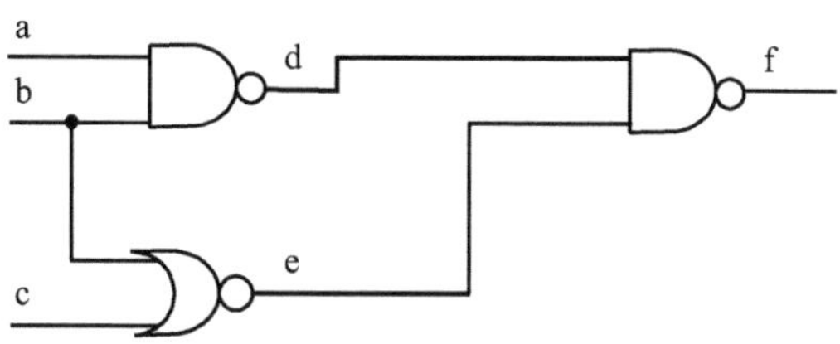

Fig. 2.8 Example of a simulated circuit

However, with the noted interpretation of the physical meaning of f_l, the choice of specific rules requires its own models for calculating f_l, since different values are obtained in each case Q_l^{φ}. As we will see after considering other interpretations of the physical meaning of f_l, there are also variants when the above rules are equivalent. When determining f_l as the correcting value (2.16), the models for calculating f_l, are obtained as follows:

1. The rules for calculating elementary logic functions are selected (Table 2.3).
2. Voltages $U_{\text{in}1}$, $U_{\text{in}2}$, …, U_{in} are given to the inputs of the basic logic cell (TTL (Fig. 1.13), ECL (Fig. 1.14), CMOS (Fig. 1.15), etc.).
3. By simulation at the circuit level, the output voltage of the basic logic cell is determined for a given DF value: U_{out}^{E}.
4. $U_{\text{in}i,\ i=1,\ 2,\ \ldots,\ n}$ is transformed into $Q_{\text{in}i,\ i=1,\ 2,\ \ldots,\ n}$ according to the formula (2.10).
5. Using the selected rules, the value Q_l^{φ} is calculated.
6. Q_l^{φ} it transformed into U_{out}^{M} according to formula (2.11).
7. Using the formulas (2.12)–(2.16) (the accuracy of calculations is given), the value f_l is determined for a given value of DF.

Then the experiments are repeated for other values of DF and the input voltages. To explain the operation of the logic block, consider the following example. Let the simple circuit shown in Fig. 2.8. be simulated.

Suppose that the influence of only one (external) DF is taken into account. It is the irradiation flux (V_1) that varies with time according to the function $D = f_1(t)$ (Fig. 2.9).

Suppose there are models of dependencies $f_l = f_2(D)$ of the NAND gate for different values of Q_l (Fig. 2.10). For simplicity, as an example, the models f_l obtained experimentally are given only for two values of Q_l: 0 and 0, 86.

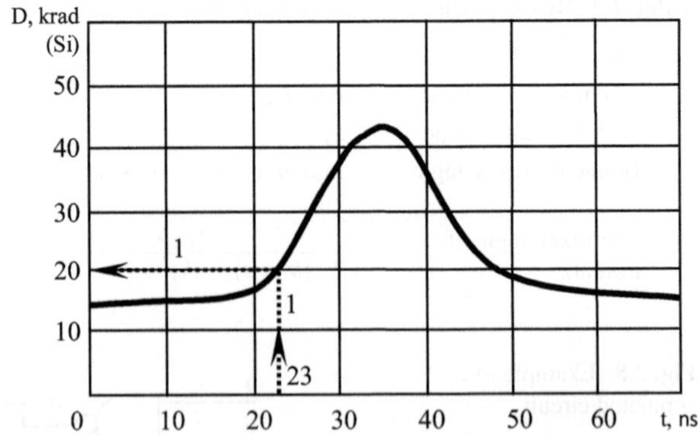

Fig. 2.9 The change in the absorbed dose of radiation in time: $D = f_1(t)$

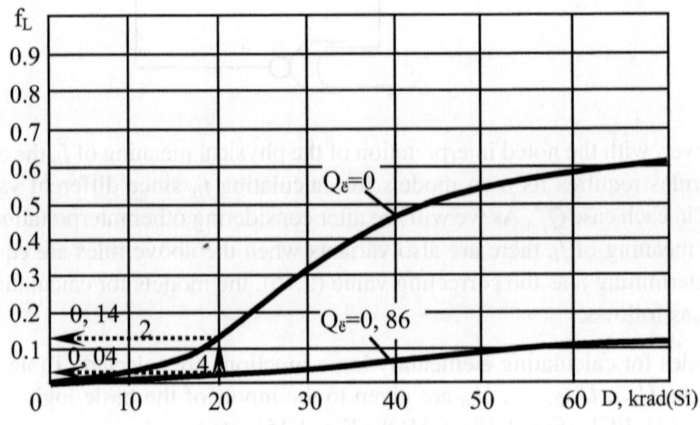

Fig. 2.10 Models of dependencies $f_l = f_2(D)$ of the NAND gate

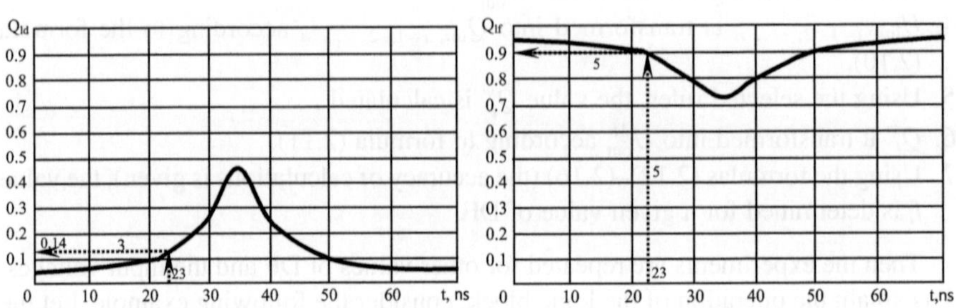

Fig. 2.11 Changes in the states of the nodes *d* and *f*

For simplicity, it is also assumed that for an OR cell, always $f_l = 0$ (which, as a rule, does not correspond to reality). Figure 2.11 shows the simulation results of the circuit in Fig. 2.8 with the change in the absorbed dose of radiation shown in Fig. 2.9 and when the input combination abc 110 is applied.

It is assumed that the input combination does not change during the considered time interval. The simulation was performed using minimax logic. The dependencies in Fig. 2.10 also correspond to this logic. At node *e*, a constant signal 1 is obtained, since $b = 1$, $c = 0$ and, according to the noted logic, $Q_{le} = \max\{1, 0\} = 1$, and according to RSE: $Q_{le} = Q_{le} + f_{l\ \text{or}} = 1 + 0 = 1$. Therefore, the diagram of the change in $Q_{l\mathring{a}}$ is not given here.

For clarity, the arrows show the sequence of actions when finding the states of nodes *d* and *f* at time $t_{\text{sim}} = 23\text{ns}$:

1. Since $Q_a = 1$, $Q_b = 1$, we obtain $Q_{ld} = 1 - \min\{1, 1\} = 0$.
2. According to function $D = f_1(t)$, for $t = 23$ ns, the following is found: $D = 20$ krad (Fig. 2.9, arrow 1).
3. According to the dependence $f_l = f_2(D)$, for $Q_l = 0$ and $D = 20$ krad, the following is obtained: fdd = 0.14 (Fig. 2.10, arrow 2).
4. According to RSE, the following is obtained: $Q_{ld} = Q_{ld} = f_{ld} = 0 = 0.14 = 0.14$ (Fig. 2.11, arrow 3).
5. Since $Q_d = 0.14$ and $Q_e = 1$, the following is obtained: $Q_{lf} = 1 - \min\{1;0.14\} =$ 0, 86.
6. According to the dependence $f_l = f_2(D)$, for $Q_l = 0.86$ and $D = 20$ krad, the following is obtained: $f_{lf} = 0.04$ (Fig. 2.10, arrow 4).
7. According to RSE, the following is obtained: $Q_{lf} = Q_{lf} + f_{lf} = 0.86 + 0.04 = 0.9$ (Fig. 2.11, arrow 5).

In the same way, the values of Q_{ld} and Q_{lf} at other moments of simulation are obtained. The strategy for selecting the simulation moments is given in Chap. 3. Here, to simplify the example, one can still assume that the t_{sim} changes with a constant step Δt_{sim}.

Uncertainty (fuzzy) of the node state with respect to any of the steady-state values (0 or 1):

$$f_l^0 = Q_l^{\text{DF}} - 0, \tag{2.18}$$

$$f_l^1 = Q_l^{\text{DF}} - 1. \tag{2.19}$$

In this case, the values of the input ($Q_{\text{in}1}$, $Q_{\text{in}2}$, …, $Q_{\text{in}n}$) and output (Q_{out}) variables of the model also show the degree of uncertainty with respect to one of the stationary values. Analysis of possible variants of calculation rules for logical functions (Table 2.3) showed that in case of coincidence of RSE with the chosen method of calculations, all the logic is equivalent. But in each case the procedure for determining the value of f_l is changing. For example, if the minimax logic is used to calculate the Q_l values and one of the arguments is f_l, then f_l, is determined by the method given above, with the only difference that in step 5, the formulas of the first row of Table 2.3 are used to calculate Q_{out}. To explain the essence of the logic block in case of representation of model variables by fuzzy values, one can consider an example of simulating the same circuit (Fig. 2.8) under the same conditions. Only the radiation effect is taken into account (Fig. 2.9). A constant combination abc

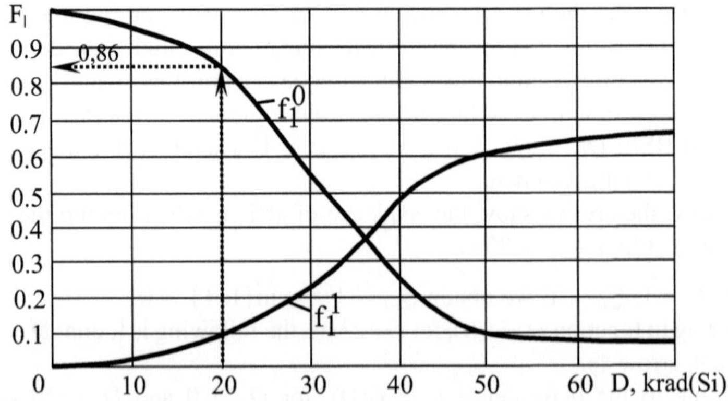

Fig. 2.12 Models of dependencies $f_l = f_2(D)$ of NAND gate

110 is applied to the input of the circuit. In this case, in comparison with the previous one, only the function $f_l = f_2(D)$ changes (Fig. 2.12).

At the moment $t_{sim} = 23$ ns, as in the previous case, the value $D = 20$ krad. Since $Q_a = Q_b = 1$, it follows from the minimax logic that $Q_{ld} = 1 - \min\{1, 1\} = 0$. Therefore, a f_l^0 value is used for substitution in the RSE. Since node *d* is the output of the NAND cell, $Q_{ld} = 1 - \min\ (1;0.86) = 0.14$. In the same way, it is possible to determine the states of other nodes at other moments of time.

Other interpretations of the physical meaning of f_l are also possible: probabilistic, when f_l is represented by the probability distribution of the values of the changes in the state of the node of digital circuits in the result of DF effect, etc.

2.2.2 Inertia block of MGE

The initial information for the inertia block [281] of MGE (Fig. 2.5) are:

1. The value of $Q_l \in [a, b]$ calculated by the logic block.
2. The values of timing parameters of the gate $f_{ti,\ i=1,\ 2,\ \ldots,\ k}$, defined in the GME (Fig. 1.119), with consideration of all DF effects at the current simulation moment and the states of other gates of the circuit.

On the basis of such input data, the inertia block generates the signals Q_{out} and Q^n_{out}.

A characteristic feature of simulation of digital circuits, with consideration of DF effects, is the possibility of forming signals with an arbitrary amplitude and duration (i.e. form) in all (including internal) nodes of the device. Therefore, the choice of the list of timing parameters of the model should facilitate simulation of the phenomena typical for such cases: gradual quenching of the signal when passing through a certain type of circuit (Fig. 1.72), movement of a series of pulses (Fig. 1.73), incomplete switching of digital cells, etc.

In [236–241] with the help of experiments at the circuit-level simulation, a justification was made for selecting the composition of timing parameters of gates for mixed-mode gate-circuit-level simulation. Since the mixed-mode simulation is characterized by the same signal features at the nodes of digital circuits as in case of considering DF effects, it is possible to use similar timing parameters in the MGE. However, the peculiarity of the MCE (Sect. 1.4) is the variability of timing parameters during the simulation of transient processes, which has a significant effect on the rules of the inertiality of the model. The developed rules are given below.

The set of timing parameters of the gates model proposed in [239] includes the following:

1. Δt_{min}—the minimum duration of the input signal of the gate, at which changes of the output state of the element is already observed (Fig. 2.13).
2. Δt_b—the time interval between the moment of state change at the input of the gate (t_0) and the beginning of the rise (or fall) of the signal at the output (t_b, Fig. 2.14).

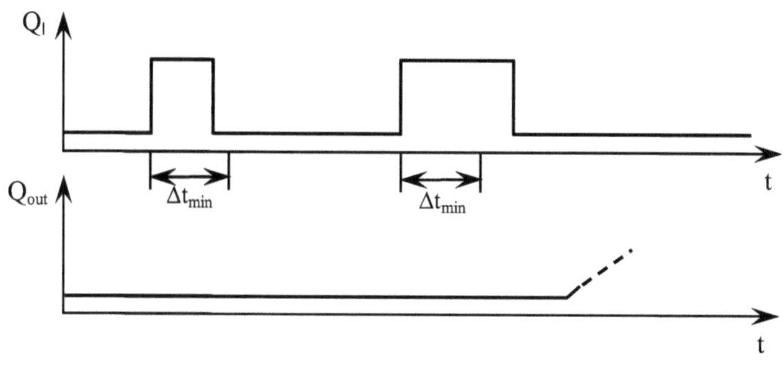

Fig. 2.13 Definition of Δt_{min}

Fig. 2.14 Definition of Δt_b

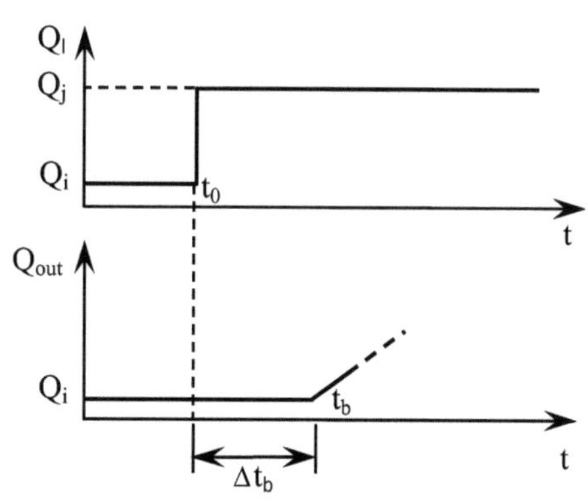

Fig. 2.15 Definition of Δt_s

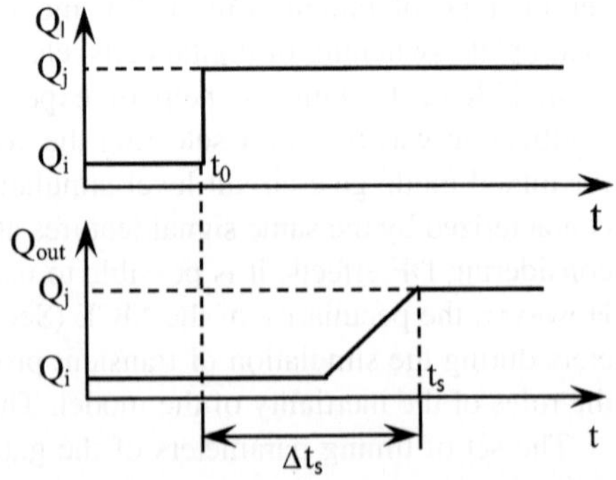

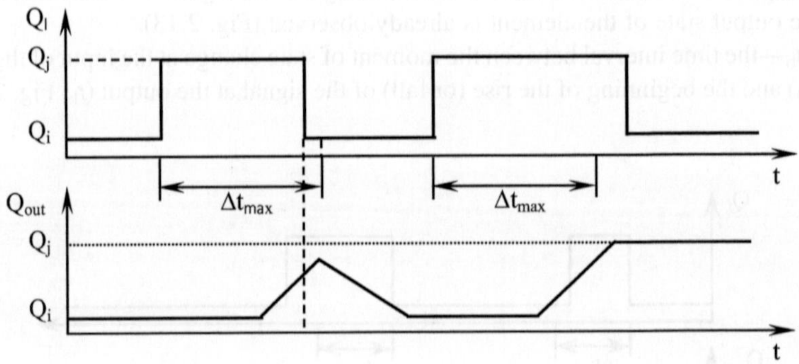

Fig. 2.16 Definition of Δt_{max}

3. Δt_s—the time interval between the change in state at the input of the gate (t_0) and the setup of the signal at the output (t_s, Fig. 2.15.)
4. Δt_{max}—the maximum duration of the input signal, at which the output of the gate is not fully switching (Fig. 2.16).

It was shown in [239] that:

1. The mentioned parameters have different values for switching "0" → "1" and "1" → "0", i.e. $\Delta t_{min}^{01} \neq \Delta t_{min}^{10}$, $\Delta t_b^{01} \neq \Delta t_b^{10}$, $\Delta t_c^{01} \neq \Delta t_s^{10}$, $\Delta t_{max}^{01} \neq \Delta t_{max}^{10}$.
2. The following ratio between timing parameters takes place:

$$0 < \Delta t_{min} < \Delta t_b < \Delta t_{max} < \Delta_{ts}. \tag{2.20}$$

The fact is that for durations of the input signal t_{in} belonging to some range $\Delta t_{min} < t_{in} < \Delta t_b$, incomplete switching in a chain of sequentially connected gates is observed (Fig. 1.104). As a result, the slope of signal changes, which leads to a gradual elimination of pulses in the chain (Fig. 1.75). For the values $t_{in} < \Delta t_{min}$, even the first gate of the chain does not operate. Numerical experiments have also stated that $\Delta t_{min} < \Delta t_b$ even in case when the number of gates in the chain is equal to one.

$\Delta t_{\max}$ does not coincide with Δt_s, either, since at $\Delta t_{\max} < t_{\text{in}} < \Delta t_s$ all the gates of chains are fully switching.

However, in the gate model proposed in [239], the following significant simplifications were made:

1. It is assumed that $0 < \Delta t_{\min} = \Delta t_b < \Delta t_{\max} = \Delta t_s$.
2. Timing parameters are defined for switching only between the states "0" and "1".
3. It is accepted that Δt_b, $\Delta t_s = \text{const}$ in the entire simulation time interval $0 \ldots T_{\text{sim}}$.

In the proposed inertia block it is assumed that:

1. Relation (2.20) takes place.
2. Timing parameters of a digital cell $\Delta t_{\min}^{Q_iQ_j}, \Delta t_b^{Q_iQ_j}, \Delta t_{\max}^{Q_iQ_j}, \Delta t_s^{Q_iQ_j}$ for switching between arbitrary states Q_i and Q_j: are taken into account: $a \leq Q_i \leq b$, $a \leq Q_j \leq b$, $Q_i < Q_j$.
3. The parameters $\Delta t_{\min}^{Q_iQ_j}, \Delta t_b^{Q_iQ_j}, \Delta t_{\max}^{Q_iQ_j}, \Delta t_s^{Q_iQ_j}$, defined in the current (lth) moment of simulation t_{siml}, can have other values already at the next (($l + 1$)th) time point of calculation $t_{\text{siml}+1}$. Due to the above assumptions, as will be seen from further consideration of the behavior of the inertia block in all possible situations, MGE is obtained, satisfying all the requirements formulated in Sect. 1.3.

The inertia block at each current simulation moment t_{sim} can be in one of the following states:

1. The steady state, when the transient processes in the gate under the influence of the last change Q_l (due to changes in the input signals $Q_{\text{in}1}, Q_{\text{in}2}, \ldots, Q_{\text{inn}}$) have completely ended before the moment t_{sim}.
2. The state of a single excitation, if processes occur inside the gate leading to a change in the signal Q_{out} after the moment t_{sim}.
3. The state of the summation of excitations, when new changes in the input signals or DF, confirming or vice versa, planned switching of the output of the gate are coming on the gates in the single excitation mode.

If the gate is in the steady state (Fig. 2.17), then $Q_l = Q_i = \text{const}$. The variable Q_{int}, simulating the degree of development of transient processes inside the gate, takes the value 0. The meaning of this value is that the gate is in a "quiet" state. $Q_{\text{out}} = Q_{\text{out}}^b = Q_l = Q_i = \text{const}$. $U_{\text{out}} = U_i = \text{const}$. U_i is determined by the formula (2.11) by substituting the value of the $Q_{\text{out}}^b = Q_i$. The gate leaves the steady state only when Q_l changes. This, in turn, can occur only as a result of a change in f_l or $Q_{\text{in}i,\ i=1,2,\ \ldots,\ n}$. If Q_l is constant, the steady-state mode will remain in the inertia block even if the timing parameters of the element change as a result of DF effects. As will be seen from the consideration of the behavior of the model in other cases, even changes in the values of timing parameters alone can cause the transmission of MGE to other operation modes. However, not all changes of f_l or $Q_{\text{in}i}$ can lead to an increase or decrease in the value of Q_l (the output of the gates from the steady-state mode). This also depends on the signal values on the other inputs of the circuit and

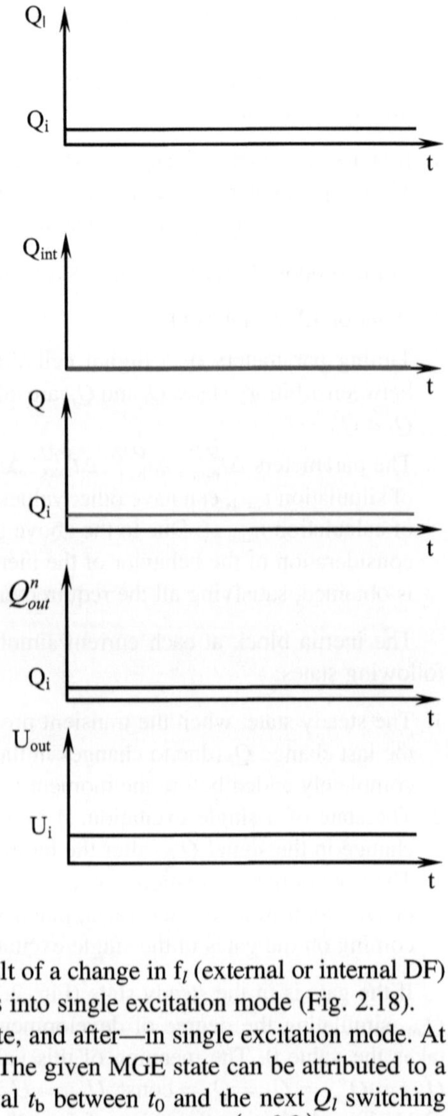

Fig. 2.17 The steady state of the gate

the function being executed. If, as a result of a change in f_l (external or internal DF) or Q_{ini} Q_l should change, the MGE goes into single excitation mode (Fig. 2.18).

Until time t_0, MGE was in steady state, and after—in single excitation mode. At time t_0, Q_l was switched from Q_i to Q_j. The given MGE state can be attributed to a single excitation class if the time interval t_b between t_0 and the next Q_l switching exceeds the duration of the transient processes inside the MGE$\left(\Delta t_s^{Q_iQ_j}\right)$. Otherwise, summation of the excitations takes place.

The state of a single excitation of the gate is divided into four stages associated with the transition of the model time through the boundaries of each of the timing parameters of the gate, starting from the moment t_0:

$$t_0 \leq t_{\text{sim}} \leq t_{\min}^{Q_iQ_j} = t_0 + \Delta t_{\min}^{Q_iQ_j}.$$

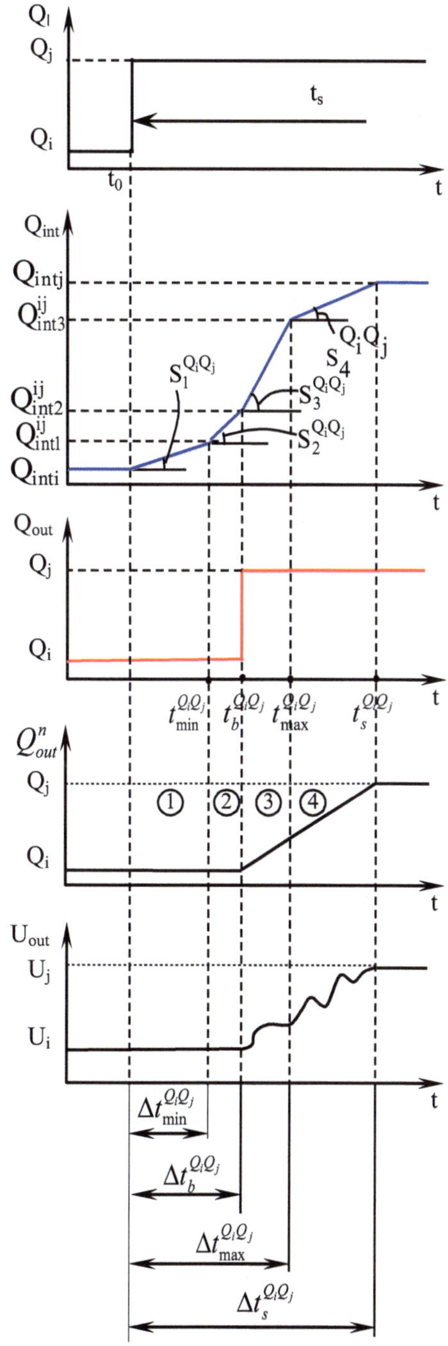

Fig. 2.18 Single excitation mode

$$t_{\min}^{Q_iQ_j} \leq t_{\mathrm{sim}} \leq t_{\mathrm{b}}^{Q_iQ_j} = t_0 + \Delta t_{\mathrm{b}}^{Q_iQ_j}.$$

$$t_{\mathrm{b}}^{Q_iQ_j} \leq t_{\mathrm{sim}} \leq t_{\max}^{Q_iQ_j} = t_0 + \Delta t_{\max}^{Q_iQ_j}.$$

$$t_{\max}^{Q_iQ_j} \leq t_{\mathrm{sim}} \leq t_{\mathrm{s}}^{Q_iQ_j} = t_0 + \Delta t_{\mathrm{s}}^{Q_iQ_j}.$$

The transition of the MGE from one switching stage to another is related to the overlap of the variable Q_{int} of a certain threshold (for example, transition from the first stage to the second is associated with the overlap of the threshold Q_{int1}^{ij}) and the change in the rate of increase in Q_{int} (or decrease if $Q_j < Q_i$). It is not difficult to guess that in case of a linear change of Q_{int} during the entire duration of one stage, the corresponding velocities can be determined as follows:

$$S_1^{Q_iQ_j} = \frac{Q_{\mathrm{int1}}^{ij} - Q_{\mathrm{int}i}}{\Delta t_{\min}^{Q_iQ_j}}, \tag{2.21}$$

$$S_2^{Q_iQ_j} = \frac{Q_{\mathrm{int2}}^{ij} - Q_{\mathrm{int1}}^{ij}}{\Delta t_{\mathrm{b}}^{Q_iQ_j} - \Delta t_{\min}^{Q_iQ_j}}, \tag{2.22}$$

$$S_3^{Q_iQ_j} = \frac{Q_{\mathrm{int3}}^{ij} - Q_{\mathrm{int2}}^{ij}}{\Delta t_{\max}^{Q_iQ_j} - \Delta t\acute{i}}{}^{Q_iQ_j}, \tag{2.23}$$

$$S_4^{Q_iQ_j} = \frac{Q_{\mathrm{int}j} - Q_{\mathrm{int3}}^{ij}}{\Delta t_{\mathrm{s}}^{Q_iQ_j} - \Delta t_{\max}^{Q_iQ_j}}. \tag{2.24}$$

The degree of influence of the linear approximation of the function $Q_{\mathrm{int}}(t)$ during one switching stage on the overall accuracy of calculations is considered in Sect. 2.4. It is shown that this factor slightly reduces the overall accuracy of simulation results. However, with such an approximation, as seen from further analysis of the essence of the MGE, the threshold values $Q_{\mathrm{int1}}^{ij}, Q_{\mathrm{int2}}^{ij}, Q_{\mathrm{int3}}^{ij}$ are only illustrative and are not included in final formulas for determining the critical moments of the change in the input signal of the gate for both standard and nonstandard input disturbances. Otherwise, there would be a problem of determining the marked threshold values. The solution to this problem is difficult because of the need to conduct a huge number of experimental studies of basic logic cells.

The physical meaning of the stages of a single excitation is also connected with the fact that if during one stage (for example, the first one) at the time t_1 a change in Q_l occurs in the opposite direction ($Q_j \rightarrow Q_i$), then, due to the inertia of the opposite switching, the time of the beginning of the drop Q_{int} t_2 ($t_2 > t_1$, Fig. 2.19) will fall into either the current stage, or, as a maximum, the next (for example, the second).

Other transitions (for example, t_2 going to the third stage) are excluded. The latter is confirmed by the results of circuit simulation using SPICE [74, 75] of basic gates in Figs. 1.13–1.15. Thus, if the time of inertia of the reverse switching is denoted by $\widetilde{\Delta t}$, then the following conditions are observed in the MGE inertia block:

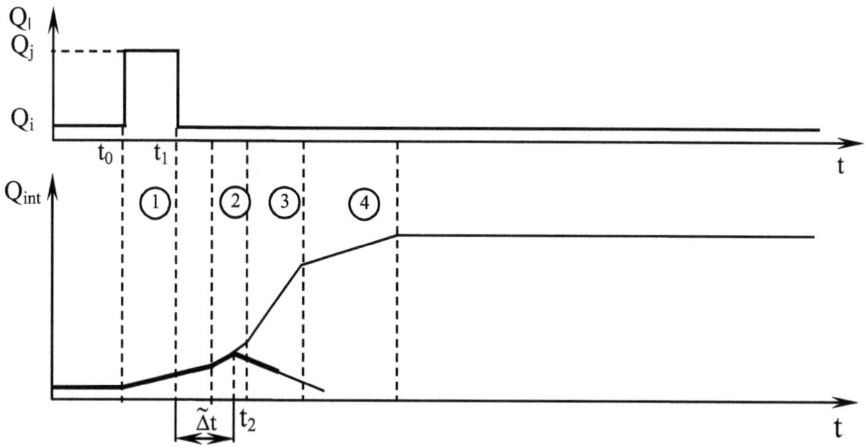

Fig. 2.19 Inertia of the reverse switching of the MGE

$$\text{If } t_0 \le t_1 \le t_{\min}^{Q_iQ_j}, \quad \text{then } t_2 \le t_{\text{b}}^{Q_iQ_j}; \tag{2.25}$$

$$\text{If } t_{\min}^{Q_iQ_j} \le t_1 \le t_{\text{b}}^{Q_iQ_j}, \quad \text{then } t_2 \le t_{\max}^{Q_iQ_j}; \tag{2.26}$$

$$\text{If } t_{\text{b}}^{Q_iQ_j} \le t_1 \le t_{\max}^{Q_iQ_j}, \quad \text{then } t_2 \le t_{\text{s}}^{Q_iQ_j}; \tag{2.27}$$

$$\text{If } t_{\max}^{Q_iQ_j} \le t_1 \le t_{\text{y}}^{Q_iQ_j}, \quad \text{then } t_2 = t_{\text{s}}^{Q_iQ_j}. \tag{2.28}$$

In addition, the value $\widetilde{\Delta}t$ at the end point of the current stage must coincide with the duration of the next one, i.e.

$$\widetilde{\Delta}t\left(t_{\min}^{Q_iQ_j}\right) = \Delta t_{\text{b}}^{Q_iQ_j} - \Delta t_{\min}^{Q_iQ_j}, \tag{2.29}$$

$$\widetilde{\Delta}t\left(t_{\text{b}}^{Q_iQ_j}\right) = \Delta t_{\max}^{Q_iQ_j} - \Delta t_{\text{b}}^{Q_iQ_j}, \tag{2.30}$$

$$\widetilde{\Delta}t\left(t_{\max}^{Q_iQ_j}\right) = \Delta t_{\text{s}}^{Q_iQ_j} - \Delta t_{\max}^{Q_iQ_j}, \tag{2.31}$$

$$\widetilde{\Delta}t\left(t_{\text{s}}^{Q_iQ_j}\right) = 0, \tag{2.32}$$

If one takes a linear approximation of the function $\widetilde{\Delta}t(t_1)$ during each stage, then it will have the form of polyline 1, shown in Fig. 2.20

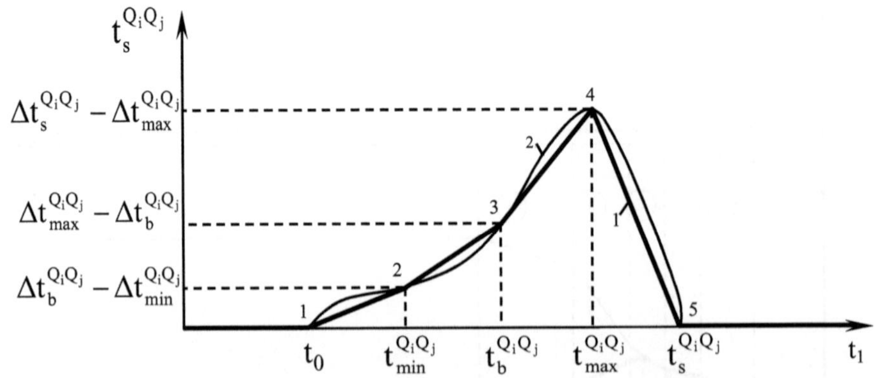

Fig. 2.20 The dependence of the inertia of the reverse switching of the gate from the moment of arrival of the opposite signal

The analytical description of this broken line is as follows:

$$\widetilde{\Delta t} = \begin{cases} \dfrac{\left(\Delta t_b^{Q_iQ_j} - \Delta t_{min}^{Q_iQ_j}\right)\cdot(t-t_0)}{\Delta t_{min}^{Q_iQ_j}}, & \text{if } t_0 \le t \le t_{min}^{Q_iQ_j} \\ \Delta t^{Q_iQ_j} - \Delta t_{min}^{Q_iQ_j} + \dfrac{\left(\Delta t_{max}^{Q_iQ_j} - 2\cdot\Delta t_b^{Q_iQ_j} + \Delta t_{min}^{Q_iQ_j}\right)\cdot\left(t - t_{min}^{Q_iQ_j}\right)}{\Delta t_b^{Q_iQ_j} - \Delta t_{min}^{Q_iQ_j}}, & \text{if } t_{min}^{Q_iQ_j} \le t \le t_b^{Q_iQ_j} \\ \Delta t_{max}^{Q_iQ_j} - \Delta t_b^{Q_iQ_j} + \dfrac{\left(\Delta t_s^{Q_iQ_j} - 2\cdot\Delta t_{max}^{Q_iQ_j} + \Delta t_b^{Q_iQ_j}\right)\cdot\left(t - t_b^{Q_iQ_j}\right)}{\Delta t_{max}^{Q_iQ_j} - \Delta t_b^{Q_iQ_j}}, & \text{if } t_b^{Q_iQ_j} \le t \le t_{max}^{Q_iQ_j} \\ \dfrac{\left(\Delta t_s^{Q_iQ_j} - \Delta t_{max}^{Q_iQ_j}\right)\cdot\left(t_s^{Q_iQ_j} - t\right)}{\left(\Delta t_s^{Q_iQ_j} - \Delta t_{max}^{Q_iQ_j}\right)}, & \text{if } t_{max}^{Q_iQ_j} \le t \le t_s^{Q_iQ_j} \\ 0, \quad \text{if } t < t_0 \text{ or } t > t_s^{Q_iQ_j} & \end{cases} \tag{2.33}$$

The noted dependence can be obtained by simulating standard basic logic cells at the circuit level. However, it should be noted that the experimental dependence (curve 2 in Fig. 2.20) must necessarily pass through points 1 … 5. This statement is due to the fact that points 1 ….. 5 are obtained from the definitions of timing parameters $\Delta t_{min}^{Q_iQ_j}, \Delta t_b^{Q_iQ_j}, \Delta t_{max}^{Q_iQ_j}, \Delta t_s^{Q_iQ_j}$. If on separate areas (1–2, 2–3, 3–4, 4–5) the experimental curve 2 deviates significantly from the linear dependence of Fig. 2.20, then it is possible to apply more accurate approximation formulas and create a library of models providing different accuracy of calculations.

The single excitation mode of the MCE in case of reverse switching (switching $Q_lQ_j \rightarrow Q_i$) is similar to that described above. However, in general case:

$$Q^{ji}_{\text{int1}} \neq Q^{ij}_{\text{int3}}, \tag{2.34}$$

$$Q^{ji}_{\text{int2}} \neq Q^{ij}_{\text{int2}}, \tag{2.35}$$

$$Q^{ji}_{\text{int3}} \neq Q^{ij}_{\text{int1}}, \tag{2.36}$$

Another qualitative feature of the MCE inertia block in single excitation mode is that in the time interval $t_0 \ldots t_s^{Q_iQ_j}$ there can be changes (recalculations) of timing parameters of the model. Such a recalculation is performed at each time point of the model computation $t_{\text{sim}i}$. The mechanisms for selecting $t_{\text{sim}i}$ are discussed in Chap. 3. Here, for the time being, it can be stated that in the interval $t_0 \ldots t_s^{Q_iQ_j}$ several Q_{int} corrections can occur (Fig. 2.21). Figure 2.21 as an example, shows three dependences of $Q_{\text{int}}(t)$:

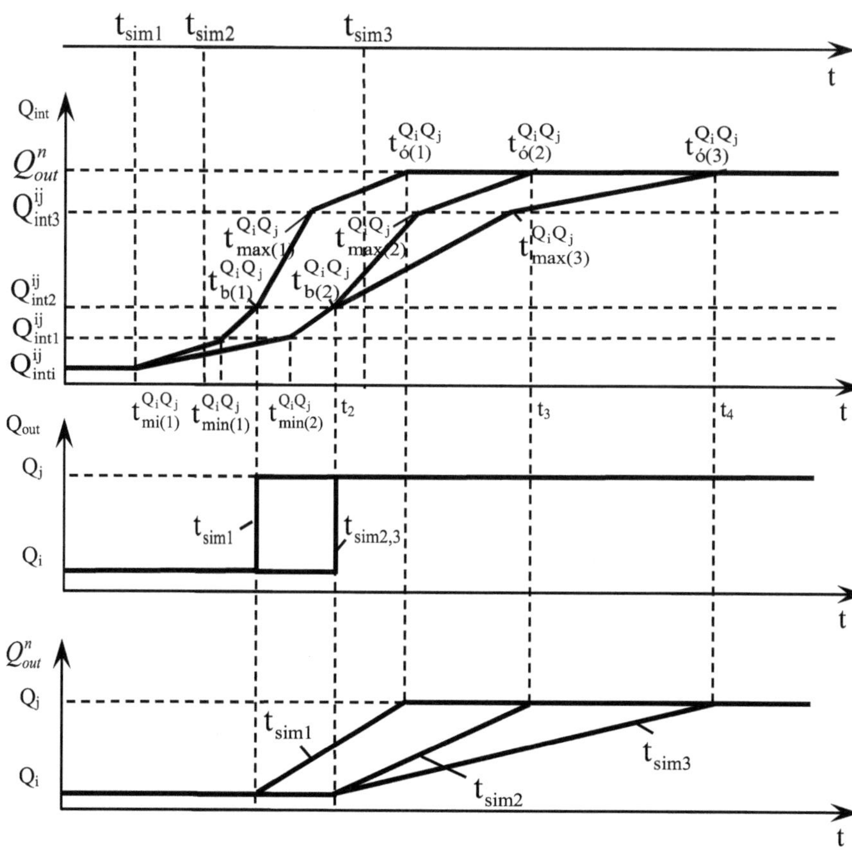

Fig. 2.21 Influence of timing parameter corrections on the operation of MGE

1. $t_{\text{sim1}} = t_0$ i.e. the first dependence is calculated at the moment of switching of the input signal Q_l in the direction $Q_i \to Q_j$. In this case, Q_{int} reaches the corresponding threshold values at the moments $t_{\min(1)}^{Q_iQ_j}, t_{\text{b}(1)}^{Q_iQ_j}, t_{\max(1)}^{Q_iQ_j}, t_{\text{s}(1)}^{Q_iQ_j}$.
2. $t_0 \leq t_{\text{sim2}} \leq t_{\min(1)}^{Q_iQ_j}$, i.e. the second, already corrected $Q_{\text{int}}(t)$ dependence was obtained during the first stage of a single excitation. All overlaps of threshold values change: $t_{\min(2)}^{Q_iQ_j}, t_{\text{b}(2)}^{Q_iQ_j}, t_{\max(2)}^{Q_iQ_j}, t_{\text{s}(2)}^{Q_iQ_j}$. Therefore, Q_{out} changes (the switching time of the output signal t_2), and Q_{out}^{n} (the duration of the rise and fall of the output signal is $t_3 - t_2$).
3. $t_{\text{b}(2)}^{Q_iQ_j} \leq t_{\text{sim3}} \leq t_{\max(2)}^{Q_iQ_j}$. In this case, the correction of timing parameters will occur during the third stage, when already $Q_{\text{int}}(t) > Q_{\text{int2}}^{ij}$, i.e. Q_{out} switches.

Therefore, only the timing parameters $\Delta t_{\max}^{Q_iQ_j}$ and $\Delta t_{\text{s}}^{Q_iQ_j}$ change (the moments of overlapping the thresholds $t_{\max(3)}^{Q_iQ_j}, t_{\text{c}(3)}^{Q_iQ_j}$) which only leads to a tightening (or vice versa) of rise and fall of the output signal $Q_{\text{out}}^{b}(t_4 - t_2)$.

It should also be noted that the need to reverse the model time due to the correction of timing parameters of gates is excluded. A simple justification for what has been said is illustrated in Fig. 2.22 by the example of the first stage of a single excitation.

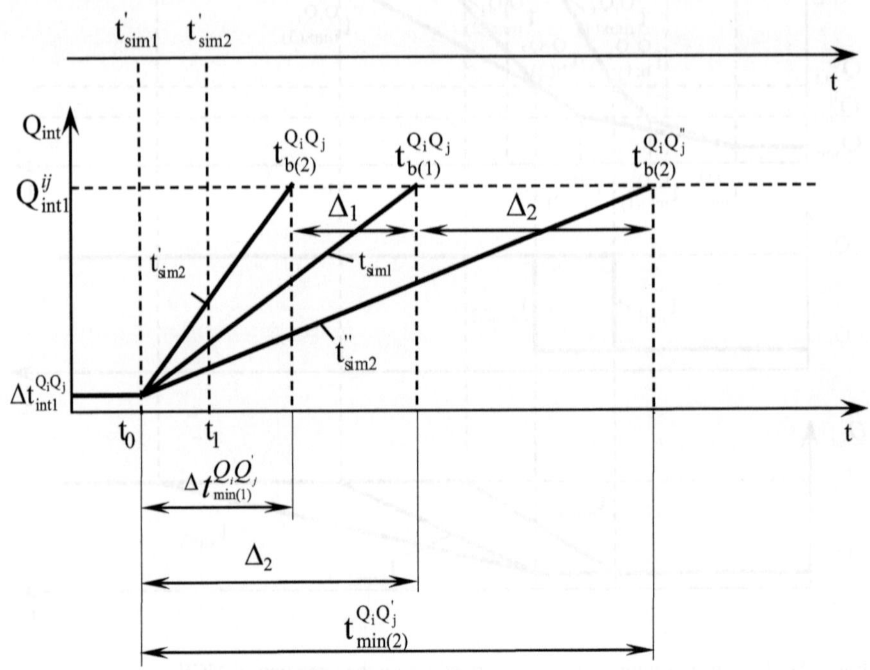

Fig. 2.22 Correction of timing parameters of the gate

Suppose, at the time of simulation $t_{sim1} = t_0$, single excitation is recorded. At this point, the MGE timing models produce a value $\Delta t_{min}^{Q_iQ_j} = \Delta t_{min0}^{Q_iQ_j}$. Therefore, the expected moment $Q_{int}(t)$ of crossing the threshold value $t_{b(1)}^{Q_iQ_j} = t_{sim1} + \Delta t_{sim0}^{Q_iQ_j}$ is calculated by the inertia unit, then the next time $t_{mod2} = t_1$, arriving at the first stage of single excitation, is calculated by the simulation time calculation algorithm $t_0 \leq t_{sim2} \leq t_{b(1)}^{Q_iQ_j}$. Correction of all timing parameters of the MGE, including Δ $t_{min}^{Q_iQ_j}$occurs. The new value can be either less $\left(\Delta t_{sim1}^{Q_iQ_j}\right)$ or greater $\left(\Delta t_{sim2}^{Q_iQ_j}\right)$ than the old value, i.e. the moment of overlapping the variable $Q_{int}(t)$ of the threshold value Q_{int1}^{ij} can be shifted both to the left by value $\Delta_1 = \Delta t_{min0}^{Q_iQ_j} - \Delta t_{min1}^{Q_iQ_j}$, and to the right by $\Delta_2 = \Delta t_{min2}^{Q_iQ_j} - \Delta t_{min0}^{Q_iQ_j}$. The shift to the right is of no interest from the considered point of view. If there is a left shift by value Δ_1, then in any case the overlapping point of the threshold value $t_{b2}^{Q_iQ_j'}$ should be to the right of t_1, because as a result of the change before the moment $t_1 Q_{int}(t)$ it has not reached the value Q_{int1}^{ij} yet, and additional (nonzero) time is needed for this. Hence it follows that if the next simulation time instant falls into the *i*th stage of a single excitation ($i = 1, 2, 3, 4$)), then MGE recognizes changes only of those timing parameters, which determine the duration of only the current and next ($\geq i$) stages. This statement, of course, also holds for states of summation of excitations.

Thus, the inertia block reacts to changes of DF values and corrects the output signal during single excitation. In other words, Q_{int} "follows" after changes of DF and can dynamically change its form during the time range $t_0 \ldots t_s^{Q_iQ_j}$. And the length of time $t_0 \ldots t_s^{Q_iQ_j}$ also has a variable character. Changes in $Q_{int}(t)$ form can occur more than once, during each of the stages and can shift the overlaps of threshold values not only to the right but also to the left, in accordance with the above-mentioned condition.

As an example, Fig. 2.23 shows the dynamics of changing the form $Q_{int}(t)$ at different moments of time (t_{sim2}, t_{sim3}, t_{sim4}) during the third stage of excitation. The upper graph shows the values of timing parameters given by the GME (Fig. 1.119) at different moments of simulation. It is clear that the changes Q_{int} will affect the signal Q_{out}^{b}accordingly.

The state of the summation of the excitations arises in case when the moment t_1 of reverse switching Q_1 occurs before the completion of a single excitation, i.e. $t_1 \leq t_s^{Q_iQ_j}$ takes place. Since, t_1 as can occur during each of the four stages of single excitation, four variants are possible, respectively.

Variant 1. The moment of the reverse switching Q_l falls on the first stage of a single excitation, i.e. $t_0 \leq t_1 \leq t_{min}^{Q_iQ_j}$. From the moment $t_1 Q_{int}(t)$ continues to increase during $\widetilde{\Delta t}$, determined by the formula (2.33) (case $t_0 \leq t_1 \leq t_{min}^{Q_iQ_j}$), and only at the moment t_2, $Q_{int}(t)$ drop begins. Depending on the ratio of the moments $t_2 = t_1 + \widetilde{\Delta t}$ and $t_{min}^{Q_iQ_j}$, $Q_{int}(t)$ may not reach (if $t_2 \leq t_{min}^{Q_iQ_j}$) or pass (or else) the threshold value Q_{int1}^{ij}. Accordingly, the gate does not pass or manage to pass to the second stage of excitation. Therefore, two sub-variants are possible, shown respectively in Figs. 2.24 and 2.25.

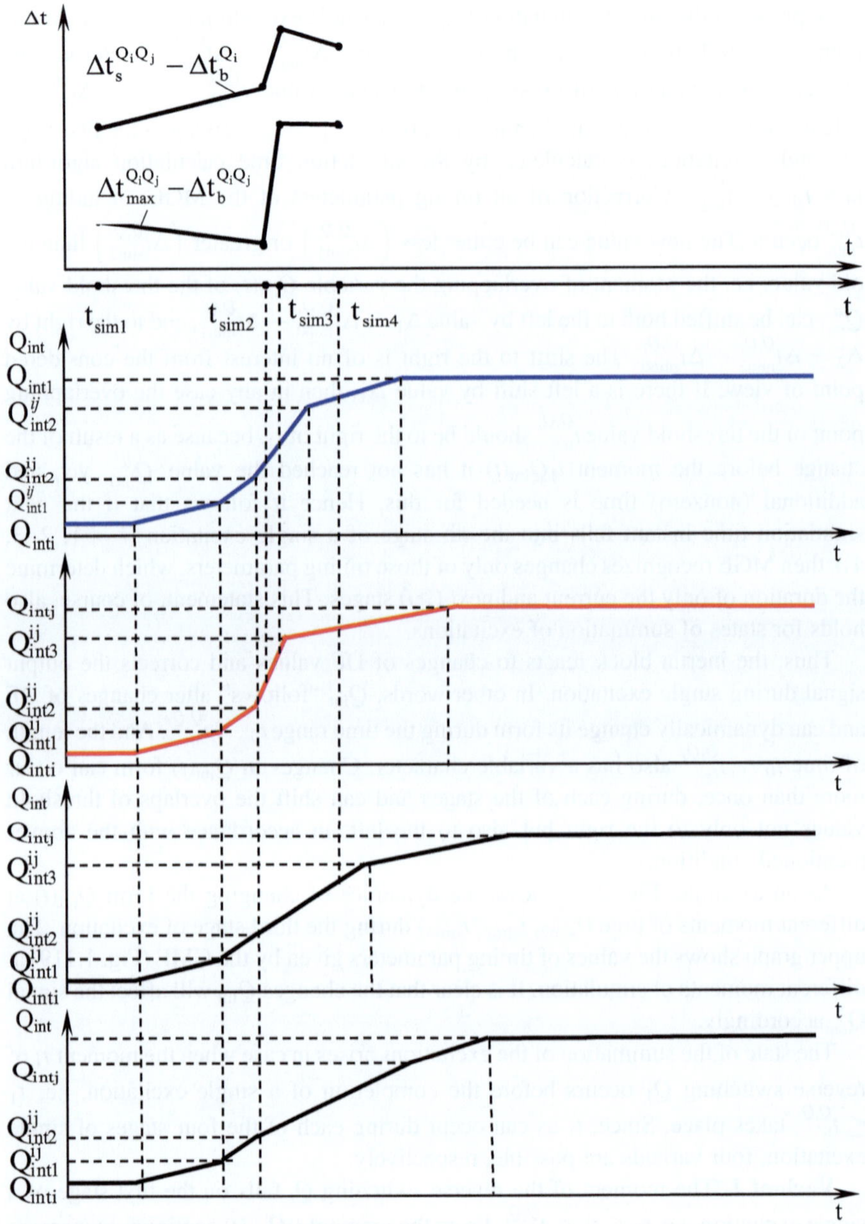

Fig. 2.23 Dynamics of change of Q_{int}

Sub-variant 1.1. (Fig. 2.24). $Q_{int}(t)$, increasing, does not reach the threshold value, Q_{int1}^{ij}, i.e. $t_2 = t_1 + \widetilde{\Delta} t \leq t_{min}^{Q_iQ_j} = t_0 + \Delta t_{min}^{Q_iQ_j}$. This condition takes the following form after substituting the value of $\widetilde{\Delta} t$ from formula (2.33) and the corresponding transformations:

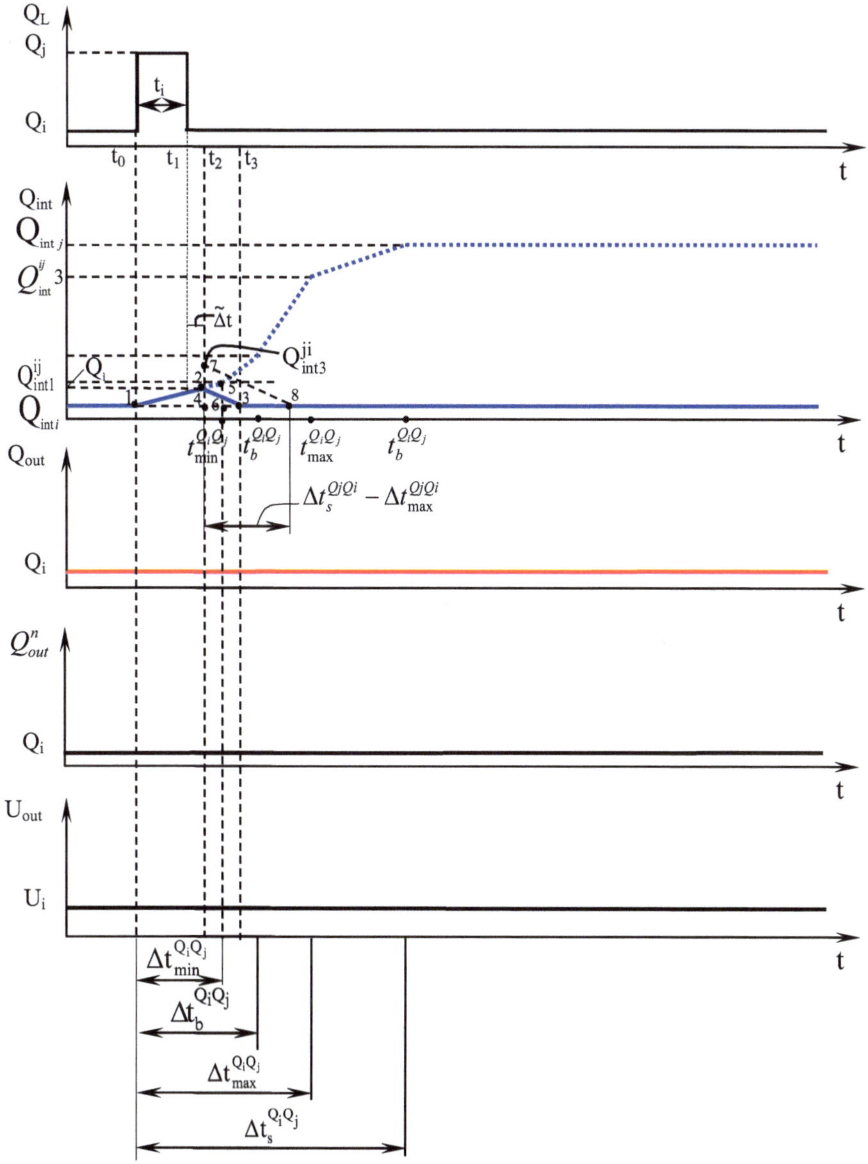

Fig. 2.24 Summation of excitations in the first stage (sub-variant 1)

$$t_u = t_1 - t_0 \leq \frac{\left(\Delta t_{\min}^{Q_iQ_j}\right)^2}{\Delta t_{\mathrm{b}}^{Q_iQ_j}}. \tag{2.37}$$

Calculate the coordinates of the characteristic points (1–3) of the behavior $Q_{\rm int}(t)$, expressing them in terms of known values. Here for the purpose of demonstrating the

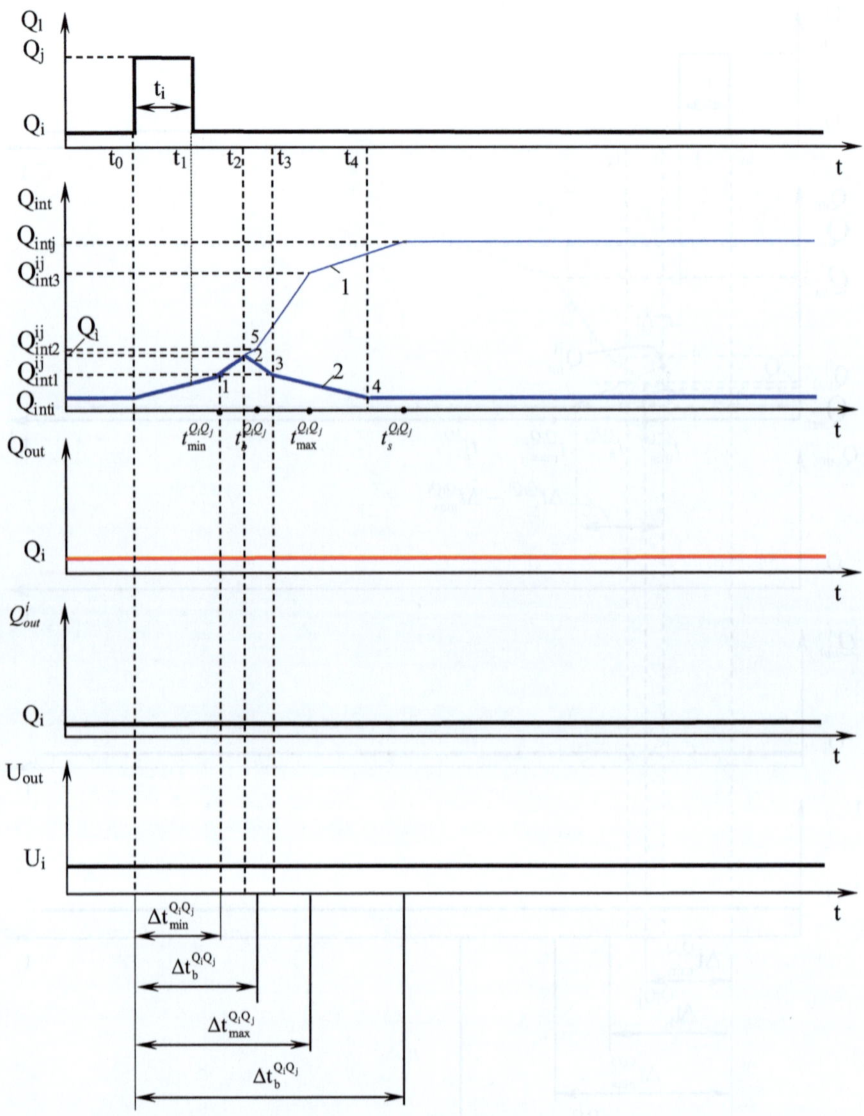

Fig. 2.25 Summation of the excitations in the first stage (sub-variant 2)

method of obtaining model relations, a detailed step of constructing formulas is given. For other variants of behavior, the $Q_{\text{int}}(t)$ formulas of the model will be presented in ready form.

The coordinates of point 1 are known $\left(t = t_0; Q = Q_{\text{int}_i} i\right)$. Define the coordinates of point 2, i.e. t_2 and Q_{M}. The moment of time t_2 when $Q_{\text{int}}(t)$, increasing, reaches its maximum value Q_{M}, is defined as:

$$t_2 = t_1 + \widetilde{\Delta t} = t_1 + \frac{\left(\Delta t_{\mathrm{b}}^{Q_iQ_i} - \Delta t_{\min}^{Q_iQ_i}\right) \cdot (t_1 - t_0)}{\Delta t_{\min}^{Q_iQ_i}}. \tag{2.38}$$

After simplifications, this is obtained:

$$t_2 = t_1 \cdot \frac{\Delta t_{\mathrm{b}}^{Q_iQ_i}}{\Delta t_{\min}^{Q_iQ_i}} + t_0 \cdot \left(1 - \frac{\Delta t_{\mathrm{b}}^{Q_iQ_i}}{\Delta t_{\min}^{Q_iQ_i}}\right). \tag{2.39}$$

The value Q_{M} can be defined as follows:

$$Q_i = Q_{\mathrm{int}i} + \Delta Q, \tag{2.40}$$

where ΔQ is the value on which $Q_{\mathrm{int}}(t)$ increases during $t_0 \ldots t_2$. ΔQ can be calculated from the similarity of triangles 1–2–4 and 1–5–6:

$$\frac{\Delta Q}{Q_{\mathrm{int1}}^{ij} - Q_{\mathrm{int}i}} = \frac{t_2 - t_0}{t_{\min}^{Q_iQ_j} - t_0}. \tag{2.41}$$

It is clear (2.39) that

$$t_2 - t_0 = \frac{\Delta t_{\mathrm{b}}^{Q_iQ_i}}{\Delta t_{\min}^{Q_iQ_i}} \cdot (t_1 - t_0). \tag{2.42}$$

From the definition $\Delta t_{\min}^{Q_iQ_j}$ it follows that $t_{\min}^{Q_iQ_j} - t_0 = \Delta t_{\min}^{Q_iQ_j}$. Substituting these values in (2.41), then the value ΔQ (2.40), the following is obtained:

$$Q_i = Q_{\mathrm{int}i} + \left(Q_{\mathrm{int1}}^{ij} - Q_{\mathrm{int}i}\right) \cdot \frac{\Delta t_{b}^{Q_iQ_i}}{\left(\Delta t_{\min}^{Q_iQ_i}\right)^2} \cdot (t_1 - t_0). \tag{2.43}$$

For boundary values of the difference $t_1 - t_0$ (relation 2.37), the following is obtained:

(a) If, $t_1 = t_0$ then, $t_2 = t_0, Q_i = Q_{\mathrm{int}i}$ i.e. point 2 coincides with 1;

(b) If, $t_1 - t_0 = \dfrac{\left(\Delta t_{\min}^{Q_iQ_i}\right)^2}{\Delta t_{\mathrm{b}}^{Q_iQ_i}}$, $t_2 = t_0 + \Delta t_{\min}^{Q_iQ_i}$, $Q_i = Q_{\mathrm{int1}}^{ij}$ i.e. point 2 coincides with 5.

From the coordinates of point 3, only t_3 needs to be calculated, since $Q = Q_{\mathrm{int}i}$. The drop rate $Q_{\mathrm{int}}(t)$ in fragment 2–3 corresponds to the speed of the fourth stage of reverse switching Q_1 from Q_j to Q_j (fragment 7–8). And the value Q_{int3}^{ji}, as stated above, in general case may not coincide with Q_{int1}^{ij}. From the similarity of triangles 2–3–4 and 7–8–4, this can be written:

$$\frac{t_3 - t_2}{\Delta t_s^{Q_jQ_i} - \Delta t_{\max}^{Q_jQ_i}} = \frac{\Delta Q}{Q_{\text{int3}}^{ji} - Q_{\text{int}i}}. \tag{2.44}$$

Substituting the values t_2 from (2.39), ΔQ from (2.43), after the corresponding algebraic transformations, the following is obtained:

$$t_3 = t_0 + (t_1 - t_0) \cdot \frac{\Delta t_N^{Q_iQ_j}}{\Delta t_{\min}^{Q_iQ_j}} \cdot \left(1 + \frac{\Delta t_s^{Q_jQ_i} - \Delta t_{\max}^{Q_jQ_i}}{\Delta t_{\min}^{Q_iQ_j}} \cdot \frac{Q_{\text{int1}}^{ij} - Q_{\text{int}i}}{Q_{\text{int3}}^{ji} - Q_{\text{int}i}}\right). \tag{2.45}$$

Thus, in the first sub-variant during $(t_1 - t_0) \cdot \frac{\Delta t_b^{Q_iQ_j}}{\Delta t_{\min}^{Q_iQ_j}} \cdot \left(1 + \frac{\Delta t_s^{Q_jQ_i} - \Delta t_{\max}^{Q_jQ_i}}{\Delta t_{\min}^{Q_iQ_j}}\right) \cdot \frac{Q_{\text{int1}}^{ij} - Q_{\text{int}i}}{Q_{\text{int3}}^{ji} - Q_{\text{int}i}}$, the gate is in an excited state, and if new excitations do not arrive until the moment t_3 (due to a change in value Q_1 or timing parameters of the gate), then the element goes into a steady state Q_i. Since during the time interval $t_0 \ldots t_3 Q_{\text{int}}(t)$ does not reach the threshold value Q_{int2}^{ij} (all the time is less than even Q_{int1}^{ij}), there are no changes at the output of the MGE, i.e. $Q_{\text{out}} = Q_{\text{out}}^n = Q_i = \text{const}$, $U_{\text{out}} = U_i = \text{const}$.

Sub-variant 1.2. $Q_{\text{int}i}(t)$, increasing, passes through the threshold value, Q_{int1}^{ij}, i.e. $t_{\min}^{Q_iQ_i} = t_0 + \Delta t_{\min}^{Q_iQ_i} \le t_2 = t_1 + \widetilde{\Delta} t \le t_b^{Q_iQ_i} = t_0 + \Delta t_b^{Q_iQ_i}$ (Fig. 2.25).

This condition after the transformations takes the following form:

$$\frac{\left(\Delta t_{\min}^{Q_iQ_j}\right)^2}{\Delta t_b^{Q_iQ_j}} \le t_s = t_1 - t_0 \le \Delta t_{\min}^{Q_iQ_j}. \tag{2.46}$$

Calculate the coordinates of the following points (Fig. 2.25): 1—overlap of the threshold value Q_{int1}^{ij} at the rise $Q_{\text{int}}(t)$, 2—reaching the $Q_{\text{int}}(t)$ maximum value of Q_i; 3—overlaps of the threshold value Q_{int3}^{ij} when decreasing $Q_{\text{int}}(t)$; 4—completion of transient processes, i.e. $Q_{\text{int}}(t)$ achieves its stationary value $Q_{\text{int}i}$.

The coordinates of point 1 can be easily determined, since according to the definition of the parameter $\Delta t_{\min}^{Q_iQ_j}$, increasing, reaches a value $Q_{\text{int}i}$ at a moment of time $t_{\min}^{Q_iQ_j} = t_0 + \Delta t_{\min}^{Q_iQ_j}$.

The value t_2 is defined similarly to the previous sub-variant (2.39). In this case, the calculation formula changes:

$$Q_i = Q_{\text{int1}}^{ij} + \Delta Q. \tag{2.47}$$

Here ΔQ differs from the analogous value of formula (2.40), since in this case $\Delta Q = Q_i - Q_{\text{int1}}^{ij}$. Therefore, the following can be written:

$$\frac{\Delta Q}{Q_{\text{int2}}^{ij} - Q_{\text{int1}}^{ij}} = \frac{t_2 - t_{\min}^{Q_iQ_j}}{t_b^{Q_iQ_j} - t_{\min}^{Q_iQ_j}}. \tag{2.48}$$

After substituting the corresponding values and simplifications, the following is obtained:

$$Q_M = Q_{\mathrm{int1}}^{ij} + \left(Q_{\mathrm{int2}}^{ij} - Q_{\mathrm{int1}}^{ij}\right) \cdot \frac{\Delta t_{\mathrm{b}}^{Q_iQ_j} \cdot (t_1 - t_0) - \left(\Delta t_{\mathrm{min}}^{Q_iQ_j}\right)^2}{\Delta t_{\mathrm{min}}^{Q_iQ_j} \cdot \left(\Delta t_{\mathrm{b}}^{Q_iQ_j} - \Delta t_{\mathrm{min}}^{Q_iQ_j}\right)}. \tag{2.49}$$

At boundary values of the difference$t_1 - t_0$ (relation 2.46) the following is obtained:

(a) If $t_1 - t_0 = \frac{\left(\Delta t_{\mathrm{min}}^{Q_iQ_j}\right)^2}{\Delta t_{\mathrm{b}}^{Q_iQ_j}}$, then $t_2 = t_0 + \Delta t_{\mathrm{min}}^{Q_iQ_j}$, $Q_i = Q_{\mathrm{int1}}^{ij}$, i.e. point 2 coincides with 1;

(b) If $t_1 = t_0 + \Delta t_{\mathrm{min}}^{Q_iQ_j}$, then $t_2 = t_1 + \Delta t_{\mathrm{b}}^{Q_iQ_j}$, $Q_i = Q_{\mathrm{int2}}^{ij}$, i.e. point 2 coincides with 1.

One of the coordinates of point 3 is known: $Q = Q_{\mathrm{int1}}^{ij}$, and the time t_3 is calculated, as in the previous case:

$$t_3 = t_1 \cdot \frac{\Delta t_{\mathrm{b}}^{Q_iQ_j}}{\Delta t_{\mathrm{min}}^{Q_iQ_j}} + t_0 \cdot \left(1 - \frac{\Delta t_{\mathrm{b}}^{Q_iQ_j}}{\Delta t_{\mathrm{min}}^{Q_iQ_j}}\right) + \frac{Q_{\mathrm{int2}}^{ij} - Q_{\mathrm{int1}}^{ij}{}^{ji}}{Q_{\mathrm{int}i}\,2} - Q_{\mathrm{int1}}^{ji}$$
$$\cdot \frac{\Delta t_{\mathrm{b}}^{Q_iQ_j} \cdot (t_1 - t_0) - \left(\Delta t_{\mathrm{min}}^{Q_iQ_j}\right)^2}{\Delta t_{\mathrm{b}}^{Q_iQ_j} \cdot \left(\Delta t_{\mathrm{b}}^{Q_iQ_j} - \Delta t_{\mathrm{min}}^{Q_iQ_j}\right)} \cdot \left(\Delta t_{\mathrm{max}}^{QjQ_i} - \Delta t_{\mathrm{b}}^{Q_jQi}\right). \tag{2.50}$$

The value Q of point 4 is also known: $Q_{\mathrm{int}i}$. Since transient processes occur between points 3 and 4, which are characteristic of the fourth stage of reverse switching, then:

$$t_4 = t_3 + \Delta t_{\mathrm{s}}^{Q_jQ_i} - \Delta t_{\mathrm{max}}^{Q_jQ_i}. \tag{2.51}$$

Thus, in the second sub-variant, during the time interval $t_0\ldots t_4$, the gate is in an excited state, and if new excitations do not arrive until the moment t_4, the element again goes into a steady stateQ_i. Since during the time interval $t_0\ldots t_4$, $Q_{\mathrm{int}}(t)$ does not reach the threshold value Q_{int2}^{ij} again (although during the interval $t_1\ldots t_3$ it crosses the threshold Q_{int1}^{ij}), the constancy of the values remains at the output of the MGE, as in the previous sub-variant.

In the first variant of the summation of excitations, too, after the moment t_1, changes in the DF can occur. If new timing points of simulation arise between the moments t_1 and the completion of the transient processes because of them (t_3 in Fig. 2.24 and t_4 in Fig. 2.25), then the MGE obtains new values of timing parameters, and other coordinates of the characteristic points of transient processes are obtained by recalculating the given formulas. Accordingly, the formula $Q_{\mathrm{int}}(t)$ changes. Due to such changes, mutual transitions between the two sub-variants are also possible. For example, at the moment $t_{\mathrm{sim0}} = t_1$ (Fig. 2.26), excitation is recorded in

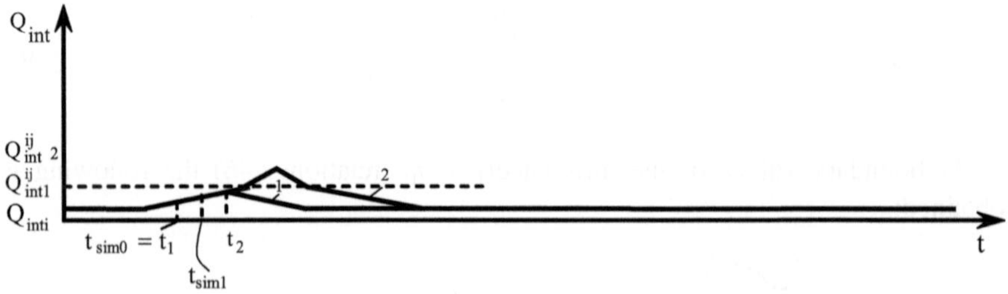

Fig. 2.26 Dynamics of change in Q_{int}

accordance with the rules of sub-variant 1.1 (line 1), and at the time $t_1 < t_{\text{sim1}} < t_2$ because of the recalculation of timing parameters the form $Q_{\text{int}}(t)$ (line 2) changes.

In fact, transition to the second sub-variant occurs, since in case of line 2, Q_{int} manages to pass the threshold value Q^{ij}_{int1}. A reverse transition is also possible. Although the output of the MGE is constant in all cases, the mentioned transitions must be taken into account. As will be seen from further consideration, if Q_i switches again before the completion of transient processes, the characteristic points of future switching are determined with higher accuracy.

However, due to the influence of DF, a more significant change of $Q_{\text{int}}(t)$ can occur when it already manages to pass through the following threshold value: Q^{ij}_{int2}. In this case, a second variant of the summation of the excitations arises, described below. This is a cardinal transition, since Q_{out} manages to switch to a state Q_j.

Variant 2. The moment of time of the reverse switching Q_l goes to the second stage of a single excitation, i.e. $t^{Q_iQ_j}_{\min} \leq t_1 \leq t^{Q_iQ_j}_{\text{b}}$ (Fig. 2.27). As in the previous variant, beginning from the moment t_1, $Q_{\text{int}}(t)$ continues to grow during time $\widetilde{\Delta}t$. However, in this case $\widetilde{\Delta}t$ is determined according to the second branch $\left(t^{Q_iQ_j}_{\min} \leq t_1 \leq t^{Q_iQ_j}_{\text{b}}\right)$ of formula (2.33), and at the moment $t_2 = t_1 + \widetilde{\Delta}t$, the drop of $Q_{\text{int}}(t)$ begins. The main difference of this variant from the previous one is that in any case the time point t_2 falls in the range $t^{Q_iQ_j}_{\text{b}} \ldots t^{Q_iQ_j}_{\max}$.

This follows from the formula (2.33): $\widetilde{\Delta}t\left(t = t^{Q_iQ_j}_{\min}\right) = \Delta t^{Q_iQ_j}_{\text{b}} - \Delta t^{Q_iQ_j}_{\min}$ (the beginning of the marked interval), and $\widetilde{\Delta}t\left(t = t^{Q_iQ_j}_{\max}\right) = \Delta t^{Q_iQ_j}_{\max} - \Delta t^{Q_iQ_j}_{\text{b}}$ (the end of the interval). Therefore, during certain times $Q_{\text{int}}(t) > Q^{ij}_{\text{int2}}$, and at the output there is an impulse. Define the coordinates of the characteristic points of the change $Q_{\text{int}}(t)$: 1—overlaps of the threshold value Q^{ij}_{int2} in case of the increase in $Q_{\text{int}}(t)$; 2—reaching the maximum value Q_i; 3—overlaps of the threshold value Q^{ji}_{int2} when decreasing $Q_{\text{int}}(t)$; 4—overlaps of the threshold value Q^{ji}_{int3} when decreasing $Q_{\text{int}}(t)$; 5—setup of a stationary value $Q_{\text{int}}(t)$.

The coordinates of point 1 are known: $t = t^{Q_iQ_j}_{\text{b}} = t_0 + \Delta t^{Q_iQ_j}_{\text{b}}$ and $Q = Q^{ij}_{\text{int2}}$. Define the coordinates of point 2:

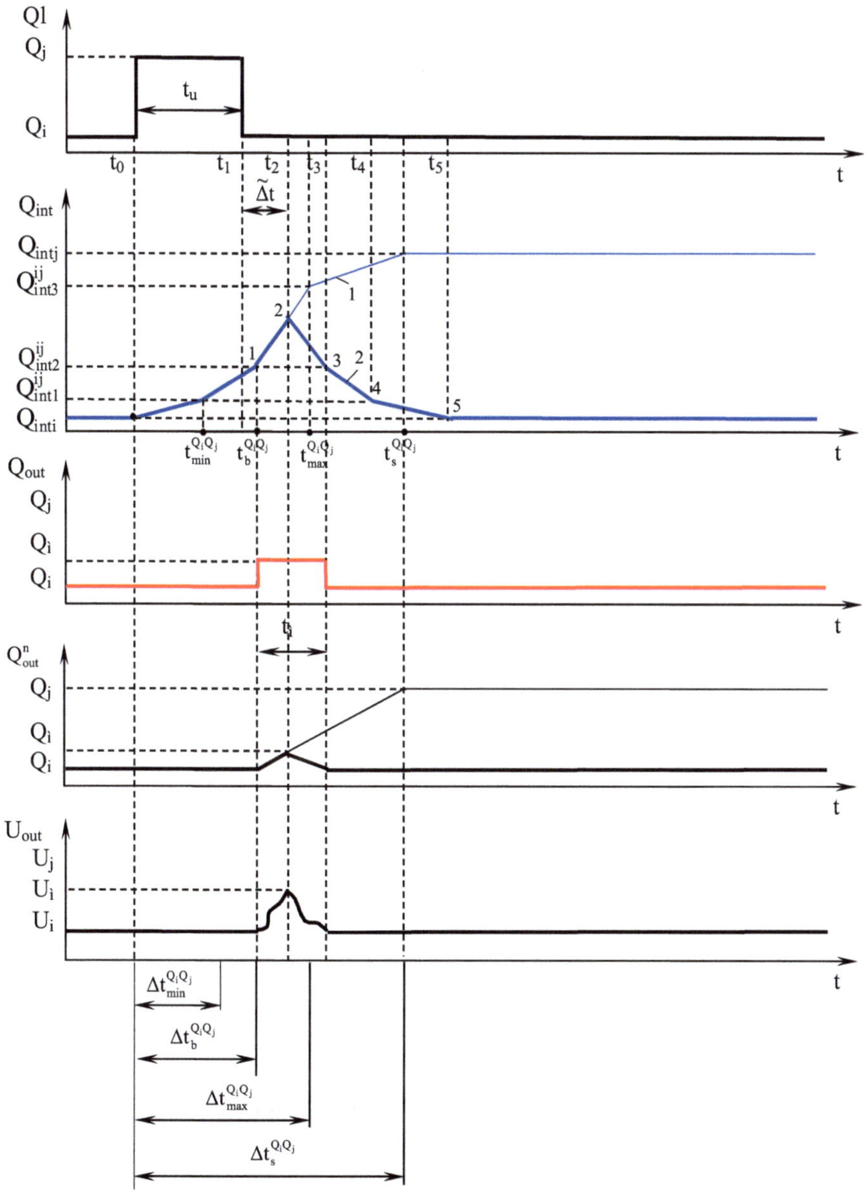

Fig. 2.27 Summation of excitations in the second stage

$$Q_i = Q^{ij}_{\text{int2}} + \Delta Q, \tag{2.52}$$

$$\text{where}\quad \frac{\Delta Q}{Q^{ij}_{\text{int3}} - Q^{ij}_{\text{int2}}} = \frac{t_2 - t_{\text{b}}^{Q_iQ_j}}{t_{\max}^{Q_iQ_j} - t_{\text{b}}^{Q_iQ_j}}. \tag{2.53}$$

Substituting the corresponding values, this is obtained:

$$Q_i = Q^{ij}_{\text{int2}} + \left(Q^{ij}_{\text{int3}} - Q^{ij}_{\text{int2}}\right) \cdot \frac{t_1 - t_0 - \Delta t^{Q_iQ_j}_{\min}}{\Delta t^{Q_iQ_j}_{\text{b}} - \Delta t^{Q_iQ_j}_{\min}}, \tag{2.54}$$

$$\begin{aligned} t_2 &= t_1 + \widetilde{\Delta} t \\ &= t_1 + \Delta t^{Q_iQ_j}_{\text{b}} - \Delta t^{Q_iQ_j}_{\min} + \frac{\left(\Delta t^{Q_iQ_j}_{\max} - 2 \cdot \Delta t^{Q_iQ_j}_{\text{b}} + \Delta t^{Q_iQ_j}_{\min}\right) \cdot \left(t_1 - t^{Q_iQ_j}_{\min}\right)}{\Delta t^{Q_iQ_j}_{\text{b}} - \Delta t^{Q_iQ_j}_{\min}}. \end{aligned} \tag{2.55}$$

For boundary values t_1, the following is obtained:

(a) When $t_1 = t^{Q_iQ_j}_{\min}, t_2 = t_1 + \Delta t^{Q_iQ_j}_{\text{b}} - \Delta t^{Q_iQ_j}_{\text{mini}}$, and $Q_i = Q^{ij}_{\text{int2}}$, i.e. point 2 merges with 1.

(b) When $t_1 = t^{Q_iQ_j}_{\text{b}}, t_2 = t_1 + \Delta t^{Q_iQ_j}_{\max} - \Delta t^{Q_iQ_j}_{\text{b}}$, and $Q_i = Q^{ij}_{\text{int3}}$, i.e. point 2 coincides with 3.

The value at $Q_{\text{int}}(t)$ point 3 is Q^{ji}_{int2}. As for time t_3, it can be defined as follows:

$$\frac{\Delta Q}{Q^{ji}_{\text{int1}} - Q^{ji}_{\text{int2}}} = \frac{t_3 - t_2}{t\acute{i}} {}^{Q_iQ_j} - t^{Q_iQ_j}_{\min}. \tag{2.56}$$

After substituting the values, this is obtained:

$$\begin{aligned} t_3 = t_1 + \Delta t^{Q_iQ_j}_{\text{b}} - \Delta t^{Q_iQ_j}_{\min} + \frac{\left(\Delta t^{Q_iQ_j}_{\max} - 2 \cdot \Delta t^{Q_iQ_j}_{\text{b}} + \Delta t^{Q_iQ_j}_{\min}\right) \cdot \left(t_1 - t^{Q_iQ_j}_{\min}\right)}{\Delta t^{Q_iQ_j}_{\text{b}} - \Delta t^{Q_iQ_j}_{\min}} \\ + \left(t_1 - t_0 - \Delta t^{Q_iQ_j}_{\min}\right) \cdot \frac{Q^{ij}_{\text{int3}} - Q^{ij}_{\text{int2}}}{Q^{ji}_{\text{int1}} - Q^{ji}_{\text{int2}}}. \end{aligned} \tag{2.57}$$

On fragment 3–4, a normal third stage of reverse switching occurs. Therefore, for point 4, $Q = Q^{ji}_{\text{int3}}$, and $t_4 = t_3 + \Delta t^{Q_jQ_i}_{\max} - \Delta t^{Q_jQi}_{n}$. And on fragment 4–5, the reverse switching continues $Q_{\text{int}}(t)$. The coordinates of point 5 are as follows: $Q = Q_{\text{int}i}$, $t_5 = t_4 + \Delta t^{Q_jQ_i}_{\text{c0}} - \Delta t^{Q_jQi}_{\max}$.

In the time interval between points 1 and 3, the value $Q_{\text{int}}(t)$ is greater than the threshold value Q^{ij}_{int2}. Therefore, an incomplete pulse is formed at the output of the MGE. Its amplitude can be determined as follows:

$$\begin{aligned} Q_i = Q_i + \left(Q_j - Q_i\right) \cdot \frac{\left(\Delta t^{Q_iQ_j}_{\max} - 2 \cdot \Delta t^{Q_iQ_j}_{\text{b}} + \Delta t^{Q_iQ_j}_{\min}\right) \cdot \left(t_1 - t^{Q_iQ_j}_{\min}\right)}{\left(\Delta t^{Q_iQ_j}_{\text{b}} - \Delta t^{Q_iQ_j}_{\min}\right) \cdot \left(\Delta t^{Q_iQ_j}_{\text{s}} - t_1 + t_0\right)} \\ - \frac{\Delta t^{Q_iQ_j}_{\min}}{\Delta t^{Q_iQ_j}_{\text{s}} - t_1 + t_0}, \end{aligned} \tag{2.58}$$

and duration:

$$t_{\mathrm{M}} = \frac{\left(\Delta t_{\max}^{Q_iQ_j} - 2 \cdot \Delta t_{\mathrm{b}}^{Q_iQ_j} + \Delta t_{\min}^{Q_iQ_j}\right) \cdot \left(t_1 - t_{\min}^{Q_iQ_j}\right)}{\Delta t_{\mathrm{b}}^{Q_iQ_j} - \Delta t_{\min}^{Q_iQ_j}} - \Delta t_{\min}^{Q_iQ_j} + \left(t_1 - t_0 - \Delta t_{\min}^{Q_iQ_j}\right) \cdot \frac{Q_{\mathrm{int3}}^{ij} - Q_{\mathrm{int2}}^{ij}}{Q_{\mathrm{int1}}^{ji} - Q_{\mathrm{int2}}^{ji}}. \tag{2.59}$$

Thus, in the second variant of summation of excitations at the output of the gates, an incomplete pulse is obtained, the duration and amplitude of which largely depend on the location of t_2 in the interval $t_{\mathrm{b}}^{Q_iQ_j} \ldots t_{\max}^{Q_iQ_j}$, which in turn depends on the moment t_1, i.e. duration of the input signal. Actually, the phenomenon of incomplete triggering of the gates and the dependence of the output signal parameters on the input signal within certain limits are simulated.

If, in the time interval $t_1 \ldots t_{\mathrm{c}}^{Q_iQ_j}$, the timing parameters of the gates are recalculated due to changes in the DF values, this can lead to phenomena of two types:

1. Change in the duration and amplitude of the output signal (Fig.2.28).
2. Transition to the third variant of the summation of excitations.

Variant 3. The moment of the reverse switching time Q_l falls on the third stage of excitation (Fig. 2.29), i.e. $t_{\mathrm{b}}^{Q_iQ_j} \leq t_1 \leq t_{\max}^{Q_iQ_j}$. Starting from the moment $t_1 Q_{\mathrm{int}}(t)$, it continues to increase to the value $Q_{\mathrm{int}\grave{i}}$ for a time $\widetilde{\Delta t}$determined by the formula (2.33) (the third case).

Then the drop of $Q_{\mathrm{int}}(t)$ begins. Since at point 2 in any case $Q_{\mathrm{int1}}^{ji} \leq Q_{\mathrm{int}}(t) \leq Q_{\mathrm{int}}^{j}$, the reverse switching $Q_{\mathrm{int}}(t)$ starts from the middle of the first stage. Define the coordinates of the characteristic switching points: $Q_{\mathrm{int}}(t)$ 1—the overlap of the threshold value Q_{int3}^{ij} during the rise in $Q_{\mathrm{int}}(t)$; 2—beginning of drop in $Q_{\mathrm{int}}(t)$; 3–5—overlaps of threshold values Q_{int1}^{ji}, Q_{int2}^{ji}, Q_{int3}^{ji}, with decreasing $Q_{\mathrm{int}}(t)$; 6—completion of transient processes.

The coordinates of point 1 are as follows: $t = t_{\max}^{Q_iQ_j} = t_0 + \Delta t_{\max}^{Q_iQ_j}, \quad Q = Q_{\mathrm{int3}}^{ij}$.

The moment of time to reach $Q_{\mathrm{int}}(t)$ its maximum value is determined as follows:

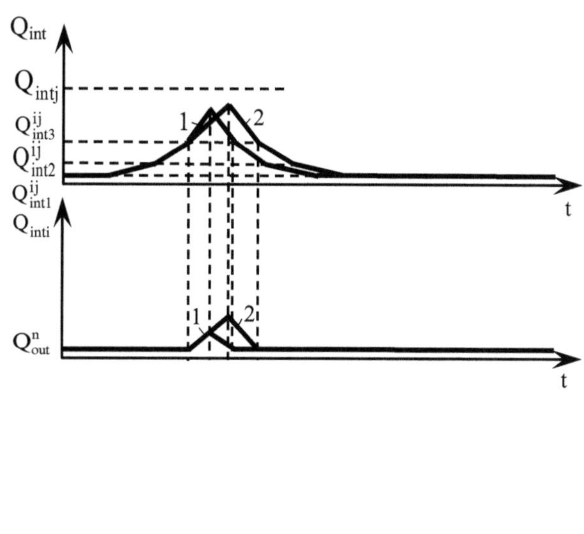

Fig. 2.28 Dynamics of change of Q_{int}

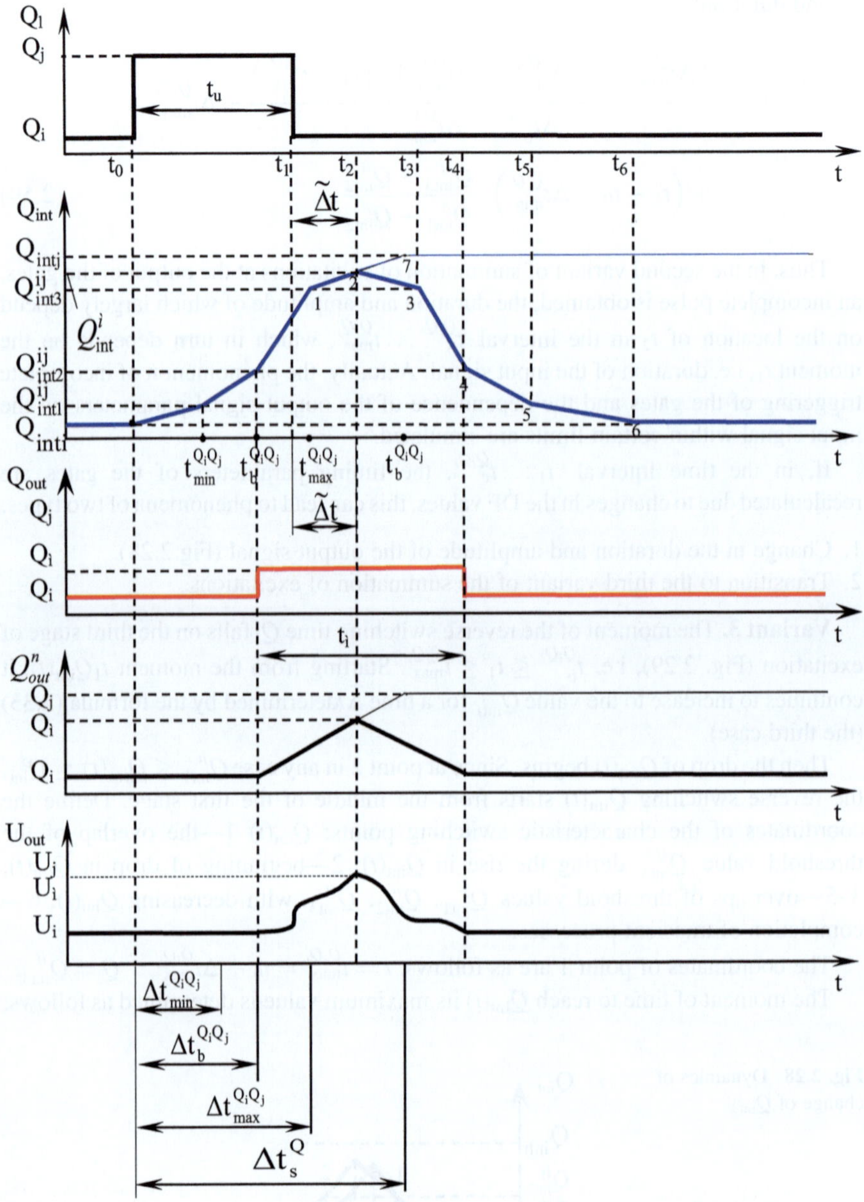

Fig. 2.29 Summation of excitations in the third stage

$$
\begin{aligned}
t_2 &= t_1 + \widetilde{\Delta} t \\
&= t_1 + \Delta t_{\max}^{Q_iQ_j} - \Delta t_{\mathrm{b}}^{Q_iQ_j} \\
&\quad + \frac{\left(\Delta t_{\mathrm{b}}^{Q_iQ_j} - 2 \cdot \Delta t_{\max}^{Q_iQ_j} + \Delta t_{\mathrm{b}}^{Q_iQ_j}\right) \cdot \left(t_1 - t_0 - \Delta t_{\mathrm{b}}^{Q_iQ_j}\right)}{\Delta t_{\max}^{Q_iQ_j} - \Delta t_{\mathrm{b}}^{Q_iQ_j}},
\end{aligned} \tag{2.60}
$$

The maximum value $Q_{\mathrm{int}\grave{i}}i$ is defined as:

$$
Q_{\mathrm{int}\grave{i}} = Q_{\mathrm{int3}}^{ij} + \Delta Q, \tag{2.61}
$$

where ΔQ is calculated as follows:

$$
\frac{\Delta Q}{Q_{\mathrm{int}j} - Q_{\mathrm{int3}}^{ij}} = \frac{t_2 - t_{\max}^{Q_iQ_j}}{\Delta t_{\mathrm{b}}^{Q_iQ_j} - t_{\max}^{Q_iQ_j}}. \tag{2.62}
$$

Substituting the corresponding values and simplifying, this is obtained:

$$
Q_{\mathrm{int}\grave{i}} = Q_{\mathrm{int3}}^{ij} + \left(Q_{\mathrm{int}j} - Q_{\mathrm{int3}}^{ij}\right) \cdot \frac{t_1 - t_0 - \Delta t_{\mathrm{b}}^{Q_iQ_j}}{\Delta t_{\max}^{Q_iQ_j} - \Delta t_{\mathrm{b}}^{Q_iQ_j}}. \tag{2.63}
$$

For boundary values t_1, there are the following:

(a) If $t_1 = t_0 + \Delta t_{\mathrm{b}}^{Q_iQ_j}$, then $Q_{\mathrm{int}\grave{i}} = Q_{\mathrm{int3}}^{ij}$, i.e. point 2 coincides with 1;
(b) If $t_1 = t_0 + \Delta t_{\max}^{Q_iQ_j}$, then $Q_{\mathrm{int}\grave{i}} = Q_{\mathrm{int}j}$, i.e. point 2 coincides with 7.

Point 3: $Q = Q_{\mathrm{int1}}^{ji}$

$$
t_3 = t_2 + \frac{Q_{\mathrm{int}j} - Q_{\mathrm{int3}}^{ij}}{Q_{\mathrm{int}j} - Q_{\mathrm{int1}}^{ji}} \cdot \frac{\left(t_1 - t_0 - \Delta t_{\mathrm{b}}^{Q_iQ_j}\right) \cdot \Delta t_{\min}^{QjQ_i}}{\Delta t_{\max}^{Q_iQ_j} - \Delta t_{\mathrm{b}}^{Q_iQ_j}}. \tag{2.64}
$$

Point 4: $Q = Q_{\mathrm{int2}}^{ji}$, $t = t_3 + \Delta t_{\mathrm{b}}^{Q_jQ_i} - \Delta t_{\min}^{Q_jQ_i}$.
Point 5: $Q = Q_{\mathrm{int3}}^{ji}$, $t = t_3 + \Delta t_{\max}^{Q_jQ_i} - \Delta t_{\min}^{Q_jQ_i}$.
Point 6: $Q = Q_i$, $t = t_3 + \Delta t_{\mathrm{s}}^{Q_jQ_i} - \Delta t_{\min}^{Q_jQ_i}$.

The duration of the output pulse will be: $t_i = t_4 - t_{\mathrm{b}}^{Q_iQ_j}$. By substituting the values, there is:

$$
\begin{aligned}
t_i &= t_1 - t_0 + \Delta t_{\max}^{Q_iQ_j} - 2 \cdot \Delta t\acute{i}^{\;Q_iQ_j} \\
&\quad + \frac{\left(\Delta t_{\mathrm{s}}^{Q_iQ_j} - 2 \cdot \Delta t_{\max}^{Q_iQ_j} + \Delta t\acute{i}^{\,Q_iQ_j}\right) \cdot \left(t_1 - t_0 - \Delta t_{\mathrm{b}}^{Q_iQ_j}\right)}{\Delta t_{\max}^{Q_iQ_j} - \Delta t_{\mathrm{b}}^{Q_iQ_j}} + + \frac{Q_{\mathrm{int}j} - Q_{\mathrm{int3}}^{ij}}{Q_{\mathrm{int}j} - Q_{\mathrm{int1}}^{ji}} \\
&\quad \cdot \frac{\left(t_1 - t_0 - \Delta t_{\mathrm{b}}^{Q_iQ_j}\right) \cdot \Delta t_{\min}^{Q_iQ_j}}{\Delta t_{\max}^{Q_iQ_j} - \Delta t_{\mathrm{b}}^{Q_iQ_j}}.
\end{aligned} \tag{2.65}
$$

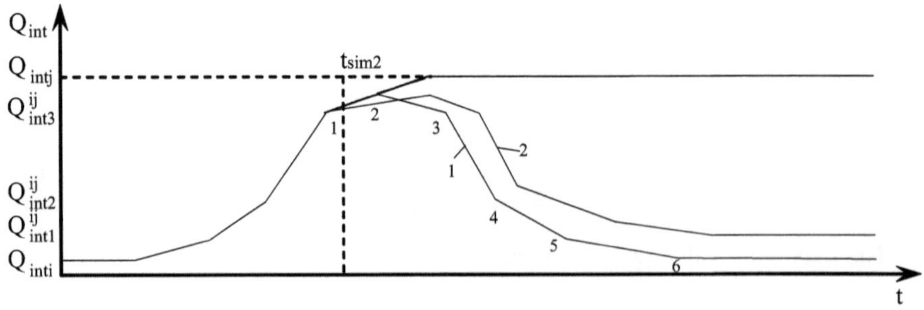

Fig. 2.30 Dynamics of change of Q_{int}

The amplitude of the output pulse is:

$$Q_i = Q_i + (Q_j - Q_i) \cdot \frac{t_1 - t_0 - \Delta t_{\text{b}}^{Q_iQ_j}}{\Delta t_{\text{s}}^{Q_iQ_j} - \Delta t_{\text{b}}^{Q_iQ_j}} \cdot \frac{\Delta t_{\text{s}}^{Q_iQ_j} - \Delta t_{\max}^{Q_iQ_j}}{\Delta t_{\max}^{Q_iQ_j} - \Delta t_{\text{b}}^{Q_iQ_j}} + \frac{\Delta t_{\max}^{Q_iQ_j} - \Delta t_{\text{b}}^{Q_iQ_j}}{\Delta t_{\text{s}}^{Q_iQ_j} - \Delta t_{\text{b}}^{Q_iQ_j}}. \tag{2.66}$$

The value U_i is determined by the formula (2.11) by substituting the value Q_i. If, after the moment of time t_1, the timing parameters of the gate are recalculated (for example, at the moment t_{sim2}, as in Fig. 2.30), then the timing coordinates of points 2 . . . 7 and the value $Q_{\text{int}i}$ of point 2 will change, if $t_{\text{sim2}} < t_2$. The broken t_1 line 1 is replaced by 2 (Fig. 2.30), which can affect the duration and amplitude of the output pulse.

Variant 4. The moment of time of the reverse switching Q_l goes to the fourth stage of a single excitation (Fig. 2.31), i.e. $t_{\max}^{Q_iQ_j} \le t_1 \le t_{\text{s}}^{Q_iQ_j}$. Starting from the moment $t_1 Q_{\text{int}}(t)$ continues to increase to a value $Q_{\text{int}j}$ during $\widetilde{\Delta t}$ determined by formula (2.33) (fourth case). Then the drop of $Q_{\text{int}}(t)$ begins. It follows from (2.33) that, regardless of the location of the timing point t_1 in the interval $\left[t_{\max}^{Q_iQ_j}, t_{\text{s}}^{Q_iQ_j}\right]$, $Q_{\text{int}}(t)$ will reach a value $Q_{\text{int}j}$. This also follows from the definition of timing parameter $\Delta t_{\text{s}}^{Q_iQ_j}$.

Therefore, at point 1 in any case $Q_{\text{int}}(t) = Q_{\text{int}j}$, and the normal reverse switching begins $Q_{\text{int}}(t)$. Define the coordinates of the characteristic switching points $Q_{\text{int}}(t)$: 1—reaching the maximum value $Q_{\text{int}j}$, 2–4—overlaps of the threshold values Q_{int1}^{ji}, Q_{int2}^{ji}, Q_{int3}^{ji} with decreasing $Q_{\text{int}}(t)$; 5—completion of transient processes.

The coordinates of point 1 are as follows: $t_2 = t\acute{o}^{Q_iQ_j}$, $\quad Q = Q_{\text{int}j}$.

The coordinates of the remaining points are as follows:

Point 2: $Q = Q_{\text{int1}}^{ji}$, $\quad t = t_3 = t_{\text{s}}^{Q_iQ_j} + \Delta t_{\min}^{Q_jQ_i} = t_0 + \Delta t_{\text{s}}^{Q_iQ_j} + \Delta t_{\min}^{Q_jQ_i}$.
Point 3: $Q = Q_{\text{int2}}^{ji}$, $\quad t = t_4 = t_{\text{s}}^{Q_iQ_j} + \Delta t_{\text{b}}^{Q_jQ_i} = t_0 + \Delta t_{\text{s}}^{Q_iQ_j} + \Delta t_{\text{s}}^{Q_jQ_i}$.
Point 4: $Q = Q_{\text{int3}}^{ji}$, $\quad t = t_5 = t_{\text{s}}^{Q_iQ_j} + \Delta t_{\max}^{Q_jQ_i} = t_0 + \Delta t_{\text{s}}^{Q_iQ_j} + \Delta t_{\max}^{Q_jQ_i}$.

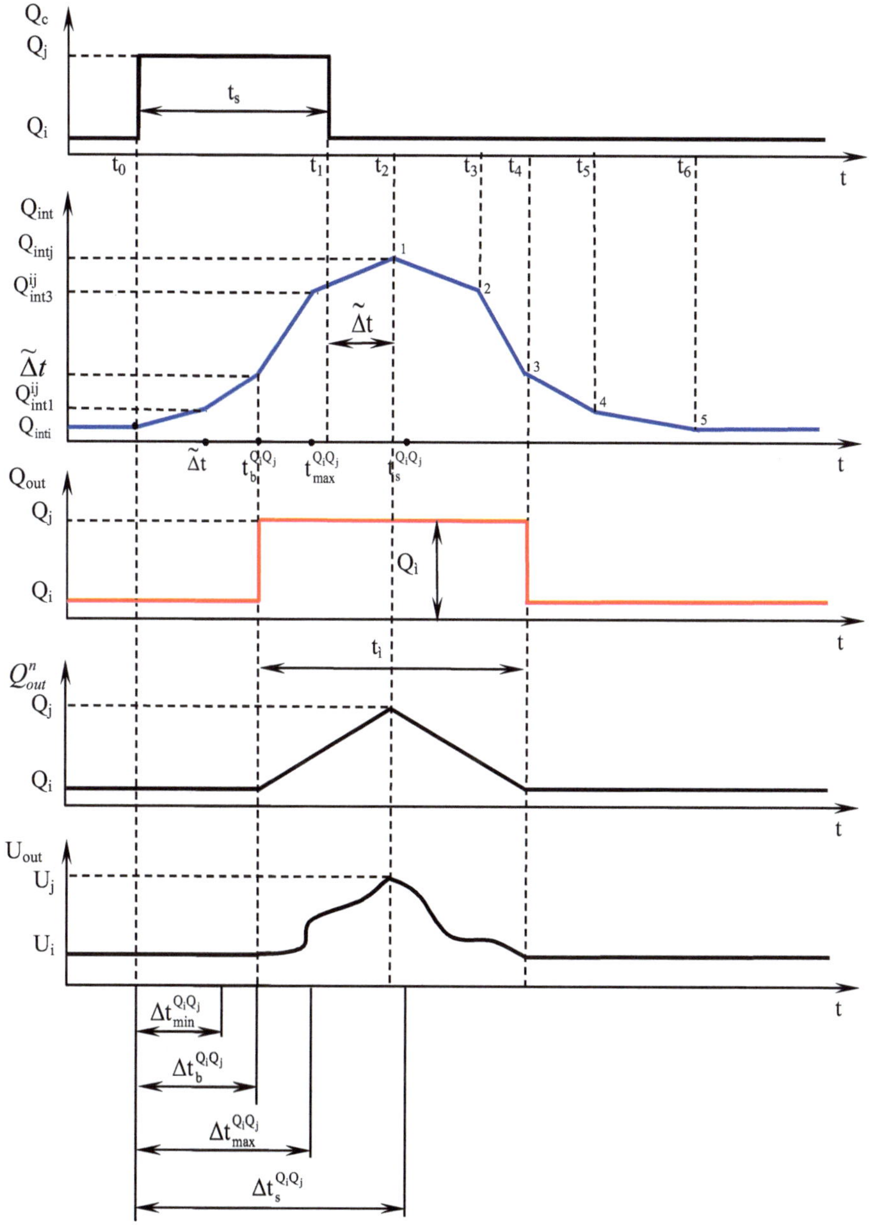

Fig. 2.31 Summation of excitations in the fourth stage

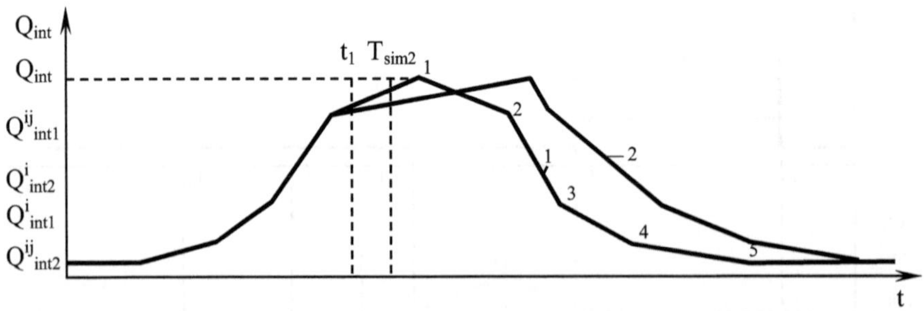

Fig. 2.32 Dynamics of variation of Q_{int}

Point 5: $Q = Q_{\mathrm{int}i}$, $t = t_6 = t_{\mathrm{s}}^{Q_iQ_j} + \Delta t_{\mathrm{s}}^{Q_jQ_i} = t_0 + \Delta t_{\mathrm{s}}^{Q_iQ_j} + \Delta t_{\mathrm{s}}^{Q_jQ_i}$.
The duration of the output pulse will be equal to:

$$\begin{aligned} t_i &= t_4 - t_{\mathrm{b}}^{Q_iQ_j} = t_0 + \Delta t_{\mathrm{s}}^{Q_iQ_j} + \Delta t_{\mathrm{b}}^{Q_jQ_i} - t_0 - \Delta t\acute{i}^{Q_iQ_j} \\ &= \Delta t_{\mathrm{s}}^{Q_iQ_j} + \Delta t_{\mathrm{b}}^{Q_jQ_i} - \Delta t_{\mathrm{b}}^{Q_iQ_j}. \end{aligned} \tag{2.67}$$

The amplitude of the output pulse will be:

$$Q_{\grave{i}} = Q_j - Q_i. \tag{2.68}$$

The value $U_{\grave{i}}$ can be determined by the formula (2.11), using $Q_{\grave{i}}$ from (2.68).

Thus, a pulse with a maximum amplitude is obtained. Changes in timing parameters of the gate after a moment of time t_1 (for example, at the moment t_{sim2}, as shown in Fig. 2.32) can only be displaced by timing coordinates of the points 1 … 5 (broken line 2 in Fig. 2.32). This will only change the duration of the output pulse.

When considering the summation of the excitations, it was assumed that on the gate, which is in a single excitation mode, a counter-directional switching of the input value Q_l may occur.

However, before the end of the action of the counter-directional switching, the signal can again return to its original state, etc. Thus Q_l can have multiple rapid changes, while the gate has not yet managed to react to previous effects. In this case, with the next change of Q_l, at the time t_1 of this switching, the state of the gate is considered, i.e. value $Q_{\mathrm{int}}(t_1)$ and direction of its change. Depending on this, according to formula (2.33), the interval of time $\widetilde{\Delta} t$ during which $Q_{\mathrm{int}}(t_1)$ continues its previous switching is calculated, and at the moment $t_1 + \widetilde{\Delta} t$, $Q_{\mathrm{int}}(t_1)$, beginning with the value$Q_{\mathrm{int}}\left(t_1 + \widetilde{\Delta} t\right)$, reacts to the changes of Q_l. For example, from Fig. 2.33 it can be seen that as a result of switching Q_l at moments t_0 and t_1, $Q_{\mathrm{int}}(t)$ should have changed according to polyline 1. However, at time t_2, when the transient processes as a result of the previous effects of Q_l have not yet terminated, Q_l again switches back to the state Q_j. Therefore, during $\widetilde{\Delta} t$, determined by the rules described above, $Q_{\mathrm{int}}(t)$ changes in broken line 1. Then $Q_{\mathrm{int}}(t)$ continues to grow already in polyline 2. If Q_l does not change again, then the output of the gate turns out to be fully switched. However, the time of this switching t_f is already different from the normal one (when Q_l switches only at the moment t_0).

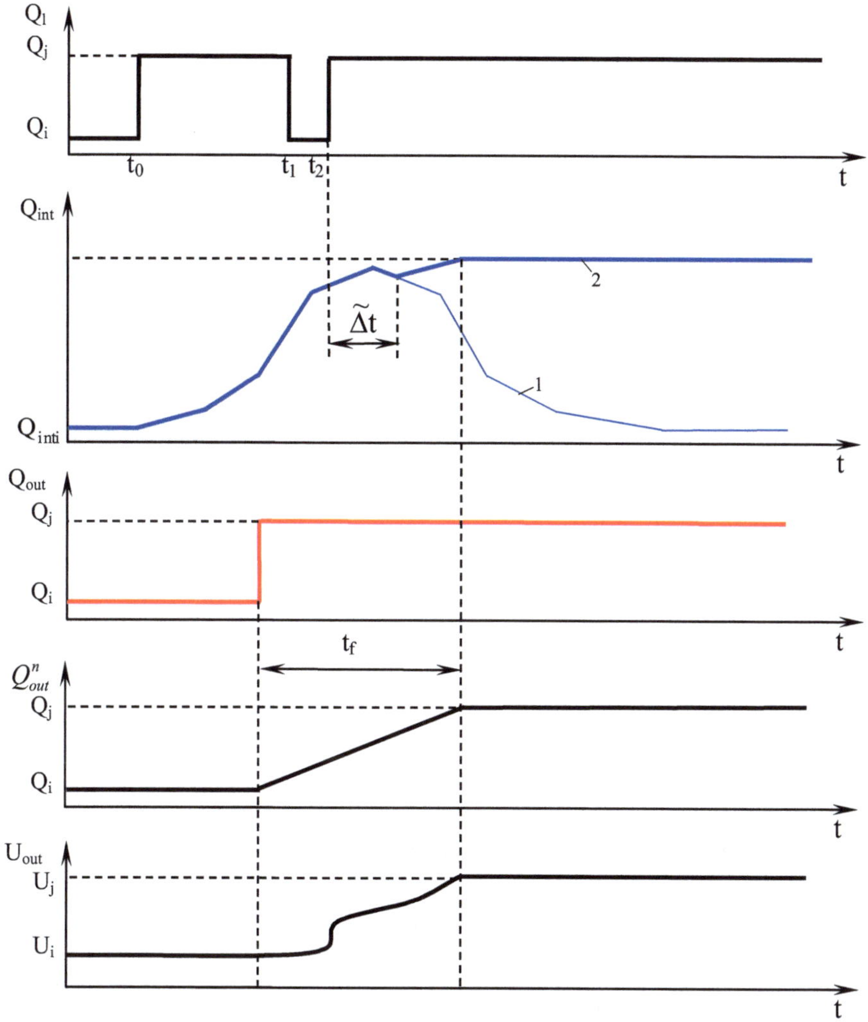

Fig. 2.33 Multiple, rapid changes of Q_l

2.3 Determination of Timing Parameters of the MCE

(a) The methodology to determine timing parameters of the MCE using their definitions

The timing parameters of the proposed model of the MCE are not standard [2, 77, 78]. However, they can be determined using simple experiments using the SPICE circuit-level simulation tool [74, 75]. Computer experiments are carried out under normal conditions of operation of gates (room temperature, stationary radiation

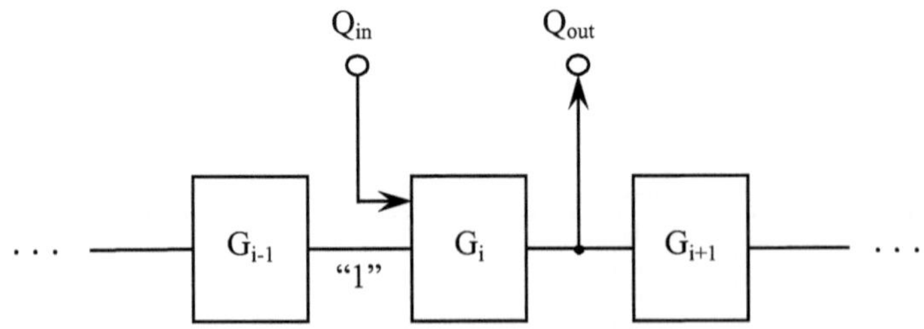

Fig. 2.34 Scheme for determining timing parameters of the gate

background, average load, etc.) In particular, to ensure the average load, the operation of gates is considered as part of a chain of consecutively connected gates (Fig. 1.104). And in such a chain, the behavior of not the first gate is studied, but of the gate with the number exceeding the quantity of gates, after which the parameters of the signals are practically setup. Experiments at the circuit level show that when a rectangular voltage jump $U_0 \rightarrow U_1$ is applied to the input of a circuit (Fig. 1.104), the difference between the values of timing parameters of the ith and $(i + 2)$th gate (gate inverts the signal) does not exceed 5% for $i \geq 7$. Therefore, the input and output signals of the gates, whose number is $i \geq 7$ are simulated. Thus, the scheme for conducting the experiments can be represented as in Fig. 2.34.

Such a constant voltage is given to the input of the gate chain ($U0$, if i is even, and $U1$—otherwise), so that the first input of Gi has a constant signal "1" ($U1$).

If it is necessary to determine the timing parameters of the Gi when switching its output from the Qi state to Qj, then the input signal Qin at the second input Gi switches from b–a–Qi to b–a–Qj (with minimax logic). The stresses corresponding to the states b–a–Qi and b–a–Qj are calculated using formula (2.11). The calculation of a specific timing parameter is carried out using its definition. For example, in order to calculate $\Delta t_{\min}^{Q_iQ_j}$ (Fig. 2.13), a pulse "$b - a - Qi$" → "$b - a - Qj$" → "$b - a - Qi$" with a duration at which constant value Qi is preserved in the output of the gate. Then the duration of the input pulse gradually (the step of increasing the duration depends on the necessary accuracy of determining the timing parameter $\Delta t^{Q_iQ_j}$) increases. At a certain value of the duration of the input pulse at the output of the gate, changes in the output voltage are observed. This duration will be the value of the timing parameter $\Delta t_{\min}^{Q_iQ_j}$. In approximately the same way, using the definitions of other parameters (Figs. 2.14–2.16), it is possible to measure the values $\Delta t_{\mathrm{b}}^{Q_iQ_j}$, $\Delta t_{\max}^{Q_iQ_j}$, $\Delta t_{\mathrm{s}}^{Q_iQ_j}$. The timing parameters defined in this way will be denoted $\Delta t_{\min 1}^{Q_iQ_j}$, $\Delta t_{\mathrm{b}1}^{Q_iQ_j}$, $\Delta t_{\max 1}^{Q_iQ_j}$, $\Delta t_{\mathrm{s}1}^{Q_iQ_j}$.

(b) The methodology to determine timing parameters of the MCE, based on measuring signals of different gates

To improve the accuracy of determining the timing parameters $\Delta t_{\mathrm{b}}^{Q_iQ_j}$ and $\Delta t_{\mathrm{s}}^{Q_iQ_j}$ and to reduce the number of experiments with SPICE [74, 75], another method of calculation is proposed. Timing parameters, determined by this method, are marked as index 2.

Since the gate in the chain in Fig. 1.104 inverts the input signal, if at the moment t_0, switching "Q_i" → "Q_j" is given to the input of the chain, the duration of which is longer than the time of setting up processes in the first gate, the following is obtained:

$$\begin{aligned} t_{e1} &= t_0, \\ t_{b1} &= t_0 + \Delta t_b^{b-a-Q_i, b-a-Q_j}, \\ t_{s1} &= t_0 + \Delta t_s^{b-a-Q_i, b-a-Q_j}, \end{aligned} \tag{2.69}$$

where t_{e1}, t_{b1}, and t_{s1} are respectively the instants of the excitation time, the beginning of the change and the setup of the signal for the first gates.

The excitation of the second gates will coincide with the moment of switching of the first gate;

$$\begin{aligned} t_{e2} &= t_{b1} = t_0 + \Delta t_b^{b-a-Q_i, b-a-Q_j}, \\ t_{b2} &= t_{e2} + \Delta t_b^{Q_iQ_j} = t_0 + \Delta t_b^{b-a-Q_i, b-a-Q_j} + \Delta t_b^{Q_iQ_j}, \\ t_{s2} &= t_{e2} + \Delta t_s^{Q_iQ_j} = t_0 + \Delta t_b^{b-a-Q_i, b-a-Q_j} + \Delta t_s^{Q_iQ_j}. \end{aligned} \tag{2.70}$$

The excitation of the third gate will coincide with the moment of switching of the second gate, i.e.

$$\begin{aligned} t_{e3} &= t_{b2} = t_0 + \Delta t_b^{b-a-Q_i, b-a-Q_j} + \Delta t_b^{Q_iQ_j}, \\ t_{b3} &= t_{e3} + \Delta t_b^{b-a-Q_i, b-a-Q_j} = t_0 + 2 \cdot \Delta t_b^{b-a-Q_i, b-a-Q_j} + \Delta t_b^{Q_iQ_j}, \\ t_{s3} &= t_{e3} + \Delta t_s^{b-a-Q_i, b-a-Q_j} = t_0 + \Delta t_b^{b-a-Q_i, b-a-Q_j} + \Delta t_b^{Q_iQ_j} + \Delta t_s^{b-a-Q_i, b-a-Q_j}. \end{aligned} \tag{2.71}$$

In the same way, for the ith gate, it can be written:

$$\begin{aligned} t_{en} &= t_0 + \left[\frac{i}{2}\right] \cdot \Delta t_b^{b-a-Q_i, b-a-Q_j} + \left[\frac{i-1}{2}\right] \cdot \Delta t_b^{Q_iQ_j}, \\ t_{bn} &= t_0 + \left[\frac{i+1}{2}\right] \cdot \Delta t_b^{b-a-Q_i, b-a-Q_j} + \left[\frac{i}{2}\right] \cdot \Delta t_b^{Q_iQ_j}, \\ t_{sn} &= t_0 + \left[\frac{i}{2}\right] \cdot \Delta t_b^{b-a-Q_i, b-a-Q_j} + \left[\frac{i-1}{2}\right] \cdot \Delta t_b^{Q_iQ_j} \\ &\quad + \left(1 - (-1)^i\right) \cdot \frac{\Delta t_s^{b-a-Q_i, b-a-Q_j}}{2} + \left(1 - (-1)^{i+1}\right) \cdot \frac{\Delta t_s^{Q_iQ_j}}{2}. \end{aligned} \tag{2.72}$$

Here, the notation $[x]$ denotes the integer part (without rounding) of the number x.

Measuring the moments of time t_b and t_s at the outputs of the mth (t_{bm}, t_{sm}) and nth (t_{bn}, t_{sn}) gates ($n > m$) and substituting these values in (2.72), this is obtained:

$$\Delta t_{\rm b}^{Q_iQ_j} = \frac{\left[\frac{n+1}{2}\right]\cdot(t_{{\rm b}m}-t_0)+\left[\frac{m+1}{2}\right]\cdot(t_0-t_{{\rm b}n})}{\left[\frac{m}{2}\right]\cdot\left[\frac{n+1}{2}\right]-\left[\frac{n}{2}\right]\cdot\left[\frac{m+1}{2}\right]}, \quad (2.73)$$

$$\Delta t_{\rm s}^{Q_iQ_j} = \frac{2\cdot A}{\left(1+(-1)^{m+1}\right)\cdot(1-(-1)^n)-\left(1+(-1)^{n+1}\right)\cdot(1-(-1)^m)},$$

$$\text{where} A = \left(1-(-1)^b\right)\cdot t_{sm}-(1-(-1)^m)\cdot t_{sn}$$

$$-(1-(-1)^n)\cdot\left(t_0+\left[\frac{m}{2}\right]\cdot\Delta t_{\rm b}^{b-a-Q_i,b-a-Q_j}\right.$$

$$++\left[\frac{m-1}{2}\right]\cdot\Delta t_{\rm b}^{Q_iQ_j})+(1-(-1)^m)\cdot$$

$$\left(t_0+\left[\frac{n}{2}\right]\cdot\Delta t_{\rm b}^{b-a-Q_i,b-a-Q_j}+\left[\frac{n-1}{2}\right]\cdot\Delta t_{\rm b}^{Q_iQ_j}\right). \quad (2.74)$$

Formulas (2.73) and (2.74) allow determining the timing parameters $\Delta t_{\rm b_2}^{Q_iQ_j}$ and $\Delta t_{\rm s_2}^{Q_iQ_j}$ with higher accuracy than $\Delta t_{\rm b_1}^{Q_iQ_j}$ and $\Delta t_{\rm s_1}^{Q_iQ_j}$. This is because the increase in the difference *m-n* automatically leads to a division of the measurement error by a larger number. This reduces the proportion of the measurement error to the value of timing parameter of one gate. In addition, the second method for determining timing parameters requires only a single simulation of the gate chain at the circuit level. However, in this case only the values of the parameters $\Delta t_{\rm b_2}^{Q_iQ_j}$ and $\Delta t_{\rm s_2}^{Q_iQ_j}$ are obtained. The values of the parameters $\Delta t_{\rm min_2}^{Q_iQ_j}$ and $\Delta t_{\rm max_2}^{Q_iQ_j}$ cannot be determined by the second method because of the lack of information about them in the output graphs of the voltage changes in the nodes of the gate chain. However, the values of $\Delta t_{\rm min_2}^{Q_iQ_j}$ and $\Delta t_{\rm max_2}^{Q_iQ_j}$ can be adjusted after determining $\Delta t_{\rm b_2}^{Q_iQ_j}$ and $\Delta t_{\rm s_2}^{Q_iQ_j}$:

$$\Delta t_{\rm min_2}^{Q_iQ_j} = \Delta t_{\rm min_1}^{Q_iQ_j}\cdot\frac{\frac{\Delta t_{\rm b_2}^{Q_iQ_j}}{\Delta t_{\rm b_1}^{Q_iQ_j}}+\frac{\Delta t_{\rm s_2}^{Q_iQ_j}}{\Delta t_{\rm s_1}^{Q_iQ_j}}}{2}, \quad (2.75)$$

$$\Delta t_{\rm max_2}^{Q_iQ_j} = \Delta t_{\rm max_1}^{Q_iQ_j}\cdot\frac{\frac{\Delta t_{\rm b_2}^{Q_iQ_j}}{\Delta t_{\rm b_1}^{Q_iQ_j}}+\frac{\Delta t_{\rm s_2}^{Q_iQ_j}}{\Delta t_{\rm s_1}^{Q_iQ_j}}}{2}. \quad (2.76)$$

(c) The methodology to determine timing parameters of the model of the MCE through the values at full switching

To provide simulation of digital circuits with consideration of DF, a large number of sets of timing parameters $\Delta t_{\rm min}^{Q_iQ_j}$, $\Delta t_{\rm b}^{Q_iQ_j}$, $\Delta t_{\rm max}^{Q_iQ_j}$, $\Delta t_{\rm s}^{Q_iQ_j}$ for different pairs Q_i, Q_j ($a \le Q_i \le Q_j \le b$) is required from the MGE. The number of such sets depends on the permissible absolute error of determining value $Q_l : \Delta Q_l$. If the set of timing parameters for switching $Q_i \to Q_j$ is known, then for $Q_i \to Q_k$ it is necessary to define a new set of timing parameters if $Q_k - Q_j > \Delta Q_l$. Therefore, if Q_l changes in the range $[a, b]$, then the number of possible levels of Q_l will be equal to $\frac{b-a}{\Delta Q_l} + 1$, and the number of possible transitions $Q_i \to Q_j$ between these levels $-\left(\frac{b-a}{\Delta Q_l} + 1\right) \cdot \frac{b-a}{\Delta Q_l}$. It is clear that even for small values ofΔQ_l, the number of sets of timing parameters can be a huge number. Therefore, the following procedure is proposed for determining the set of timing parameters when switching Q_l from state Q_i to Q_j.

1. For a standard basic logic cell (Figs. 1.13–1.15), by one of the above written methodologies, a set of timing parameters Δt_{min}, Δt_{b}, Δt_{max}, Δt_{s} for switching "0" → "1" and "1" → "0" is determined.
2. For the possible values of the range $Q_i \ldots Q_j$, as well as the location of the interval $[Q_i, Q_j]$ inside $[a, b]$ (i.e. difference $Q_i - a$) for specific values Q_i and Q_j experimentally (using SPICE) , the following dependencies are obtained:
$$\frac{\Delta t_{\mathrm{min}}^{Q_iQ_j}}{\Delta t_{\mathrm{min}}^{01}} = f_1(Q_j - Q_i), \quad \frac{\Delta t_{\mathrm{b}}^{Q_iQ_j}}{\Delta t_{\mathrm{b}}^{01}} = f_2(Q_j - Q_i), \quad \frac{\Delta t_{\mathrm{max}}^{Q_iQ_j}}{\Delta t_{\mathrm{max}}^{01}} = f_3(Q_j - Q_i),$$
$$\frac{\Delta t_{\mathrm{s}}^{Q_iQ_j}}{\Delta t_{\mathrm{s}}^{01}} = f_4(Q_j - Q_i) \quad \text{etc., as well as} \quad \frac{\Delta t_{\mathrm{min}}^{Q_iQ_j}}{\Delta t_{\mathrm{min}}^{01}} = F_1(Q_i - a),$$
$$\frac{\Delta t_{\mathrm{b}}^{Q_iQ_j}}{\Delta t_{\mathrm{b}}^{01}} = F_2(Q_i - a), \frac{\Delta t_{\mathrm{max}}^{Q_iQ_j}}{\Delta t_{\mathrm{max}}^{01}} = F_3(Q_i - a), \frac{\Delta t_{\mathrm{s}}^{Q_iQ_j}}{\Delta t_{\mathrm{s}}^{01}} = F_4(Q_i - a).$$
3. The dependencies f_1, f_2, f_3, f_4 and F_1, F_2, F_3, F_4 are approximated by the nth degree polynomial. In Chap. 3, the dependences, obtained experimentally for the basic logic cells in Figs. 1.13–1.15 are illustrated. It is shown that to ensure the necessary accuracy of the calculation of the entire circuit, it is often sufficient to approximate these dependencies even by linear or quadratic functions.
4. If the gate is simulated, the timing parameters ($\Delta t_{\mathrm{min}}^{01}$, $\Delta t_{\mathrm{b}}^{01}$, $\Delta t_{\mathrm{max}}^{01}$, $\Delta t_{\mathrm{s}}^{01}$, $\Delta t_{\mathrm{min}}^{10}$, $\Delta t_{\mathrm{b}}^{10}$, $\Delta t_{\mathrm{max}}^{10}$, $\Delta t_{\mathrm{s}}^{10}$) of which differ from the standard basic logic cell (for which the experiments were performed), then to obtain each function f_i or F_i often only two-three (depending on the degree of the approximation formula) computer experiments, aimed to calculate the coefficients of the approximation formula are given. Moreover, the form of the formula for the same circuit-level base remains constant. Thus, using a small number of experiments, approximate formulas f_i and F_i are obtained.
5. First, with the help of f_i and then F_i the timing parameters of the simulated gate are specified.

For clarity, the following example is given. Suppose, for the TTL basic logic cell (Fig. 1.13), it is experimentally setup that the dependences $\frac{\Delta t_{\mathrm{min}}^{Q_iQ_j}}{\Delta t_{\mathrm{min}}^{01}} = f_1(Q_j - Q_i)$, $\frac{\Delta t_{\mathrm{b}}^{Q_iQ_j}}{\Delta t_{\mathrm{b}}^{01}} = f_2(Q_j - Q_i)$, $\frac{\Delta t_{\mathrm{max}}^{Q_iQ_j}}{\Delta t_{\mathrm{max}}^{01}} = f_3(Q_j - Q_i)$ and $\frac{\Delta t_{\mathrm{s}}^{Q_iQ_j}}{\Delta t_{\mathrm{s}}^{01}} = f_4(Q_j - Q_i)$ have a quadratic form. It is clear that with the $Q_j - Q_i = 0$, the value of the corresponding ratio should be equal to 0, and with the $Q_j - Q_i = 1$, the same value should be equal to 1. Therefore, the indicated dependencies should be characterized by a function of the type $a \cdot (Q_j - Q_i)^2 + (1 - a) \cdot (Q_j - Q_i)$, where $0 < \grave{a} < 1$, and to determine the coefficient a of each of the dependences, it is necessary to conduct only one computer experiment at specific value of the difference $Q_j - Q_i$. Suppose, after carrying out such experiments, the following dependencies are obtained:

$$\begin{aligned} \frac{\Delta t_{\min}^{Q_iQ_j}}{\Delta t_{\min}^{01}} &= 0.42 \cdot (Q_j - Q_i)^2 + 0.58 \cdot (Q_j - Q_i), \\ \frac{\Delta t_{\mathrm{b}}^{Q_iQ_j}}{\Delta t_{\mathrm{b}}^{01}} &= 0.36 \cdot (Q_j - Q_i)^2 + 0.64 \cdot (Q_j - Q_i), \\ \frac{\Delta t_{\max}^{Q_iQ_j}}{\Delta t_{\max}^{01}} &= 0.61 \cdot (Q_j - Q_i)^2 + 0.39 \cdot (Q_j - Q_i), \\ \frac{\Delta t_{\mathrm{s}}^{Q_iQ_j}}{\Delta t_{\mathrm{s}}^{01}} &= 0.56 \cdot (Q_j - Q_i)^2 + 0.44 \cdot (Q_j - Q_i). \end{aligned} \tag{2.77}$$

For example, for a TTL basic logic cell, computer experiments have established that for a particular value of the difference $Q_j - Q_i = c$ ($\tilde{n} \le 1$), if the interval $[Q_i, Q_j]$ shifts within the range [0, 1] (i.e. first $Q_i = 0$, $Q_j = c$,then $Q_i = \Delta Q(\Delta Q \ll 1)$, $Q_j = \Delta Q + c$, $Q_i = 2 \cdot \Delta Q$, $Q_j = 2 \cdot \Delta Q + c$ etc., and at the end $Q_i = 1 - c$, $Q_j = 1$), then the dependencies $\frac{\Delta t_{\min}^{Q_iQ_j}}{\Delta t_{\min}^{01}} = F_1(Q_i)$, $\frac{\Delta t_{\mathrm{b}}^{Q_iQ_j}}{\Delta t_{\mathrm{b}}^{01}} = F_2(Q_i)$, $\frac{\Delta t_{\max}^{Q_iQ_j}}{\Delta t_{\max}^{01}} = F_3(Q_i)$ and $\frac{\Delta t_{\mathrm{s}}^{Q_iQ_j}}{\Delta t_{\mathrm{s}}^{01}} = F_4(Q_i)$ have a linear form: $(1 - b \cdot Q_i) \cdot \frac{\Delta t^{Q_iQ_j}}{\Delta t^{01}}$. In this case, for each of the marked dependencies, it is sufficient to perform only one computer experiment to determine the value of the coefficient b. Suppose, after carrying out such experiments, for example, for, $Q_j - Q_i = c = 0.3$ the following dependencies are obtained:

$$\begin{aligned} \frac{\Delta t_{\min}^{Q_iQ_j}}{\Delta t_{\min}^{01}} &= (1 - 0.06354Q_i)\frac{\Delta t_{\min}^{Q_iQ_j}}{\Delta t_{\min}^{01}}, \\ \frac{\Delta t_{\mathrm{b}}^{Q_iQ_j}}{\Delta t_{\mathrm{b}}^{01}} &= (1 - 0.08976Q_i)\frac{\Delta t_{\mathrm{b}}^{Q_iQ_j}}{\Delta t_{\mathrm{b}}^{01}}, \\ \frac{\Delta t_{\max}^{Q_iQ_j}}{\Delta t_{\max}^{01}} &= (1 - 0.09095Q_i)\frac{\Delta t_{\max}^{Q_iQ_j}}{\Delta t_{\max}^{01}}, \\ \frac{\Delta t_{\mathrm{s}}^{Q_iQ_j}}{\Delta t_{\mathrm{s}}^{01}} &= (1 - 0.10944Q_i)\frac{\Delta t_{\mathrm{s}}^{Q_iQ_j}}{\Delta t_{\mathrm{s}}^{01}}. \end{aligned} \tag{2.78}$$

The computer experiments also determined the values of timing parameters of the TTL cell when switching "0" → "1" : $\Delta t_{\min}^{01} = 0.1978\,\text{ns}$, $\Delta t_{\mathrm{b}}^{01} = 0.4576$ ns, $\Delta t_{\max}^{01} = 0.9674\,\text{ns}$, $\Delta t_{\mathrm{s}}^{01} = 1.1677\,\text{ns}$.

Suppose, during simulation, it was found out that *Ql* should switch from the state 0.5 to 0.9, i.e. $Qi = 0.5$ and $Qj = 0.9$. First, from the formulas (2.77), define (substitute $Qj - Qi = 0.4$) that $\frac{\Delta t_{\text{min}}^{Q_iQ_j}}{\Delta t_{\text{min}}^{01}} = 0.2992$, $\frac{\Delta t_{\text{b}}^{Q_iQ_j}}{\Delta t_{\text{b}}^{01}} = 0.3136$, $\frac{\Delta t_{\text{max}}^{Q_iQ_j}}{\Delta t_{\text{max}}^{01}} = 0.2536$ and $\frac{\Delta t_{\text{s}}^{Q_iQ_j}}{\Delta t_{\text{s}}^{01}} = 0.2656$. Then, using the formulas (2.78), correct the obtained relations:

$$\frac{\Delta t_{\text{min}}^{Q_iQ_j}}{\Delta t_{\text{min}}^{01}} = 0.2897, \frac{\Delta t_{\text{b}}^{Q_iQ_j}}{\Delta t_{\text{b}}^{01}} = 0.2995, \frac{\Delta t_{\text{max}}^{Q_iQ_j}}{\Delta t_{\text{max}}^{01}} = 0.2421, \frac{\Delta t_{\text{s}}^{Q_iQ_j}}{\Delta t_{\text{s}}^{01}} = 0.2416.$$

The finally corrected values of timing parameters of the gate when switching Q_l from 0.5 to 0.9 have this form: $\Delta t_{\text{min}}^{0.5;0.9} = 0.0573\,\text{ns}$, $\Delta t_{\text{b}}^{0.5;0.9} = 0.13705\,\text{ns}$, $\Delta t_{\text{max}}^{0.5;0.9} = 0.2342\,\text{ns}$, $\Delta t_{\text{s}}^{0.5;0.9} = 0.28212\,\text{ns}$.

(d) The methodology to determine the timing parameters of the MCE through reference data

The determination of timing parameters of the MGE when *Ql Qi Qj* switches through parameter values at full ("0" → "1" or "1" → "0") switching, allows expressing $\Delta t_{\text{min}}^{Q_iQ_j}$, $\Delta t_{\text{b}}^{Q_iQ_j}$, $\Delta t_{\text{max}}^{Q_iQ_j}$, $\Delta t_{\text{s}}^{Q_iQ_j}$ also through the reference parameters of digital cells. It is known [2, 77, 78] that the reference timing parameters of digital cells are the following (Fig. 2.35):

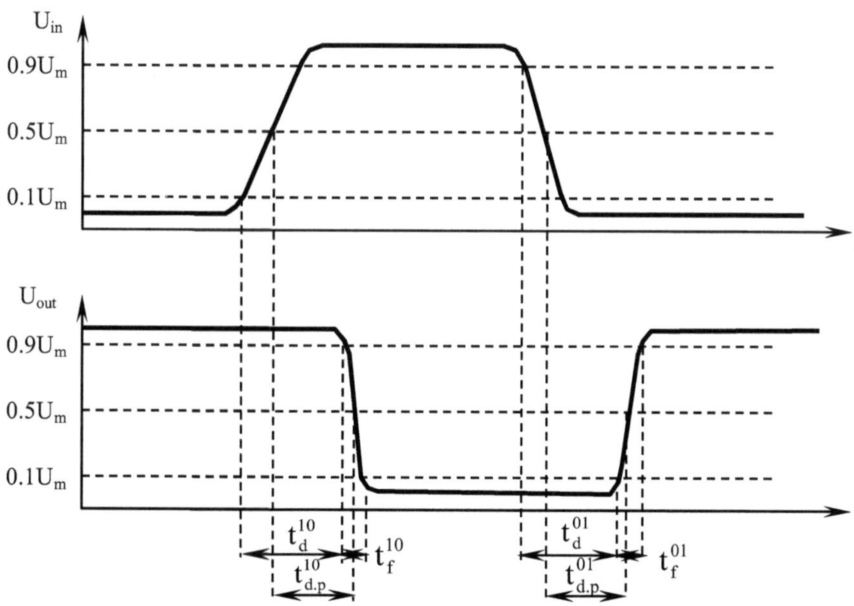

Fig. 2.35 Reference timing parameters of a digital cell

1. $t^{10}_{\text{p.d}}, t^{01}_{\text{p.d}}$—respectively, propagation delay times when the gate switches "0" → "1" and "1" → "0". These are the time intervals between the moments of crossing the input (U_{in}) and output (U_{out}) voltages of the threshold value $0.5U_{\text{m}}$.
2. t^{10}_{d}, t^{01}_{d}—respectively, the delay times of on and off switching when the gate switches "0" → "1" and "1" → "0". These are the time intervals between the crossing of U_{in} and U_{out}, respectively, the threshold values $0.1U_{\text{m}}$ and $0.9U_{\text{m}}$.
3. $t_{\text{f}}, t_{\text{r}}$—respectively, the duration of rise and fall of the output signal when the gate switches "0" → "1" and "1" → "0". These are the time intervals between the moments of crossing of U_{out} threshold values $0.1U_{\text{m}}$ and $0.9U_{\text{m}}$.

In [239], relations were obtained that allow relating the values of timing parameters Δt^{01}_{b}, Δt^{10}_{b}, Δt^{01}_{b}, Δt^{10}_{b} to the reference data:

$$\Delta t^{01}_{\text{b}} = t^{01}_{\text{d}} - 0.125t^{01}_{\text{f}}, \quad \Delta t^{10}_{\text{b}} = t^{10}_{\text{d}} - 0.125t^{10}_{\text{f}}, \quad \Delta t^{01}_{\text{s}} = t^{01}_{\text{d}} + 1.125t^{01}_{\text{f}}, \quad \Delta t^{10}_{\text{b}} = t^{10}_{\text{d}} + 1.125t^{10}_{\text{f}} \tag{2.79}$$

or through another set of reference parameters:

$$\Delta t^{01}_{\text{b}} = t^{01}_{\text{p.d}} - 0.625t^{01}_{\text{f}}, \quad \Delta t^{10}_{\text{b}} = t^{10}_{\text{p.d}} - 0.625t^{10}_{\text{f}}, \quad \Delta t^{01}_{\text{s}} = t^{01}_{\text{p.d}} + 0.625t^{01}_{\text{f}}. \quad \Delta t^{10}_{\text{s}} = t^{10}_{\text{p.d}} + 0.625t^{10}_{\text{f}} \tag{2.80}$$

The remaining timing parameters ($\Delta t^{01}_{\min}$, $\Delta t^{10}_{\min}$, $\Delta t^{01}_{\max}$, $\Delta t^{10}_{\max}$) can be determined using the values calculated from formulas (2.79) or (2.80), and the coefficients:

$$k_1 = \frac{\Delta t^{01}_{\min}}{\Delta t^{01}_{\text{b}}}, \; k_2 = \frac{\Delta t^{10}_{\min}}{\Delta t^{10}_{\text{b}}}, \; k_3 = \frac{\Delta t^{01}_{\max}}{\Delta t^{01}_{\text{b}}}, \; k_4 = \frac{\Delta t^{10}_{\max}}{\Delta t^{10}_{\text{b}}}.$$

As for the recalculations of timing parameters of the MGE due to deviations in the operating conditions of the gates from normal, they are performed by the models described in Chap. 3.

Thus, to determine the timing parameters of the gate $\Delta t^{Q_iQ_j}_{\min}$, $\Delta t^{Q_iQ_j}_{\text{b}}$, $\Delta t^{Q_iQ_j}_{\max}$, $\Delta t^{Q_iQ_j}_{\text{s}}$ a small number of computer experiments and calculations are required.

2.4 Research of Properties of the MCE

Analyze the properties of the developed MCE, from the point of view of its compliance with the requirements formulated in Sect. 2.2.

1. The adequacy of the proposed model of the MCE can be assessed by comparing the simulation results of separate basic logic cells of TTL (Fig. 1.13), ECL (Fig. 1.14), CMOS (Fig. 1.15), etc., obtained in different ways: by the help of MCE and circuit-level simulation of the same digital cell. However, when considering a particular digital cell, some important features of the MCE will not be revealed from the point of view of reflecting the phenomena inherent in

signal propagation through interacting within the digital circuit gates (for example, Fig. 1.72, Fig. 1.73, etc.). Therefore, the accuracy of the MCE will be evaluated by observing examples of simulating the most common typical digital circuit fragments using the proposed model and comparing them with the simulation results at the circuit level, as well as known logical models.

Consider the circuit in Fig. 1.104 first in static mode. Suppose that all the DF (*Vi*, i=1, 2, . . ., *Wj*, j=1, 2, . . .) are equal to zero. Suppose a voltage U_0 corresponding to a logic "0" is applied to the input of the circuit. Using formula (2.10), define the logic signal at the input of the first gate:

$$Q_{\text{in}1} = Q_0 = a + (b - a) \cdot \frac{U_0 - U_{\text{in}_{\min}}}{U_{\text{in}_{\max}} - U_{\text{in}_{\min}}} \tag{2.81}$$

The output logic signal G_1, which is simultaneously the input for G_2, can be determined according to the NOT function from Table 2.3:

$$Q_{\text{out}1} = Q_{\text{in}2} = Q_1 = a + b - Q_{\text{in}1} = b + (a - b) \cdot \frac{U_0 - U_{\text{in}_{\min}}}{U_{\text{in}_{\max}} - U_{\text{in}_{\min}}} \tag{2.82}$$

Similarly, the remaining signals of the digital circuit can be determined: $Q_{\text{out}2} = Q_{\text{in}3} = Q_2 = Q_{\text{in}1}$; $Q_{\text{out}3} = Q_{\text{in}4} = Q_3 = Q_{\text{in}2}$ etc. Thus, for the ith node of the circuit in Fig. 1.104, this is obtained:

$$Q_i = \frac{1-(-1)^{i+1}}{2} \cdot a + \frac{1-(-1)^{i}}{2} \cdot b + (-1)^{i+1} \cdot (a - b) \cdot \frac{U_0 - U_{\text{in}_{\min}}}{U_{\text{in}_{\max}} - U_{\text{in}_{\min}}} \tag{2.83}$$

Substituting the value of Q_i into the formula (2.11), the value of the voltage in the ith node of the consecutive chain of gates is obtained:

$$\begin{aligned} U_i = {} & U_{\text{out}_{\min}} + \left(U_{\text{out}_{\max}} - U_{\text{out}_{\min}}\right) \times \\ & \times \frac{\left(1-(-1)^{i}\right) \cdot U_{\text{in}_{\max}} - \left(1-(-1)^{i+1}\right) \cdot U_{\text{in}_{\min}} + (-1)^{i} \cdot 2U_0}{2\left(U_{\text{in}_{\max}} - U_{\text{in}_{\min}}\right)} \end{aligned} \tag{2.84}$$

Since $U_{\text{out},\, i} = \text{U}_{\text{in},\, i+1}$ (for a consecutive chain of gate), it can be assumed that $U_{\text{out}_{\max}} = U_{\text{in}_{\max}} = U_{\max}$ and $U_{\text{out}_{\min}} = U_{\text{in}_{\min}} = U_{\min}$. With this in mind, this is obtained:

$$U_i = U_{\min} + \frac{\left(1-(-1)^{i}\right) \cdot U_{\max} - \left(1-(-1)^{i+1}\right) \cdot U_{\min} + (-1)^{i} \cdot 2 \cdot U_0}{2} \tag{2.85}$$

For even i, $U_i = U_0$ is obtained, and for odd ones—$U_i = U_{\min} + U_{\max} - U_0$, which is (see Fig. 2.36) the value of U_1.

It is clear that when the voltage $U_{\text{in}} = U_1$ is applied to the input of the digital circuit (Fig. 1.104), everything will be reversed: at nodes with even numbers, the voltage U_1 is obtained, and with odd—U_0. Thus, in case standard static signals are

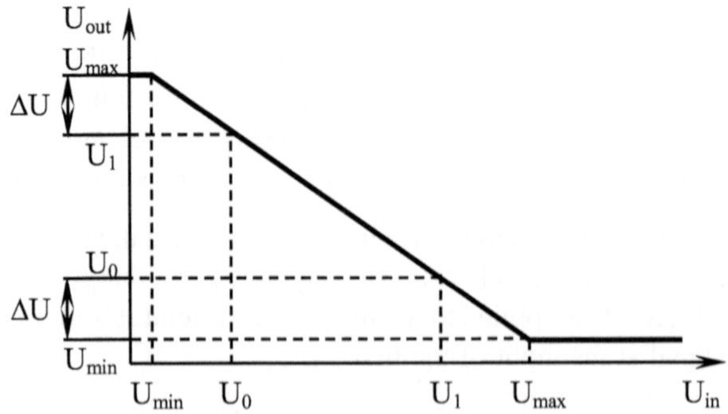

Fig. 2.36 Switching characteristic of the gate

given to the input of the gate chain, reliable simulation results are obtained. However, the same results are also given by ordinary logical models [164, 172, 174, 181, 182, 191, 196, 201, 207]. According to them, too, if Q_{in} = "0", then in nodes with even numbers the logical state is "0", and with odd numbers—"1". When converting into voltages, U_0 and U_1, respectively, are obtained. But in case of the MCE, unlike traditional models, the formula (2.85) also holds for an arbitrary input voltage U_{in} from the range $U_{min} \le U_{in} \le U_{max}$, and not only in case of U_0 or U_1:

$$U_i = U_{min} + \frac{\left(1 - (-1)^i\right) \cdot U_{max} - \left(1 - (-1)^{i+1}\right) \cdot U_{min} + (-1)^i \cdot 2 \cdot U_{in}}{2} \tag{2.86}$$

If $U_{in} = U_{min} + c(U_{max} - U_{min})$ (where $0 \le c \le 1$), i.e. an intermediate voltage between U_{min} and U_{max} is applied to the input of the circuit, then from (2.86) it follows that at nodes with even numbers, the level $U_{min} + c(U_{max} - U_{min})$ is obtained, and with odd numbers $U_{max} - c(U_{max} - U_{min})$. In particular, if $U_{in} = (U_{min} + U_{max})/2$, i.e. $c = 0.5$, then the same voltage will be set in all nodes of the circuit $U = (U_{min} + U_{max})/2$. As an example in Table 2.4 shows the values of the stresses in nodes 9 and 10 (U_9 and U_{10}) of a digital circuit in Fig. 1.104 (the circuit from Fig. 1.15 is used as gate), determined using SPICE (U_{min} = 0.0B, U_{max} = 1.0B) and the MCE for different values of c. The average relative error for U_9 is 1.41%, and for U_{10}—2.37%. A dependence of the relative error on the node number of a digital circuit is shown in Fig. 2.37. It does not exceed 3.7%. Approximately the same results are obtained when using the circuit in Figs. 1.13 (2.9%) and 1.14 (2.7%). The relative error in determining the static stresses in the flip-flop in Fig. 2.38 when using the circuit in Fig. 1.13 does not exceed 4.3%, the circuit in Fig. 1.14—3.9%, the circuits in Fig. 1.15—3.6% compared to the results obtained with the help of SPICE. Similar experiments for some circuits from the ISCAS 89 test series when implementing the basic logical cell with the help of CMOS (Fig. 1.15) showed the following results: s27 no more than 3.4%, s208—4.6%, s298—4.8%. Thus, the MCE adequately simulates the static mode of operation of the consecutive gate chain in all cases, i.e. for an arbitrary value Q_{in} from the range $[a, b]$.

Table 2.4 Estimation of the accuracy of determining static stresses at nodes 9 and 10 of the circuit in Fig. 1.104

	U_9 (V)				U_{10} (V)			
C	MCE	SPICE	Δ_i (2.12)	ε_i (2.13)	MCE	SPICE	Δ_i (2.12)	ε_i (2.13)
0.0	1.0	0.9987	0.0013	0.0013	0.0	0.0093	0.0093	–
0.1	0.9	0.8983	0.0017	0.0019	0.1	0.0911	0.0089	0.0977
0.2	0.8	0.7979	0.0021	0.0026	0.2	0.1916	0.0084	0.0438
0.3	0.7	0.6975	0.0025	0.0036	0.3	0.2210	0.0079	0.0357
0.4	0.6	0.5969	0.0031	0.0052	0.4	0.3927	0.0073	0.0185
0.5	0.5	0.4965	0.0035	0.0070	0.5	0.4933	0.0067	0.0136
0.6	0.4	0.3958	0.0042	0.0106	0.6	0.5943	0.0057	0.0096
0.7	0.3	0.2952	0.0048	0.0163	0.7	0.6952	0.0048	0.0069
0.8	0.2	0.1946	0.0054	0.0277	0.8	0.7959	0.0041	0.0052
0.9	0.1	0.0939	0.0061	0.0650	0.9	0.8968	0.0032	0.0036
1.0	0.0	0.0068	0.0068	–	1.0	0.9974	0.0026	0.0026

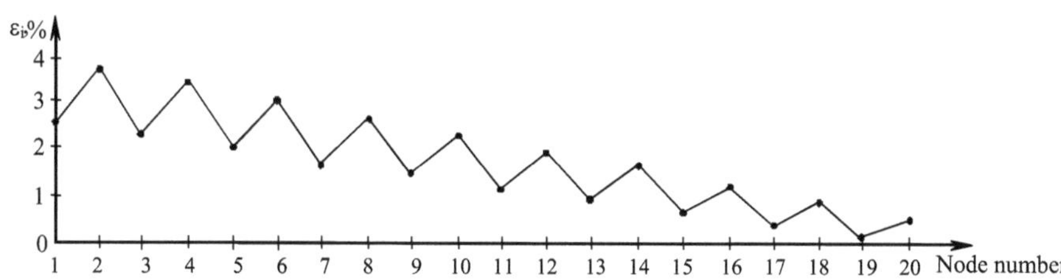

Fig. 2.37 The experimental dependence of the relative error in determining the static stresses of the circuit in Fig. 1.104 from the number of digital circuit node

Fig. 2.38 FF circuit

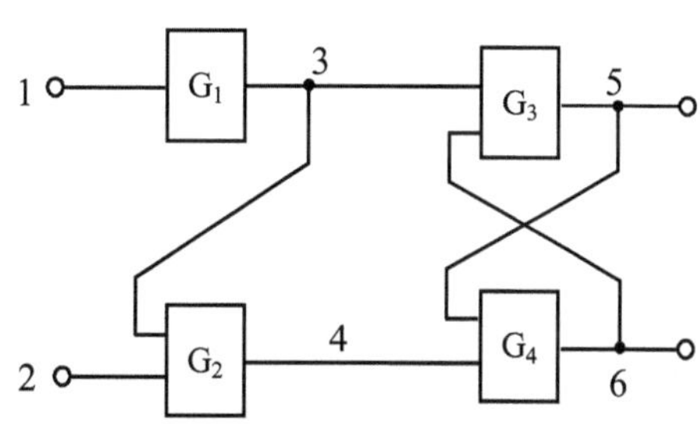

And traditional logical models, because of the limitations of their alphabets, can give adequate simulation results only for certain values of the input voltage: in case of two-valued ({0, 1}) models [164, 174, 179, 182, 192]—only for $U_{\text{in}} = U_0$ and $U_{\text{in}} = U_1$, with three-valued $(0, 1, x)$ [163, 188, 202, 233, 238, 290] only for $U_{\text{in}} = U_i$, $U_{\text{in}} = U_1$ and $U_{\text{in}} = (U_0 + U_1)/2$ (if "x" is interpreted as the average between "0" and "1" states), etc.

Unlike known logic models, the MCE is able to reflect changes in static voltage in this circuit nodes also in case of DFs. For example, if as a result of DF effects, the output voltage value of ith (for example, suppose that i-odd) gate changes by ΔU_i

and for simplicity first assume that the other gates are not affected by DF, then according to (2.86), the voltage $U = U_{\min} + c(U_{\max} - U_{\min} + \Delta U_i)$ will be obtained in the outputs of all subsequent gates of odd numbered nodes, and $U_{\max} - c(U_{\max} - U_{\min} - \Delta U_i)$—even-numbered. Of course, in all cases the condition $U_{\min} \leq U \leq U_{\max}$ must be observed. Thus, the property of the chain from Fig. 1.104 is reflected to extend the DF effect, applied to one of its component gates, through the followers of this element. This phenomenon occurs in real digital circuits [2, 282].

However, DF, naturally, affects simultaneously all the gates of the circuit. In this case, to determine the stresses in all digital circuit nodes in Fig. 1.104 it is necessary to have a dependence of the size of the deviation of the signal ΔU on the voltage value in case of the current DF value. To simplify the example, consider the case when $\Delta U = \text{const}$ and does not depend on the magnitude of the output voltage of the gate, calculated without consideration of DF. This means that a constant value of ΔU must be added to the voltage value at any node of the circuit, taking into account its sign (2.17). As seen from Fig. 1.33, ΔU for the basic TTL cell (Fig. 1.13) in case of a change in the ambient temperature at the state "1" (U^1) is positive, and in case of "0" (U^0)—negative. To simplify the considered example, in the first approximation, one can assume that $|\Delta U^1| = |\Delta U^0| = \Delta U$. Suppose that the voltage $U_{in}(U_{\min} \leq U_{in} \leq U_{\max})$ is applied to the input of the digital circuit (Fig. 1.104). The following notation is introduced:

$$
\begin{aligned}
c_0 &= \frac{U_{in} - U_{\min}}{U_{\max} - U_{\min}}, \quad c_i = \frac{U_i - U_{\min}}{U_{\max} - U_{\min}}, \quad i = 1, 3, 5, \ldots, \\
c_i &= \frac{U_{\min} + U_{\max} - U_i}{U_{\max} - U_{\min}}, \quad i = 2, 4, 6, \ldots
\end{aligned}
\tag{2.87}
$$

c_i shows the degree of closeness of the voltage at the output of G_i to $U_{\max}$ or to $U_{\min}$. For example, for nodes with odd numbers c_i shows the degree of proximity of U_i to $U_{\max}$: for $U_i = U_{\max}$ $c_i = 1$ and for $U_i = U_{\min}$ $c_i = 0$. In general case, $0 \leq c_i \leq 1$. c_0 corresponds to the input node of a digital circuit in Fig. 1.104. Suppose c_0 is closer to 0, i.e. U_{in} is closer to the voltage corresponding to the logic state "0" (U^0). Then, naturally, at the nodes of a digital circuit with odd numbers, voltages close to U^1 will be set, and in nodes with even numbers, to U^0. Using (2.86), (2.87), and also taking into account the changes in the sign of the voltage ΔU in the corresponding states, the following is obtained for the nodes:

(a) With odd numbers:

$$
U_i = U_{\max} - c_{i-1} \cdot (U_{\max} - U_{\min}) + \Delta U = U_{\max} - U_i + U_{\min} + \Delta U \tag{2.88}
$$

(b) With even numbers:

$$
\begin{aligned}
U_i &= U_{\max} - \tilde{n}_{i-1} \times (U_{\max} - U_{\min}) - \Delta U \\
&= 2 \cdot U_{\min} + U_{\max} - U_i - \Delta U
\end{aligned}
\tag{2.89}
$$

Table 2.5 The results of digital IC simulation from Fig. 1.104 with consideration of DF

i	U_i (V)		Δ_i	ε_i
	MCE	SPICE		
0	0.9	0.9000	0.0000	0.0000
1	3.4	3.4782	0.0782	0.0225
2	1.1	1.0914	0.0086	0.0079
3	3.2	3.3007	0.1007	0.0305
4	1.3	1.3548	0.0548	0.0404
5	3.0	2.9643	0.0357	0.0121
6	1.5	1.6004	0.1004	0.0627
7	2.8	2.7841	0.0159	0.0057
8	1.7	1.7694	0.0694	0.0392
9	2.6	2.6324	0.0324	0.0123
10	1.9	1.8832	0.0168	0.0089

Bearing in mind that $U_0 = U_{\text{in}}$ and assuming that the TTL cell is used as gate in a digital circuit in Fig. 1.104 (Fig. 1.13), $U_{\min} = 0.6$B, $U_{\max} = 3.5$B, $U_{\text{in}} = 0.9$B, $\Delta U = 0.2$B, the results shown in Table 2.5 are obtained.

Even with rather rough assumptions, taken in the observation of an example ($\Delta U =$ const, $|\Delta U^1| = |\Delta U^0| = \Delta U$), the average relative error compared to the results obtained with the SPICE is only 2.2%, and the maximum error is 6.27%. Such results are also obtained when considering other examples of digital circuits. From the results of Table 2.5, it is also seen that the MCE adequately reflects the property of the successive chain of the gate of a gradual decrease in the amplitude of the signal ($U_{\text{m}} = U^1 - U^0$) under the influence of DF, which occurs in real digital circuits [1].

Thus, in the static operating mode of digital circuits, the developed model of considering DF effects accurately calculates the stress level values for both standard voltage levels at digital circuit inputs (U_0, U_1) and for arbitrary values between them. Such a capability does not have known logical models of digital cells. The accuracy of calculations is close to the accuracy of calculations with SPICE and on average yields only 5% ... 8%. The accuracy of calculations of static signal levels with the help of the MCE is mainly influenced by the errors of approximation of the dependencies $Q_{\text{in}i} = f_1(U_{\text{in}i})$ and $Q_{\text{out}i} = f_2\left(Q^b_{\text{out}}\right))$ by first-order formulas (2.10) and (2.11), the error in determining $f_{\dot{e}}$ in (2.17) (in the considered example—ΔU) and the machine errors in the calculation of the gate functions (Table 2.3). However, studies show that with an increase in the degree of approximation polynomial of the functions f_1 and f_2, an insignificant gain in the accuracy of the results is attained due to a significant increase in the calculation time of the MCE. For an example from Table. 2.4, the transition to a quadratic approximation of the functions f_1 and f_2 reduces the average relative error in the determination of U_9 and U_{10} from 1.41% and 2.37%, respectively, to 1.38% and 2.29% due to an increase in the number of calculations by 34.6%. Therefore, from the practical point of view, the use of (2.10) and (2.11) seems appropriate. As for the error in determining f, it also depends on the error in approximating the experimental dependences obtained with SPICE.

In the example, this is the dependence of ΔU on the voltage in the ith node of the digital circuit at a specific value of DF. In practical programs, it is more expedient to have several models for the same dependence obtained by approximating the experimental data by polynomials of different degrees. In this case, it will also be possible to organize an adaptive simulation system [291, 293].

The MCE correctly simulates the behavior of digital cells in dynamic mode.

Consider the circuit in Fig. 1.104 for three values of the duration of the input signal t_i.

(a) $t_i > \Delta t_s^{Q_iQ_j}$ the first gate is excited at the moment of time $t_{e1} = t_0$, will switch to the moment$t_{b1} = t_{b1} + \Delta t_b^{Q_iQ_j}$, which will lead to the excitation of the second gate. i.e. $t_{e2} = t_{\acute{1}} 1 = t_0 + \Delta t\acute{i}^{\,Q_iQ_j}$. The state of the first gate is established at the time. Switching the second gate leads to the third wave excitation, and thus the signal will propagate along the gate chain.

(b) $t_s < \Delta t_b^{Q_iQ_j}$ in this case, since the second switching of the input signal will occur earlier than $t_{\min 1}$, the excitation of the first gate will be transferred to the quenching stage and after a while the transient processes will end. Thus, the MCE allows simulating such a digital cell feature as filtering of input signals, the duration of which is less than a certain value.

(c) $\Delta t_b^{Q_iQ_j} < t_u < \Delta t_s^{Q_iQ_j}$ in this case, the reverse switching of the gate will happen in a time $\Delta \tilde{t}_n^{Q_jQ_i} < \Delta t_n^{Q_jQ_i}$. If $t_1 + \Delta \tilde{t}_b^{Q_jQ_i} < t_0 + \Delta t_s^{Q_iQ_j}$, then there will be an incomplete operation of the gate, and if $t_0 + \Delta t_s^{Q_iQ_j} < t_1 + \Delta \tilde{t}_b^{Q_jQ_i} < t_0 + \Delta t_s^{Q_iQ_j} + \Delta t_b^{Q_jQ_i}$, then shortened pulse will be obtained at the output of the gate.

Consider the behavior of the MCE in the simulation of sequential circuits. Assume that the elementary FF cell (Fig. 2.38) is in a stable equilibrium state $Q5 = Qi$, $Q6 = Qj$ and at the input 1 at the time $t0$ the signal changes from Qj to Qi (Fig. 2.39). Then $G1$ is excited. The interruption of $G1$ will occur through the time $\Delta t_b^{Q_iQ_j}$ that will excite $G2$ (Fig. 2.40a). $G2$ is transferred at the time $t_0 + \Delta t_b^{Q_iQ_j} + \Delta t_b^{Q_jQ_i}$. The state at the outputs of $G1$ and $G2$ is set accordingly in $t_0 + \Delta t_s^{Q_iQ_j}$ and$t_0 + \Delta t_b^{Q_iQ_j} + \Delta t_s^{Q_jQ_i}$.

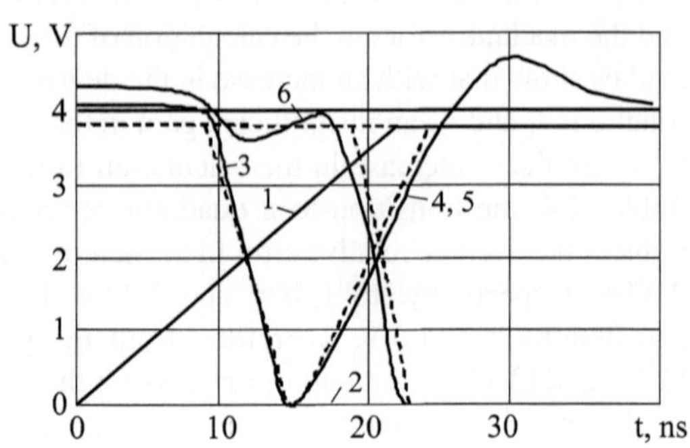

Fig. 2.39 The transient processes in the flip-flop of Fig. 2.38

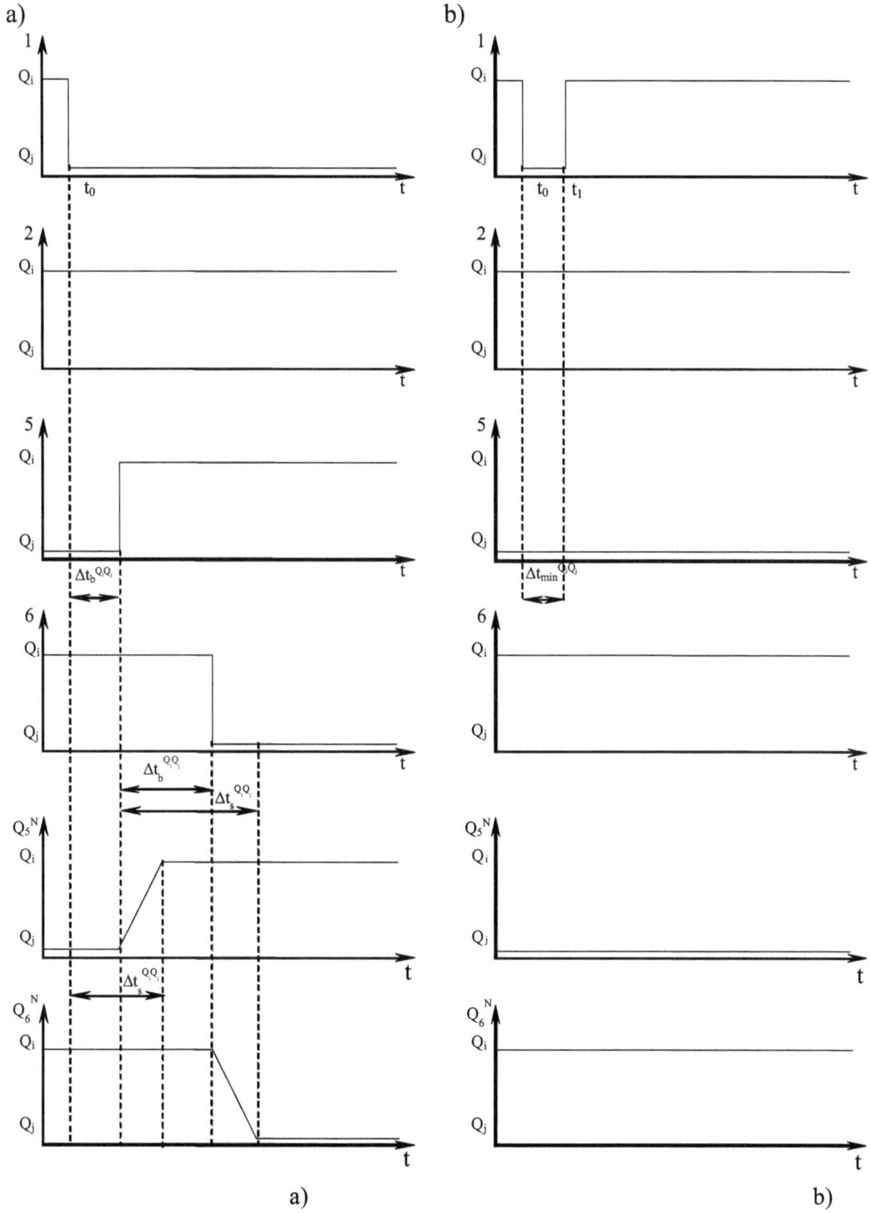

Fig. 2.40 The simulation results of FF cell from Fig. 2.38

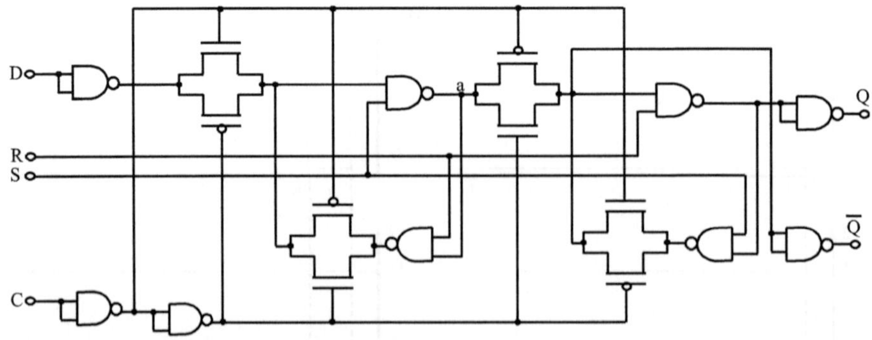

Fig. 2.41 Synchronous D FF circuit

If at a moment of time $t1$ the signal at input 1 switches back (from Qi to Qj), different results are obtained depending on the duration of the input pulse ($ti = t1 - t0$). When excitation $t_s < \Delta t_b^{Q_iQ_j}$ of $G1$ is transferred to the third stage, this element will not switch and the circuit will remain in the initial state ($Q5 = Qi$, $Q6 = Qj$) (Fig. 2.40b).

In case where $\Delta t_b^{Q_iQ_j} < t_u < \Delta t_s^{Q_iQ_j}$, the $G1$ switches back at the time $t_1 + \Delta \tilde{t}_b^{Q_jQ_i}$, and its output produces a shortened pulse. If the duration of the pulse at the output of $G1$ ($t_u = t_1 + \Delta \tilde{t}_b^{Q_jQ_i} - t_0 - \Delta t_b^{Q_iQ_j}$) is less, then $G2$ will not switch (Fig. 2.40c). If, however, tu at the output of $G1$ is greater than $\Delta t_b^{Q_jQ_i}$, but less than $\Delta t_s^{Q_jQ_i}$, then at the output of $G2$ an incomplete operation will result (Fig. 2.40d).

To illustrate the accuracy of the model, Fig. 2.39 shows the transient processes in the flip-flop (Fig. 2.38) performed on the TTL basis (Fig. 1.13) calculated with SPICE at the circuit level (solid lines) and the MCE (dashed lines) with the recalculation of Qout in Uout by (2.11).

Relative accuracy of signals reproduction with the help of MCE in comparison with the results of SPICE is 6.84%, and the accuracy of determination is t_{f5}^{01}—4.74% (Fig. 2.41).

In order to illustrate the capabilities of the MCE in a dynamic mode with consideration of DF, Fig. 2.42 shows the results of simulating a synchronous D FF (Fig. 2.41) at a temperature of 125 °C and with standard input signals (the duration of which exceeds the delay of the used gate from Fig. 1.15) with the help of SPICE—U_q^1, VCS [294]—U_q^2 and MCE—U_q^3. U_q^3 differs from U_q^1 by only 4.38%, while the relative error in determining U_q^2 exceeds 46%. The reason is that VCS does not have the capability to consider the effect of temperature.

Figure 2.43 shows the simulation results of the same circuit (Fig. 2.41) using the mentioned tools. From Fig. 2.43 it is clear that if the VCS calculates with an error of 64.7% at switching time of the output of the FF $\left(U_q^2\right)$ and produces a qualitatively incorrect result for U_a^2, then the MCE allows obtaining qualitatively accurate simulation results with a time error not exceeding 3.72%. Thus, the developed MCE correctly simulates the behavior of the digital cell with standard and nonstandard input signals, not only in static but also in dynamic modes.

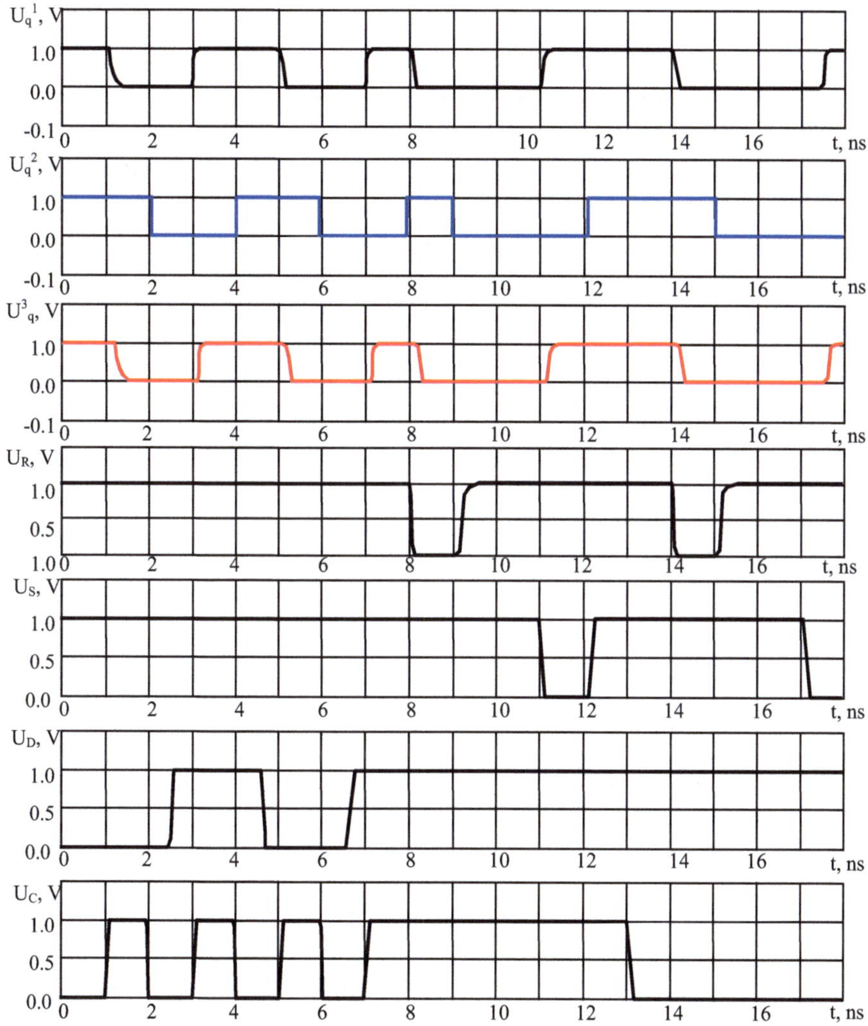

Fig. 2.42 Simulation results of a synchronous D FF at $T^0 = 125\,°\mathrm{C}$: (**a**) U_q^1—SPICE; (**b**) U_q^2—VCS; (**c**) U_q^3—ELAIS-L (using MCE)

The accuracy of calculations with the help of MCE in the dynamic mode is mainly affected by errors in the approximation of the dependencies (2.21)–(2.24), (2.33), Qbm(t), etc., errors in determining timing parameters of the MCE $\Delta t_{\min}^{Q_iQ_j}$, $\Delta t_{\mathrm{b}}^{Q_iQ_j}$, $\Delta t_{\max}^{Q_iQ_j}$, $\Delta t_{\mathrm{s}}^{Q_iQ_j}$, as well as errors in the approximation of the dependencies of timing parameters of the MCE from various DFs given in Chap. 3. However, the illustrated research shows the expediency of approximation of the dependences (2.21)–(2.24), (2.33), Qbm(t), etc. by linear function. In particular, Fig. 2.44 shows the obtained dependences of the relative error of calculations (εi) and

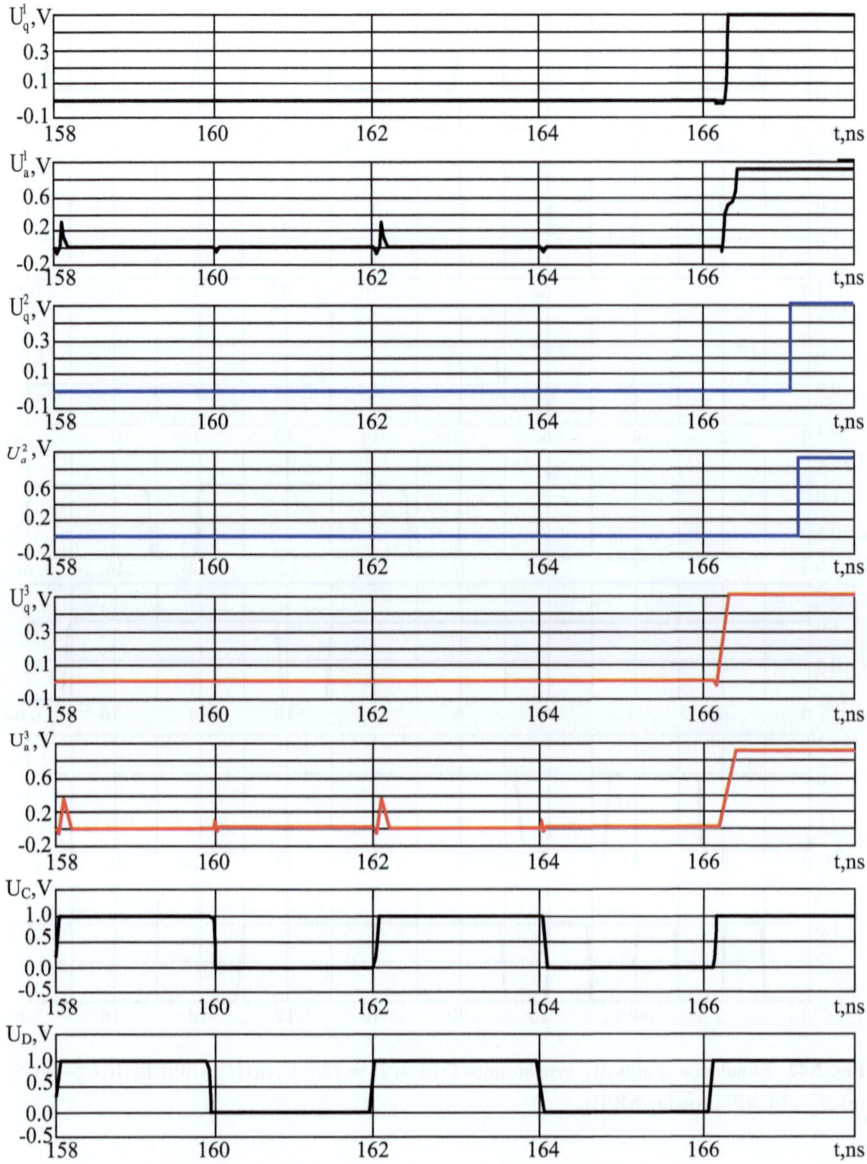

Fig. 2.43 Simulation results of a synchronous D FF at times close to the installation time: (**a**) U_q^1, U_a^1—SPICE; (**b**) U_q^2, U_a^2—VCS; (**c**) U_q^3, U_a^3—ELAIS-L (using MCE)

computer time (*t*) (Pentium IV, clock frequency—2.2 GHz) on the degree of the approximate polynomial *Q*int(*t*) when obtaining the results in Fig. 2.43. The dependencies in Fig. 2.44 clearly confirm the advisability of approximating *Q*int(*t*) by a linear function.

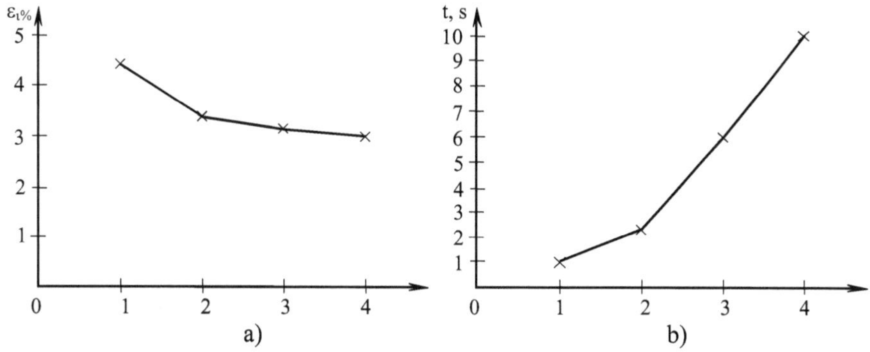

Fig. 2.44 Dependences of the relative error of calculations ε_i with the help of MCE and the expenditure of machine time (t) on the degree of approximating polynomial

The error in determining the timing parameters $\Delta t_{\text{min}}^{Q_iQ_j}$, $\Delta t_{\text{b}}^{Q_iQ_j}$, $\Delta t_{\text{max}}^{Q_iQ_j}$, $\Delta t_{\text{s}}^{Q_iQ_j}$ is regulated by applying the methodology, described in Sect. 2.3 (2.73)–(2.76). As for the error in approximating the dependencies of timing parameters of the MCE from different DFs given in Chap. 3, for the reasons for adaptive simulation [291–293] it is more appropriate to have a library of different versions of the same dependence having different accuracy and computational complexity.

2. Differentiated accuracy of the reflection of various static states and switching stages of the gate, considering their significance from the point of view of affecting other nodes of digital circuits is achieved by an accurate determination of switching times of a digital circuit (points 1–5 in Fig. 2.20) and linear approximation of the durations of the remaining switching stages of a digital circuit (fragments 1–2, 2–3, etc. in Fig. 2.20), as well as the use of different models for determining the MCE parameters, depending on the value of DF, which having different accuracy of calculations.
3. The MCE is quite universal from the point of view of all, or at least most types of digital circuits, regardless of their configuration, etc. The performed calculations of different digital circuits and the same digital circuits on different circuit-levels (Figs. 1.13–1.15) confirm this property of the MCE.
4. The MCE is compatible with models of circuit-level and provides the ability to implement mixed-mode simulation in addition to purely logical one. This is facilitated by the absence of jumplike switching at the output of the MCE, the ability to give to the input of the MCE and to receive signals from the output with nonstandard levels and timing parameters.
5. A one-to-one correspondence of its timing parameters with the standard parameters of modern digital circuits is foreseen in the MCE (Sect. 2.3).
6. In the MCE, the amount of calculations and the number of parameters are not much different from traditional analogues of the logical level due to the addition of new features.

Table 2.6 Estimation of the increase in the amount of computation with the help of MCE

Circuit name	Number of inputs	Number of outputs	Number of FFs	Number of logic gates	Simulation time (s) VCS	ELAIS-L	Increase of calculation amount (%)
s27	4	1	3	10	0.36	0.37	3.4
s208	11	2	8	96	1.08	1.12	3.8
s298	3	6	14	119	0.72	0.76	4.3
s344	9	11	15	160	0.72	0.76	4.9
s349	9	11	15	161	15.6	16.4	5.4
s382	3	6	21	158	1	1.1	5.8
s386	7	7	6	159	1.7	1.8	6.6
s400	3	6	21	162	19.2	20.6	7.5
s420	19	2	16	196	22.8	24.7	8.2
s444	3	6	21	181	30	32.6	8.8
s510	19	7	6	211	1.3	1.4	9.2
s526	3	6	21	193	20.4	22.3	9.5
s641	35	24	19	379	30	32.9	9.6
s713	35	23	19	393	42	46	9.8
s820	18	19	5	289	7.9	8.7	10
s832	18	19	5	287	39.6	43.6	10.1
s838	35	2	32	390	58	64	10.3
s953	16	23	29	395	20.4	22.5	10.5
s1196	14	14	18	529	44.4	49.1	10.7
s1238	14	14	18	508	72	79.8	10.9
s1423	17	5	74	657	61	67.8	11.1
s1488	8	19	6	653	7.4	8.2	11.2
s1494	8	19	6	647	64	71	11.4
s5378	35	49	179	2779	234	261	11.6
s13207	19	22	228	5597	794	888	11.8
s13207	31	121	669	7951	790	884	11.9
s15850	14	87	597	9772	1086	1216	12

This is confirmed by the results of a comparison (Table 2.6) of the simulation time of a series of test circuits from the ISDCS89 series using the developed tool ELAIS-L and VCS [294] for gate-level simulation and optimization of digital circuits for the characteristic transformations of signals of each digital circuit (Processor—Pentium IV, clock frequency—2.2 GHz). The increase in the amount of computations with the help of the MCE in comparison with traditional logical models on the considered series of circuits is approximately 8.9%. If one considers the fact of ensuring the accuracy of calculation results, inferior to SPICE by only 5.8%, then the goal of constructing the MCE is reached.

Chapter 3
Models for Determining the Influence of DF

3.1 Methodology for Constructing Models for Determining the Influence of DF

The adequacy of gate-level simulation of digital circuits with consideration of DF to a large extent depends on the accuracy of the model of determining DF effects (MDE) on the parameters of digital cells and the functioning of digital circuits, which must meet the requirements formulated in Sect. 1.3. The MDE is understood as the ME and PDE (Sect. 1.4). Depending on the DF (Table 1.1), the parameters of digital cells vary over a wide range (sometimes several times (Sect. 1.1)). Therefore, in order to increase the reliability of simulation results, it is necessary to take into account the dependencies of the MCE parameters on the DF. These dependencies are generally complex, nonlinear in nature and are determined mainly by the circuit-level basis of gates (TTL, ECL, CMOS, etc.), as well as specific circuit-level solutions used in the construction of gates. Determining the dependencies of the MCE parameters on the DF by experiment or by simulating it on a circuit-level is a difficult task, since a large number of experiments or many variants of computer calculations are required to remove the mentioned dependencies. Therefore, it is necessary to develop a methodology for obtaining an MDE with a minimal amount of computer time. The following ways of building MDE are possible:

1. Analytical use of the formulas obtained in fundamental research on basic logical cells [1, 2, 33, 35, 70, 72, 78, 80] and various DF [32, 38, 41–58, 86–117, 136–146]. With the help of the dependencies given in the noted works, it is possible to determine the parameters of the MCE with specific values of DF quite accurately. The disadvantage of this approach is that there are many parameters in the formulas for determining gate-level parameters given in the noted works, the method of determining which is too difficult or unknown. This requires a large number of experimental measurements. For example, in work on interconnects [7, 8, 22–26, 86–99] analytical dependencies of timing parameters of interconnects on their length are given. With their help, distortions of a useful signal due

V. Melikyan, *Simulation and Optimization of Digital Circuits*,
https://doi.org/10.1007/978-3-319-71637-4_3

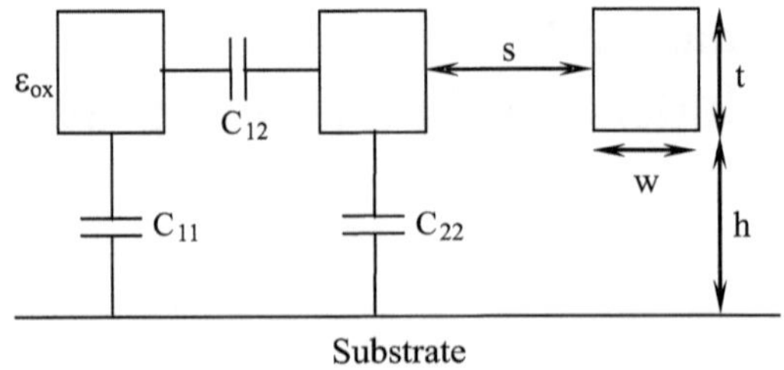

Fig. 3.1 The model for determining the specific capacitance

to the parasitic influence of interconnects introduced by connection lines depending on their length are determined quite accurately. For example, in [99] a complex analytical model of the voltage variation at the RC, interconnect output is used:

$$U(t) = \frac{V_0}{T_{\mathrm{R}}} \cdot t + \left[\frac{b^2}{2} \mathrm{erfc}\left(\frac{b^2}{\sqrt{4t}} \right) - b\sqrt{t\,\mathrm{e}^{-\frac{b^2}{4t\pi}}} \right.$$
$$\left. + (2qb + 2q^2)\mathrm{erfc}\left(\frac{b}{\sqrt{4t}} \right) - 2q^2\,\mathrm{e}^{\frac{t+qb}{q^2}}\mathrm{erfc}\left(\frac{\sqrt{t}}{q} + \frac{b}{2\sqrt{t}} \right) \right] U(t), \tag{3.1}$$

where $b = \frac{2h - l_i}{h\sqrt{R_0 C_0}}$; $q = \frac{C_{\mathrm{L}}}{C_0}\sqrt{R_0 C_0}$; $T_{\mathrm{R}} = R_0 C_0$; l_i—length; h—interconnect width; R_0 and C_0 specific resistance and interconnect capacitance; C_{L}—interconnect load capacitance. In order to determine the specific values of interconnect parameters (for example, capacitance), in turn, complex analytical expressions are used that relate their quantities to the geometric dimensions of interconnects. The model for determining the specific capacitance is presented in the form shown in Fig. 3.1.

The formulas for determining the components of capacitances are as follows:

$$\frac{C_{11}}{\varepsilon_{\mathrm{ox}}} = 1.12\frac{w}{h} + \left[0.75\left(\frac{w}{h}\right)^{0.11} + 0.68\left(\frac{t}{h}\right)^{0.49} \right]$$
$$+ \left[-0.039\left(\tfrac{w}{h}\right)^{0.092} + 0.83\left(\tfrac{t}{h}\right)^{0.028} \right] \cdot \left(1 - \mathrm{e}^{-\frac{s}{h}}\right), \tag{3.2}$$

$$\frac{C_{22}}{\varepsilon_{\mathrm{ox}}} = \frac{w}{h} + 2.24\left(\frac{w}{h}\right)^{0.0275}\left(1 - 0.85\,\mathrm{e}^{-0.62\frac{s}{h}}\right)$$
$$+ 0.32\ln\left(\frac{t}{s}\right) \cdot \left(0.15\frac{s}{h} \cdot \mathrm{e}^{-1.62\frac{t}{s}} - 0.12\,\mathrm{e}^{-0.065\frac{s}{t}}\right), \tag{3.3}$$

$$\frac{C_{12}}{\varepsilon_{\mathrm{ox}}} = \frac{t}{s} + 1.31\left(\frac{t}{h}\right)^{0.073}\left(\frac{s}{h} + 1.12\right)^{-0.81}, \tag{3.4}$$

where w, t, s, h—geometric dimensions of interconnect; $\varepsilon_{\mathrm{ox}}$ is the dielectric permittivity.

In [295], in order to improve the accuracy of calculations, the influence of the skin effect on the transient processes in interconnects was considered. In case of a mismatched line, for example, the following expressions are derived for the determination of the rise and fall and the delay of the signal:

$$t_{\mathrm{r.f.}} \approx 4.2T_1N\left(1 - \mathrm{e}^{-|k|^2/\left(1-|k|^2\right)}\right), \tag{3.5}$$

$$t_{\mathrm{d}} \approx 1.38T_1N\left(1 - \mathrm{e}^{-|k|^2/\left(1-|k|^2\right)}\right), \tag{3.6}$$

where T_1 is the propagation delay per unit length of the interconnect without taking into account the additional delay; N—multiplicity; k—the reflection coefficient.

However, in the analytical formulas for determining the interconnect delays, given in the mentioned papers, there are many parameters, the methodology of determining of which is too difficult or unknown. For example, in the above model, such parameters are N and k. In addition, the accuracy of the marked models (and, correspondingly, the expenditure of computer time) is much higher than necessary for the case of gate-level simulation with consideration of DF. In fact, the requirement for the MDE, formulated in Sect. 1.3, is not fulfilled in accordance with for the accuracy of simulation to the current character of the dependence of gate parameters on the DF value. Even such an accurate asynchronous gate model, as developed by the MCU (Sect. 2.2), yields a rather rough form of output signal in comparison with the real one, and in these conditions, the excessive accuracy of interconnect model is not justified.

2. Empirical—obtaining experimental dependences of the MCE parameters on the DF and their approximation using some known method, for example, interpolation spline functions [296]. The advantage of this approach is the controllability of the accuracy of determining the MCE parameters with the help of the MDE by changing the order of the interpolation functions. However, this approach also has a significant drawback—it requires a large number of experiments or many variants of gate simulation at the circuit level to determine the parameters of approximation of dependencies.
3. Semiempirical—combination of two previous methods: the form of the dependences of the MCE parameters on the DF is determined by analytical relations, and the approximation parameters by means of computer calculations. As will be seen in what follows, with such a semiempirical approach, as a rule, the number of experiments necessary to ensure sufficient accuracy of approximation is much less than for a purely analytical or purely empirical approach.

As an example, consider how to implement a semiempirical approach for digital circuits, performed in TTL (Fig. 1.13) basis.

On the basis of the expressions (2.79) and (2.80) relating the MCE parameters with the reference parameters, as well as using the formulas for determining the reference timing parameters of basic TTL gate (Fig. 1.13), given in [77], the following is obtained:

$$\Delta t_{\min}^{Q_iQ_j} = t_{\mathrm{com.en}}, \tag{3.7}$$

$$\Delta t_b^{Q_iQ_j} = t_{\mathrm{com.en}} - t_{\mathrm{sat}} + t_{\mathrm{res}} + 0.1 \cdot \tau_{\mathrm{en.eq}}, \tag{3.8}$$

$$\Delta t_{\max}^{Q_i Q_j} = t_{\text{com.en}} - t_{\text{sat}} + t_{\text{res}} + 0.15 \cdot \tau_{\text{en.eq}}, \tag{3.9}$$

$$\Delta t_{\text{s}}^{Q_i Q_j} = t_{\text{com.en}} - t_{\text{sat}} + t_{\text{res}} + 0.275 \cdot \tau_{\text{en.eq}}, \tag{3.10}$$

$$\Delta t_{\min}^{Q_j Q_i} = t_{\text{d.en}} + \tau_{\text{in.en}}, \tag{3.11}$$

$$\Delta t_{\text{b}}^{Q_j Q_i} = t_{\text{d.en}} + \tau_{\text{in.en}} + t_{\text{sat}} + t_{\text{d.en4}} + 0.1 \cdot \tau_{\text{en.eq}}, \tag{3.12}$$

$$\Delta t_{\max}^{Q_j Q_i} = t_{\text{d.en}} + \tau_{\text{in.en}} + t_{\text{sat}} + t_{\text{d.en4}}, \tag{3.13}$$

$$\Delta t_{s}^{Q_j Q_i} = t_{\text{in.en}} + \tau_{\text{d.en}} + t_{\text{d.en4}} - 0.1 \cdot \tau_{\text{en.eq}} + 0.125 \cdot (t_{\text{com.en}} + t_{\text{res}} + t_{\text{sat}}), \tag{3.14}$$

where $t_{\text{d.en}}$—enable delay of transistor T3; $\tau_{\text{in.en}}$—enable time constant of transistor T3; t_{sat}—time required to saturate the transistor T3; $t_{\text{d.en4}}$—enable delay of transistor T4; $\tau_{\text{en.eq}}$—equivalent value of the cut-off time constant; $t_{\text{com.en}}$—time required to reduce the voltage at the output of the gate to a level sufficient for enabling the emitter junctions of T1 and T2 load cells; t_{res}—the time of resolution of carriers from the base of transistors of T1 and T2 loading gate when enabling their emitter junctions; t_{sat}—saturation time of T1 and T2 loading gates.

In [77] the following expressions are given for determination of separate components of timing parameters included in (3.7)–(3.14):

$$t_{\text{d.en}} = R_1 \left[c_1 + c_2 + c_{\text{e3}} + \frac{c_{\text{c3}} \cdot (c_{\text{c3}} + c_n)}{c_{\text{c3}} + c_{\text{c4}} + c_n} \right] \cdot \ln \left(1 + \frac{U_{\text{nom}}^0}{E - U_{\text{cb.t1}} - U_{\text{en.t}}} \right), \tag{3.15}$$

$$\tau_{\text{in}} = \frac{c_1 + c_2}{\frac{1}{R_1} + \frac{\beta_{N4}}{R_2} + \frac{1}{r_{\text{in.e}}}}, \tag{3.16}$$

$$t_{\text{sat}} = \sqrt{0.1 \cdot \left(U_{\text{out}}^1 - U_{\text{out}}^0 \right) \cdot \frac{\tau_{\text{TN3}} \cdot (c_{\text{ce}} + c_{\text{c3}} + c_{\text{b}})}{I_{\text{bs}}}}, \tag{3.17}$$

$$t_{d.en} = \sqrt{\frac{2 \cdot U_{\text{en}} \cdot R_{\text{S}}}{I_{\text{bs}}}} \cdot (c_{\text{c3}} + c_{\text{e4}})^2 \cdot c_{\text{ce}}, \tag{3.18}$$

$$\tau_{\text{en.ec}} = \sqrt{\tau_{\text{en3}}^2 + \frac{R_2^2}{\beta_{N4}^2} \cdot \left[c_{\text{ce}} + c_{\text{b}} + \frac{c_{\text{c3}} \cdot (c_{\text{e3}} + c_3)}{c_{\text{c3}} + c_{\text{e3}} + c_3} \right]^2}, \tag{3.19}$$

$$t_{\text{com.en}} = \frac{\beta_{N4} \cdot \left(U_{\text{out}}^1 - E + I_{\text{bs}} \cdot R_1 + U_{\text{en.t1}} \right)}{3 \cdot R_2 \cdot I_{\text{bs}}} \times \left(1 + \sqrt{1 + \frac{A_1 \cdot (c_{\text{c4}} + c_{\text{c3}} + c_{\text{b}})}{(A_2 - E) \cdot (A_3 + c_{\text{c4}} + c_{\text{c3}} + c_{\text{b}})}} \right), \tag{3.20}$$

$$t_{\text{res}} = \tau_{\text{b3}} \cdot \frac{1 - \frac{I_{\text{cs.min}}}{\beta_{N3} \cdot I_{\text{bs}}}}{\frac{\beta_{N3}}{n_{\text{T}}} \cdot \left(1 - \mathrm{e}^{\frac{-t_{\text{d.in}}}{\tau_{\beta N}}} \right) - \frac{1}{2} \cdot \left[[1 - \frac{\tau_{b3}}{\tau_{\beta N3} \cdot \left(\mathrm{e}^{\frac{-t_{\text{d.in}}}{\tau_{\beta N}}} - 1 \right)} \right] \cdot \left(1 - \frac{I_{\text{cs.min}}}{\beta_{N3} \cdot I_{\text{bs}}} \right)}, \tag{3.21}$$

where

$$I_{\text{cs.min}} = \frac{E - U_{\text{ces}}}{N \cdot R_2} + n_{\text{T}} \cdot \frac{E - U_{\text{bs.t1}} - U_{\text{ces}}}{N \cdot R_1},$$

$$t_{\text{sat}} = \left(U_{\text{bs.i}} - U_{\text{ces.t1}} - 0.1 \cdot U_{\text{out}}^1 - 0.9 \cdot U_{\text{out}}^0\right) \cdot \frac{c_{\text{c3}} + \frac{c_{\text{ce}} + c_{\text{b}} + n_{\text{T}} \cdot c_1}{\beta_{N2}}}{I_{bs} \cdot \left(1 - n_{\text{T}} + \frac{R_1 \cdot \beta_{N3}}{R_2 \cdot \beta_{N2}}\right)}, \quad (3.22)$$

The following notations are used in the expressions (3.15)–(3.22): C_{ci} C_{ei}—capacitance of the collector and emitter junctions of the *i*th transistor; C_{ce}—capacitance of the collector–emitter of the transistor; β_{Ni}—the current transfer coefficient of the base of the *i*th transistor; $\tau_{\text{TN}i}$—the mean transit time of carriers in the base of the *i*th transistor; $\tau_{\text{en}i}$—the enable time constant; τ_{ai}—the accumulation time constant of the *i*th transistor; $\tau_{\beta N}$—the time constant of β_N ; E—the voltage of the power source; I_{bs}—the base current of the transistor when operating in saturation mode; R_i—the value of the *i*th resistance of the circuit; r_{ine}—input resistance of the transistor; $t_{\text{d.in}}$—the delay time for switching of emitter junctions T1 and T2; U_{out}^1, U_{out}^0 —output voltages in the states "1" and "0"; $U_{\text{en.t}}$—enable voltage of the transistor; U_{bst1}, U_{cst1}—base-collector voltage of saturated T1; $U_{\text{cb.t1}}$—a difference of potentials of collector-base T1; U_{ces}—collector–emitter voltage of a saturated transistor; $U_{\text{bs.i}}$—potential of base of saturated transistor; n_{T}—number of gate inputs; c_1 and c_2—parasitic capacitances: c_1 includes the capacitances of substrates T1 and T2, the resistances R_1 and emitters T_1 and T_2, and c_2 is the capacitance of substrate T3; U_{nom}^0, U_{nom}^1 —noise immunity of "0" and "1"; I_{bs}—the base current of the transistor in saturation mode.

Expressions (3.7)–(3.22) allow determining the analytical dependences of timing parameters of the MCE on certain DFs with a high degree of accuracy: the load capacitance C_{L}, the supply voltage E, etc. However, this requires knowledge of about 30 parameters, which makes their usage difficult. Of course, these values can be measured, but the number of measurements can become commensurable or exceed the number of experiments to determine the required dependencies. Therefore, in order to reduce the laboriousness of the process of obtaining an MDE, real curves can be approximated by simpler functions. This is done on the example of the MDE $\Delta t_{\text{b}}^{10}(C_{\text{L}})$. According to (3.12), (3.15)–(3.22), this dependence has the following form:

$$\Delta t_{\text{b}}^{10} = A_0 + A_1 \cdot \frac{A_2 + c_{\text{L}}}{A_3 + c_{\text{L}}} + A_4 \cdot \sqrt{A_5 + c_{\text{L}}} - 0.1 \cdot \sqrt{\frac{A_5 \cdot A_1^2}{c_{\text{L}}} + (A_6 + c_{\text{L}})^2 \cdot A_7}. \quad (3.23)$$

The expression (3.23) represents an MDE $\Delta t_{\text{b}}^{10}(C_{\text{L}})$, the parameters of which are the coefficients A_0–A_7. If the parameters of expressions (3.15)–(3.22) are known, then these coefficients can be determined through them:

$$A_0 = R_1 \cdot (c_1 + c_2 + c_{e3}) \cdot \ln\left[1 + \frac{U^0_{nom}}{E - U_{cb.t1} - U_{en.t}}\right] + \frac{c_1 + c_2}{\frac{1}{R_1} + \frac{\beta_{N4}}{R_2} + \frac{1}{r_{in.e}}} + \sqrt{\frac{2 \cdot U_{en.t} \cdot R_3}{I_{bs}} \cdot (c_{c3} + c_{e4}) \cdot c_{ce}}, \quad (3.24)$$

$$A_1 = R_1 \cdot c_{\hat{e}2} \cdot \ln\left(1 + \frac{U^0_{nom}}{E - U_{cb.t1} - U_{en.t}}\right), \quad (3.25)$$

$$A_2 = c_{c3}, \quad (3.26)$$

$$A_3 = c_{c2} + c_{c3}, \quad (3.27)$$

$$A_4 = \sqrt{0.1 \cdot (U^1_{out} - U^0_{out}) \cdot \frac{\tau_{TN3}}{I_{bs}}}, \quad (3.28)$$

$$A_5 = \frac{\tau_{en3}}{R_2}, \quad (3.29)$$

$$A_6 = c_{c4} + \frac{c_{c3} \cdot (c_{e3} + c_2)}{c_{c3} + c_{e3} + c_2}, \quad (3.30)$$

$$A_7 = \frac{R_2^2}{\beta^2_{N4}}. \quad (3.31)$$

As already noted, the determination of coefficients in expressions (3.24)–(3.31) is difficult because of the large number of nonstandard parameters in them. Since the number of coefficients is 8, then, if Δt^{10}_bat 8 values of C_L are known, $\Delta t^{10}_b(C_L)$MDE MDE can be obtained. To do this, experimentally or by gate simulation at the circuit level it will be necessary to determine Δt^{10}_b for 8 values of C_L.

To make it easy, it is possible to make a piecewise approximation of the dependence $\Delta t^{10}_b(C_L)$. If $C_L \ll C_{ce} + C_{cd}$, then expression (3.23) can be expanded in a series and confined to the second term of this series:

$$\Delta t^{10}_b = A_0 + A_1 \cdot C_L. \quad (3.32)$$

In this case, the absolute error does not exceed the value of the third term of the series, which is insignificant for practical circuits (Fig. 1.13). If, however, these relations take place:

$$C_L \gg c_{c3} + c_{c4}, \quad C_L \gg \tau_{en3}, \quad (3.33)$$

then the dependence $\Delta t^{10}_b(c_L)$can be approximated by the following function:

$$\Delta t^{10}_b = A_0 + A_1 \cdot \sqrt{c_L} + A_2 \cdot c_L, \quad (3.34)$$

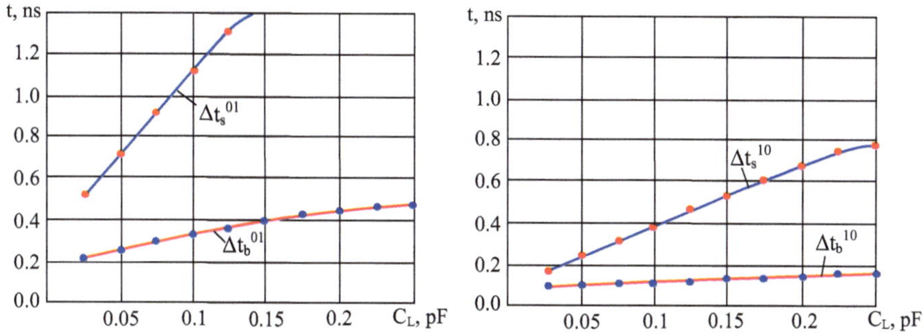

Fig. 3.2 Dependencies of model parameters on load capacitance

i.e. in the first case, two, and in the second case, three experiments are required to obtain the MDE $\Delta t_{\mathrm{b}}^{10}(c_{\mathrm{L}})$. For the circuit in Fig. 1.13 the approximation obtained as a result of computer experiment of the dependence $\Delta t_{\mathrm{b}}^{10}(c_{\mathrm{L}})$given above is produced, and the following MDE is obtained:

$$\Delta t_{\mathrm{b}}^{10} = \begin{cases} 0.17 + 1.6 \cdot c_{\mathrm{L}} & \text{when } c_{\mathrm{L}} \le 0.1\,\mathrm{pF}, \\ -0.2251 + 2.386 \cdot \sqrt{c_{\mathrm{L}}} - 1.99 \cdot c_{\mathrm{L}} & \text{when } c_{\mathrm{L}} > 0.1\,\mathrm{pF}. \end{cases} \tag{3.35}$$

Relative error (ε), defined in the form:

$$\varepsilon = \frac{\Delta t_{\mathrm{ne}}^{10} - \Delta t_{\mathrm{nm}}}{\Delta t_{\mathrm{ne}}^{10}} \cdot 100\%, \tag{3.36}$$

does not exceed 3% (Fig. 3.2) in the range of changes c_{L} from 0 to 0.25 pF. Here $\Delta t_{\mathrm{ne}}^{10}$ is the value of $\Delta t_{\mathrm{b}}^{10}$ obtained by simulating at the device level; Δt_{nm}—with the help of the expression (3.35).

The analysis of expressions (3.15)–(3.22) similarly shows that the dependences $\Delta t_{\mathrm{c}}^{10}(c_{\mathrm{L}})$, $\Delta t_{\mathrm{b}}^{01}(c_{\mathrm{L}})$, $\Delta t_{\mathrm{s}}^{01}(c_{\mathrm{L}})$ can be approximated by the following functions:

$$\Delta t_{\mathrm{b}}^{10} = \begin{cases} A_0 + A_1 \cdot c_i & \text{at } c_{\mathrm{L}} \ll c_{c3} + c_{c4} \text{ and } c_{\mathrm{L}} \ll \tau_{\mathrm{en}3} \\ A_0 + A_1 \cdot \sqrt{c_{\mathrm{L}}} \end{cases}, \tag{3.37}$$

in other cases;

$$\Delta t_{\mathrm{b}}^{01} = A_0 + A_1 \cdot c_{\mathrm{L}}, \tag{3.38}$$

$$\Delta t_{\mathrm{s}}^{01} = A_0 + A_1 \cdot c_{\mathrm{L}}. \tag{3.39}$$

For the circuit in Fig. 1.13 the coefficients of approximation are defined:

$$\Delta t_{\mathrm{s}}^{10} = \begin{cases} 0.316667 + 8.13333 \cdot c_{\mathrm{L}} & \text{when } c_{\mathrm{L}} \le 0.1\,\mathrm{pF}, \\ -15.8668 - 94.3284 \cdot \sqrt{c_i} - 128.792 \cdot c_{\mathrm{L}} & \text{when } c_{\mathrm{L}} > 0.1\,\mathrm{pF}, \end{cases} \quad (3.40)$$

$$\Delta t_{\mathrm{b}}^{10} = 0.0825 + 0.35 \cdot c_{\mathrm{L}}, \quad (3.41)$$

$$\Delta t_{\mathrm{s}}^{10} = -0.01333 + 3.46667 \cdot c_{\mathrm{L}}. \quad (3.42)$$

As the comparison of the approximation results with the curves obtained by simulating at the device level shows (Fig. 3.2), the relative error for $\Delta t_{\mathrm{s}}^{10}$ does not exceed 6%, for $\Delta t_{\mathrm{b}}^{01}$—2%, for $\Delta t_{\mathrm{s}}^{01}$—3%

Thus, with the help of a semiempirical technique, it is possible to construct various MDEs, meeting the requirements formulated in Sect. 1.3.

Below are the developed MDE of the total dose of radiation D, the ambient temperature T, the supply voltage E, the rate of change of the input signal S, the duration of the input pulse t_{in}, the amplitude of the input signal U_{min}, the number of loads N for TTL (1.13), ECL (1.14), and CMOS (1.15) of standard basic logic cells.

3.1.1 *The Basic Logic Cell TTL (Fig. 1.13; Tables 3.1, 3.2, 3.3, 3.4, 3.5, 3.6, 3.7)*

Table 3.1 MDE of the total radiation dose D[a]

MCE parameter	MDE	Relative error (%)
$\Delta t_{\mathrm{min}}^{01}$	$0.0001D^2 + 0.0006D + 0.061$	6.48
$\Delta t_{\mathrm{b}}^{01}$	$0.0004D^2 - 0.0021D + 0.51875$ at $D \le 60$ $0.00053D^2 - 0.02453D + 2.125$ at $D > 60$	4.42
$\Delta t_{\mathrm{max}}^{01}$	$0.0001D^2 + 0.0089D + 0.41875$	0.82
$\Delta t_{\mathrm{s}}^{01}$	$0.00008D^2 + 0.0067D + 0.08$ at $D \le 40$ $-0.00031D + 0.5$ at $40 < D \le 60$ $0.00027D^2 + 0.02297D + 0.875$ at $D > 60$	4.87
$\Delta t_{\mathrm{min}}^{10}$	$0.0008D^2 - 0.0768D - 0.2375$	3.95
$\Delta t_{\mathrm{b}}^{10}$	$-0.01D - 0.0375$ at $D \le 20$ $0.0018D^2 - 0.1881D + 2.8086$ at $D > 20$	6.46
$\Delta t_{\mathrm{max}}^{10}$	$0.00042D^2 + 0.0083D$ at $D \le 60$ $0.1D - 4$ at $D > 60$	3.82
$\Delta t_{\mathrm{s}}^{10}$	$0.015D + 0.3375$ at $D \le 20$ $0.1122D - 2.125$ at $D > 20$	5.45
f_1^0	$-0.0003D^2 + 0.0028D + 1$ at $D \le 45$ $-0.00028D^2 + 0.0456D - 1.041$ at $D > 45$	6.17
f_1^1	$-0.002D^2 + 0.077D + 1$ at $D \le 45$ $0.00016D^2 - 0.027D + 1.3455$ at $D > 45$	6.42

[a]In functions $\Delta t = f(D)$ Δt is measured in ns, D in ns/krad, in $f_1 = f(D)$ f_1 is a dimensionless value, D is measured in 1/krad. Range of change D = 0–100 krad

Table 3.2 MDE of ambient temperature T

MCE parameter	MDE	Relative error (%)
Δt_{min}^{01}	$0.00008T + 0.0698$ at $T \leq 60$ $-0.000029T^2 + 0.00373T - 0.05375$ at $T > 60$	2.45
Δt_{b}^{01}	$-0.00008T^2 - 0.00077T + 0.206$	5.42
Δt_{max}^{01}	$-0.00015T^2 - 0.00023T + 0.9514$	1.89
Δt_{s}^{01}	$-0.00002T^2 - 0.00023T + 1.0514$	1.81
Δt_{min}^{10}	$-0.000006T^2 - 0.00052T + 0.0255$	6.28
Δt_{b}^{10}	$-0.00048T + 0.039$ at $T \leq 40$ $0.00003T^2 + 0.0052T$+0.1925 at $40 < T \leq 80$ $-0.00184T + 0.1375$ at $T > 80$	5.37
Δt_{max}^{10}	$-0.00009T^2 - 0.00077T + 0.3969$ at $T \leq 40$ $-0.00297T + 0.47$ at $T > 40$	3.31
Δt_{s}^{10}	$-0.00009T^2 + 0.001T + 0.4569$ at $T \leq 60$ $-0.003T + 0.525$ at $T > 60$	0.91
f_1^0	$0.000017^2 - 0.003T + 0.055$	3.37
f_1^1	$-0.000037^2 + 0.0028T + 0.955$	2.82

Table 3.3 MDE of supply voltage E

MCE parameter	MDE	Relative error (%)
Δt_{min}^{01}	$0.01361E^2 - 0.26E + 1.865$	6.50
Δt_{b}^{01}	$0.0053E^2 - 0.11E + 0.515$	3.12
Δt_{max}^{01}	$0.0053E^2 - 0.11E + 0.455$	6.08
Δt_{s}^{01}	$0.01361E^2 - 0.26E + 1.965$	3.51
Δt_{min}^{10}	$0.01016E^2 - 0.1219E + 0.2905$ at $E \leq 7$ $-0.015E + 0.04$ at $E > 7$	6.38
Δt_{b}^{10}	$0.0055E^2 - 0.75E + 0.232$ at $E \leq 7$ $-0.015E + 0.08$ at $E > 7$	6.29
Δt_{max}^{10}	$0.0105E^2 - 0.234E + 1.12125$	3.81
Δt_{s}^{10}	$0.0105E^2 - 0.234E + 1.35125$	3.79
f_1^0	$0.00625E^2 - 0.1E + 0.34375$ at $E \leq 7$ $-0.015E + 0.04$ at $E > 7$	2.41
f_1^1	$-0.00208E^2 + 0.06667E + 0.71875$	0.32

Table 3.4 MDE of slope of the input signal S

MCE parameter	MDE	Relative error (%)
Δt_{min}^{01}	$-0.0005S + 0.03$	0.47
Δt_{b}^{01}	$0.00002S^2 - 0.00089S + 0.0615$	4.00
Δt_{max}^{01}	$-0.00039S^2 - 0.02589S + 1.2023$	2.15
Δt_{s}^{01}	$-0.0017S^2 - 0.055S + 1.41542$	1.80
Δt_{min}^{10}	$-0.00011S^2 + 0.002527S - 0.05369$	1.11
Δt_{b}^{10}	$-0.00064S^2 + 0.0159S - 0.06708$ at $S \leq 20$ $0.00009S^2 + 0.00656S + 0.0875$ at $S > 20$	5.81
Δt_{max}^{10}	$-0.0185S + 0.70625$ at $S \leq 40$ $-0.0004S^2 - 0.03344S + 0.8425$ at $S > 40$	2.85
Δt_{s}^{10}	$-0.0185S + 0.77625$ at $S \leq 60$ $-0.00009S^2 + 0.00344S + 0.5125$ at $S > 60$	1.54
f_1^0	$0.000075S^2 + 0.002S + 1$	0.55
f_1^1	$-0.000875S^2 - 0.0375S$	3.92

Table 3.5 MDE of duration of the input pulse t_{in}

MCE parameter	MDE	Relative error (%)
Δt_{min}^{01}	$0.1458t_{in}^2 - 0.19375t_{in} - 0.0075$ at $t_{in} \leq 0.9$ $-0.21354t_{in}^2 + 0.75625t_{in} - 0.57141$ at $t_{in} > 0.9$	4.31
Δt_{b}^{01}	$0.4558t_{in}^2 - 0.19375t_{in} + 0.0525$ at $t_{in} \leq 0.9$ $-0.21354t_{in}^2 + 0.75625t_{in} - 0.51141$ at >0.9	3.12
Δt_{max}^{01}	$-0.554t_{in}^2 + 1.8588t_{in} - 0.4753$	5.09
Δt_{s}^{01}	$-0.554t_{in}^2 + 1.8588t_{in} - 0.3753$	4.81
Δt_{min}^{10}	$0.0363t_{in}^2 - 0.09884t_{in} + 0.0014$	5.2
Δt_{b}^{10}	$-0.0417t_{in} + 0.0075$ at $t_{in} \leq 0.6$ $0.125t_{in}^2 - 0.2542t_{in} + 0.09$ at $0.6 < t_{in} \leq 1.2$ $0.039t_{in}^2 - 0.12t_{in} + 0.0525$ at $t_{in} > 1.2$	5.11
Δt_{max}^{10}	$-0.069t_{in}^2 + 0.404t_{in} - 0.1$ at $t_{in} \leq 1.2$ $-0.0792t_{in} + 0.19$ at $t_{in} > 1.2$	4.22
Δt_{s}^{10}	$-0.1227t_{in}^2 + 0.484t_{in} - 0.0892$	1.32
f_1^0	$-1.16667t_{in}^2 + 2.71667t_{in} - 0.51$ at $t_{in} \leq 0.9$ $0.008333t_{in} + 0.9825$ at $t_{in} > 0.9$	6.11
f_1^1	$-0.0111t_{in} + 0.02333$	0

Table 3.6 MDE of amplitudes of the input signal U_{min}

MCE parameter	MDE	Relative error (%)
Δt^{01}_{min}	$-0.2875\ U_{min} + 0.8025$ at $U_{min} \leq 3.0$ $-0.05\ U_{min} + 0.09$ at $\mathrm{U}_{min} \leq 3.0$	1.57
Δt^{01}_{b}	$-0.05625\ U_{min}^2 - 0.0625\ U_{min} + 0.69375$ at $U_{min} \leq 3.0$ $-0.08\ U_{min}^2 + 0.59\ U_{min} - 1.05$ at $U_{min} \leq 3.0$	6.25
Δt^{01}_{max}	$-0.3\ U_{min} + 2.3$ at $U_{min} \leq 3.0$ $0.095\ U_{min}^2 - 1.06\ U_{min} + 3.725$ at $U_{min} \leq 3.0$	1.03
Δt^{01}_{s}	$0.0906\ U_{min}^2 - 1.025\ \mathrm{U}_{min} + 3.7594$	1.43
Δt^{10}_{min}	$-0.1125\ U_{min} + 0.3625$ at $U_{min} \leq 3.0$ $0.0281\ U_{min}^2 - 0.30938\ U_{min} + 0.7$ at $U_{min} \leq 3.0$	6.45
Δt^{10}_{b}	$0.125\ U_{min}^2 - 0.7125S + 1.1375$ at $U_{min} \leq 3.0$ $0.0281\ U_{min}^2 - 0.3094\ U_{min} + 0.8$ at $U_{min} \leq 3.0$	6.39
Δt^{10}_{max}	$-0.5625\ U_{min} + 2.3125$ at $U_{min} \leq 2.0$ $0.0906\ U_{min}^2 - 1.0156\ U_{min} + 2.856$ at $U_{min} > 2.0$	0.81
Δt^{10}_{s}	$0.11797\ U_{min}^2 - 1.3094\ U_{min} + 3.7414$	0.29
f^0_1	$-0.375\ \mathrm{U}_{min}^2 + 0.45\ U_{min} - 0.2125$	2.12
f^1_1	$-0.0025\ \mathrm{U}_{min}^2 + 0.005\ U_{min} + 0.0375$	3.34

Table 3.7 MDE of the number of loads N

MCE parameter	MDE	Relative error (%)
Δt^{01}_{min}	$0.004256\ N^2 - 0.00745\ N + 0.02494$	1.36
Δt^{01}_{b}	$0.004256\ N^2 - 0.00745\ N + 0.08494$	1.36
Δt^{01}_{max}	$0.011875\ N^2 - 0.01688\ N + 0.924$ at $N \leq 6$ $0.1\ N - 0.65$ at $N > 6$	1.74
Δt^{01}_{s}	$0.5031\ N^2 - 2.5019\ N + 3.0175$ at $N \leq 3$ $-0.073\ N^2 + 1.1926\ N - 2.884$ at $N \leq 3$	6.45
Δt^{10}_{min}	$-0.00068\ N^2 + 0.02\ N - 0.2094$	2.93
Δt^{10}_{b}	$0.00375\ N^2 - 0.0125\ N - 0.06$ at $N \leq 5$ $-0.00833\ N^2 + 0.12\ N - 0.4225$ at $N > 5$	6.42
Δt^{10}_{max}	$0.02667\ N + 0.25333$ at $N \leq 4$ $-0.01\ N^2 - 0.18\ N - 0.2$ at $N > 4$	1.25
Δt^{10}_{s}	$0.0097\ N^2 - 0.00917\ N + 0.37$ at $N \leq 6$ $-0.016\ N^2 + 0.34375\ N - 0.8775$ at $N > 6$	0.63
f^0_1	$-0.0952\ N^2 + 0.014\ N + 0.995$	2.14
f^1_1	$0.0065\ N^2 - 0.016\ N + 0.0095$	1.47

3.1.2 Basic Logical Cell ECL (Fig. 1.14; Tables 3.8, 3.9, 3.10, 3.11, 3.12, 3.13, 3.14)

Table 3.8 MDE of total radiation dose *D*

MCE parameter	MDE	Relative error (%)
Δt^{01}_{min}	$0.00005D^2 - 0.00033D - 0.14875$	1.2
Δt^{01}_{b}	$-0.00018D^2 + 0.00725D - 0.09875$ at $D \leq 40$ $0.00009D^2 - 0.00094D - 0.08125$ at $D > 40$	6.41
Δt^{01}_{max}	$-0.0004D^2 + 0.006927D + 0.23875$	0.62
Δt^{01}_{s}	$0.00003D^2 + 0.00475D + 0.33875$ at $D \leq 40$ $0.000792D + 0.552083$ at $D > 40$	2.13
Δt^{10}_{min}	$-0.00015D^2 + 0.005875D - 0.095$ at $D \leq 60$ $-0.00024D^2 - 0.044313D - 2.075$ at $D > 60$	6.34
Δt^{10}_{b}	$-0.0002D^2 + 0.002875D - 0.045$ at $D \leq 40$ $0.00067D^2 - 0.07991D + 2.17125$ at $40 < D \leq 80$ $-0.00144D - 0.14375$ at $D > 80$	6.49
Δt^{10}_{max}	$0.00009D^2 + 0.001797D + 0.255$ at $D \leq 80$ $0.002938D + 0.10625$ at $D > 80$	0.31
Δt^{10}_{s}	$0.000209D^2 - 0.00694 + 0.305$ at $D \leq 60$ $0.000388D^2 - 0.06681D + 3.256$ at $D > 60$	5.44
f^0_1	$-0.001067D^2 - 0.02033D$ at $D \leq 30$ $0.0012D^2 - 0.104D + 2.39$ at $30 < D \leq 50$ $0.0002D + 0.18$ at $D > 50$	6.17
f^1_1	$-0.00167D^2 + 0.00283D + 1$ at $D \leq 30$ $-0.0008D^2 + 0.076D - 1.21$ at $30 < D \leq 50$ $0.0002D + 0.58$ at $D > 50$	6.42

Table 3.9 MDE of ambient temperature *T*

MCE parameter	MDE	Relative error (%)
Δt^{01}_{min}	$-0.000006T^2 - 0.00069T - 0.095$ at $T \leq 20$ $-0.000003T^2 + 0.000235T - 0.011675$ at $T > 20$	1.42
Δt^{01}_{b}	$-0.00004T^2 + 0.001828T + 0.21766$	0.48
Δt^{01}_{max}	$0.000015T^2 + 0.002T + 0.255$ at $T \leq 20$ $-0.000003T^2 + 0.001575T + 0.27075$ at $T > 20$	0.51
Δt^{01}_{s}	$-0.00003T^2 - 0.00113T - 0.0175$ at $T \leq 0$ $-0.00064T - 0.0775$ at $T > 0$	0.98
Δt^{10}_{min}	$-0.00001T^2 - 0.00063T - 0.0375$ at $T \leq 40$ $-0.000004T^2 - 0.00014T - 0.05$ at $40 < T \leq 80$ $-0.00064T - 0.03687$ at $T > 80$	2.69
Δt^{10}_{b}	$-0.00005T^2 + 0.000898T + 0.18625$	093
Δt^{10}_{max}	$-0.00001T^2 + 0.000625T + 0.2375$ at $T \leq 40$ $-0.000003T^2 + 0.00071T + 0.232$ at $T > 40$	0.38
Δt^{10}_{s}	$0.00005T^2 - 0.0025T + 1.0925$	1.24
f^0_1	$-0.00014T^2 + 0.013125T + 0.70625$	1.86

Table 3.10 MDE supply voltage E

MCE parameter	MDE	Relative error (%)
Δt_{min}^{01}	$0.0092E^2 - 0.1356E + 0.305$	3.80
Δt_{b}^{01}	$0.02797E^2 - 0.29438E + 0.67266$ at $E \leq 7$ $0.0525E^2 - 0.885E + 3.605$ at $E > 7$	5.12
Δt_{max}^{01}	$0.01528E^2 - 0.25542E + 1.229$	4.60
Δt_{s}^{01}	$0.048281E^2 - 0.57563E + 2.071$ at $E \leq 7$ $0.0525E^2 - 0.91E + 4.205$ at $E > 7$	1.56
Δt_{min}^{10}	$0.0014E^2 - 0.1417E - 0.00125$	3.60
Δt_{b}^{10}	$0.001389E^2 - 0.01417E + 0.03875$	5.30
Δt_{max}^{10}	$0.009167E^2 - 0.1575E + 0.86125$	0.72
Δt_{s}^{10}	$0.015278E^2 - 0.25583E + 1.301$	3.60
f_1^0	$-0.048E^2 - 0.7417E - 0.71$	1.87
f_1^1	$0.16875E - 1.31875$	2.41

Table 3.11 MDE of the slope of the input signal S

MCE parameter	MDE	Relative error (%)
Δt_{min}^{01}	$0.000425S^2 - 0.0165S + 0.0125$	3.18
Δt_{b}^{01}	$-0.02975S + 0.1125$ at $S \leq 5$ $0.00033S^2 - 0.01375S + 0.02417$ at $S > 5$	4.52
Δt_{max}^{01}	$0.001775S^2 - 0.0595S + 0.7375$	0.79
Δt_{s}^{01}	$0.002525S^2 - 0.092S + 1.1875$	1.54
Δt_{min}^{10}	$0.00195S^2 - 0.0325S - 0.06$ at $S \leq 10$ $0.0011S^2 - 0.03675S + 0.1875$ at $S > 10$	2.57
Δt_{b}^{10}	$0.00195S^2 - 0.0325S + 0.1$ at $S \leq 10$ $-0.00375S + 0.0075$ at $S > 10$	3.38
Δt_{max}^{10}	$-0.015S + 0.55$ at $S \leq 10$ $0.013S^2 - 0.05525S + 0.8225$ at $S > 10$	0.89
Δt_{s}^{10}	$0.002338S^2 - 0.08638S + 1.1$	3.02
f_1^0	$0.00005S^2 + 0.00005S + 1$	1.21
f_1^1	$0.004595S + 0.0001$	0.23

Table 3.12 MDE of the duration of the input pulse $t_{u_{in}}$

MCE parameter	MDE	Relative error (%)
Δt_{min}^{01}	$0.026667t_{in}^2 - 0.15417t_{in} + 0.0775$	6.18
Δt_{b}^{01}	$0.03111t_{in}^2 - 0.18t_{in} + 0.1775$	6.21
Δt_{max}^{01}	$0.3469t_{in} + 1.0825$ at $t_{in} \leq 2$ $0.2025t_{in}^2 - 1.33625t_{in} + 2.25225$ at $t_{in} > 2$	0.68
Δt_{s}^{01}	$-0.06417t_{in}^2 - 0.2629t_{in} + 1.2025$ at $t_{in} \leq 1.5$ $0.0575t_{in}^2 - 0.591t_{in} + 1.421$ at $t_{in} > 1.5$	034
Δt_{min}^{10}	$-0.2167t_{in}^2 - 0.00167t_{in} + 0.22$	3.29
Δt_{b}^{10}	$-0.022t_{in}^2 - 0.016727$ - 0.27	2.54
Δt_{max}^{10}	$-0.069t_{in}^2 - 0.075t_{u_{in}} + 0.97$	1.66
Δt_{s}^{10}	$-0.027t_{in}^2 - 0.285t_{in} + 1.27$	1.51
f_1^0	$0.0444t_{in}^2 - 0.26667t_{in} + 0.4$	6.11
f_1^1	$-0.0444t_{in}^2 + 0.26667t_{in} + 0.6$	0.97

Table 3.13 MDE of amplitudes of the input signal $U_{\min}$

MCE parameter	MDE	Relative error (%)
$\Delta t^{01}_{\min}$	$-0.2875\ U_{\min} + 0.8025$ at $U_{\min} \leq 3.0$ $-0.05\ U_{\min} + 0.09$ at $U_{\min} \leq 3.0$	1.57
$\Delta t^{01}_{\mathrm{b}}$	$-0.05625\ U_{\min}{}^{2} - 0.0625\ U_{\min} + 0.69375$ at $U_{\min} \leq 3.0$ $-0.08\ U_{\min}{}^{2} + 0.59\ U_{\min} - 1.05$ at $U_{\min} \leq 3.0$	6.25
$\Delta t^{01}_{\max}$	$-0.3\ U_{\min} + 2.3$ at $U_{\min} \leq 3.0$ $0.095\ U_{\min}{}^{2} - 1.06\ U_{\min} + 3.725$ at $U_{\min} \leq 3.0$	1.03
$\Delta t^{01}_{\mathrm{s}}$	$0.0906\ U_{\min}{}^{2} - 1.025\ U_{\min} + 3.7594$	1.43
$\Delta t^{10}_{\min}$	$-0.1125\ U_{\min} + 0.3625$ at $U_{\min} \leq 3.0$ $0.0281\ U_{\min}{}^{2} - 0.30938\ U_{\min} + 0.7$ at $U_{\min} \leq 3.0$	6.45
$\Delta t^{10}_{\mathrm{b}}$	$0.125\ U_{\min}{}^{2} - 0.7125S + 1.1375$ at $U_{\min} \leq 3.0$ $0.0281\ U_{\min}{}^{2} - 0.3094\ \mathrm{U}_{\min} + 0.8$ at $U_{\min} \leq 3.0$	6.39
$\Delta t^{10}_{\max}$	$-0.5625\ U_{\min} + 2.3125$ at $U_{\min} \leq 2.0$ $0.0906\ U_{\min}{}^{2} - 1.0156\ U_{\min} + 2.856$ at $\mathrm{U}_{\min} > 2.0$	0.81
$\Delta t^{10}_{\mathrm{s}}$	$0.11797\ U_{\min}{}^{2} - 1.3094\ U_{\min} + 3.7414$	0.29
f^{0}_{1}	$-0.375\ U_{\min}{}^{2} + 0.45\ U_{\min} - 0.2125$	2.12
f^{1}_{1}	$-0.0025\ U_{\min}{}^{2} + 0.005\ U_{\min} + 0.0375$	3.34

Table 3.14 MDE of the number of loads N

MCE parameter	MDE	Relative error (%)
$\Delta t^{01}_{\min}$	$0.00228\ N^2 + 0.01889\ N - 0.18256$	0.51
$\Delta t^{01}_{\mathrm{b}}$	$-0.00313\ N^2 + 0.0231\ N - 0.12625$ at $N \leq 4$ $0.005156\ N^2 - 0.055\ N + 0.055$ at $N > 4$	3.54
$\Delta t^{01}_{\max}$	$0.0051\ N^2 - 0.01317\ N + 0.214$	0.51
$\Delta t^{01}_{\mathrm{s}}$	$0.010\ N^2 - 0.018\ N + 0.314$	1.09
$\Delta t^{10}_{\min}$	$0.00287\ N^2 + 0.0027\ N - 0.17$	6.04
$\Delta t^{10}_{\mathrm{b}}$	$0.0002\ N^2 + 0.026\ N - 0.094$ at $N \leq 4$ $0.0089\ N^2 - 0.077\ N + 0.18$ at $N > 4$	6.44
$\Delta t^{10}_{\max}$	$0.00796\ N^2 - 0.00022\ N + 0.1998$	1.71
$\Delta t^{10}_{\mathrm{s}}$	$0.00796\ N^2 - 0.00022\ N + 0.2998$	0.57
f^{0}_{1}	$-0.073\ N^2 + 0.543\ N - 0.97$ at $N \leq 4$ $0.079\ N^2 - 0.7025\ N + 1.58$ at $N > 4$	2.14
f^{1}_{1}	$0.005\ N^2 - 0.085\ N + 1.16$	1.47

3.1.3 Basic Logical Cell CMOS (Fig. 1.15; Tables 3.15, 3.16, 3.17, 3.18, 3.19, 3.20, 3.21)

Comparison of the results obtained with these models (solid curves), with the data obtained by simulating at the device level by SPICE (dots) shows (Figs. 3.3, 3.4, 3.5, 3.6, 3.7, 3.8, 3.9, 3.10, 3.11, 3.12, 3.13, 3.14, 3.15, 3.16, 3.17, 3.18, 3.19, 3.20, 3.21,

Table 3.15 MDE of the total radiation dose D

MCE parameter	MDE	Relative error (%)
Δt^{01}_{min}	$0.00053D^2 - 1.49375D - 62.5$	3.91
Δt^{01}_{b}	$-0.03125D - 32.5$ at $D \leq 40$ $0.051D^2 - 8.729D + 233.75$ at $D > 40$	3.05
Δt^{01}_{max}	$0.02385D^2 + 0.577D + 62.5$	0.63
Δt^{01}_{s}	$0.0203D^2 + 0.71875D + 92.5$ at $D \leq 40$ $-0.084D^2 + 16.0625D - 353.75$ at $D > 40$	2.85
Δt^{10}_{min}	$0.0327D^2 - 4.996D - 41.25$	3.61
Δt^{10}_{b}	$0.075D^2 – 6.6875D - 11.25$ at $D \leq 60$ $0.0297D^2 - 5.78D + 97.5$ at $D > 60$	4.80
Δt^{10}_{max}	$-0.02438D^2 + 6.71125D + 114.5631$	2.18
Δt^{10}_{s}	$-0.01938D^2 + 6.4625D + 151.25$	1.52
f^0_1	$-0.00004D^2 - 0.00275D + 1$ at $D \leq 60$ $0.00005D^2 - 0.0105D + 1.15$ at $D > 60$	2.41
f^1_1	$0.00009D^2 - 0.00258D + 0.05$	3.23

Table 3.16 MDE of ambient temperature T

MCE parameter	MDE	Relative error (%)
Δt^{01}_{min}	$0.00134T^2 + 0.0875T - 13.2344$	5.09
Δt^{01}_{b}	$0.001146T^2 + 0.01458T - 0.125$ at $T \leq 20$ $0.001458T^2 - 0.1667T + 3.375$ at $20 < T \leq 80$ $-0.00984T^2 + 1.6844T - 72.375$ at $T > 80$	5.78
Δt^{01}_{max}	$0.001494T^2 + 0.084512T + 85.898$	0.60
Δt^{01}_{s}	$0.028125T + 106.125$ at $T \leq 40$ $0.0044T^2 - 0.4844T + 128.25$ at $T > 40$	0.48
Δt^{10}_{min}	$0.000371T^2 - 0.05781T - 7.906$	4.38
Δt^{10}_{b}	$-0.00027T^2 - 0.05781T + 3.125$ at $T \leq 40$ $-0.00422T^2 + 0.509T - 13.25$ at $T > 40$	6.12
Δt^{10}_{max}	$-0.00139T^2 + 0.4578T + 115.53$	0.53
Δt^{10}_{s}	$-0.00139T^2 + 0.4578T + 125.5313$	0.53
f^0_1	$-0.00041T^2 + 0.00125T + 0.7$ at $T \leq 40$ $-0.00003T^2 - 0.00125T + 0.15$ at $T > 40$	2.83
f^1_1	$0.000009T^2 + 0.001875T + 0.9875$	1.67

Table 3.17 MDE of supply voltage E

MCE parameter	MDE	Relative error (%)
Δt^{01}_{min}	$15.5833E^2 - 58.125E + 20.16667$	1.41
Δt^{01}_{b}	$18.9167E^2 - 71.125E + 46.8333$	4.80
Δt^{01}_{max}	$39.75E^2 - 146.875E + 188.5$	3.48
Δt^{01}_{s}	$39.75E^2 - 166.875E + 238.5$	6.13
Δt^{10}_{min}	$-16.25E^2 + 9.375E + 3.125$	5.12
Δt^{10}_{b}	$-16.25E^2 - 9.375E + 23.125$	6.31
Δt^{10}_{max}	$-10.4167E^2 - 9.375E + 123.5417$	0.74
Δt^{10}_{s}	$-68.75E + 220.625$	0.55
f^0_1	$0.0333E^2 + 0.05E + 0.86667$	3.12
f^1_1	$0.6E^2 - 1.1E+0.5$	2.43

Table 3.18 MDE of the slope of the input signal *S*

MCE parameter	MDE	Relative error (%)
Δt_{min}^{01}	$-0.00107S^2 + 0.329S - 27.56$	4.98
Δt_{b}^{01}	$-0.0185S^2 + 0.439S - 18.58$ at $S \le 130$ $-0.0024S^2 + 0.8227S - 59164$ at $S > 130$	5.73
Δt_{max}^{01}	$-0.00016S^2 - 0.8099S + 26.435$	1.21
Δt_{s}^{01}	$-0.00061S^2 + 0.6716S + 47.72$	149
Δt_{min}^{10}	$-0.00089S^2 + 0.305S - 12.715$	1.11
Δt_{b}^{10}	$-0.00089S^2 + 0.305S - 14.715$	6.27
Δt_{max}^{10}	$-0.00201S^2 + 0.859S + 57.3597$	2.85
Δt_{s}^{10}	$-0.00022S^2 + 0.6495S + 69.2767$	1.57
f_1^0	$0.000075S^2 + 0.002S + 1$	0.55
f_1^1	$-0.000875S^2 - 0.0375S$	3.92

Table 3.19 MDE of the duration of the input pulse $t_{u_{in}}$

MCE parameter	MDE	Relative error (%)
Δt_{min}^{01}	$-0.0001t_{in}^2 - 0.00958t_{in} - 18.125$	0.67
Δt_{b}^{01}	$-0.00075t_{in}^2 + 0.05t_{in} - 8.125$ at $t_{in} \le 100$ $-0.00024t_{in}^2 + 0.049375t_{in} - 13.125$ at $t_{in} > 100$	1.73
Δt_{max}^{01}	$-0.00076t_{in}^2 + 0.36625t_{in} + 33.125$	0.91
Δt_{s}^{01}	$-0.00052t_{in}^2 + 0.302917t_{in} + 48.125$	1.37
Δt_{min}^{10}	$0.00004t_{in}^2 - 0.0475t_{in} - 15.5$	5.2
Δt_{b}^{10}	$0.0004t_{in}^2 - 0.0475t_{in} - 5.5$ at $t_{in} \le 100$ $0.000263t_{in}^2 - 0.155t_{in} + 6.625$ at $t_{in} > 100$	5.11
Δt_{max}^{10}	$0.000739t_{in}^2 - 0.11583t_{in} + 59.5$	4.22
Δt_{s}^{10}	$0.361667t_{in} + 69.5$ at $t_{in} \le 150$ $-0.00039t_{in}^2 + 0.2596t_{in} + 93.625$ at $t_{in} > 150$	1.32
f_1^0	$0.000003t_{in}^2 + 0.000167t_{in} + 0.15$	0.12
f_1^1	$0.000003t_{in} - 0.00027 + 0.9$	3.21

3.22, 3.23, 3.24, 3.25, 3.26, 3.27, 3.28, 3.29, 3.30, 3.31, 3.32, 3.33, 3.34, 3.35, 3.36, 3.37, 3.38, 3.39, 3.40, 3.41, 3.42, 3.43, 3.44, 3.45, 3.46, 3.47, 3.48, 3.49, 3.50, 3.51, 3.52, 3.53, 3.54, 3.55, 3.56, 3.57, 3.58, 3.59, 3.61, 3.62, 3.63, 3.64, 3.65) that the relative error does not exceed 6.5%. The use of approximating formulas of higher order is inexpedient, since the experiments performed show that in this case the picture is similar to that in Fig. 2.44, i.e. an increase in the order of the approximation polynomial gives a small gain in accuracy with a significant increase in the machine resources necessary for the calculations for these models.

Table 3.20 MDE of amplitudes of the input signal $U_{\min}$

MCE parameter	MDE	Relative error (%)
$\Delta t^{01}_{\min}$	$23.8889\ U_{\min}^2 - 47.083\ U_{\min} - 0.625$	4.54
$\Delta t^{01}_{\mathrm{b}}$	$-31.25\ U_{\min}^2 + 15.625\ U_{\min} + 19.375$ at $U_{\min} \le 1$ $150\ U_{\min}^2 - 395.5\ U_{\min} + 251.25$ at $U_{\min} > 1$	5.02
$\Delta t^{01}_{\max}$	$73.8889\ U_{\min}^2 - 299.583\ U_{\min} + 290.625$	2.20
$\Delta t^{01}_{\mathrm{s}}$	$-168.75\ U_{\min}^2 + 15.625\ U_{\min} + 300.625$ at $U_{\min} \le 1$ $250\ U_{\min}^2 - 822.5\ U_{\min} + 688.75$ at $U_{\min} > 1$	1.67
$\Delta t^{10}_{\min}$	$-53.3333\ U_{\min} + 1.25$ at $U_{\min} \le 0.75$ $-146.667\ U_{\min}^2 + 326.667\ U_{\min} - 201.25$ at $U_{\min} > 0.75$	6.14
$\Delta t^{10}_{\mathrm{b}}$	$-63.333\ U_{\min}^2 - 5.8333S + 21.25$at $U_{\min} \le 0.75$ $-146.667\ U_{\min}^2 + 326.667\ U_{\min} - 181.25$ at $U_{\min} > 0.75$	6.42
$\Delta t^{10}_{\max}$	$-42.222\ U_{\min}^2 - 75\ U_{\min} + 228.75$	1.69
$\Delta t^{10}_{\mathrm{s}}$	$33.3333\ U_{\min}^2 - 161.667\ U_{\min} + 358.75$	2.81
f^0_1	$-0.02\ U_{\min} + 0.1\ U_{\min} \le 0.5$ $0.1\ U_{\min}^2 - 0.29\ U_{\min} + 0.21$ at $U_{\min} > 0.5$	1.78
f^1_1	$9.8 \times 10^{-17}\ U_{\min}^2 + 0.06667\ U_{\min} + 0.0375$	3.56

Table 3.21 MDE of the number of loads N

MCE parameter	MDE	Relative error (%)
$\Delta t^{01}_{\min}$	$0.5\ N - 49.75$	0.00
$\Delta t^{01}_{\mathrm{b}}$	$-2.5\ N - 23.75$	0.00
$\Delta t^{01}_{\max}$	$12.5\ N + 53.75$	0.00
$\Delta t^{01}_{\mathrm{s}}$	$22.5\ N + 63.75$	0.00
$\Delta t^{10}_{\min}$	$-1.75\ N - 31$	0.00
$\Delta t^{10}_{\mathrm{b}}$	$-3.75\ N - 5$	0.00
$\Delta t^{10}_{\max}$	$23.75\ N + 85$	0.00
$\Delta t^{10}_{\mathrm{s}}$	$33.75\ N + 95$	0.00
f^0_1	$0.114\ N - 0.114$	0.00
f^1_1	$-0.00313\ N^2 - 2.8 \times 10^{-17}\ N + 1$	0.76

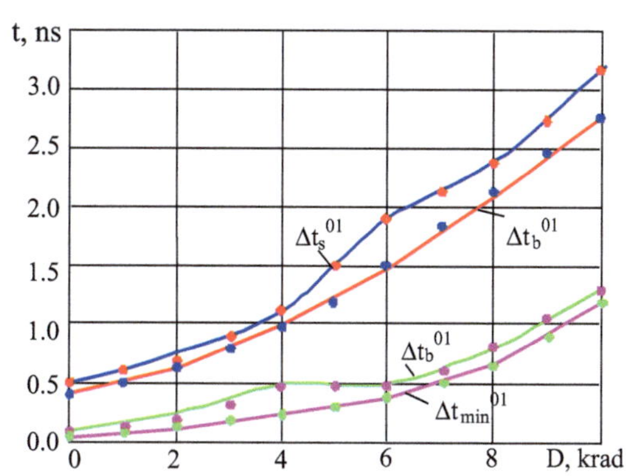

Fig. 3.3 $\Delta t^{01}_{\min}$, $\Delta t^{01}_{\mathrm{b}}$, $\Delta t^{01}_{\max}$, $\Delta t^{01}_{\mathrm{s}} = f(D)$

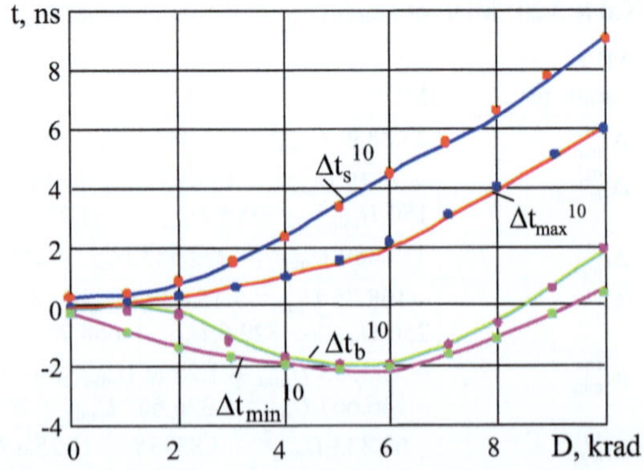

Fig. 3.4 Δt_{min}^{10}, Δt_{b}^{10}, Δt_{max}^{10}, $\Delta t_{s}^{10} = f(D)$

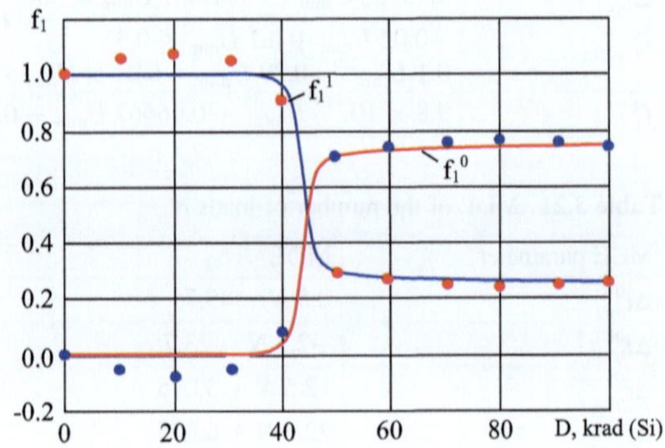

Fig. 3.5 $f_1 = f(D)$

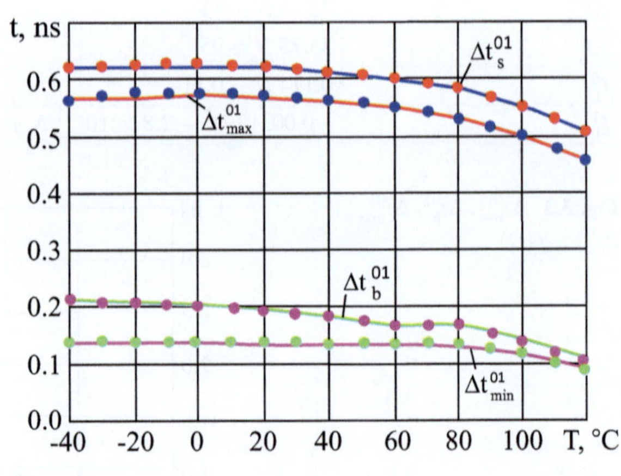

Fig. 3.6 Δt_{min}^{01}, Δt_{b}^{01}, Δt_{max}^{01}, $\Delta t_{s}^{01} = f(T)$

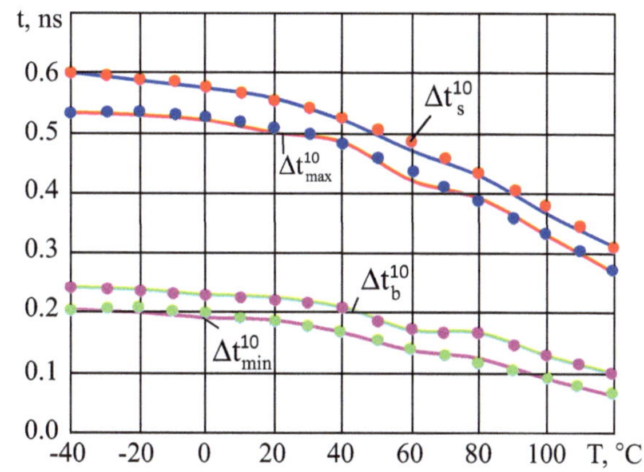

Fig. 3.7 Δt_{min}^{10}, Δt_b^{10}, Δt_{max}^{10}, $\Delta t_s^{10} = f(T)$

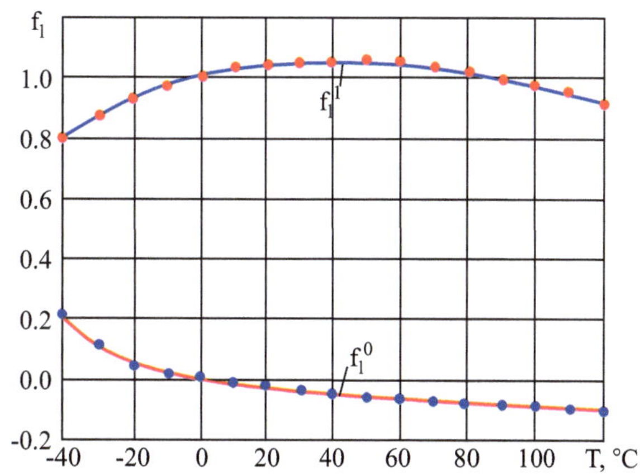

Fig. 3.8 $f_1 = f(T)$

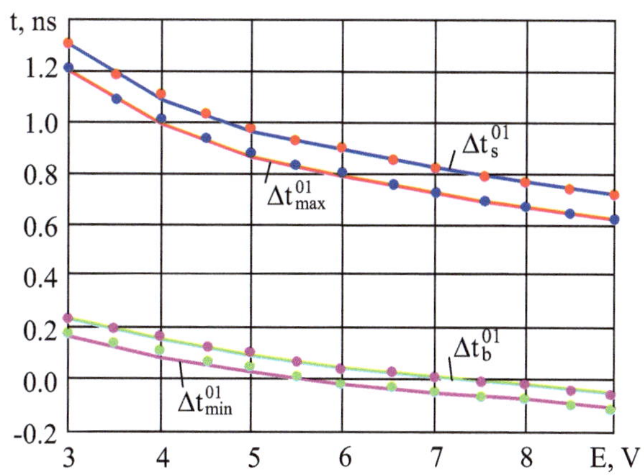

Fig. 3.9 Δt_{min}^{01}, Δt_b^{01}, Δt_{max}^{01}, $\Delta t_s^{01} = f(E)$

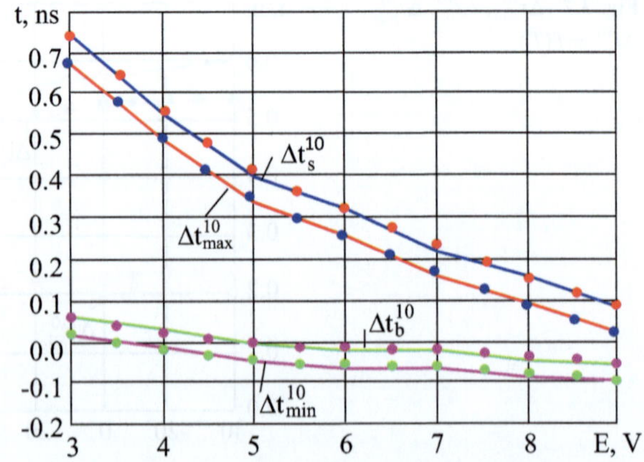

Fig. 3.10 Δt_{min}^{10}, Δt_{b}^{10}, Δt_{max}^{10}, $\Delta t_{s}^{10} = f(E)$

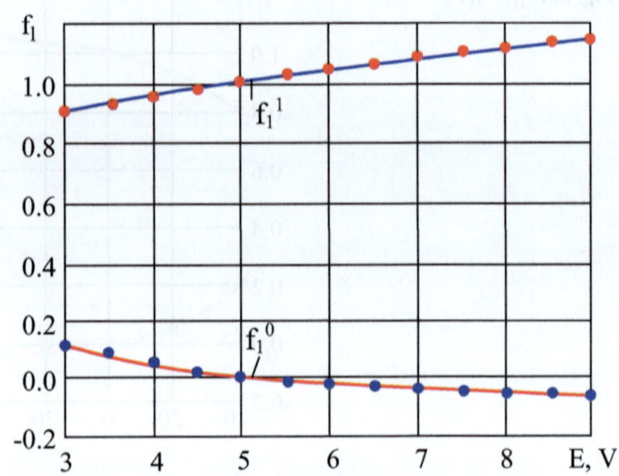

Fig. 3.11 $f_l = f(E)$

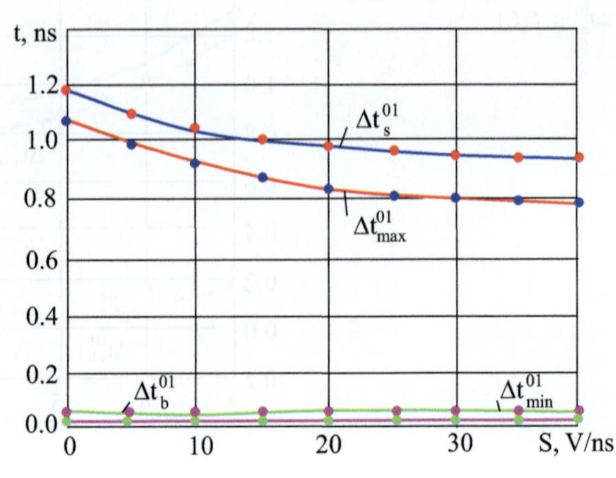

Fig. 3.12 Δt_{min}^{01}, Δt_{b}^{01}, Δt_{max}^{01}, $\Delta t_{s}^{01} = f(S)$

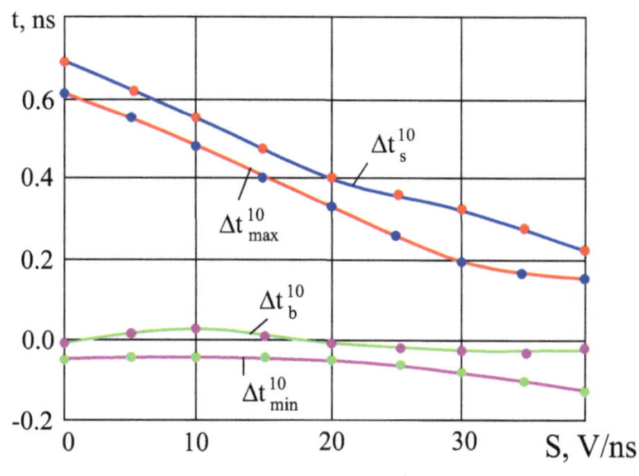

Fig. 3.13 Δt_{min}^{10}, Δt_{b}^{10}, Δt_{max}^{10}, $\Delta t_{s}^{10} = f(S)$

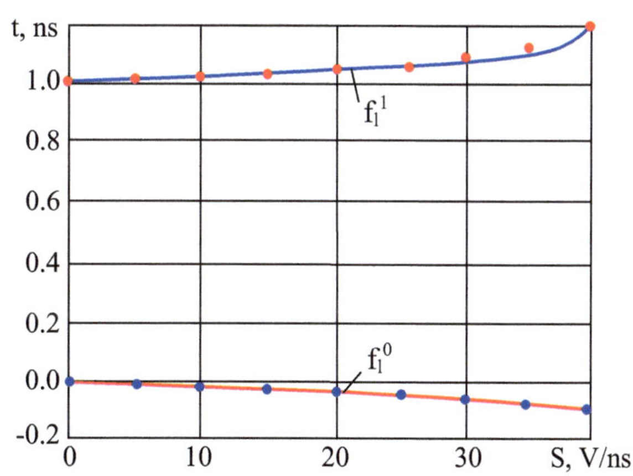

Fig. 3.14 $f_l = f(S)$

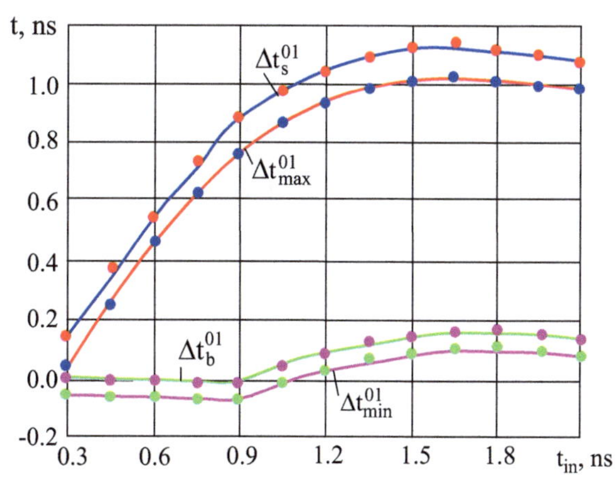

Fig. 3.15 Δt_{min}^{01}, Δt_{b}^{01}, Δt_{max}^{01}, $\Delta t_{s}^{01} = f(t_{in})$

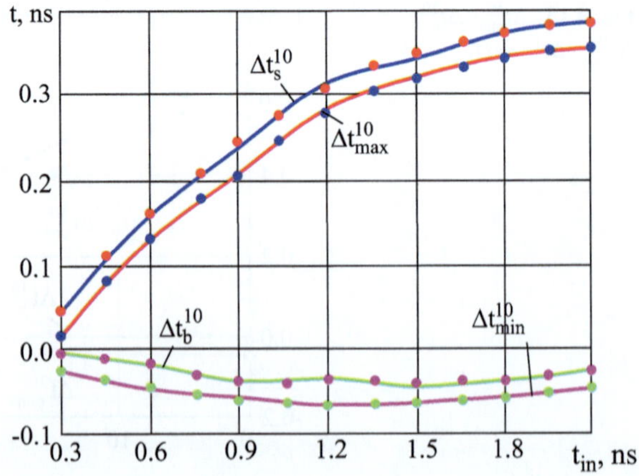

Fig. 3.16 $\Delta t_{\min}^{10}$, Δt_{b}^{10}, $\Delta t_{\max}^{10}$, $\Delta t_{s}^{10} = f(t_{in})$

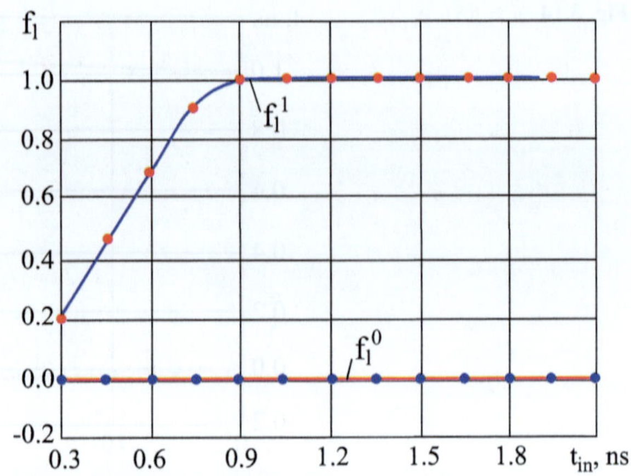

Fig. 3.17 $f_l = f(t_{in})$

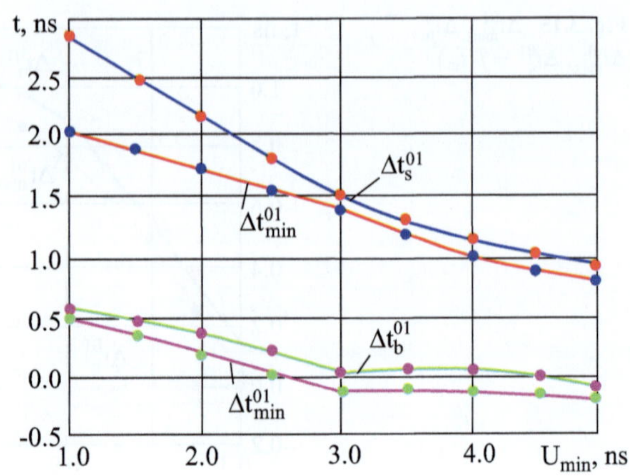

Fig. 3.18 $\Delta t_{\min}^{01}$, Δt_{b}^{01}, $\Delta t_{\max}^{01}$, $\Delta t_{s}^{01} = f(U_{\min})$

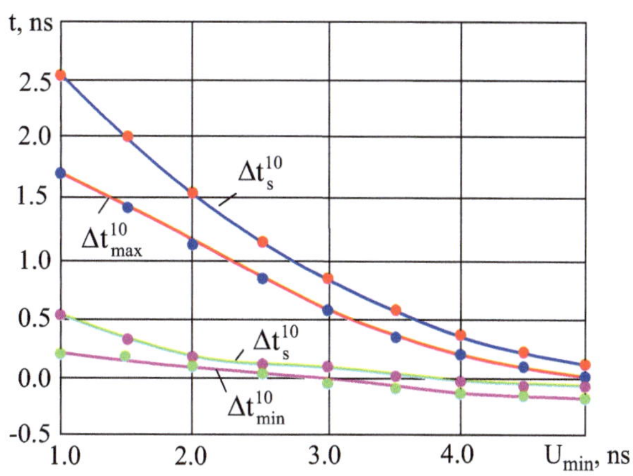

Fig. 3.19 $\Delta t_{\min}^{10}$, Δt_{b}^{10}, $\Delta t_{\max}^{10}$, $\Delta t_{s}^{10} = f(U_{\min})$

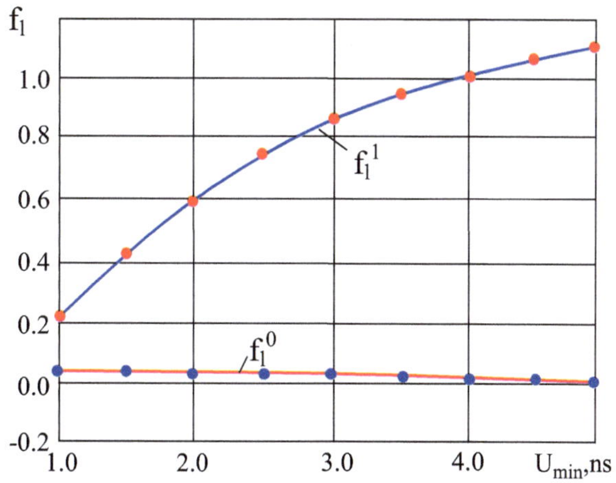

Fig. 3.20 $f_l = f(U_{\min})$

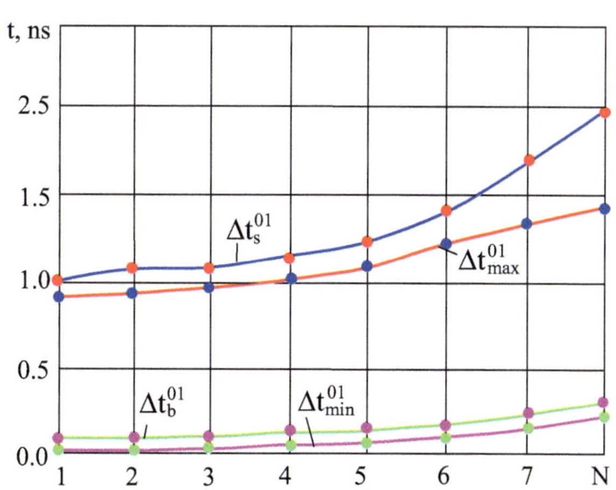

Fig. 3.21 $\Delta t_{\min}^{01}$, Δt_{b}^{01}, $\Delta t_{\max}^{01}$, $\Delta t_{s}^{01} = f(N)$

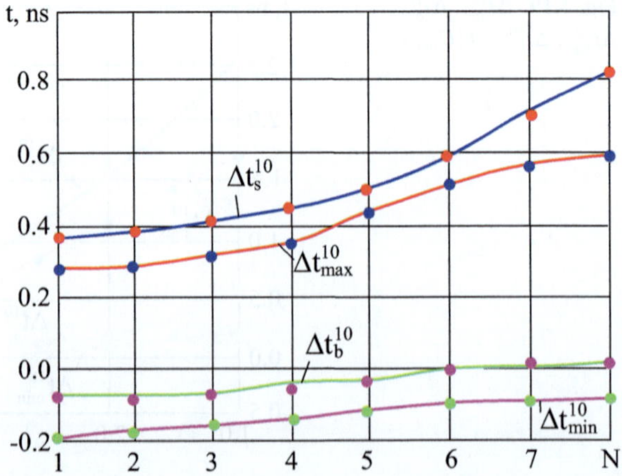

Fig. 3.22 $\Delta t_{\min}^{10}$, $\Delta t_{\mathrm{b}}^{10}$, $\Delta t_{\max}^{10}$, $\Delta t_{\mathrm{s}}^{10} = f(N)$

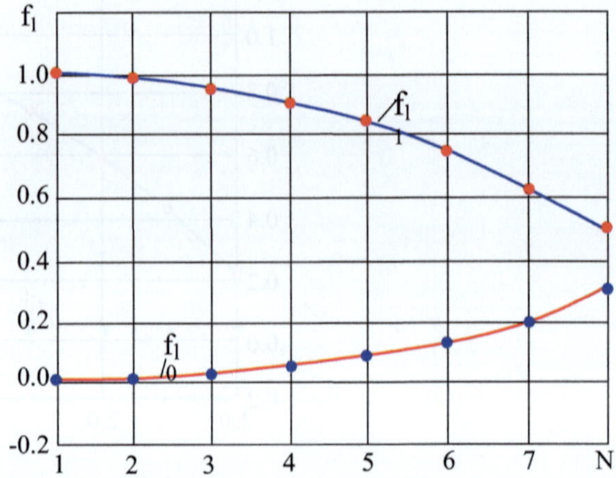

Fig. 3.23 $f_{\mathrm{l}} = f(N)$

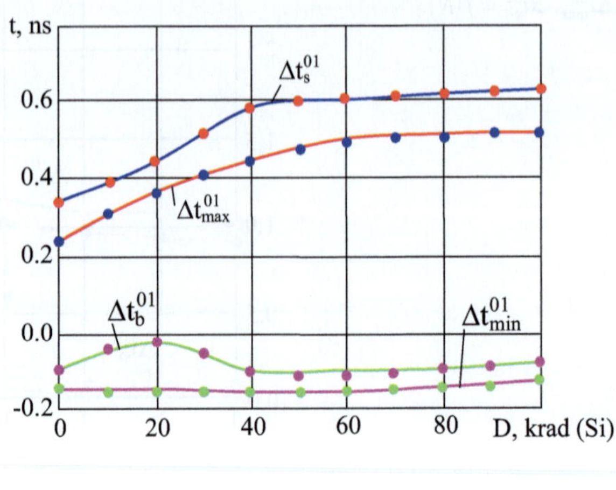

Fig. 3.24 $\Delta t_{\min}^{01}$, $\Delta t_{\mathrm{b}}^{01}$, $\Delta t_{\max}^{01}$, $\Delta t_{\mathrm{s}}^{01} = f(D)$

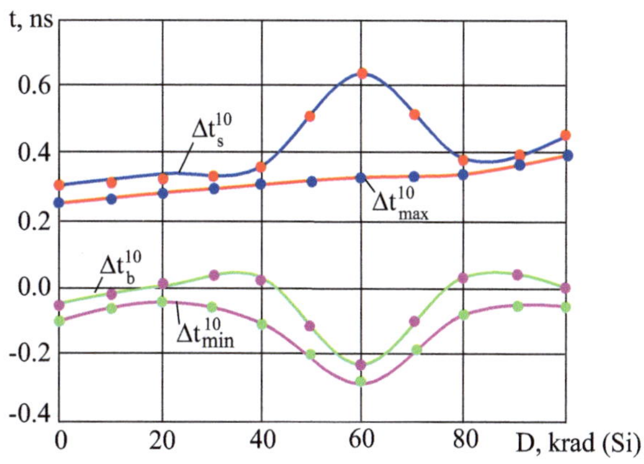

Fig. 3.25 Δt_{min}^{10}, Δt_b^{10}, Δt_{max}^{10}, $\Delta t_s^{10} = f(D)$

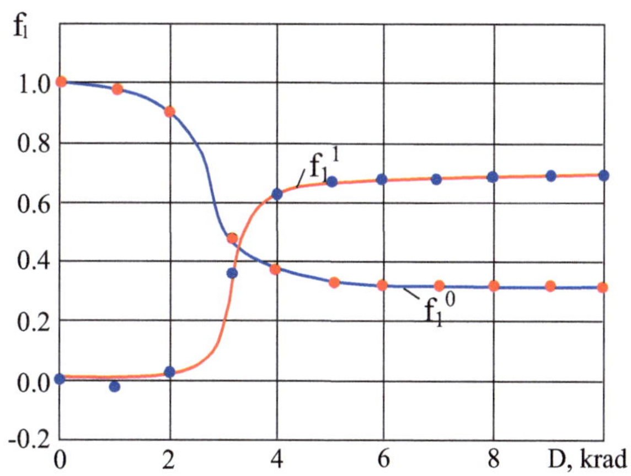

Fig. 3.26 $f_1 = f(D)$

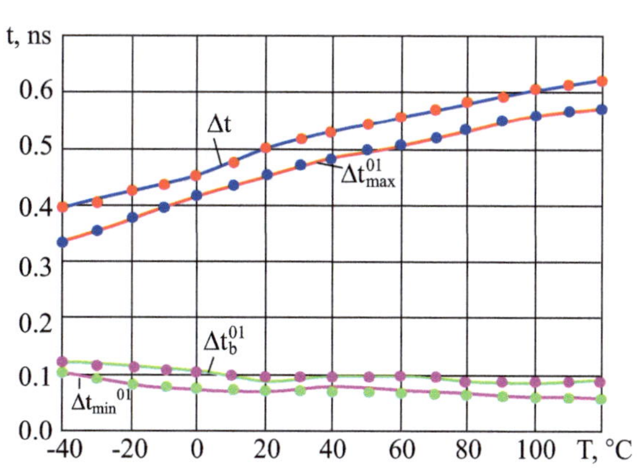

Fig. 3.27 Δt_{min}^{01}, Δt_b^{01}, Δt_{max}^{01}, $\Delta t_s^{01} = f(T)$

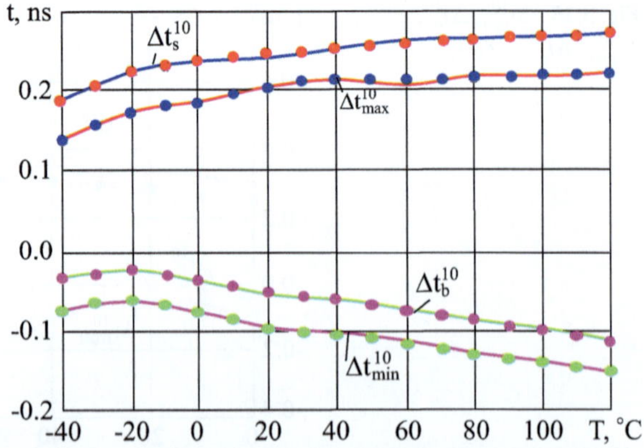

Fig. 3.28 $\Delta t_{\min}^{10}$, Δt_{b}^{10}, $\Delta t_{\max}^{10}$, $\Delta t_{s}^{10} = f(T)$

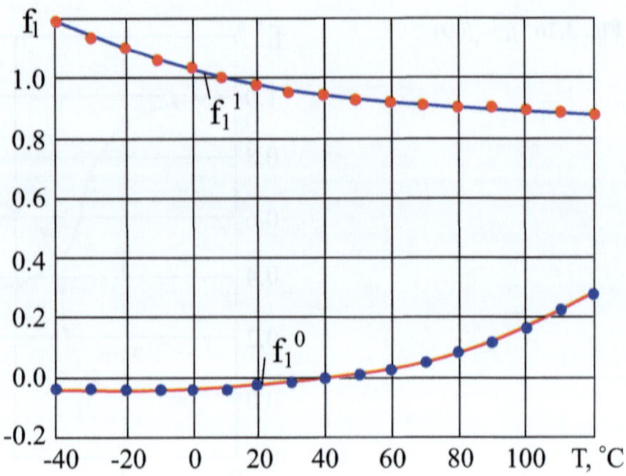

Fig. 3.29 $f_l = f(T)$

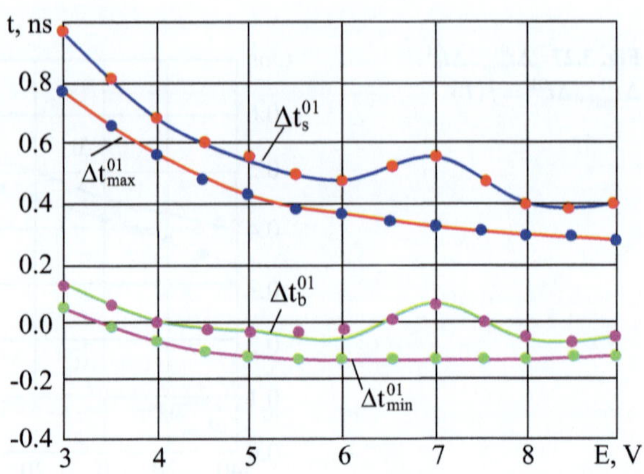

Fig. 3.30 $\Delta t_{\min}^{01}$, Δt_{b}^{01}, $\Delta t_{\max}^{01}$, $\Delta t_{s}^{01} = f(E)$

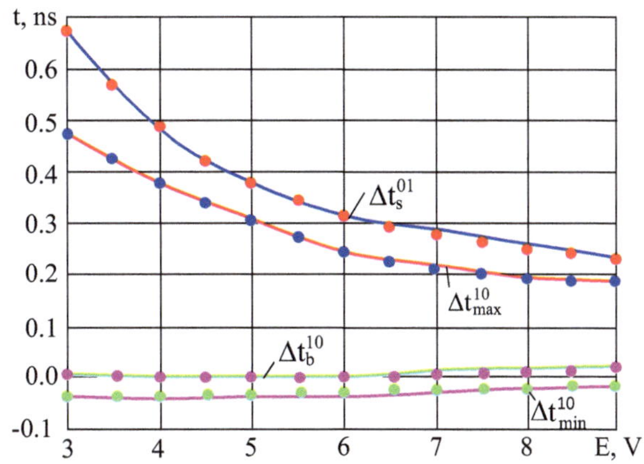

Fig. 3.31 Δt_{min}^{10}, Δt_b^{10}, Δt_{max}^{10}, $\Delta t_s^{10} = f(E)$

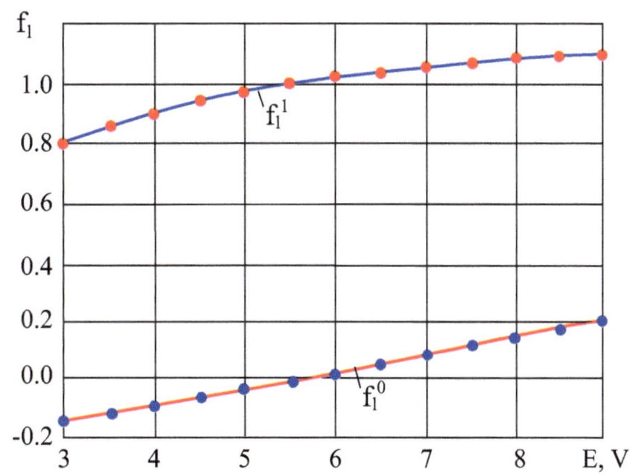

Fig. 3.32 $f_l = f(E)$

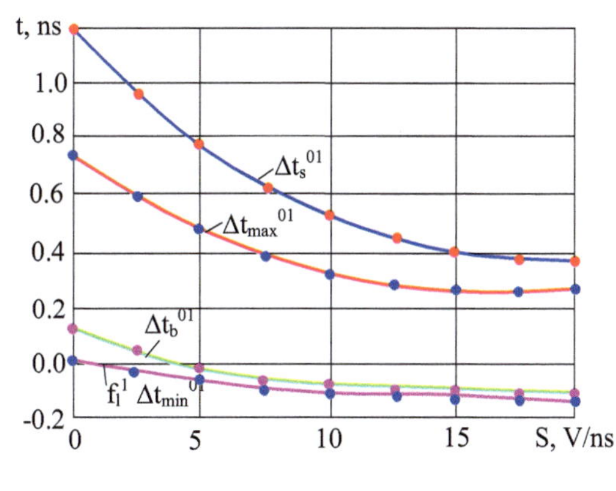

Fig. 3.33 Δt_{min}^{01}, Δt_b^{01}, Δt_{max}^{01}, $\Delta t_s^{01} = f(S)$

Fig. 3.34 Δt_{min}^{10}, Δt_b^{10}, Δt_{max}^{10}, $\Delta t_s^{10} = f(S)$

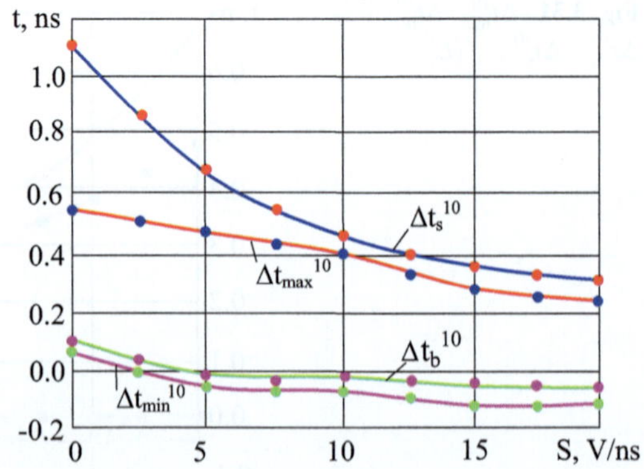

Fig. 3.35 $f_l = f(S)$

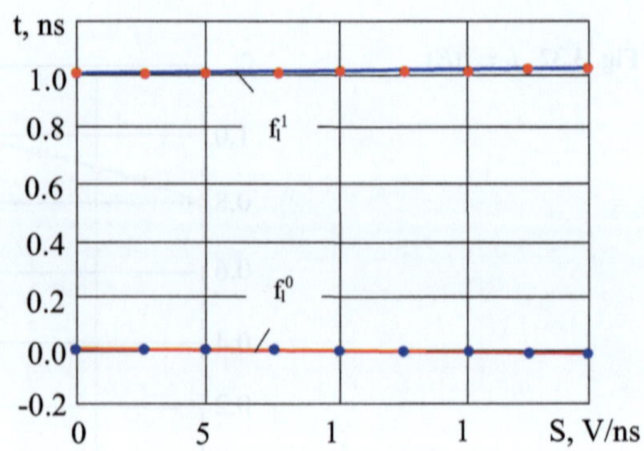

Fig. 3.36 Δt_{min}^{01}, Δt_b^{01}, Δt_{max}^{01}, $\Delta t_s^{01} = f(t_{in})$

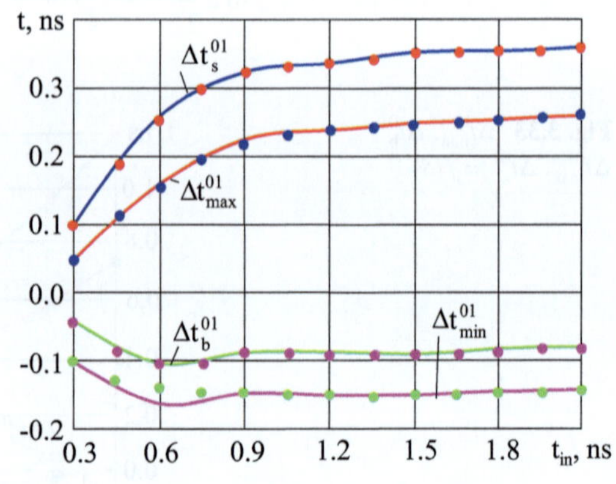

Fig. 3.37 Δt_{min}^{10}, Δt_{b}^{10}, Δt_{max}^{10}, $\Delta t_{s}^{10} = f(t_{in})$

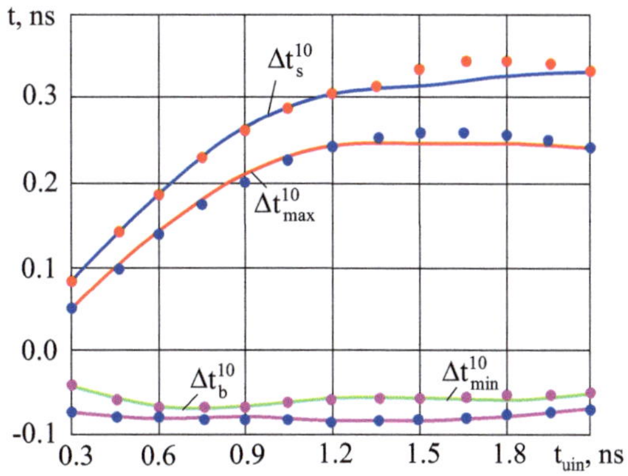

Fig. 3.38 $f_1 = f(t_{in})$

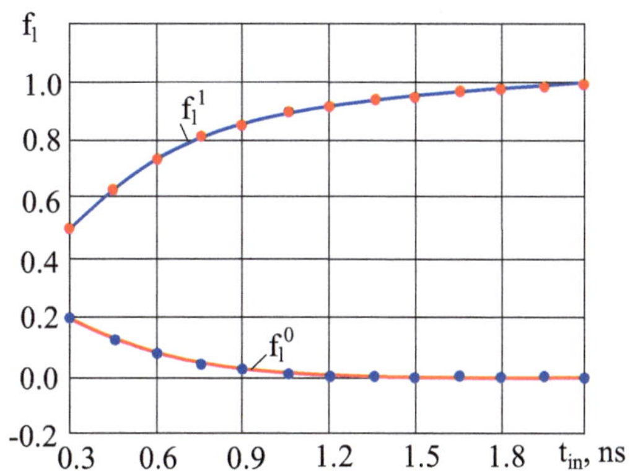

Fig. 3.39 Δt_{min}^{01}, Δt_{b}^{01}, Δt_{max}^{01}, $\Delta t_{s}^{01} = f(U_{min})$

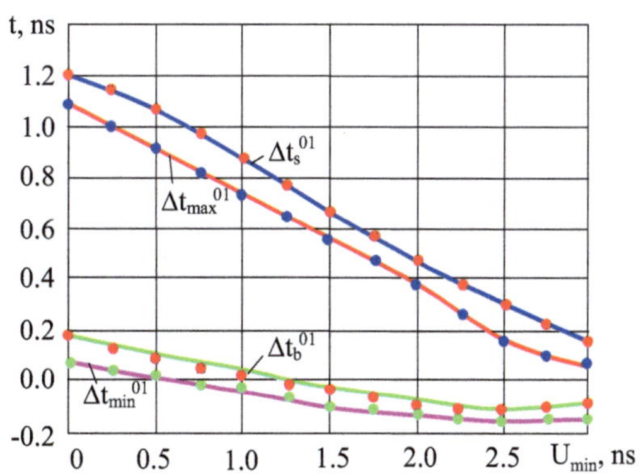

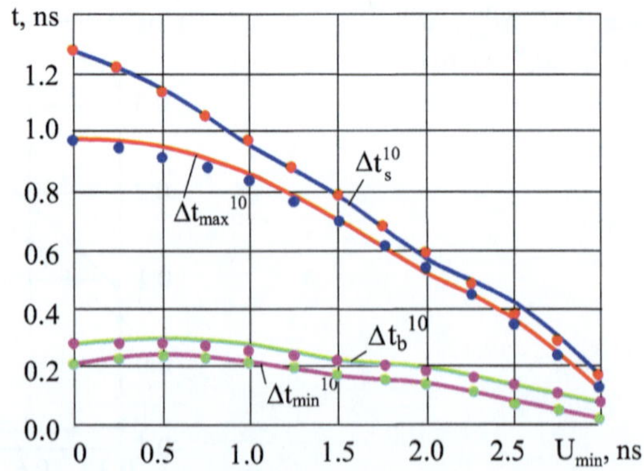

Fig. 3.40 $\Delta t_{\min}^{10}$, $\Delta t_{\mathrm{b}}^{10}$, $\Delta t_{\max}^{10}$, $\Delta t_{\mathrm{s}}^{10} = f(U_{\min})$

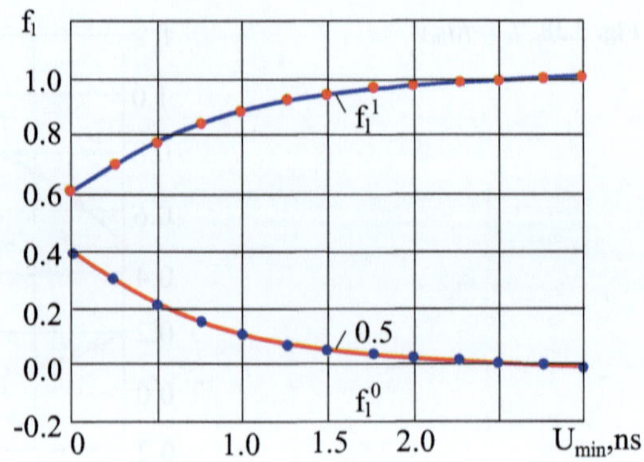

Fig. 3.41 $f_{\mathrm{l}} = f(U_{\min})$

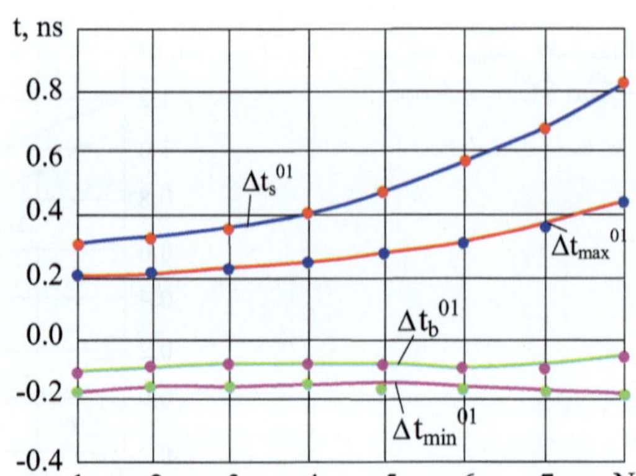

Fig. 3.42 $\Delta t_{\min}^{01}$, $\Delta t_{\mathrm{b}}^{01}$, $\Delta t_{\max}^{01}$, $\Delta t_{\mathrm{s}}^{01} = f(N)$

Fig. 3.43 Δt_{min}^{10}, Δt_b^{10}, Δt_{max}^{10}, $\Delta t_s^{10} = f(N)$

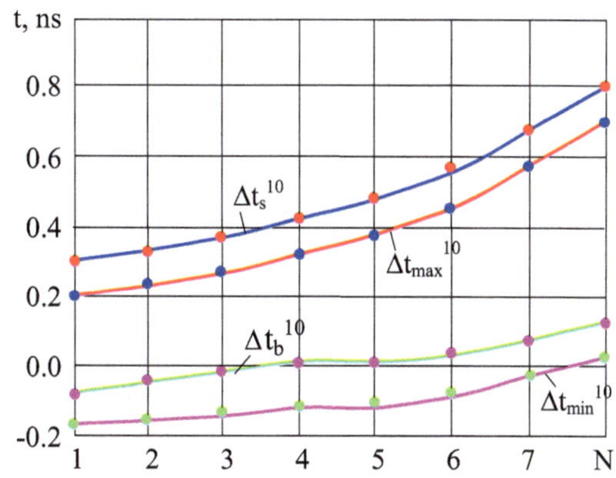

Fig. 3.44 $f_1 = f(N)$

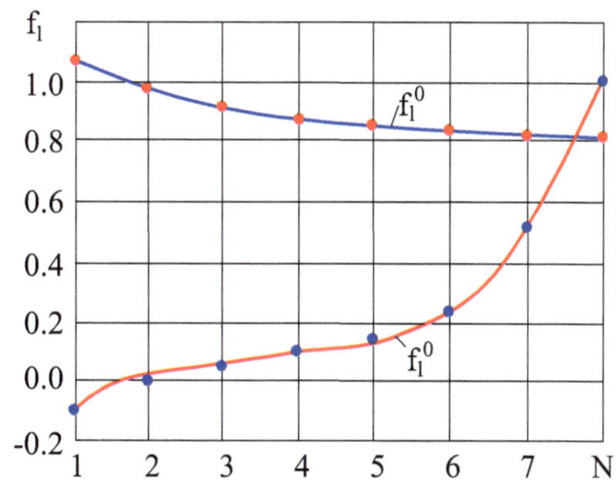

Fig. 3.45 Δt_{min}^{01}, Δt_b^{01}, Δt_{max}^{01}, $\Delta t_s^{01} = f(D)$

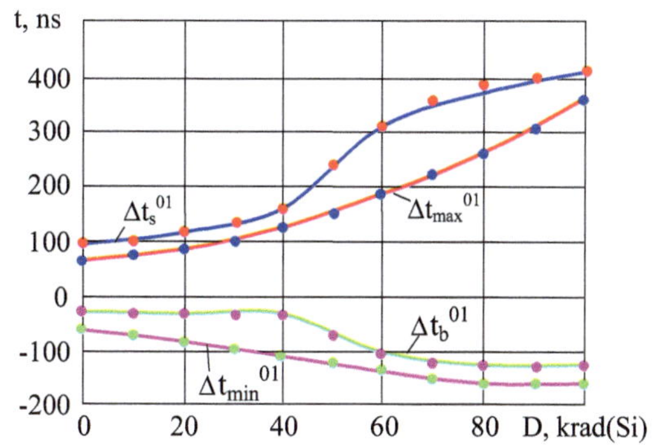

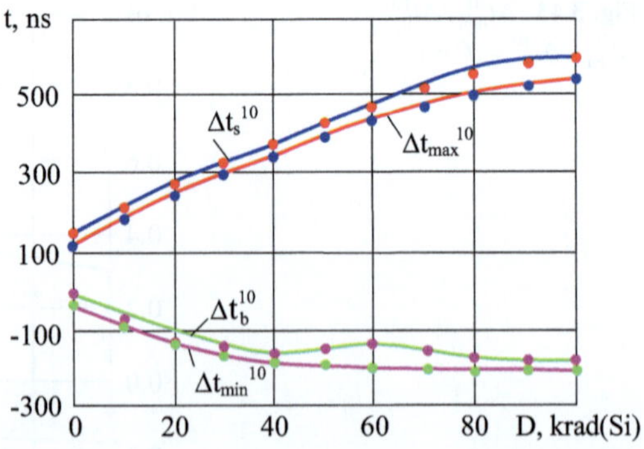

Fig. 3.46 $\Delta t^{10}_{\min}$, $\Delta t^{10}_{\mathrm{b}}$, $\Delta t^{10}_{\max}$, $\Delta t^{10}_{\mathrm{s}} = f(D)$

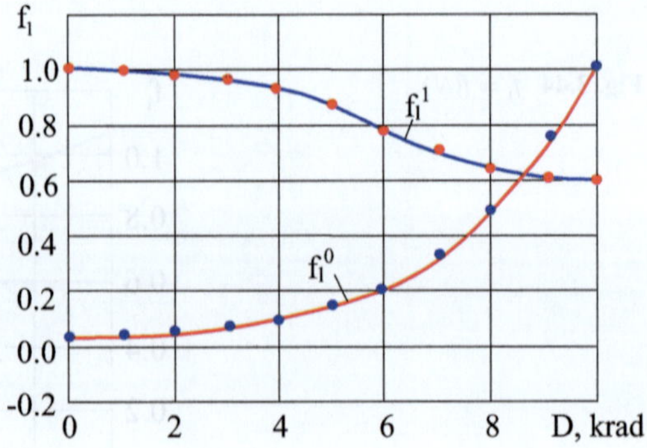

Fig. 3.47 $f_{\mathrm{l}} = f(D)$

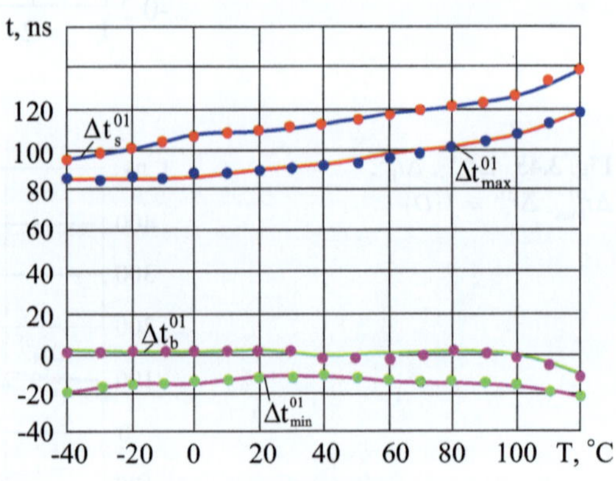

Fig. 3.48 $\Delta t^{01}_{\min}$, $\Delta t^{01}_{\mathrm{b}}$, $\Delta t^{01}_{\max}$, $\Delta t^{01}_{\mathrm{s}} = f(T)$

Fig. 3.49 Δt_{min}^{10}, Δt_b^{10}, Δt_{max}^{10}, $\Delta t_s^{10} = f(T)$

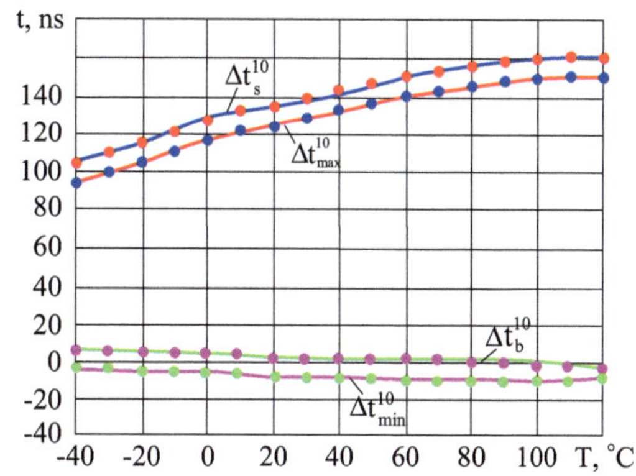

Fig. 3.50 $f_l = f(T)$

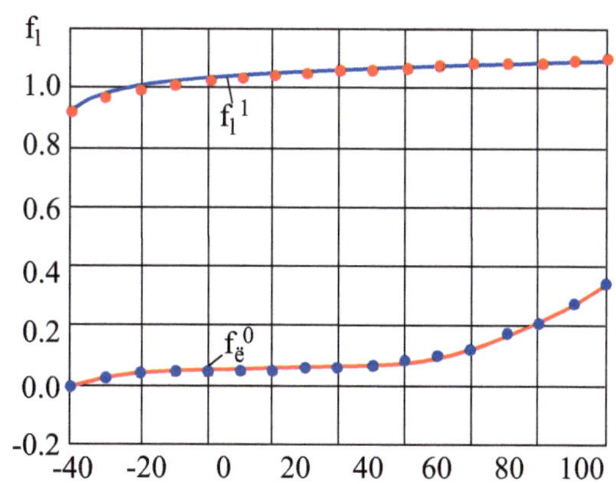

Fig. 3.51 Δt_{min}^{01}, Δt_b^{01}, Δt_{max}^{01}, $\Delta t_s^{01} = f(E)$

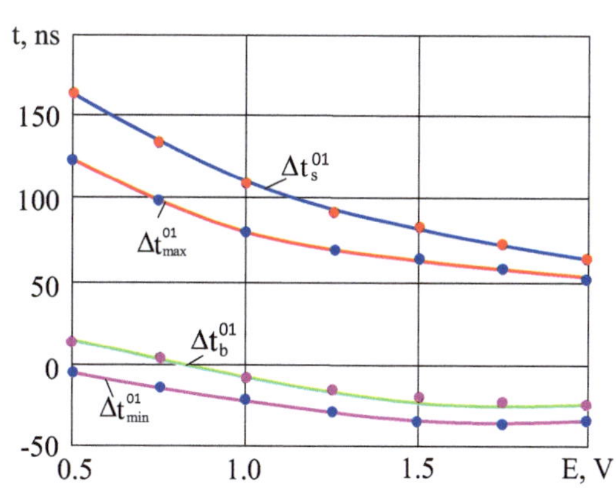

Fig. 3.52 Δt^{10}_{min}, Δt^{10}_{b}, Δt^{10}_{max}, $\Delta t^{10}_{s} = f(E)$

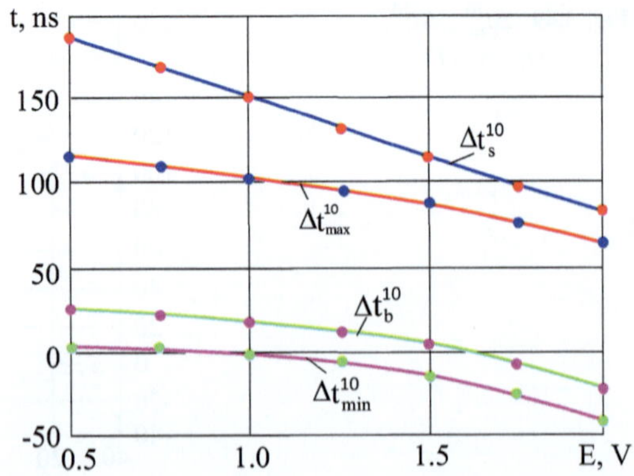

Fig. 3.53 $f_l = f(E)$

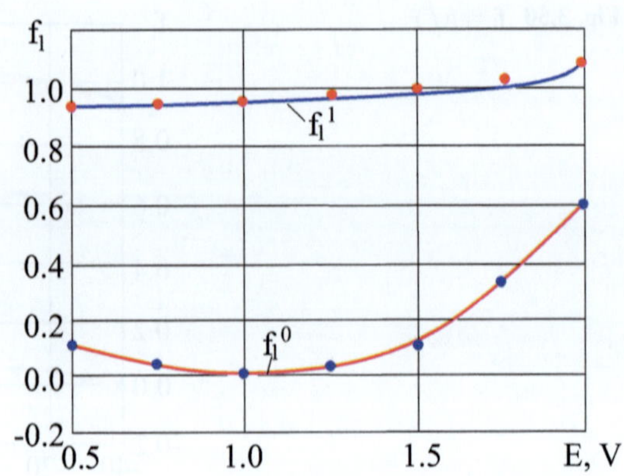

Fig. 3.54 Δt^{01}_{min}, Δt^{01}_{b}, Δt^{01}_{max}, $\Delta t^{01}_{s} = f(S)$

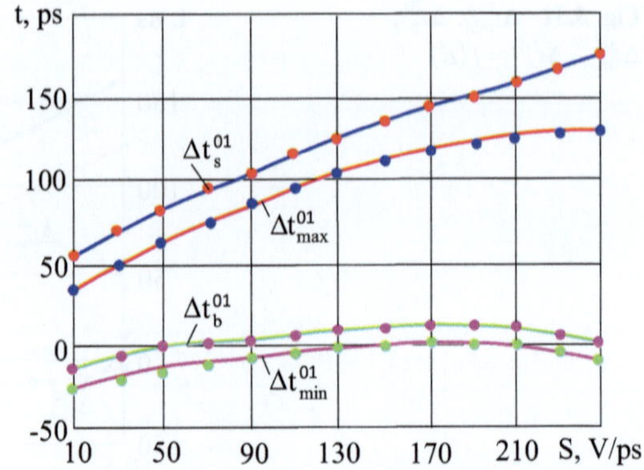

Fig. 3.55 Δt_{min}^{10}, Δt_b^{10}, Δt_{max}^{10}, $\Delta t_s^{10} = f(S)$

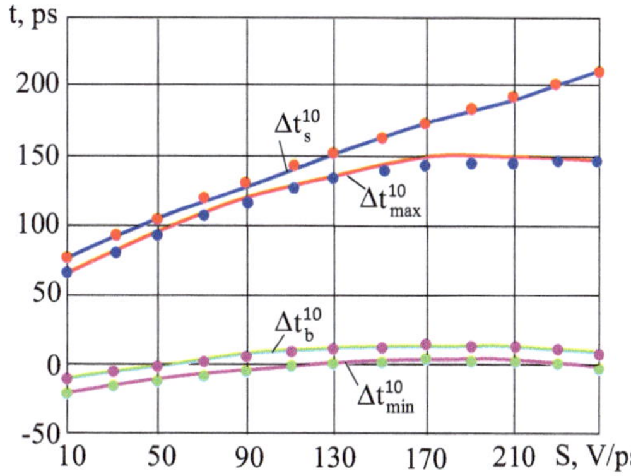

Fig. 3.56 $f_l = f(S)$

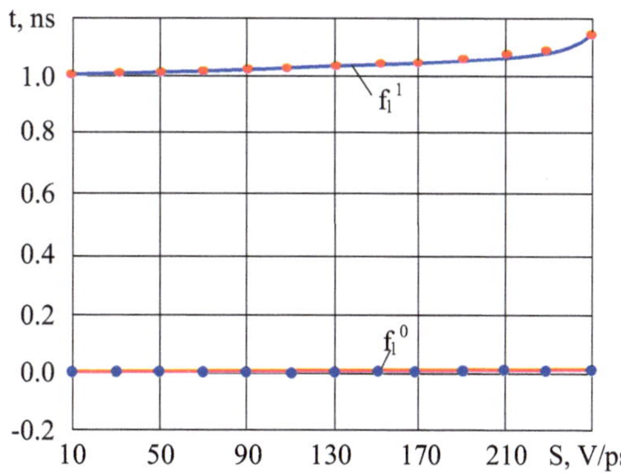

Fig. 3.57 Δt_{min}^{01}, Δt_b^{01}, Δt_{max}^{01}, $\Delta t_s^{01} = f(t_{in})$

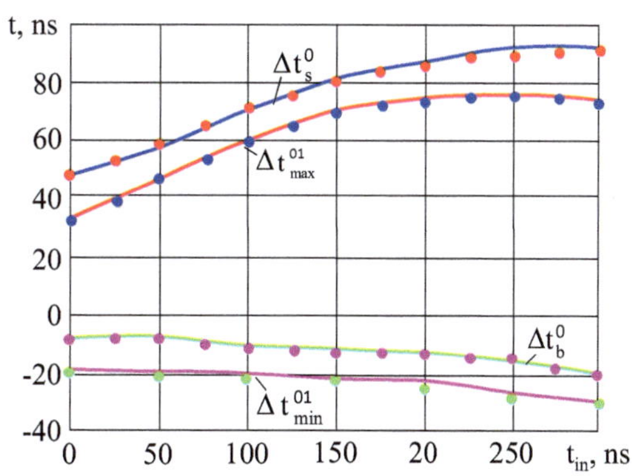

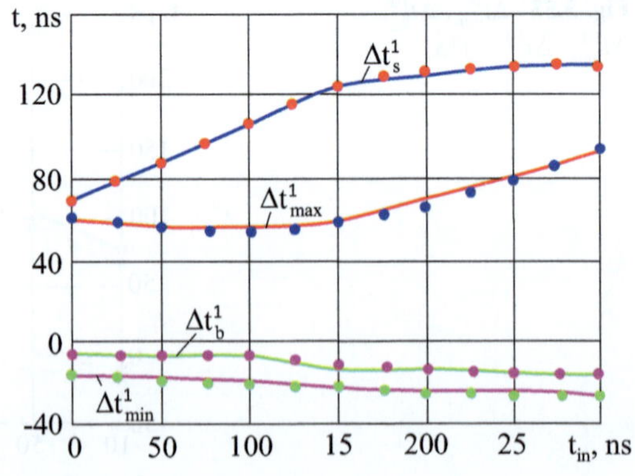

Fig. 3.58 $\Delta t_{\min}^{10}$, $\Delta t_{\mathrm{b}}^{10}$, $\Delta t_{\max}^{10}$, $\Delta t_{\mathrm{s}}^{10} = f(t_{\mathrm{in}})$

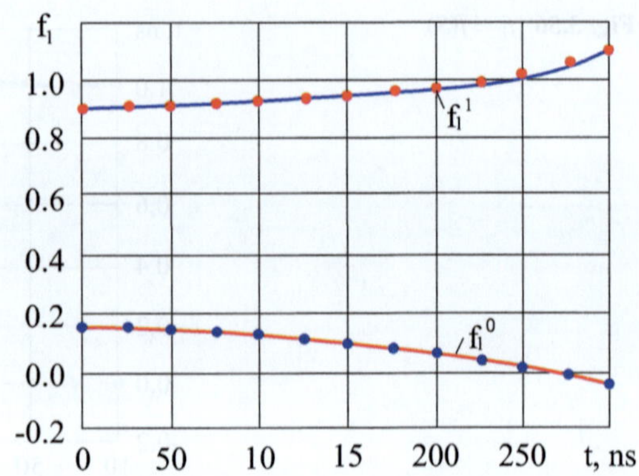

Fig. 3.59 $f_l = f(t_{in})$

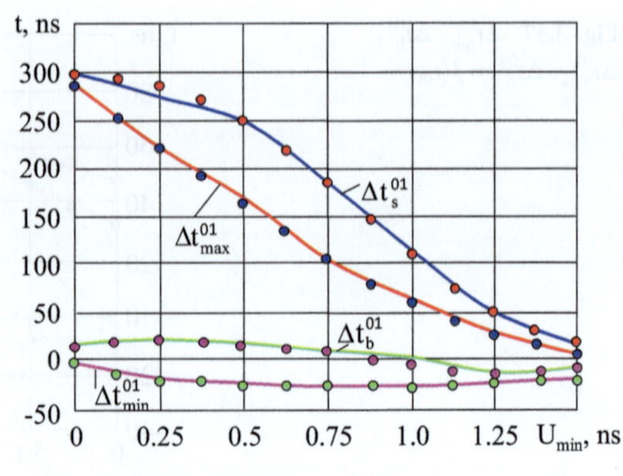

Fig. 3.60 $\Delta t_{\min}^{01}$, $\Delta t_{\mathrm{b}}^{01}$, $\Delta t_{\max}^{01}$, $\Delta t_{\mathrm{s}}^{01} = f(U_{\min})$

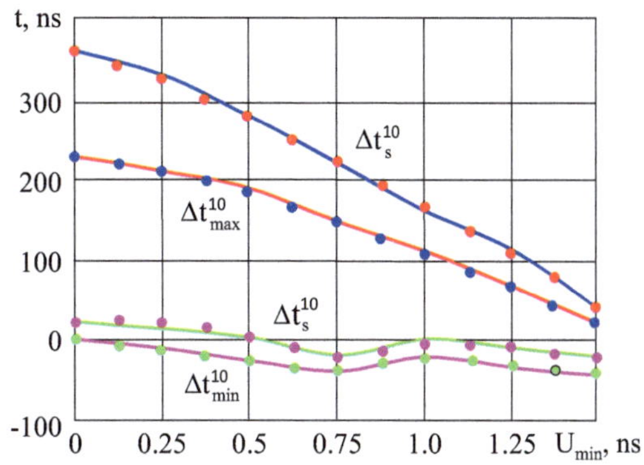

Fig. 3.61 $\Delta t_{\min}^{10}$, $\Delta t_{\mathrm{b}}^{10}$, $\Delta t_{\max}^{10}$, $\Delta t_{\mathrm{s}}^{10} = f(U_{\min})$

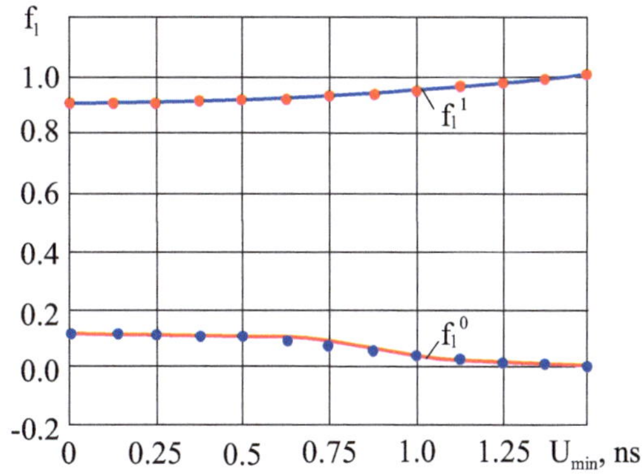

Fig. 3.62 $f_{\mathrm{l}} = f(U_{\min})$

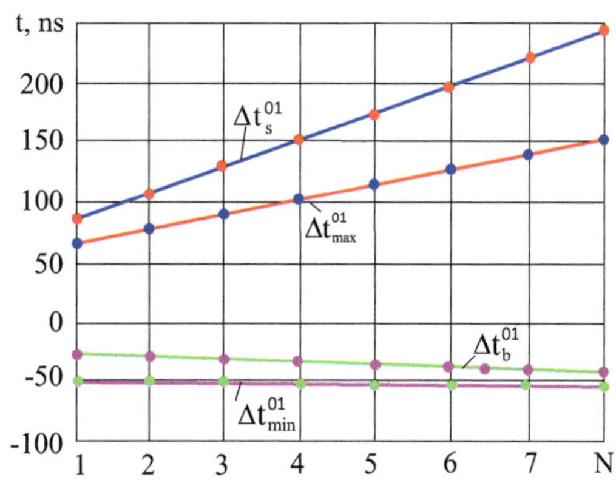

Fig. 3.63 $\Delta t_{\min}^{01}$, $\Delta t_{\mathrm{b}}^{01}$, $\Delta t_{\max}^{01}$, $\Delta t_{\mathrm{s}}^{01} = f(N)$

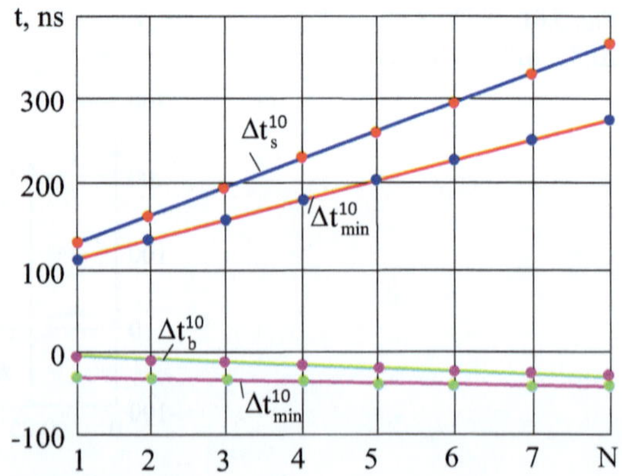

Fig. 3.64 Δt_{min}^{10}, Δt_b^{10}, Δt_{max}^{10}, $\Delta t_s^{10} = f(N)$

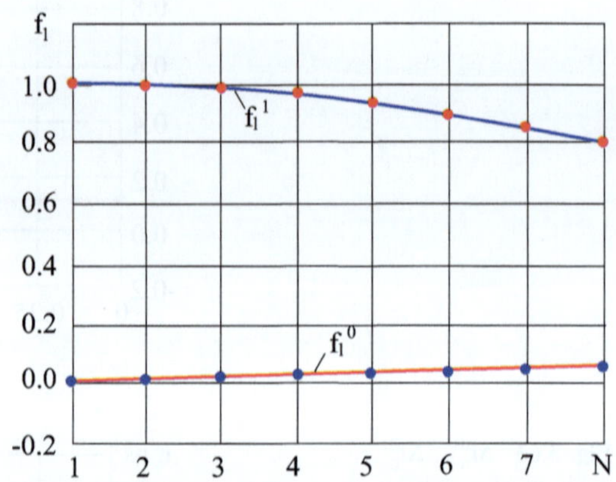

Fig. 3.65 $f_l = f(N)$

3.1.4 *Models of Interconnects*

Below are the results [100–103] of MDE interconnects obtained on the basis of the mentioned machine experiments for timing parameters of the developed MCE (Δt_{min}, Δt_b, Δt_{max}, Δt_s) in case of switching "0" → "1" and "1" → "0" for basic logic TTL cells (Fig. 1.13), ECL (Fig. 1.14) and CMOS (Fig. 1.15). RC and RLC representation of interconnects are considered, l_M is the interconnection length.

The MDEs were obtained by the following methodology:

1. The digital circuit from Fig. 1.104 is observed and timing diagrams of voltages at nodes 1–*n*, when the input signal changes from the state "0" to "1" (Fig. 1.105 for TTL) are obtained with the help of SPICE (Table 3.22).

Table 3.22 MDE interconnects for TTL technology

Interconnect model	Parameter (ns)	MDE interconnect (in formulas the dimension is Δt—ns, l_M—ns/cm)	l_M (cm)
RC	Δt^{01}_{min}	$1.3266\, l_M^2 + 0.937\, l_M - 0.0095$ $6.333\, l_M^2 - 4.7167\, l_M + 1.175$	0.0005–0.35 0.35–0.75
	Δt^{01}_{b}	$4.5249\, l_M^2 - 0.1554\, l_M + 0.10001$ $4.6667\, l_M^2 - 2.6333\, l_M + 0.95$	0.0005–0.36 0.36–0.75
	Δt^{01}_{max}	$8.8132\, l_M^2 - 0.5139\, l_M + 0.2003$ $-13.3333\, l_M^2 + 15.6667\, l_M - 2.75$ $12.5714\, l_M^2 - 13.9143\, l_M + 5.5643$	0.0005–0.33 0.33–0.48 0.48–0.75
	Δt^{01}_{s}	$7.6156\, l_M^2 + 0.478\, l_M + 0.2998$ $-17.7778\, l_M^2 + 19.1111\, l_M - 3.1111$ $11.4286\, l_M^2 - 12.2857\, l_M + 5.2857$	0.0005–0.36 0.36–0.5 0.5–0.75
	Δt^{10}_{min}	$-1.1976\, l_M^2 + 0.992\, l_M - 0.0005$ $0.5\, l_M + 0.025$	0.0005–0.35 0.35–0.75
	Δt^{10}_{b}	$-1.629\, l_M^2 + 1.4294\, l_M + 0.0993$ $0.5\, l_M + 0.225$	0.0005–0.33 0.33–0.75
	Δt^{10}_{max}	$-2.3951\, l_M^2 + 1.984\, l_M + 0.199$ $1.25\, l_M + 0.1625$	0.0005–0.37 0.37–0.75
	Δt^{10}_{s}	$-3.5927\, l_M^2 + 2.976\, l_M + 0.2985$ $1.25\, l_M + 0.4625$	0.0005–0.36 0.36–0.75
RLC	Δt^{01}_{min}	$-0.2685\, l_M^2 + 0.6684\, l_M + 0.2497$	0.0005–0.75
	Δt^{01}_{b}	$-0.269\, l_M^2 + 0.869\, l_M + 0.2996$	0.0005–0.75
	Δt^{01}_{max}	$0.2642\, l_M^2 + 0.9358\, l_M + 0.3495$	0.0005–0.75
	Δt^{01}_{s}	$0.1297\, l_M^2 + 1.3703\, l_M + 0.3993$	0.0005–0.75
	Δt^{10}_{min}	$0.4008\, l_M + 0.0998$ $-1.6\, l_M^2 + 2.4\, l_M - 0.3$	0.0005–0.25 0.25–0.75
	Δt^{10}_{b}	$0.4008\, l_M + 0.0498$ $-1.2\, l_M^2 + 1.9\, l_M - 0.25$	0.0005–0.26 0.26–0.75
	Δt^{10}_{max}	$4.0027\, l_M^2 - 1.0027\, l_M + 0.3005$	0.0005–0.75
	Δt^{10}_{s}	$4.0027\, l_M^2 - 1.0027\, l_M + 0.2005$	0.0005–0.75

2. The amount of gates is determined, after which the waveform at the output of the next gate is set with sufficient accuracy (Table 3.23). To determine the timing parameters of the MCE, 0.1 and 0.9 signal amplitude levels were used. Since the basic TTL cell (Fig. 1.13) inverts the signal, the signals between even and odd nodes are compared separately to determine the absolute and relative errors. As seen from Table 3.23, after the sixth gate, the error in determining the parameters Δt_b and Δt_s does not exceed 6.1%. Therefore, an interconnect model is included between the marked gates.
3. *n*-section RC or RLC interconnect model is connected between the sixth and seventh (for basic logic cell TTL) gates. The specific *R* and *C* values of one section were obtained using parasitic extraction tool [294] and based on the comparison of signals at nodes a and b, the values of the MCE parameters for a given *n* were obtained. By changing *n* (that is, l_M), the table dependences are

Table 3.23 Errors in determining timing parameters of the MDE interconnect

Node number	Timing parameter	Absolute value (ns)	Absolute error (ns)	Relative error (%)
1	Δt_b^{10}	0.0435	–	–
	Δt_s^{10}	0.1954	–	–
2	Δt_b^{01}	0.2071	–	–
	Δt_s^{01}	1.1677	–	–
3	Δt_b^{10}	0.0576	0.0141	24.48
	Δt_s^{10}	0.2678	0.0724	27.04
4	Δt_b^{01}	0.2733	0.0662	24.22
	Δt_s^{01}	1.3270	0.1593	12.00
5	Δt_b^{10}	0.0595	0.0019	3.19
	Δt_s^{10}	0.2817	0.0139	4.93
6	Δt_b^{01}	0.2911	0.0178	6.11
	Δt_s^{01}	1.3737	0.0467	3.40
7	Δt_b^{10}	0.0604	0.0009	1.49
	Δt_s^{10}	0.2866	0.0049	1.71
8	Δt_b^{01}	0.2964	0.0053	1.79
	Δt_s^{01}	1.3925	0.0188	1.35

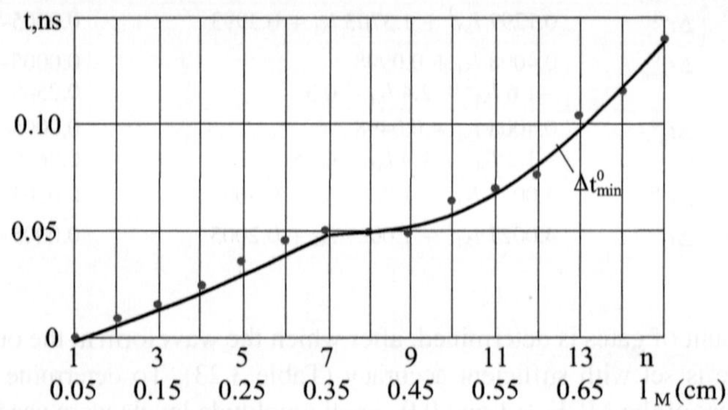

Fig. 3.66 Dependence $\Delta t_{\min}^{01} = f(l_M)$

obtained $\Delta t_{\min}^{01}, \Delta t_b^{01}, \Delta t_{\max}^{01}, \Delta t_s^{01}, \Delta t_{\min}^{10}, \Delta t_b^{10}, \Delta t_{\max}^{10}, \Delta t_s^{10} = f(l_M)$. These are the experimental points in Fig. 3.66 for the example of the dependence $\Delta t_{\min}^{01} = f$ (l_M) (column 3 of Table 3.24).

4. By means of approximation of tabular dependences, the MDE interconnects (solid line in Fig. 3.66), included in Table 3, are obtained. 3.22.

Table 3.24 Dependence $\Delta t_{\min}^{01} = f(l_M)$

n	l_M (cm)	$\Delta t_{\min}^{01}$ by experiment (ns)	$\Delta t_{\min}^{01}$ by model (ns)	Absolute error (ns)	Relative error (%)
1	0.05	0	0	0	0
2	0.10	0.100	0.116499	0.016499	14.2
3	0.15	0.160	0.179933	0.019933	11.1
4	0.20	0.250	0.250000	0	0
5	0.25	0.345	0.326700	0.018300	5.6
6	0.30	0.450	0.410033	0.039967	9.8
7	0.30	0.500	0.500000	0	0
8	0.40	0.510	0.501667	0.008333	1.7
9	0.45	0.500	0.535000	0.035000	6.5
10	0.50	0.630	0.600000	0.030000	5
11	0.55	0.700	0.696667	0.003333	0.5
12	0.60	0.750	0.825000	0.075000	9.1
13	0.65	1.050	0.985000	0.065000	6.6
14	0.70	1.154	1.176667	0.022667	1.9
15	0.75	1.400	1.400000	0	0

Similarly, the MDE interconnects for ECL (Table 3.25) and CMOS (Table 3.26) were obtained. The results of experiments similar to Fig. 3.66 and Table 3.24, for briefness are not given.

The relative error of these MCEs varies between 3.4% and 6.8% with a change in the interconnect length l_M in the range from 5 to 7500 μm.

To consider the effect of interconnects, it is proposed [103] to automatically add a Logical buffer MCE_{LB} between sequentially connected MCE_G with the parameters calculated by the MDE from Sect. 3.1.4 (Fig. 3.67a) or several MCE_{LB} with long interconnects (Fig. 3.67b). The amount of MCE_{LB} on the interconnect is roughly defined as $[\Delta t_M/\Delta t_{MCE}]$, where Δt_M is the value of the interconnection parameter calculated from the models in Sect. 3.1.4, and Δt_{MCE} are the timing parameters of the MCE for the gate of a digital circuit. In fact, with such a choice of the number of LBs on the interconnect interval, it is ensured that the parameters of the gate and LB are commensurate in order to improve the accuracy of calculations, since the MDE from Sect. 3.1.4 was obtained under close conditions for the operation of the gate to the LB.

3.1.5 MDE Noise in Power Rails

Below are the MDE noise of power rails on the parameters of the developed MDE for basic logic cells TTL (Fig. 1.13), ECL (Fig. 1.14), and CMOS (Fig. 1.15), obtained on the basis of computer experiments.

Table 3.25 MDE interconnects for ECL technology

Interconnect	Parameter (ns)	MDE interconnect (in formulas the dimension is Δt—ns, l_M—ns/cm)	l_M (cm)
RC	Δt^{01}_{min}	$6.29\, l_M^2 - 0.774\, l_M + 0.2004$ $1.32\, l_M + 0.04$	0.0005–0.51 0.51–0.75
	Δt^{01}_{b}	$6.8635\, l_M^2 - 1.1181\, l_M + 0.10056$ $1.12\, l_M - 0.01$	0.0005–0.49 0.49–0.75
	Δt^{01}_{max}	$-1.3538\, l_M^2 + 1.54155\, l_M + 1.299$ $1.61\, l_M + 1.22$ $0.7619\, l_M^2 + 0.2476\, l_M + 1.6857$	0.0005–0.26 0.26–0.51 0.51–0.75
	Δt^{01}_{s}	$-1.3538\, l_M^2 + 1.54155\, l_M + 1.499$ $1.599\, l_M + 1.41$ $0.7619\, l_M^2 + 0.2476\, l_M + 1.8857$	0.0005–0.25 0.25–0.5 0.5–0.75
	Δt^{10}_{min}	$0.3206\, l_M - 0.3998$ $0.7619\, l_M^2 + 0.2476\, l_M + 1.8857$	0.0005–0.36 0.36–0.75
	Δt^{10}_{b}	$0.4008\, l_M - 0.4998$ $0.7619\, l_M^2 + 0.2476\, l_M + 1.8857$	0.0005–0.36 0.36–0.75
	Δt^{10}_{max}	$2.4003\, l_M^2 - 0.2003\, l_M + 1.1001$	0.0005–0.75
	Δt^{10}_{s}	$2.2401\, l_M^2 - 0.08\, l_M + 0.9$	0.0005–0.75
RLC	Δt^{01}_{min}	$0.515\, l_M + 0.4197$ $0.1667\, l_M^2 - 0.1917\, l_M + 0.3625$	0.0005–0.35 0.35–0.75
	Δt^{01}_{b}	$0.7153\, l_M + 0.1996$ $-1.1333\, l_M^2 + 1.49667\, l_M + 0.215$	0.0005–0.34 0.34–0.75
	Δt^{01}_{max}	$2.7766\, l_M^2 - 2.06\, l_M + 1.601$ $-0.55\, l_M + 1.4125$	0.0005–0.33 0.33–0.75
	Δt^{01}_{s}	$0.9609\, l_M^2 - 1.1922\, l_M + 1.8006$ $-0.7775\, l_M + 1.7721$	0.0005–0.35 0.35–0.75
	Δt^{10}_{min}	$-0.8117\, l_M^2 + 1.1429\, l_M + 0.3994$	0.0005–0.75
	Δt^{10}_{b}	$-1.169\, l_M^2 + 1.411\, l_M + 0.199$	0.0005–0.75
	Δt^{10}_{max}	$-0.8534\, l_M + 1.8$ $5.613\, l_M^2 - 6.225\, l_M + 2.711$	0.0005–0.37 0.37–0.75
	Δt^{10}_{s}	$-1.087\, l_M + 1.6$ $6.413\, l_M^2 - 7.305\, l_M + 3.271$	0.0005–0.36 0.36–0.75

In Table 3.27, $f^1_{ls}, f^0_{ls}, f^1_{ld}, f^0_{ld}$ were obtained by recalculations of approximated experimental dependencies $U^1_{ns} = f(l_r)$, $U^0_{ns} = f(l_r)$, $U^1_{ld} = f(l_r)$, $U^0_{bd} = f(l_r)$, according to (2.10).

To carry out the experiments, the circuit shown in Fig. 3.69 is used.

The values $R_O = 10\ \Omega$, $C_O = 0.1$ pF and $L_O = 0.8$ nHz, obtained by parasitic extraction tool [294] were used.

The circuit in Fig. 3.69 was simulated by SPICE. A signal switching from one logic state to another (for example, "0" → "1") was applied to the input of the gate, source of noise, and the received signal was considered at the output of the gate, receiver of noise. For example, Fig. 3.70 (the circuit in Fig. 1.13 was used as a gate) shows the forms of signal changes at nodes a, b, c, d for real power rail parameters. The dashed lines show the signal received at node d with an ideal power rail.

Table 3.26 MDE interconnects for CMOS technology

Interconnect	Parameter (ns)	MDE interconnect (in formulas the dimension is Δt—ns, l_M—ns/cm)	l_M (cm)
RC	Δt^{01}_{min}	$-0.1916\ l_M^2 + 0.1587\ l_M - 0.00008$ $0.08\ l_M + 0.004$	0.0005–0.36 0.36–0.75
	Δt^{01}_{b}	$-0.2607\ l_M^2 + 0.2287\ l_M + 0.01589$ $0.08\ l_M + 0.036$	0.0005–0.35 0.35–0.75
	Δt^{01}_{max}	$-0.3822\ l_M^2 + 0.3174\ l_M + 0.0318$ $0.2\ l_M + 0.026$	0.0005–0.35 0.35–0.75
	Δt^{01}_{s}	$-0.5748\ l_M^2 + 0.4762\ l_M + 0.0477$ $0.2\ l_M + 0.074$	0.0005–3600 3600–0.75
	Δt^{10}_{min}	$0.304\ l_M^2 + 0.0994\ l_M - 0.00005$ $0.88\ l_M^2 - 0.588\ l_M + 0.17$	0.0005–0.35 0.35–0.75
	Δt^{10}_{b}	$0.724\ l_M^2 - 0.0249\ l_M + 0.016$ $0.7467\ l_M^2 - 0.4213\ l_M + 0.152$	0.0005–0.36 0.36–0.75
	Δt^{10}_{max}	$1.4101\ l_M^2 - 0.0822\ l_M + 0.032$ $-2.1333\ l_M^2 + 2.5067\ l_M - 0.44$ $2.0114\ l_M^2 - 2.2263\ l_M + 0.8903$	0.0005–0.35 0.35–0.49 0.49–0.75
	Δt^{10}_{s}	$1.2185\ l_M^2 + 0.0765\ l_M + 0.048$ $-2.8444\ l_M^2 + 3.05778\ l_M - 0.4978$ $1.8286\ l_M^2 - 1.9657\ l_M + 0.8457$	0.0005–0.37 0.37–0.51 0.51–0.75
RLC	Δt^{01}_{min}	$0.12024\ l_M + 0.01494$ $-0.36\ l_M^2 + 0.57\ l_M - 0.075$	0.0005–0.24 0.24–0.75
	Δt^{01}_{b}	$0.12024\ l_M + 0.02994$ $-0.48\ l_M^2 + 0.72\ l_M - 0.09$	0.0005–0.25 0.25–0.75
	Δt^{01}_{max}	$1.2008\ l_M^2 - 0.3008\ l_M + 0.06015$	0.0005–0.75
	Δt^{01}_{s}	$1.2008\ l_M^2 - 0.3008\ l_M + 0.09015$	0.0005–0.75
	Δt^{10}_{min}	$-0.0805\ l_M^2 + 0.2005\ l_M + 0.0749$	0.0005–0.75
	Δt^{10}_{b}	$-0.0807\ l_M^2 + 0.2607\ l_M + 0.0899$	0.0005–0.75
	Δt^{10}_{max}	$0.0793\ l_M^2 + 0.2807\ l_M + 0.1049$	0.0005–0.75
	Δt^{10}_{s}	$0.0389\ l_M^2 + 0.411\ l_M + 0.1198$	0.0005–0.75

By comparing two signals for node d, the above-described noise parameters were determined.

Approximation of tabular dependences obtained using similar experiments for various n (or, what is the same, l), the noise models (solid line in Fig. 3.71), included in Table 3.27 were obtained.

For brevity, the results of the remaining experiments are not given.

The relative error of the true MDE varies within 4.7% ... 7.2% when the distance along the power supply varies between 100 and 7500 μm between the source of noise and the gate—receiver (Fig. 3.69). Knowing the dependence of f_{lj} on l_{rj} for the ith gate in the form of an MDE from Table 3.27, it is possible to determine the total noise value f_{lj} at the output of some GLE$_i$ digital circuit. Number the gate in the order of their proximity to the power source, considering the GLE$_i$ closest. Then

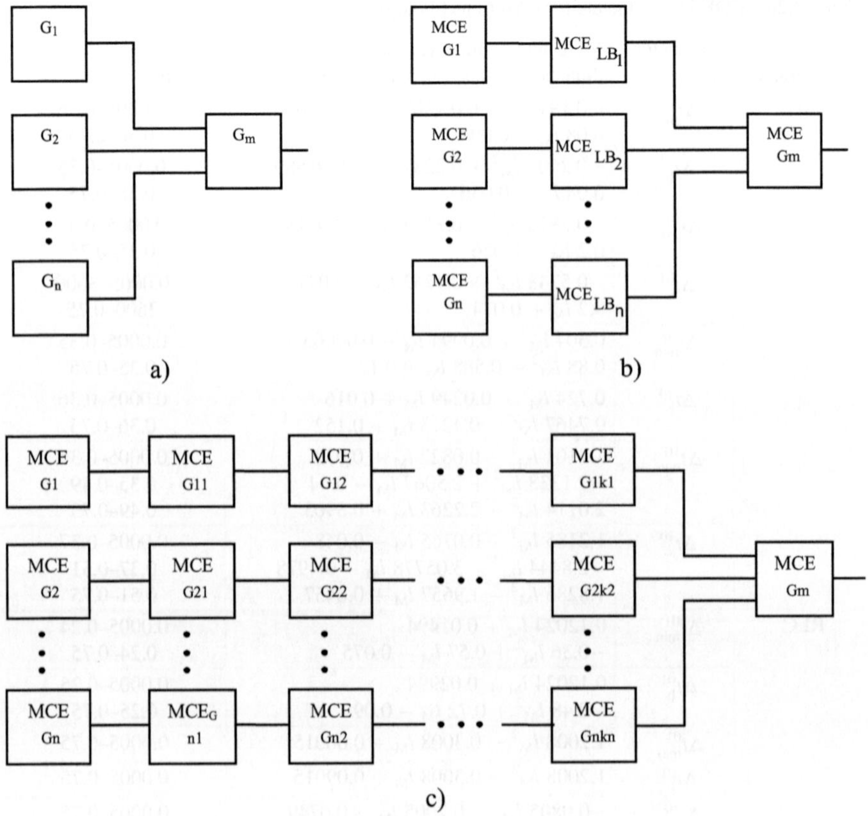

Fig. 3.67 Consideration of the effect of interconnects: (**a**) an example of a digital circuit; (**b**) digital circuit with addition of LB; (**c**) digital circuit with addition of several LBs

$$f_{\mathrm{In}} = \left[a + (b-a)\cdot\frac{U_{\mathrm{nj}} - U_{\mathrm{outmin}}}{U_{\mathrm{outmax}} - U_{\mathrm{outmin}}}\right]\cdot\sum_{j=1}^{N} q_{\mathrm{nj}}$$

$$+\sum_{j=1}^{i-1}\left[a + (b-a)\cdot\frac{U_{\mathrm{nj}} - U_{\mathrm{outmin}}}{U_{\mathrm{outmax}} - U_{\mathrm{outmin}}}\right]\cdot q_{\mathrm{nj}}, \tag{3.43}$$

$$U_{\mathrm{nj}} = U_{\mathrm{outmin}} + (U_{\mathrm{outmax}} - U_{\mathrm{outmin}})\cdot\frac{f_{\mathrm{lj}} - a}{b-a}, \tag{3.44}$$

where $q_{\mathrm{nj}} = 0$ when there is no noise in the GLE_j at a given moment of noise, and $q_{\mathrm{nj}} = 1$—in the presence.

In fact, (3.43) is a PDE for noise (Sect. 1.4). As for the PDE (1.23) for other DFs (D, E, T, etc.), they can be obtained as follows:

Table 3.27 MDE noise in power rails

Basic logic cell	Parameter (ns)	MDE (in formulas for t, the dimension is Δt—ns, and l_r—ns/cm, for f_1 the dimension l_r—1/cm)	l_r (cm)
TTL	f_{ls}^1	$-1.04\ l_r^2 + 1.5\ l_r - 0.21$	0.0005–0.75
	f_{ls}^0	$-0.4\ l_r^2 + 0.62\ l_r - 0.09$	0.0005–0.75
	f_{ld}^1	$-0.4333\ l_r + 0.4883$ $-2.6667\ l_r^2 + 2.6667\ l_r - 0.45$	0.0005–0.55 0.55–0.75
	f_{ld}^0	$-0.55\ l_r + 0.4375$ $0.4444\ l_r^2 - 1.06667\ l_r + 0.58$	0.0005–0.45 0.45–0.75
	Δt_{minn}^{01}	$0.144\ l_r^2 - 0.244\ l_r + 0.137$	0.0005–0.75
	Δt_{bn}^{01}	$0.168\ l_r^2 - 0.246\ l_r + 0.171$	0.0005–0.75
	Δt_{maxn}^{01}	$0.16\ l_r^2 - 0.36\ l_r + 0.58$	0.0005–0.75
	Δt_{sn}^{01}	$-0.22\ l_r + 0.355$	0.0005–0.75
	Δt_{minn}^{10}	$-0.08\ l_r + 0.15$	0.0005–0.75
	Δt_{bn}^{10}	$-0.08\ l_r^2 - 0.02\ l_r + 0.16$	0.0005–0.75
	Δt_{maxn}^{10}	$-0.16\ l_r + 0.34$	0.0005–0.75
	Δt_{sn}^{10}	$-0.24\ l_r^2 + 0.1\ l_r + 0.31$	0.0005–0.75
ECL	f_{ls}^1	$0.154\ l_r^2 - 0.233\ l_r + 0.156$	0.0005–0.75
	f_{ls}^0	$0.178\ l_r^2 - 0.224\ l_r + 0.182$	0.0005–0.75
	f_{ld}^1	$-0.17\ l_r + 0.33$	0.0005–0.75
	f_{ld}^0	$-0.22\ l_r^2 + 0.1\ l_r + 0.321$	0.0005–0.75
	Δt_{minn}^{01}	$-0.1826\ l_r^2 + 0.1647\ l_r - 0.0125$ $0.18\ l_r + 0.025$	0.0005–0.35 0.35–0.75
	Δt_{bn}^{01}	$-0.2517\ l_r^2 + 0.2347\ l_r + 0.0055$ $0.18\ l_r + 0.045$	0.0005–0.45 0.45–0.75
	Δt_{maxn}^{01}	$-0.381\ l_r^2 + 0.317\ l_r + 0.0318$	0.0005–0.75
	Δt_{sn}^{01}	$1.094\ l_r^2 - 1.353\ l_r + 0.4797$	0.0005–0.75
	Δt_{minn}^{10}	$-0.12\ l_r + 0.3$	0.0005–0.75
	Δt_{bn}^{10}	$-0.147\ l_r + 0.5$	0.0005–0.75
	Δt_{maxn}^{10}	$-0.162\ l_r + 0.7$ $7.76333\ l_r^2 - 11.1478\ l_r + 4.3939$	0.0005–0.45 0.45–0.75
	Δt_{sn}^{10}	$-0.222\ l_r + 0.9$ $2.513\ l_r^2 - 3.979\ l_r + 2.0816$	0.0005–0.55 0.55–0.75
CMOS	f_{ls}^1	$-0.089\ l_r + 0.14$ $-0.5\ l_r^2 + 0.4\ l_r + 0.02125$	0.0005–0.45 0.45–0.75
	f_{ld}^0	$-0.364\ l_r + 0.12$ $1.88\ l_r^2 - 2.844\ l_r + 1.0955$	0.0005–0.55 0.55–0.75
	f_{ld}^1	$-0.107\ l_r^2 + 0.2137\ l_r + 0.0399$	0.0005–0.75
	f_{ld}^0	$-0.0535\ l_r^2 + 0.1068\ l_r + 0.0199$	0.0005–0.75
	Δt_{minn}^{01}	$0.154\ l_r^2 - 0.064\ l_r + 0.057$	0.0005–0.75
	Δt_{bn}^{01}	$0.17\ l_r^2 - 0.09\ l_r + 0.093$	0.0005–0.75
	Δt_{maxn}^{01}	$0.159\ l_r^2 + 0.04\ l_r + 0.14$	0.0005–0.75

(continued)

Table 3.27 (continued)

Basic logic cell	Parameter (ns)	MDE (in formulas for t, the dimension is Δt—ns, and l_r—ns/cm, for f_1 the dimension l_r—1/cm)	l_r (cm)
	Δt^{01}_{sn}	$0.081\ l_r + 0.07$	0.0005–0.75
	Δt^{10}_{minn}	$-0.079\ l_r + 0.15$	0.0005–0.75
	Δt^{10}_{bn}	$-0.079\ l_r^2 - 0.18\ l_r + 0.06$	0.0005–0.75
	Δt^{10}_{maxn}	$0.159\ l_r + 0.18$	0.0005–0.75
	Δt^{10}_{sn}	$-0.239\ l_r^2 + 0.38\ l_r + 0.17$	0.0005–0.75

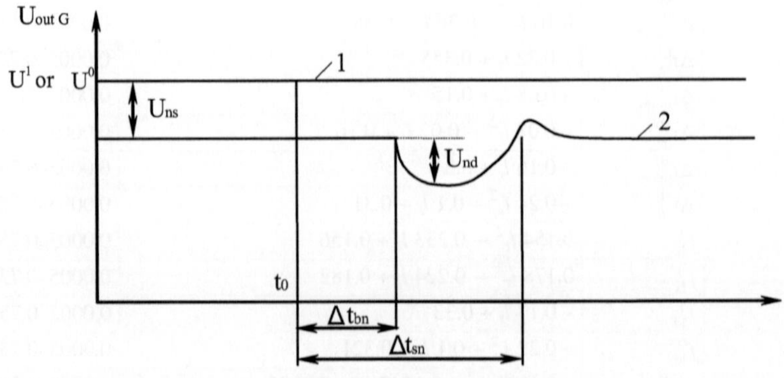

Fig. 3.68 Definitions of noise parameters. 1—voltage in the output the gate-receiver of the noise in case of an ideal power rail; 2—voltage in the output the gate-receiver of the noise taking into account the real parameters of the supply circuit; t_0—switching moment of the gate-source of noise

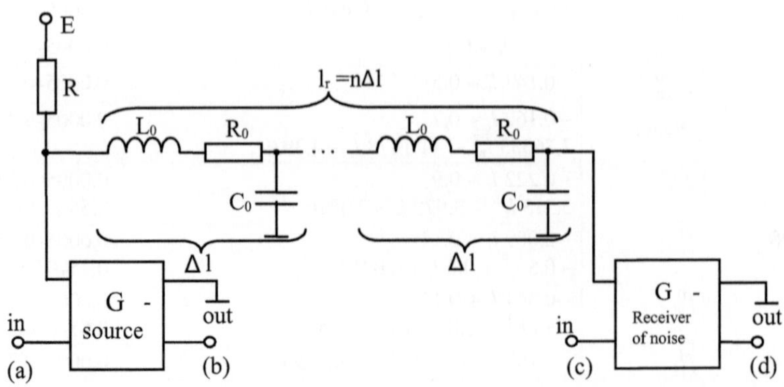

Fig. 3.69 Scheme of conducting experiments. L_r is the length of the supply chain between the gate-source of noise and gate-receiver; Δl is the length of the elementary segment of the power supply between two gates (0.05 cm); n is the number of elementary segments; R_0, L_0, C_0—resistance, inductance, and capacitance of power rails

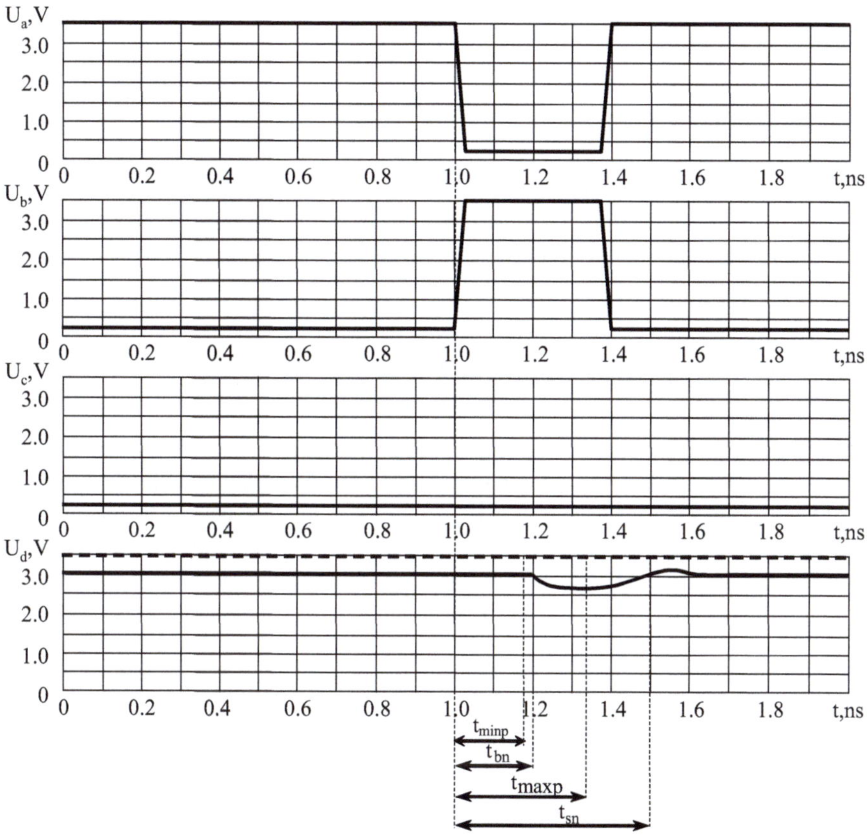

Fig. 3.70 Signal changes at the nodes a, b, c, d of the circuit in Fig. 3.69

Fig. 3.71 Dependence $U^1_{bn} = l_m$

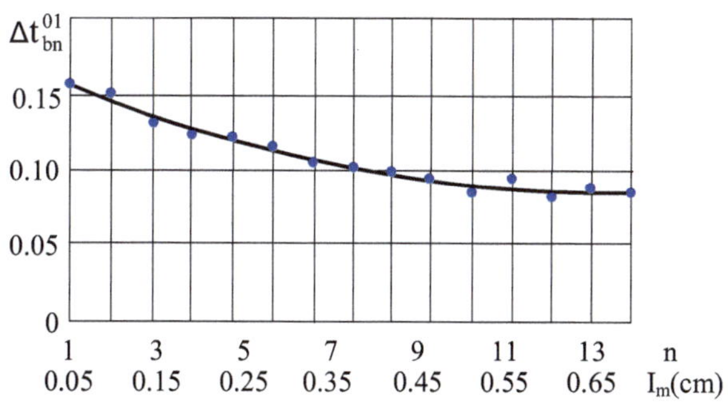

1. For each DF v_i or w_j for a specific DF η value, the values of the MCE parameters are determined (for example, P_i = MDE (η), P_i can be either f_1 or any of timing parameters of MCE) according to the developed MDE (Sects. 3.1.1–3.1.3). In fact, the values $P_{ij,j} = 1, 2, \ldots, n$ are obtained, where n is the number of different types of DF.

2. The weight coefficients a_{ij} (1.23) are defined as follows:

$$a_{ij} = \frac{P_{ij}(b-a)}{P_{\max}}, \tag{3.45}$$

where $P_{\max} = \max\{P_{ij}\}$.

The described technique for constructing models for determining the influence of DF satisfies the simulation of a digital circuit with consideration of DF.

3.2 Models of Noise in Power Lines

One of the important types of DF functioning of a digital circuit is the noise in the power rails (Sect. 1.1.2) of digital ICs. For a gate-level simulation with consideration of DF, it is necessary to have an MDE also for a given type of DF meeting the requirements of Sect. 1.3. Such requirements are met by semiempirical MDEs (Sect. 3.1.5), obtained on the basis of approximation of the results of circuit-level simulation, for the noise in power rails. However, in practical design of digital ICs, it is also often required to reflect the physical–topological parameters in the MDE. The use of models based on consideration of concentrated parasitic parameters of each section of the network of power circuits of digital ICs (Fig. 1.96) and furthermore, the distributed representation [93, 109], as well as on solving systems of differential equations (sometimes with partial derivatives) [22] in case of gate-level simulation, it is unacceptable (Sect. 1.3) because of the inadmissible inconvenience of calculations. Below is an example of the developed [140] MDE for noise in power rails, simultaneously considering the physical–topological features of power rails of modern digital ICs and providing simulation with a speed, characteristic of gate-level simulation. The MDE is based on consideration of simultaneous switching N gates from the state "1" to "0" since simulation results by SPICE (when using CMOS cells from Fig. 1.15 as a gate) confirm that it is in this case the noise is the largest (Fig. 3.72).

A typical noise signal (Fig. 3.73) [14] shows that there are clearly identifiable three stages associated with the different modes of operation of the transistors of the circuit in Fig. 1.15. Therefore, to construct the model, the circuit in Fig. 3.72 is replaced [145] by different equivalent circuits corresponding to separate stages of noise signal (Fig. 3.74). For each of these equivalent circuits (the stages of the noise signal), one can determine voltage drop on the total inductance (L_c) included between the nodes a_i and E (noise on the power rail) or between the b_i nodes and the ground (noise on the ground rail). For simplicity, further only the noise on the power rail will be spoken about. In [145], L_c is defined as:

$$\frac{1}{L_c} = \sum_{i=1}^{n} \frac{1}{L_{pri}^{e}}, \tag{3.46}$$

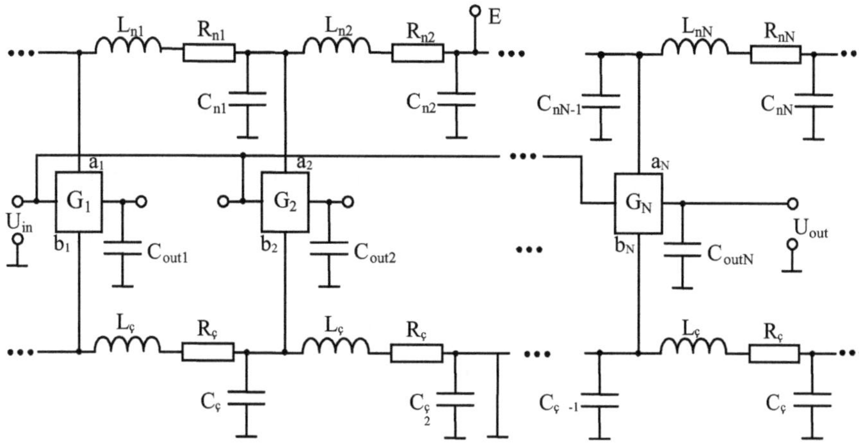

Fig. 3.72 Scheme of simultaneously switching N gates

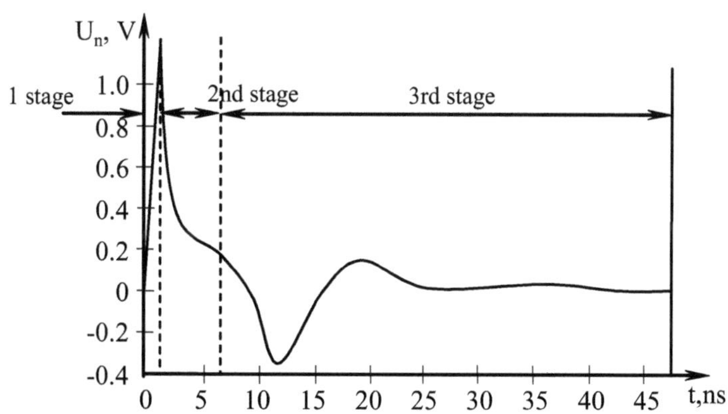

Fig. 3.73 Three stages of the noise signal

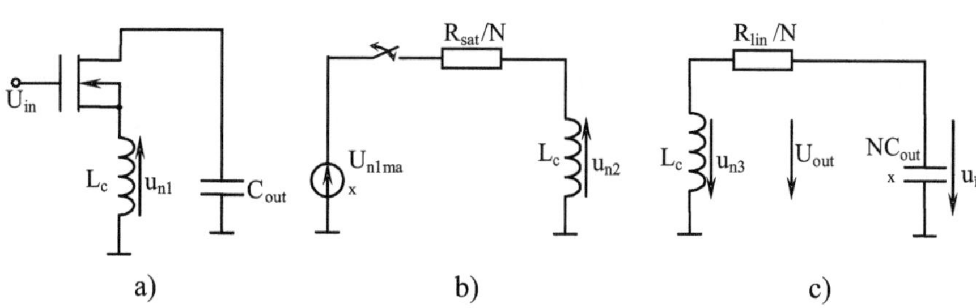

Fig. 3.74 Equivalent circuits of separate stages of the noise signal

where L_{pri}^{e} is the equivalent inductance of the ith power supply fragment. In its turn,

$$L_{\text{pri}}^{e} = \sum_{i=1}^{n} L_{\text{pri}j} \cdot \frac{I'_{\text{pr}j}}{I'_{\text{pri}}} - \sum_{i=1}^{m} L_{\text{pri,sg}j}^{e} \cdot \frac{I'_{\text{sg}j}}{I'_{\text{sg}i}}. \tag{3.47}$$

Here I'_{pr} and I'_{sg}, respectively, are the rates of change of currents at power outputs and at signal outputs; n—number of power outputs; m—number of signal outputs; $L_{\text{pri}j}$ and $L_{\text{pri,sg}j}$ are the corresponding mutual inductances. Naturally, the value of L_{c} depends on the topology of digital IC.

In the first stage of the noise signal generation (Fig. 3.73), it is assumed that the voltage grows linearly. The characteristic values of this stage are the maximum values of the voltage and the duration of the rise of the noise signal.

In the equivalent circuit of the first stage of the noise signal (Fig. 3.74a), the transistor shown generically reflects the properties of all switching transistors, C_{out} combines all the capacitances of switching gate. The voltage on L_{c} in the first stage changes as follows:

$$U_{\text{s1}}(t) = L_{\text{c}} \cdot N \cdot \frac{\mathrm{d}I_{\text{sat}}(t)}{\mathrm{d}t}, \tag{3.48}$$

where N is the number of simultaneously switching gate; $I_{\text{sat}}(t)$ is the current of a saturated transistor. The law of changing $I_{\text{sat}}(t)$ can be obtained using the transistor model (for example, BSIM3 [33]).

$$\begin{aligned} U_{\text{n1max}} &= L_{\text{c}} \cdot N \cdot w \cdot C_{\text{ox}} \cdot U_{\text{sat}} \cdot \left(\frac{U_{\text{in}}}{T} - \frac{U_{\text{n1max}}}{T - t_{\text{d}}} \right) \\ &\times \frac{(2E_{\text{sat}}l + U_{\text{in}} - U_{\text{n1max}} - U_{\text{th}}) \cdot (U_{\text{in}} - U_{\text{n1max}} - U_{\text{th}})}{(E_{\text{sat}}l + U_{\text{in}} - U_{\text{n1max}} - U_{\text{th}})^2}. \end{aligned} \tag{3.49}$$

Here, U_{n1max} is the maximum value of the voltage of the noise signal during the first stage; l is the length; w is the channel width of the transistor; C_{ox} is the specific value of the capacitance of the oxide layer; E_{sat}—the electric field intensity in the saturated transistor channel; U_{th}—transistor threshold voltage; U_{sat}—the voltage at the output of a saturated transistor. Expression (3.49) holds for the time interval [0, T], where T is the rise time of U_{in} to the maximum value; t_{d}—rise time of U_{in} to the value of U_{th}, defined as $t_{\text{d}} = \frac{U_{\text{th}}}{E_{\text{n}}} \cdot t_{\text{r}}$; E—the value of the supply voltage; t_{r}—the rise time of U_{in} to U_{sat}. During the second stage, U_{n} decreases according to a law close to the exponential function. At this stage, the n-channel transistor operates in saturation mode. Therefore, it can be assumed that during this stage the channel resistance of this transistor is small. Since the n-channel transistors of parallel connected N gates are connected in parallel, the equivalent resistance of the second stage of R_{e2} is calculated as $R_{\text{e2}} = \frac{R_{\text{sat}}}{N}$, where R_{sat} is the channel resistance of one saturated transistor. To determine the noise voltage in the second stage, the following relation is used:

$$U_{\text{n1max}} = R_{\text{e2}} \cdot I(t) + L_{\text{c}} \cdot \frac{\mathrm{d}I(t)}{\mathrm{d}t} \tag{3.50}$$

where U_{n1max} is the maximum value of the noise voltage during the first stage. From (3.50) this can be determined:

$$I(t) = \frac{U_{\text{n1max}}}{R_{\text{e2}}} \cdot \left(1 - \exp^{-\frac{t}{T_{\text{L}}}}\right), \tag{3.51}$$

where the time constant T_{L} is defined as $\frac{L_{\text{c}}}{R_{\text{e2}}}$. Using the Kirchhoff law, one can obtain a formula describing the variation of the noise voltage during the second stage:

$$U_{\text{n2}}(t) = U_{\text{n1max}} - R_{\text{e2}} \cdot I(t) = U_{\text{n1max}} \cdot \exp^{-\frac{t}{T_{\text{L}}}}. \tag{3.52}$$

Since during the third stage the noise signal has an oscillatory character, it can be described by the oscillation frequency and the attenuation coefficient of the parallel RLC oscillating circuit from Fig. 3.74c.

In this circuit, R_{lin} is the resistance of the channel when the transistor operates in the linear area of the characteristic. While in the circuit in Fig. 3.72 all transistors are connected in parallel, the equivalent resistance R_{e3} in this case can be defined as R_{lin}/N. The equivalent capacitance of the load in the third stage is defined as $C_{\text{e3}} = N \cdot C_{\text{out}}$, where C_{out} is the load capacitance of one gate.

The law of change of the noise voltage during the third stage can be determined by solving the following differential equation:

$$L_{\text{c}} \cdot C_{\text{e3}} \cdot \frac{\mathrm{d}^2 U_{\text{c}}(t)}{\mathrm{d}t^2} + R_{\text{e3}} \cdot C_{\text{e3}} \cdot \frac{\mathrm{d}U_{\text{c}}(t)}{\mathrm{d}t} + U_{\text{c}}(t) = 0. \tag{3.53}$$

Depending on the value of the attenuation coefficient, defined as:

$$\partial = \frac{R_{\text{e2}} \cdot \sqrt{L_{\text{c}} \cdot C_{\text{e3}}}}{2L_{\text{c}}}, \tag{3.54}$$

three different solutions of equation are possible (3.53):

(a) $\partial > 1$ (the case of rapid damping):

$$U_{\text{n3}}(t) = \frac{U_{\text{sat}}}{2\sqrt{\left(\partial^2 - 1\right) \cdot L_{\text{c}} \cdot C_{\text{e3}}}} \times \left\{\sqrt{L_{\text{c}} \cdot C_{ý}3} \cdot \left[\left(\partial + \sqrt{\partial^2 - 1}\right) \cdot \exp^{\frac{-t}{\partial + \sqrt{\partial^2 - 1}}} - \left(\partial - \sqrt{\partial^2 - 1}\right) \cdot \exp^{\frac{-t}{\partial - \sqrt{\partial^2 - 1}}}\right]\right\}; \tag{3.55}$$

(b) ∂ is equivalent to 1 (the case of characteristic attenuation):

$$U_{n3}(t) = U_{sat} \cdot \left(1 - \frac{1}{\sqrt{L_c \cdot C_{e3}}}\right) \cdot \exp^{\frac{-t}{\sqrt{L_c \cdot C_{e3}}}}; \tag{3.56}$$

(c) $\partial < 1$ (the case of slow damping):

$$U_{n3}(t) = U_{sat} \cdot \exp^{-\frac{R_{e3} \cdot t}{2L_c}} \cdot \cos\left[\sqrt{\frac{1-\partial^2}{L_c \cdot C_{e3}}} \cdot t - \operatorname{arctg}\left(R_{e3}\sqrt{\frac{C_{e3}}{L_c(1-\partial)^2}}\right)\right]. \tag{3.57}$$

Combining all three steps and using (2.10), one obtains the following MDE of the noise of power rails, taking into account the physical–topological parameters:

$$f_{lb} = \begin{cases} 0 & t < t_d \\ \left[a + (b-a) \cdot \dfrac{U_{n1\max} - U_{\text{outmin}}}{U_{\text{outmax}} - U_{\text{outmin}}}\right] \cdot \dfrac{t - t_3}{t_r - t_d} & t_d \le t \le t_{tr} \\ \left[a + (b-a) \cdot \dfrac{U_{n1\max} - U_{\text{outmin}}}{U_{\text{outmax}} - U_{\text{outmin}}}\right] \cdot \exp^{-\frac{t - t_3 - t_c}{T_L}} - |k| & t_r \le t \le t^* \\ \left[\dfrac{a + (b-a) \cdot \dfrac{U_{sat} - U_{\text{outmin}}}{U_{\text{outmax}} - U_{\text{outmin}}}}{2\sqrt{\left(\partial^2 - 1\right) \cdot L_c \cdot C_{e3}}}\right] \times & \\ \times \left\{\sqrt{L_c \cdot C_{e3}} \cdot \left[\left(\partial + \sqrt{\partial^2 - 1}\right) \cdot \exp^{\frac{-t}{\partial + \sqrt{\partial^2 - 1}}} \right.\right. & t > t^*, \quad \partial > 1 \\ \left.\left. - \left(\partial - \sqrt{\partial^2 - 1}\right) \cdot \exp^{\frac{-t}{\partial - \sqrt{\partial^2 - 1}}}\right]\right\} & \\ \left[a + (b-a) \cdot \dfrac{U_{sat} - U_{\text{outmin}}}{U_{\text{outmax}} - U_{\text{outmin}}}\right] \cdot \left(1 - \dfrac{1}{\sqrt{L_c \cdot C_{e3}}}\right) \cdot \exp^{\frac{-t}{\sqrt{L_c \cdot C_{e3}}}} & t > t^*, \quad \partial = 1 \\ \left[a + (b-a) \cdot \dfrac{U_{sat} - U_{\text{outmin}}}{U_{\text{outmax}} - U_{\text{outmin}}}\right] \cdot \exp^{-\frac{R_{e3} \cdot t}{2L_c}} \times & \\ \times \cos\left[\sqrt{\dfrac{1-\partial^2}{L_c \cdot C_{e3}}} \cdot t - \operatorname{arctg}\left(R_{e3}\sqrt{\dfrac{C_{e3}}{L_c(1-\partial)^2}}\right)\right] & t > t^*, \quad \partial < 1 \end{cases} \tag{3.58}$$

where a and b are the levels between which MCE gate switches (Sect. 2.1); U_{outmin} and U_{outmax}—the minimum and maximum values of the voltages at the output of the gate when switching from the state "1" and "0"; t_c—the additional time shift during the second stage of the noise signal, defined as follows:

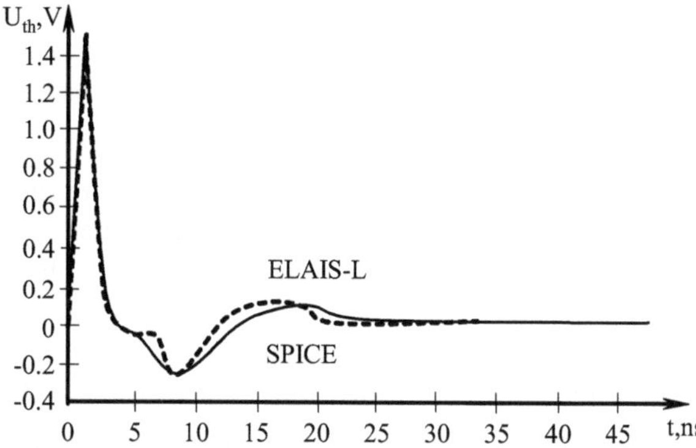

Fig. 3.75 Comparison of simulation results of the circuit on Fig. 3.4

$$t_c = T_L \cdot \ln \cdot \left[1 + \frac{k}{a + (b - a) \cdot \frac{U_{n1max} - U_{outmin}}{U_{outmax} - U_{outmin}}} \right]; \tag{3.59}$$

k—an additional voltage shift during the second stage, defined as:

$$k = \left| \min \left[a + (b - a) \cdot \frac{U_{n1max} - U_{outmin}}{U_{outmax} - U_{outmin}} \right] \cdot \exp^{-\frac{t}{T_L}} \right|, \tag{3.60}$$

U_{n1max} is defined from (3.49).

The model described in (3.58) is the MDE of noise in power rails, with consideration of physical–topological parameters. All the necessary parameters of this MDE are divided into four groups: physical constants, transistor parameters, circuit parameters, and parameters determined by simulation (for example, R_{sat}, R_{lin}). The latter can be determined with a single simulation of the gate with the help of SPICE.

To evaluate the accuracy of the developed MDE, the circuit in Fig. 3.72 at the device level has been simulated by SPICE ($N = 8$, $C_{out} = 50$ nf, $L_c = 10$ nHz, $t_r = 1$ ns) and developed by the MDE using the automated system of gate-level simulation of digital circuits with consideration of DF [140], ELAIS-L. The comparison results are shown in Fig. 3.75.

The average relative error in calculations using the developed MDE is only 4.8%, while the simulation time for ELAIS-L is about 148 times smaller.

3.3 Calculation and Optimization of the Parameters of Models for Determining the Influence of DF

The problem of calculation and optimization of parameters (COP) of the MDE DF is one of the main one in the development of the automated system of gate-level simulation and optimization of digital circuits, since the simulation results also largely depend on the accuracy of determining the parameters of the MDE (Sect. 3.2).

In the system of gate-level simulation and optimization of digital circuits with consideration of DF, the COP program of the MDE DF is an independent subsystem. This subsystem solves the problem of the COP of the MDE in two aspects:

1. The input data are the electrical parameters or characteristics of a separate gate, presented at the circuit-level detalization (for example, U^0_{nom}, C_{e3}, β_{N3} etc. in the MDE (3.23), or physical–topological parameters (as in the MDE from Sect. 3.2)), and output parameters are the parameters of the MDE (for example, (3.24)–(3.31)).
2. The input data is the parameters of the MDE (for example, (3.24)–(3.31)), and the output parameters are the parameters of the MCE (for example, (3.23)).

In the COP of the MDE DF, the target function $F(\vec{P})$ ($\vec{P}$—vector of the parameters of the MDE) should reflect the proximity of the MDE parameters to the experimental data. The problem of the COP of the MDE DF is multicriteria, and the methods of $F(\vec{P})$ formation by separate $C_j(\vec{P})$ criteria differ from the traditional optimization methods [297]. If, for example, the timing parameters of the MCE Δt_j are calculated, the $C_j(\vec{P})$ criteria will have the following form:

$$C_j(\vec{P}) = \frac{\Delta t_{je}(\vec{P}) - \Delta t_{jm}(\vec{P})}{\Delta_{\Delta t_j}}, \tag{3.61}$$

where $\Delta t_{jm}(\vec{P})$ is the value of the timing parameter of the MCE obtained by the MDE, a $\Delta t_{je}(\vec{P})$ is the experiment for a given value of the vector $\vec{P}$; $\Delta_{\Delta t_j}$—an absolute error for the value Δt_j.

If the MDE reflects the dependence of some timing parameter Δt_j on the parameter MCE $A_i(\Delta t_j = \mathrm{MDE_c}(A_i))$ and the errors are specified, and for $\Delta_{\Delta t_j}$ and Δ_{A_i}, respectively Δt_j and A_i, then the criterion of the proximity of the point $(\Delta t_{jm}, A_{im})$ can be introduced to the dependence $\Delta t_j = \mathrm{MDE_c}(A_i)$ as the shortest distance between them in a normed space of variables $\Delta \bar{t}_j = \dfrac{\Delta t_j}{\Delta_{\Delta t_j}}$ and $\bar{A}_i = \dfrac{A_i}{\Delta_{A_i}}$.

$$l_{\mathrm{m}} = \frac{\left|\Delta t_{jm} - \mathrm{MDE_c}(A_{i0}) - (A_{im} - A_{i0}) \cdot \mathrm{MDE_{co}}\right|}{\sqrt{\Delta^2_{\Delta t_i} + \Delta^2_{A_i}(\mathrm{MDE_{co}})^2}}, \tag{3.62}$$

where $\mathrm{MDE}'_{\mathrm{co}}$ is the derivative of the function of determining the influence of the DF at $A_i = A_{i0}$; A_{i0} coordinate of overlapping the function $\Delta \bar{t}_j(\bar{A}_i)$ and perpendicular to it in a normed space passing through $(\Delta \bar{t}_m, \bar{A}_{im})$. When $\mathrm{MDE}'_{\mathrm{co}} = 0$, $A_{i\mathrm{f}} = A_{im}$ is obtained, and l_{m} is determined by the error with respect to Δt_j: $l_{\mathrm{m}} = \left|\Delta t_{jm} - \mathrm{MDE_c}(A_{im})\right| / \Delta_{\Delta t_j} \Delta t_j$. When $\mathrm{MDE}'_{\mathrm{co}} \to \infty$, $\mathrm{MDE_c}(A_{i0}) = \mathrm{MDE_c}(A_{im})$

and l_{m} is determined by the error in A_i. In general case, if $l_{\text{m}} \leq \frac{\Delta_{\Delta t_j}+\Delta_{A_i}\left|\text{MDE}'_{\text{co}}\right|}{\sqrt{\Delta^2_{\Delta t_j}+\Delta^2_{A_i}\left(\text{MDE}'_{\text{co}}\right)^2}}$, then the error of the COP is within acceptable limits, i.e. point (Δt_{jm}, A_{im}) is inside the range $\pm\Delta t_j$, $\pm\Delta_{A_j}$, from one of the points of dependence $\Delta t_j = \text{MDE}_{\text{c}}(A_i)$. Thus, the criterion for the quality of the COP of the MDE DF can be defined as:

$$F_{\text{m}} = \frac{\left|\Delta t_{jm} - \text{MDE}_{\text{c}}(A_{i0}) - (A_{im} - A_{i0}) \cdot \text{MDE}'_{\text{co}}\right|}{\Delta_{\Delta t_i} + \Delta_{A_i} \cdot \left|\text{MDE}'_{\text{co}}\right|}. \tag{3.63}$$

In this case, the determination of the coordinate A_{i0} is required. If A_{i0} is so close to A_{im}, that the relation $\text{MDE}_{\text{c}}(A_{im}) = \text{MDE}_{\text{c}}(A_{i0}) + (A_{im} - A_{i0})\text{MDE}'_{\text{co}}$ takes place, then (3.63) is transformed into the following form:

$$F_{\text{m}} = \frac{\left|\Delta t_{jm} - \text{MDE}_{\text{c}}(A_{im})\right|}{\Delta_{\Delta t_j} + \Delta_{A_i} \cdot \left|\text{MDE}'_{\text{co}}\right|}, \tag{3.64}$$

where MDE_{ct} is the derivative at the point $A_i = A_{im}$. Then the objective function for $\Delta t_j = \text{MDE}_{\text{c}}(A_i)$will be defined as:

$$F = \max_m F_{\text{m}} = \max_m \frac{\left|\Delta t_{jm} - \text{MDE}_{\text{c}}(A_{im})\right|}{\Delta_{\Delta t_j} + \Delta_{A_i} \cdot \text{MDE}'_{\text{co}}}. \tag{3.65}$$

This definition of the objective function F, in contrast to [297–299], based on the absolute and relative errors between Δt_{jm} and $\text{MDE}_{\text{c}}(A_{im})$, also takes into account the slope of $\text{MDE}_{\text{c}}(A_i)$ and therefore more accurately characterizes the closeness of the point (Δt_{jm}, A_{im}) to the dependence $\Delta t_j = \text{MDE}_{\text{c}}(A_i)$.

To solve the problem of COP models of electronic devices, in particular when solving the widely discussed problem of the COP model of the third-level CMOS transistor, various optimization methods are used: Gauss, steepest descent, Levenberg-Margurd, Hook-Jeeves, etc. [297]. Consider the expediency of using them in the COP of the MDE DF.

Since the first two methods require a large number of additional experimental data, they are completely ineffective. In addition, the values of the parameters in these methods are calculated by localizing the curve of some external characteristic of the transistor, which usually does not provide the required accuracy of the results.

In the works based on the use of the other listed optimization methods, the requirements for the accuracy of calculations and the costs of machine time are not always provided, either. For this reason, new approaches to improving the accuracy of calculations and reducing the costs of machine resources have recently been proposed.

For example, in [298] an optimization method is presented, the essence of which is as follows. It is assumed that $\vec{X} = (x_1, x_2, \ldots, x_n)t$ is the vector of the parameters of the marked model that are present in its equations (n is the total number of parameters). The drain current is represented in the form of:

$$\begin{aligned} I_\mathrm{d} &= f(L, W, U_\mathrm{DS}, U_\mathrm{GS}, U_\mathrm{BS}, x_1, x_2, \ldots, x_n) \\ &= f(L, W, U_\mathrm{DS}, U_\mathrm{GS}, U_\mathrm{BS}, \vec{X}), \end{aligned} \tag{3.66}$$

where L and W are respectively the length and width of the channel of a CMOS transistor, U_DS, U_GS, U_BS—the voltage between the outputs of the transistor; $\vec{X} = \{x_1, x_2, \ldots, x_n\}$ is the vector of the model parameters.

The drain current values of the transistor $\mathrm{I_D}$ are determined in two ways: experimental (I_M):

$$I_\mathrm{M} = (I_\mathrm{m1}, I_\mathrm{m2}, \ldots, I_\mathrm{mi}, \ldots, I_\mathrm{mk})^t \tag{3.67}$$

and also using the model (I_D):

$$I_\mathrm{D} = (I_\mathrm{d1}, I_\mathrm{d2}, \ldots, I_\mathrm{di}, \ldots, I_\mathrm{dk})_t = (f_1, f_2, \ldots, f_i, \ldots, f_k) \tag{3.68}$$

for k values of the arguments of the function (3.66), where f_i is $f(L_i, W_i, V_{\mathrm{DS}i}, V_{\mathrm{GS}i}, V_{\mathrm{BS}i}, \vec{X})$.

The relative error between I_D and I_M is expressed through the residual vector $R(\vec{X})$:

$$R(\vec{X}) = \left(1 - \frac{f_1}{I_{m_1}}, 1 - \frac{f_2}{I_{m_2}}, \ldots, 1 - \frac{f_k}{I_{m_k}}\right)^t. \tag{3.69}$$

The objective function is defined as:

$$\mathrm{OBJ}(\vec{X}) = R^t(\vec{X}) \cdot R(\vec{X}) = \sum_{i=1}^{k} \left(1 - \frac{f_i}{I_{m_i}}\right)^2. \tag{3.70}$$

The indicator of comparability of experimental and model values is introduced $\mathrm{CFERR} = \left[\frac{\mathrm{OBJ}(\vec{X}^*)}{K}\right]^{\frac{1}{2}}$, where $\vec{X}^*$ is the point of the global minimum $\mathrm{OBJ}(\vec{X})$. Mathematically $\vec{X}^*$ is obtained either by minimizing $\mathrm{OBJ}(\vec{X})$, or by determining the gradient zero, i.e.

$$\mathrm{OBJ}(\vec{X}) = 2J^t(\vec{X}) \cdot R(\vec{X}) = 0, \tag{3.71}$$

where $J(\vec{X})$—the Jacobi matrix.

Elements $J(\vec{X})$ are defined as: $(-(\partial f_i/\partial x_i)/I_{mi})$, $i=1,2,\ldots, k$, $j=1,2,\ldots,n$.

For given initial values of the parameters ($X0$), R() is approached by means of the Taylor series, i.e. equations (3.71) are represented by a Gaussian iteration:

$$J^t(X^m)J(X^m)\Delta X^m = -J^t(X^m)R(X^m), \tag{3.72}$$

where m is the iteration number; ΔX^m is the mth enhancement vector. The Gaussian iteration, applied in this work, has the following advantages: the rate of convergence is quadratic, and only first-order partial derivatives are required.

However, the lack of such an approach is well known: a strong dependence of the rate of convergence on the accuracy of the choice of initial values.

In [299] another method of approximating the curves was used, lying in the fact that an error function is compiled, with the minimization of which the calculated and measured characteristics are consistent. The following error function was used:

$$E(\vec{P}) = \sum_{(W_i,L_i)} \left[\sum \frac{| I_{\text{ds}}(\vec{P} - I_{\text{ds}}) |}{\max(I_{\text{ds}}, I_{\text{dmin}})} \right], \tag{3.73}$$

where $I_{\text{ds}}(\vec{P})$ is the value calculated using the model; I_{dmin} is the value of the current below which the absolute error is used instead of the relative error; $\vec{P} = [\text{VTO}, \text{NSUB}, \text{PHI}\ldots]$—the vector of the calculated model parameters; (W_i, L_i)—the width and length of the transistor respectively; $\vec{P} = P_{\min}$—the value at which $E(\vec{P})$ is minimal.

To determine $P_{\min}$, a modification of the simplex method is used, which differs from the gradient method in the thing that not one parameter changes in the error function, but a multiparameter search is performed while changing the values of the whole set of parameters. Various geometric operations are performed with the aim of directing the vector of parameters towards the optimum point.

The algorithm proposed in [299] has several obvious advantages: comparative simplicity of calculations, a conveniently implemented algorithm, small program volumes, a small number of preset parameters.

However, the algorithm also has a number of significant drawbacks: the dependence of all the coordinates of the simplex vertices on the same scale factor α, the decrease in step size when hit in an area with a narrow "ravine" or "ridge," etc., the low speed of the algorithm operation due to the lack of information from previous iterations.

The simulation of the noted and also other optimization methods for solving the problem of the COP of the MDE DF showed that, from the point of view of reliability and efficiency, they do not meet the requirements for constructing systems for gate-level simulation and optimization of digital circuits with consideration of DF. Therefore, the algorithm [300, 301], given in Sect. 3.3.1 and oriented at the solution of the COP problem of the MDE DF, is proposed, which is a modification of the Rosenbrock algorithm [302].

3.3.1 Description of the COP Algorithm of the MDE DF

step 1:	the initial point $P^{(0,0)}$is chosen in such a way that the condition $a_j(P^{(0,0)}) \geq \delta$is performed for $j = 1,2, \ldots m$. The accuracy parameter >0 is selected. (in practical calculations it was assumed that $=10^{-4}$, $\delta = 10^{-4}$).
step 2:	$V_i^{(0)} = e_i$ for $i = 1,2, \ldots ,n$.
step 3:	$f_j = F(P^{(0,0)})$ for all $j = 1,2, \ldots ,m$.
step 4:	$\hat{e} = 0$, $\mu = 0$;
step 5:	$S_i^{(0,0)} = 0.1$; $d_i^{(k)} = 0$ and $V_i^{(k)} = -1$ for all $i = 1,2, \ldots ,n$
step 6:	$l = 0$; $i = 1$;
step 7:	$P' = P^{(k,n\cdot l+i-1)} + S_i^{(k,l)} \cdot V_i^{(k)}$ is formed.
step 8:	if $F(P') > F(P^{(k,n \cdot 1 + i - 1)}$, go to step 16, otherwise: $F' = F(P')$ and $i = 1$.
step 9:	if $a_i(P) \leq 0$, go to step 18.
step 10:	if $a_i(P) \geq \delta$, then $F' = F(P')$ and go to step 13.
step 11:	Calculation of F by (3.65).
step 12:	if $F' < F(P^{(k,n \cdot l + i - 1)})$, go to step 16.
step 13:	if $i < m$, then $i = i + 1$ and go to step 9.
step 14:	$P^{(k,n\cdot l+i)} = P'$, $S_i^{(k,l+1)} = 3 \cdot S_i^{(k,l)}$ and $d_i^{(k)} = d_i^{(k)} + S_i^{(k,l)}$,
step 15:	if $V_i^{(k)} = -1$, then $V_i^{(k)} = 0$, go to step 18.
step 16:	$P^{(k,n\cdot l+i)} = P^{(k,n\cdot l+i-1)}$, $S_i^{(k,l+1)} = -\frac{1}{2} S_i^{(k,l)}$
step 17:	if $V_i^{(k)} = 0$, then $V_i^{(k)} = 1$.
step 18:	if $V_i^{(k)} = 1$ for all $i = 1,2, \ldots, m$, go to step 20;
step 19:	if $i < n$, then $i = i + 1$ and if $i = n$, then $l = l + 1$ and $i = 0$.
step 20:	$P^{(k+1,0)} = P^{(k,n\cdot l+i)} = P^{(k,0)} + \sum_{j=1}^{n} d_j^{(k)} \cdot V_j^{(k)}$.
step 21:	Formation of the vector $a_i^{(k)} = \sum_{j=1}^{n} d_j^{(k)}$ for all $i = 1,2, \ldots ,n$.
step 22:	if $F > F_{\min}$ ($F_{\min}$—user-defined minimum limit of the value of the objective function), the algorithm operation is terminated.
step 23:	Formation of a new search vector $\vec{V}_j^{(k+1)}$ for all $i = 1,2, \ldots ,n$ according to (3.75)
step 24:	$c = c + 1$, go to step 1.

3.3.2 An Example of the COP Algorithm of MDE DF

Below is an example of the operation of the COP algorithm of the MDE DF (Fig. 3.76 and Table 3.28) for two steps. At the initial point $P^{(0,0)}$, the search directions coincide with the unit vectors.

The essence of the algorithm is as follows. The number of search directions changes so that the search surface can move parallel to the axes of the rotating coordinate system. One of the axes should be oriented to the direction that is best for modification.

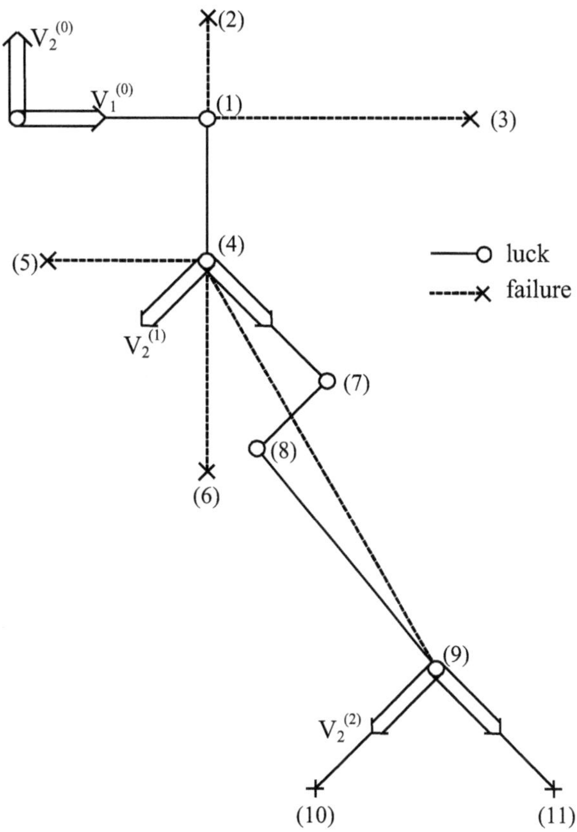

Fig. 3.76 The operation of the algorithm of the COP of the MDE DF for two iterations

Table 3.28 Algorithm of the COP of the MDE DF for two steps

			Value of parameters		Step size		
n	Step number k	Step number $kl + i$	P_1	P_2	S_1	S_2	Step result
0	0	0	0.0	9.0	2	2	
1	0	1	2.0	9.0	2		+
2	0	2	2.0	11.0		2	–
3	0	3	8.0	9.0	6		–
4	0	4	2.0	8.0		−1	+
5	0	5	−1.0	8.0	−3		–
6	0	6	2.0	5.0		−3	–
4	1	0	2.0	8.0	2	2	0
7	1	1	3.8	7.1	2		+
8	1	2	2.9	5.3		2	+
9	1	3	8.3	2.6	6		+
10	1	4	5.6	−2.7		6	–
11	1	5	24.4	−5.4	18		–
12	2	0	8.3	2.6	2	2	0

+: luck, –: failure, 0: return to the previous point

At the beginning, search directions are defined as unit vectors.

$$\vec{V}_i^{(0)} = \vec{e}_i, \qquad \text{for}_{i=1,2,\ldots,n}. \tag{3.74}$$

From the initial point of the parameter vector $\vec{P}^{(0,0)}$, one sample with a discrete step $\vec{S}_i^{(0,0)}, i = 1, 2, \ldots, n$ is made in each search direction. In case of success (the value of the objective function F (3.65) increases), the modified vector of variables does not change, and the step size is multiplied by a positive number $\alpha > 1$. Otherwise, the parameter vector takes its previous value, and the step size is multiplied by a negative number $-1 < \beta < 0$ (good results were obtained with $\alpha = 3$ and $\beta = -0.5$). This procedure is repeated until in each direction one luck and one failure are obtained (not necessarily sequentially). At the end of the first part of the search, the coordinate system rotates. For the above procedure, the recursive formulas of Gram and Schmidt are used:

$$\vec{V}_i^{(k+1)} = \frac{\vec{w}_i}{|\vec{w}_i|} \quad \text{for }_{i=1,2,\ldots,n}, \tag{3.75}$$

where

$$w_i = \begin{cases} a_i & \text{for}_{i=1}. \\ a_i - \sum_{j=1}^{i-1} \left(a_i^T \cdot \vec{V}_j^{(k+1)} \right) \cdot \vec{V}_j^{k+1} & \text{for }_{i=2,3,\ldots,n}, \end{cases} \tag{3.76}$$

a_i is defined as $a_i = \sum_{j=1}^{n} \vec{d}_j^{(k)} \cdot \vec{V}_j^{(k)}$ for $i = 1, 2, \ldots, n$.

Scalar values $\alpha_i^{(k)}$ show backward directed projection distances in the direction $\vec{V}_j^{(k)}$, during the kth iteration. The value of the objective function is checked after each modification of the parameter vector $\vec{P}$. In case of the modification, the objective function is corrected as follows:

$$F'(\vec{P}) = F(\vec{P}) + \sum_{j=1}^{n} \Theta_j(\vec{P}) \cdot (f_j - F(\vec{P})),$$

where

$$\Theta_j(\vec{P}) = \begin{cases} 0, & \text{if } a_j(\vec{P}) \geq \delta, \\ 3\eta - 4\eta^2 + 2\eta^3, & \text{if } 0 < a_j(\vec{P}) < \delta, \\ 1, & \text{if } \vec{a}_j(\vec{P}) \leq 0, \end{cases} \tag{3.77}$$

$$\eta = 1 - \frac{1}{\delta} \cdot a_j(\vec{P}).$$

Here f_j is the next value of the objective function belonging to the success of the search.

The main modification of the algorithm is related to the following. It is known that the Rosenbrock algorithm [302] diverges if some $d_j = (1 \leq j \leq n) = 0$. To avoid this, in practical implementations of this algorithm [297] the components a_i are re-sorted so that those d_j whose values are zero are located at the end of the array, and the optimization algorithm is applied only to the $n–q$ components, where q is the number of d_j with zero value. However, this procedure of data re-sorting significantly increases the estimated time. In the algorithm proposed in [300], the computer time for rearrangement is not wasted, even if $d_{k-1} = 0$, provided that $\sum_{j=k}^{n} d_j^2 \neq 0$. If the above condition is satisfied, then even at $d_{k-1} = 0$, using expressions (3.78)–(3.81), one can directly define $\vec{V} \rightarrow_i^{k+1}$:

$$V_1^1 = \frac{\sum_{i=1}^{n} d_i V_1^0}{\sqrt{\sum_{i=1}^{n} d_i^2}}, \tag{3.78}$$

$$w_k = \frac{d_{k-1}^2 \cdot \sum_{i=1}^{k} d_i V_1^0 - d_{k-1} V_{k-1}^0 \sum_{i=1}^{k} d_i^2}{\sum_{i=1}^{k-1} d_i^2} = \frac{a_k |a_{k-1}|^2 - a_{k-1} |a_k|^2}{|a_{k-1}|^2}, \tag{3.79}$$

$$|w_k| = d_{k-1} \sqrt{\frac{\sum_{i=1}^{k} d_i^2}{\sum_{i=1}^{k} d_i^2}} = \sqrt{|a_{k-1}|^2 - |a_k|^2} \cdot \frac{|a_k|}{|a_{k-1}|}, \tag{3.80}$$

$$V_k^1 = \frac{d_{k-1} \sum_{i=1}^{k} d_i V_i^0 - V_{k-1}^0 \sum_{i=1}^{k} d_i^2}{\sqrt{\sum_{i=1}^{k-1} d_i^2 \sum_{i=1}^{k} d_i^2}} = \frac{d_{k-1} a_k - V_{k-1}^0 |a_k|^2}{|a_{k-1}||a_k|} = \frac{a_k |a_{k-1}|^2 - a_{k-1} |a_k|^2}{|a_{k-1}||a_k| \sqrt{|a_{k-1}|^2 - |a_k|^2}}, \tag{3.81}$$

Thus, if $d_{k-1} = 0$, then $V_k^1 = -V_{k-1}^0$ is taken when the condition $\sum_{i=1}^{k} d_j^2 = 0$ is satisfied. The proof of the accuracy of (3.79)–(3.81) is given below.

If it is assumed that (3.79)–(3.81) hold for a particular value $k > 1$, then using (3.75) and (3.76), the following is obtained:

$$
\begin{aligned}
w_{k+1} &= a_{k+1} - \sum_{j=1}^{k}\left(a_{k+1}\cdot V_j^1\right)V_j^1 = a_{k+1} - \left(a_{k+1}\cdot V_1^1\right)V_1^1 - \sum_{j=2}^{k}\left(a_k + V_{k1}^1\right)V_1^1 \\
&= a_{k+1} - \left(a_{k+1}\cdot\frac{a_1}{|a_1|}\right)\frac{a_1}{|a_1|} - \sum_{j=2}^{k}\left(a_{k+1}\cdot\frac{a_j|a_{j-1}|^2 - a_{j-1}|a_j|^2}{|a_{j-1}||a_j|\sqrt{|a_{j-1}|^2-|a_j|^2}}\right) \\
&\quad\times\left(\frac{a_j|a_{j-1}|^2 - a_{j-1}|a_j|^2}{|a_{j-1}||a_j|\sqrt{|a_{j-1}|^2-|a_j|^2}}\right)
\end{aligned}
\tag{3.82}
$$

Now

$$
a_{k+1}\cdot a_1 = \sum_{i=1}^{k+1} d_i V_i^0 \cdot \sum_1 d_i^2 V_i^0 = \sum_{i=1}^{k+1} d_i^2 = |a_{k+1}|^2 \quad \text{(beginning with } k > 1\text{).} \tag{3.83}
$$

Similarly, this is obtained:

$$
a_{k+1}\cdot a_j = a_{k+1}\cdot a_{j-1} = |a_{k+1}|^2 \quad \text{(beginning with } k+1 > j\text{).} \tag{3.84}
$$

Hence:

$$
\begin{aligned}
w_{k+1} &= a_{k+1} - \frac{a_1|a_{k+1}|^2}{|a_1|^2} - \sum_{j=2}^{k}\left\{\frac{|a_{k+1}|^2\left(|a_{j-1}|^2 - |a_j|^2\right)}{|a_{j-1}|^2|a_j|^2}\right\}\times\left\{\frac{a_j|a_{j-1}|^2 - a_{j-1}|a_j|^2}{|a_{j-1}|^2-|a_j|^2}\right\} \\
&= a_{k+1} - \frac{a_1|a_{k+1}|^2}{|a_1|^2} - |a_{k+1}|^2\sum_{j=2}^{k}\frac{a_j|a_{j-1}|^2 - a_{j-1}|a_j|^2}{|a_{j-1}|^2|a_j|^2} \\
&= a_{k+1} - |a_{k+1}|^2\left\{\sum_{j=2}^{k}\left(\frac{a_j}{|a_j|^2} - \frac{|a_{j-1}|}{|a_{j-1}|^2}\right) + \frac{a_1}{|a_1|^2}\right\}
\end{aligned}
\tag{3.85}
$$

But

$$
\sum_{j=2}^{k}\left(\frac{a_j}{|a_j|^2} - \frac{|a_{j-1}|}{|a_{j-1}|^2}\right) = \sum_{j=2}^{k}\frac{a_j}{|a_j|^2} - \sum_{j=2}^{k-1}\frac{a_j}{|a_j|^2} = \frac{a_k}{|a_k|^2} - \frac{a_1}{|a_1|^2} \tag{3.86}
$$

Thus:

$$
w_{k+1} = a_{k+1} - \frac{a_k|a_{k+1}|^2}{|a_k|^2} = \frac{a_{k+1}|a_k|^2 - a_k|a_{k+1}|^2}{|a_k|^2} \tag{3.87}
$$

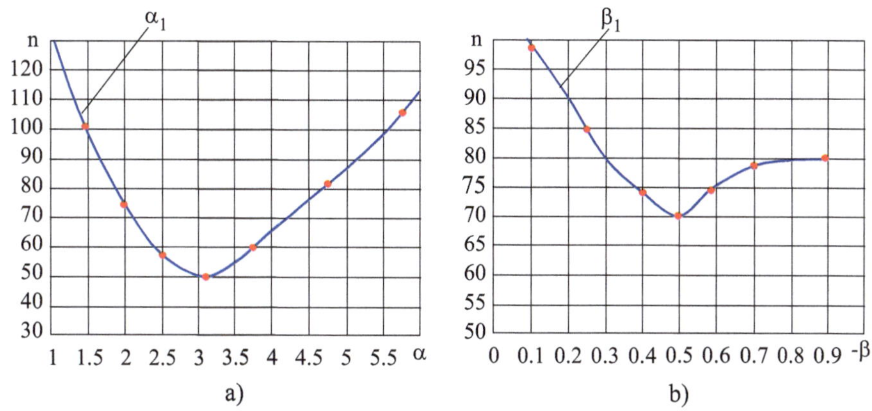

Fig. 3.77 Dependence of n from α and β for the MDE $\Delta t_{b}^{01} = f(T)$ of basic CMOS logical gate

but (3.87) is the same as (3.79) with k replaced by $(k + 1)$. Thus, if (3.79) holds for a given value of k, then it also holds for the next, larger value of k. But as (3.79) also holds for $k = 2$, then the accuracy of (3.79) is proved by the method of mathematical induction. And (3.80) and (3.81) follow from (3.79).

By eliminating the process of marked reordering, the proposed algorithm generates a significant gain (on average by 10% ... 15%) on the calculation time.

Another modification is related to the choice of parameters α and β the algorithm, which is made as follows. On the example of the COP of an MDE series from Sect. 3.1.1 to 3.1.3, the dependences of the number of steps of the operation of algorithm n fromα and β. were obtained. In Fig. 3.77 as an example, these dependencies are shown for the MDE $\Delta t_{b}^{01} = f\left(T^{0}\right)$ of the basic logical cell CMOS (Table 3.3).

Then, based on the average statistical processing of these values, the values α and β, used in the COP subsystem of the MDE DF of gate-level simulation and optimization of digital circuits ELAIS-L were determined. The convergence of the proposed algorithm is not considered here, since they are analogous to the case of the Rosenbrock algorithm [302]. In order to demonstrate the advantages of the proposed algorithm of the COP of the MDE DF, it was used to optimize the parameters of the third-level SPICE model of the CMOS transistor [300]. TOX (oxide layer thickness), GAMMA (threshold parameter of the substrate), PHI (surface potential), THETA (modulation of charge carrier mobility), and U_0 (substrate mobility) were chosen as optimized parameters of the mentioned model. When the values of the marked parameters differed from the experimental values by no more than 5%, the calculation time was approximately 28% less than in [299].

3.4 Models of Circuits with Switched Capacitors

At present, various types of analog devices for discrete signal processing, in particular, filters on switchable capacitors (SC) [303–306] are produced by integrated technology. Due to their good economic parameters (low power consumption, small size, etc.), these devices are promising.

A characteristic feature of such circuits is their duality in relation to the type of information processed and the way it is processed. On the one hand, these are analog devices, because the nature of the change in signal information (voltage, currents, charges) is continuous, on the other hand, clocked, i.e. the process of information transformation takes place at discrete instants of time (in cycles). Therefore, these devices occupy an intermediate position between analog and digital circuits.

Simulation of devices for discrete processing of analog signals of a marked type using existing simulation tools is difficult for the following reasons:

1. The continuous nature of the change in information signals excludes the possibility of using traditional gate-level simulation [163–165, 195–197, 180–200].
2. The process of processing analog signals occurs for a large number (usually several thousand) cycles, and simulation time with the help of circuit simulation tools turns out to be too large [74, 75].
3. These circuits have the property of changing in time not only the parameters but also the structure of the circuit. Research of circuits on the SC shows that they consist of elements of two types:
   (a) Capacitance with keys (Fig. 3.78a), which is connected to one of the two pairs of nodes at the beginning of each clock cycle, and switches to the other pair of nodes in the middle of the clock, thereby changing the configuration of the circuit;
   (b) Capacitance with an operational amplifier (Fig. 3.78b), which is connected to the feedback loop of the operational amplifier (OA).

The property of changing the structure of the circuit during one clock of operation causes additional difficulties for general circuit-level simulation tools [74, 75], in which such an option is not anticipated. Therefore, there is a need to develop device

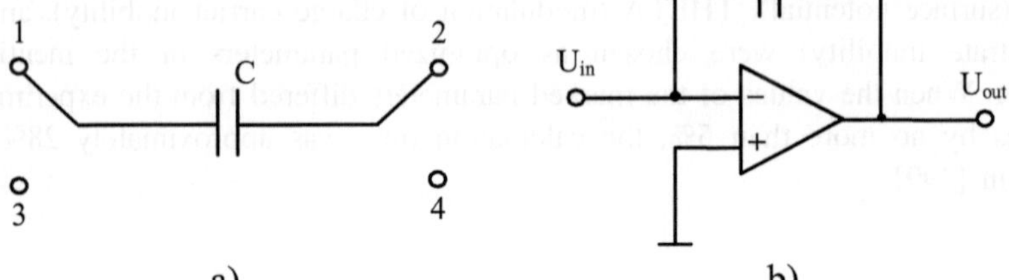

Fig. 3.78 Elements on switched capacitors

models for discrete processing of analog signals, providing simulation of these circuits with minimal expenditure of computer resources. The technique of gate-level simulation with continuous values of variables, proposed in Sect. 1.4, allows constructing similar models for SC circuits. Some approaches to construct such models are discussed below.

At present, the following approaches are known from the literature for constructing models of circuits on a SC, for a complete analysis of which is given in [307, 308]:

1. For a specific method of mathematical description of the circuit (the method of state variables, the modified method of nodal potentials, etc.), special models of circuits on the SC are developed. These models reproduce the law of variation of physical quantities (currents, voltages, charges) on the outputs of the elements of such circuits during one clock cycle (T) of operation. The values of these quantities, determined from the model for the end of the clock, are the initial data in the simulation of the next clock cycle. For example, for the state variables method, the following model is proposed in [309]: if $\vec{X}_1(t)$ and $\vec{X}_2(t)$ are state vectors for the time intervals, respectively (nT, $nT + \tau$) and ($nT + \tau$, $(n + 1)T$), then the equations of states of the output quantities are written as follows:

$$\begin{aligned} y_1(t) &= C_1 \cdot X_2(nT) + D_1 \cdot U_1(t) && \text{at } nT \le t \le nT + \tau, \\ y_2(t) &= C_2 \cdot X_1(nT + \tau^-) + D_2 \cdot U_1(t) && \text{at } nT + \tau \le t \le (n+1)T. \end{aligned} \tag{3.88}$$

Here, τ—time point belonging to the interval nT - $(n + 1)T$; n—the clock number; C_1, C_2, D_1, D_2—matrices of real coefficients that are variable in time (due to the inconsistent configuration of the circuit). A record of the form t means immediately before the time t.

2. For each possible configuration of the circuit formed at a specific SC position, a simplified equivalent circuit is constructed [308]. For example, for the circuit in Fig. 3.79a, with the two possible states of the S key, the equivalent circuits shown in Fig. 3.79b, c are obtained. Depending on the positions of the switched capacitances, a corresponding equivalent circuit is simulated.

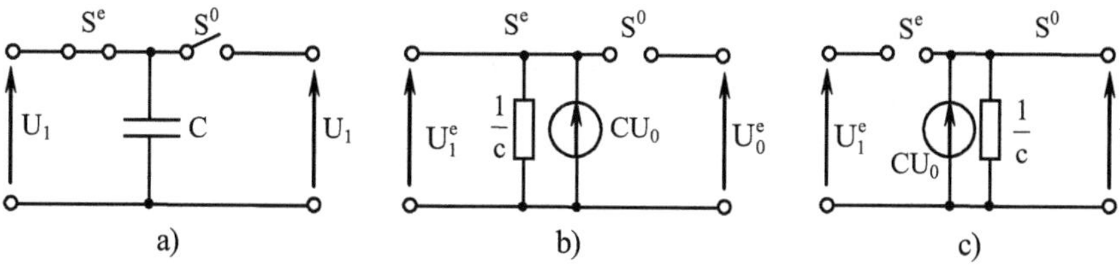

Fig. 3.79 Equivalent circuits with different key positions

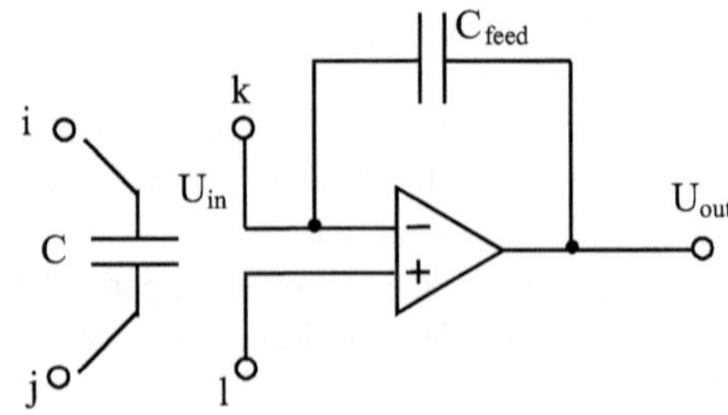

Fig. 3.80 OA with capacitance

A common drawback of the circuit models known from the literature on SC is that variables are tracked inside the clock cycle of the circuit operation, i.e.

$$h_{\max} < T \tag{3.89}$$

where $h_{\max}$ is the maximum integration step,

If one considers that the number of clock cycles of these circuits is large, then one can conclude that simulation with known models requires a large expenditure of computer time. In addition, the disadvantage of the noted approaches is that it is necessary to develop special software tools for calculating changing matrixes of coefficients or alternately switching equivalent circuits at each instant of time. One of the approaches that allow to reduce the expenditure of computer time is the development of discrete models corresponding to the switching times of capacitances [238]. Suppose that a switchable capacitance C is connected to the input of an OA with a capacitance in the feedback loop, which is connected to nodes i and j in the first half of the clock, and then switches between the input of the OA (node k) and node l (Fig. 3.80).

It will be assumed that for half the clock, all the transient processes associated with the recharging of the capacitors can be completed. Then by the end of the first half of the cycle the charge of the capacitance C will be equal to:

$$Q_{\mathrm{c}}^{S+\frac{1}{2}} = C \cdot \left(U_i^{S+\frac{1}{2}} - U_j^{S+\frac{1}{2}}\right), \tag{3.90}$$

where S is the clock number. The record of type $t^{S\,+\,1/2}$ means the value of t in the middle of $(S + 1)$th clock; U_i, U_j are the voltages in the ith and jth nodes of the circuit. In the second half of the $(SS + 1)$th cycle, the capacitance C switches to the nodes k and l. For the second half of the clock, after the voltage is set, the voltage at node k will be equal to:

$$U_k = U_{\mathrm{in}} = \frac{U_{\mathrm{out}}}{k}, \tag{3.91}$$

where U_{in}, U_{out} are respectively the input and output voltages of the OA. If it is assumed that k is sufficiently large, then $U_k \approx 0$.

Changing the charge into capacitance C in the second half of the clock will equal to:

$$\Delta Q_{\mathrm{c}} = Q_{\mathrm{c}}^{S+1} - Q_{\mathrm{c}}^{S+\frac{1}{2}} = C \cdot \left(-U_l^{S+1} - U_i^{S+\frac{1}{2}} - U_j^{S+\frac{1}{2}}\right). \tag{3.92}$$

If it is assumed that the input resistance of the OA is infinitely large, as in most papers [307–309], it can be assumed that the charge change onto the capacitance C–Q_{c} is transferred to the feedback capacitance OA (C_{feed}), i.e.

$$\Delta Q_{\mathrm{c}} = \Delta Q_{\mathrm{c.feed}}. \tag{3.93}$$

If it is assumed that the output resistance of the OA is infinitely large, then

$$\Delta Q_{\mathrm{c.feed}} = C_{\mathrm{feed}} \cdot \Delta U_{\mathrm{c.feed}}, \tag{3.94}$$

where $\Delta U_{\mathrm{c.\ feed}}$ is the increment of voltage on the capacitance C_{feed}.

From the expressions (3.92)—(3.94) this is obtained:

$$\Delta Q_{\mathrm{c}} = \frac{C}{C_{\mathrm{feed}}} - \left(U_j^{S+\frac{1}{2}} - U_i^{S+\frac{1}{2}} - U_l^{S+1}\right). \tag{3.95}$$

Taking into account that $U_{\mathrm{in}} \approx 0$, this is obtained:

$$\Delta U_{\mathrm{out}} = U_{\mathrm{c.feed}}. \tag{3.96}$$

From (3.96) the value of the output voltage of the OA at the end of the $(S + 1)$th cycle of operation is obtained:

$$U_{\mathrm{out}}^{S+1} = U_{\mathrm{out}}^{S+\frac{1}{2}} + \Delta U_{\mathrm{out}}^{S+1} = U_{\mathrm{out}}^{S+\frac{1}{2}} + \frac{C}{C_{\mathrm{feed}}} \cdot \left(U_j^{S+\frac{1}{2}} - U_i^{S+\frac{1}{2}} - U_l^{S+1}\right). \tag{3.97}$$

If the output of the OA is connected to its input through the SC, it can depend on itself. In general, several SCs can be connected to the input of the OA, and the voltage at the output of the SC at the end of the $(S + 1)$th clock cycle will be determined as follows:

$$U_{\mathrm{out}}^{S+1} = U_{\mathrm{out}}^{S+\frac{1}{2}} + \sum_p k \cdot U_{\mathrm{p}}, \tag{3.98}$$

where U_{p} is the voltage of the nodes connected to the OA SC inputs, and the matrix k can be determined from the configuration of the circuit (as in (3.97)). It follows from expression (3.98) that circuits on a SC can be simulated using algorithms similar to event-driven simulation (Sect. 4.1).

Knowing the voltage of the circuit nodes at the end of the previous clock, their new values are calculated one at a time according to (3.98). If U_{out}^{S+1} and $U_{\mathrm{out}}^{S+1/2}$ differ more than the given error in voltage ΔU, then the voltage in this node changes to

U_{out}^{S+1}. If U_{out}^{S+1} and $U_{out}^{S+1/2}$ differ less than ΔU, the remaining nodes are considered. When the voltage in a node $U^{S+1/2}$ and U^{S+1} changes, it is considered what the element in this half-clock is connected to. If it is connected to a node that is already excited, then no action is taken with the output node. If the considered element is connected to an unexcited node, then excitation is signaled for it (a signal about the inconsistency of the input and output voltages). The simulation of this cycle of the circuit operation will end when all nodes of the circuit go into an unexcited state, which indicates that the steady-state values of the voltages of all the circuit nodes for the end of the considered cycle have been obtained.

From the above, it follows that this approach is easy to implement in mixed-mode gate-level-circuit-level simulation tools [226, 231–248] with the help of some modification of LM (Fig. 1.110). In this case, the analogues of logical variables are continuous voltages U_i, and logical operations (disjunction and conjunction) are replaced by arithmetic ones (addition and multiplication).

The disadvantages of this approach are the following:

1. Simulation is done by half-clocks, and the number of cycles can be very large.
2. In the presence of feedbacks one cycle can have many events occurring in the same elements. Then, to determine the steady-state values of the stresses, a multiple solution of the same equations of the system (3.98) is required.
3. To implement this approach, it is necessary to develop special software tools that automatically determine the values of the coefficients of the matrix k (3.98) from the configuration of the circuit. Therefore, another approach is suggested for constructing integrated models of the SC circuit [218, 238]. In the proposed model, it is assumed that in order to accelerate the calculation of circuits on a SC, it is not possible to consider in detail the changes in the voltages during each cycle of the circuit operation and the unit of time as the duration of the cycle of operation of circuit T. In this case, the differential equations of the circuit can be written with respect to the voltages obtained at the end of each clock, i.e. at discrete instants of time, nT, where n is an integer number. In fact, these will be equations relative to the envelopes of transient processes, and the corresponding model of elements will give an integrated representation of the processes occurring during the cycle. With this approach, it becomes possible to predict the course of the envelopes by several cycles ahead and, as a result, simulate a circuit with an integration step equal to or greater than T.

Assume that the switching capacitor (SC) was connected to the nodes with numbers 1 and 2 at the beginning of the clock, and switched to nodes with numbers 3 and 4 in the middle of the clock (Fig. 3.78a). If it is assumed that the charge time of the capacitance is small in comparison with T and denote by $t = nT$ the end time of the clock, then the voltage on the capacitance is set equal to $U_{\tilde{n}}\ddot{i} - 1 = U_2(t - T)$ $-U_1(t - T)$ at the beginning of this clock when C is connected to nodes 1 and 2. This voltage is maintained until the capacitance switches to nodes 3 and 4, after which it changes to a value of U_{cn} and is stored this way until the end of the cycle. Proceeding from this, the following can be written: $U_{cn} = U_4(t) - U_3(t)$. Through terminals 3 and 4, during the time of the cycle, a charge equal to the integrated change in charge on the capacitance flows:

$$\Delta Q = c \cdot (U_{cn-1} - U_{cn}) = c \cdot [U_2(t-T) - U_1(t-T) - U_4(t) + U_3(t)], \quad (3.99)$$

or the average capacitance current flowing into the node 3 and flowing from the node 4 is equal to

$$I_{\text{aver}3,4} = \frac{\Delta Q}{T} = \frac{c}{T} \cdot [U_2(t-T) + U_3(t) - U_1(t-T) - U_4(t)]. \quad (3.100)$$

If the voltages $U_1(t-T)$ and $U_2(t-T)$ are expressed in terms of the values $U_1(t)$ and $U_2(t)$ and the increments of these voltages per cycle ΔU_1, ΔU_2, then expression (3.100) is written as follows:

$$I_{\text{aver}3,4}(t) = \frac{c}{T} \cdot [U_3(t) - U_4(t) + U_2(t) - U_1(t) - \Delta U_1 - \Delta U_2]. \quad (3.101)$$

Since the equations are written with respect to the voltages at the end of each clock, and the period T is taken for a unit of time, it can conditionally be assumed that these voltages are continuous functions of t. Then the increments of voltage per cycle are determined by expressions:

$$\Delta U_i = U_i' \cdot T. \quad (3.102)$$

Taking this into account, relation (3.101) takes the form of the differential equation:

$$I_{\text{aver}3,4} = \frac{c}{T} \cdot \left(U_3 - U_4 + U_2 - U_1 + TU_1' - TU_2'\right). \quad (3.103)$$

A similar consideration of the charge transferred per clock from capacitor C to nodes 1 and 2 gives the following equation describing the average current flowing through capacitor C from node 1 to node 2:

$$I_{\text{aver}1,2} = \frac{c}{T} \cdot \left(U_1 - U_2 + U_4 - U_3 + TU_3' - TU_4'\right). \quad (3.104)$$

Thus, a commutated capacitor can be represented by a four-terminal network, the mean pole currents of which are described by equations (3.103) and (3.104).

Now consider an op-amp with a feedback capacitance C_{feed} (Fig. 3.78b), to the input and output nodes of which are connected during the clock cycle of the SC. If the output impedance of the amplifier is small enough, the capacitors connected to the output do not affect the voltage U_{out}. If the input resistance of the op-amp is large, then the average current flowing from the SC through the input terminal goes to the C_{feed} and leads to a change in the charge of this capacitance per cycle:

$$\Delta Q_{\text{in}} = I_{\text{in}} \cdot T. \tag{3.105}$$

On the other hand, the change in this charge is:

$$\Delta Q_{\text{feed}} = C_{\text{feed}} \cdot (\Delta U_{\text{in}} - \Delta U_{\text{out}}), \tag{3.106}$$

where ΔU_{in}, ΔU_{out}—changes in voltages per cycle.

Since, $\Delta_{\text{out}} = -k \cdot U_{\text{in}}$and $k \gg 1$, the quantity in expression (3.106) can be neglected by ΔU_{in}. Combining (3.105) and (3.106), this is obtained:

$$I_{\text{aver.in}} = -\frac{C_{\text{feed}}}{T} \cdot \Delta U_{\text{out}}, \tag{3.107}$$

and taking into account (3.102):

$$I_{\text{aver.in}} = -C_{\text{feed}} \cdot \Delta U'_{\text{out}}. \tag{3.108}$$

Thus, an OA with a feedback capacitance C_{feed} can be represented by a three-terminal network whose input node is closed to the ground, and the output voltage is described by the differential equation (3.108).

The advantages of these models are as follows:

1. It takes no more than one step per clock. However, the step size can be larger than T, then one step spans several cycles.
2. The application of these models does not require any changes in the circuit-level simulation tools or in MA (Fig. 1.110) in mixed-mode simulation tools.
3. Models allow conducting frequency analysis, and with their help, it is possible to directly obtain amplitude-phase characteristics of envelopes, whereas previously considered models do not allow this.

Section 3.4.1 provides examples of simulation of circuits with a SC, demonstrating the advantages of the developed models, as well as the issues of improving the accuracy of these models taking into account the actual parameters of the OA.

3.4.1 Examples of the Use of Integrated Circuit Models on a SC

Figure 3.81 shows an integrator circuit on the SC. The transient response of the envelope of the output voltage in this circuit can be obtained analytically:

$$U_{\text{out}} = -U_{\text{in}} \cdot \frac{C_1}{C_2 + C_{\text{feed}}} \left[1 - \mathrm{e}^{-\frac{C_1 t}{(C_{\text{feed}} + C_2)T}} \right]. \tag{3.109}$$

Exactly the same expression is obtained if one jointly solves the equations of the models (3.103), (3.104), (3.108) and the Kirchhoff equation with respect to average

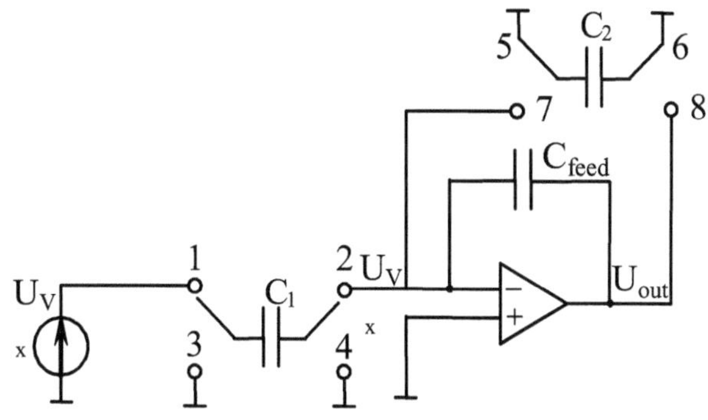

Fig. 3.81 Integrator

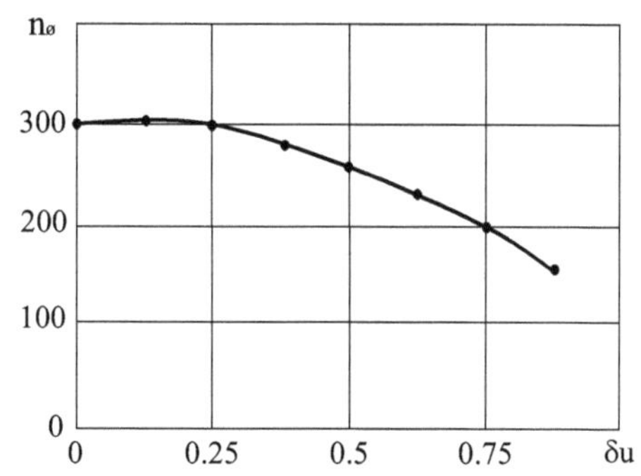

Fig. 3.82 Dependence of the number of steps on the voltage error

currents. At $C_1 = 1$ nF, $C_2 = 1$ nF, $C_{feed} = 0.01$ nF, about 300 clock cycles should be calculated to obtain the complete transient response. When using the proposed models and having the given relative stress error at step $\delta_u \leq 0.05$, this required 27 s of Pentium IV computer time (clock frequency 2.2 GHz). Calculation of the same circuit using key macromodels and an operational amplifier requires about 0.3 s per clock or 96 s on 300 clocks. When calculating at the circuit level using SPICE, this requires at least an order of more time. Thus, the use of integrated models reduces the time spent by an order in comparison with conventional macromodels. Figure 3.82 shows the dependence of the number of steps n_s, required for the calculation of 300 clock cycles, on the given relative error of stresses δ_u. From this graph it is seen that as δ_u increases above a certain limit, the number of steps becomes less than the number of clock cycles, i.e. the step size exceeds the duration of the clock cycle *T*.

Figure 3.83 shows a diagram of a more complex filter on the SC. The transient response calculated at $T = 3.9$ μs is shown in Fig. 3.84. Figure 3.84a shows the most initial area in a few clocks. The true step-by-step dependence of $U_{out}(t)$ is superimposed on the calculated envelope characteristic. Figure 3.84b illustrates a large fragment of the transient process of the envelope, which has a complex oscillatory character. Finally, Fig. 3.84c depicts a graph of the envelope of these oscillations. All three graphs are obtained for one calculation of the circuit.

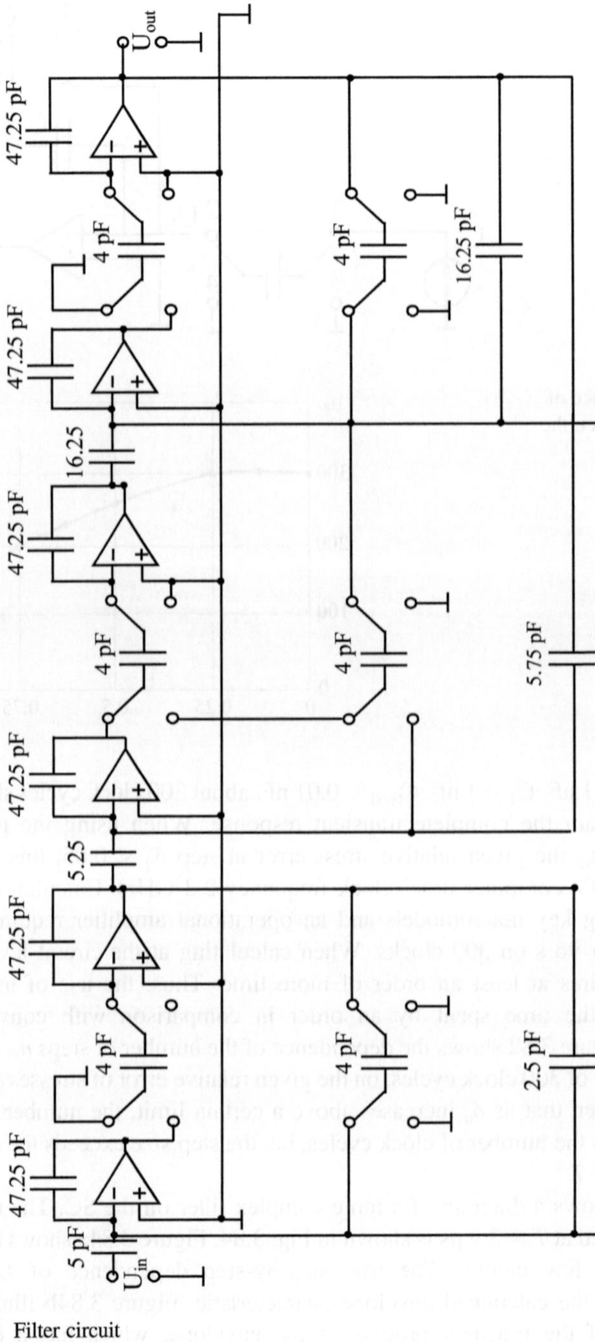

Fig. 3.83 Filter circuit

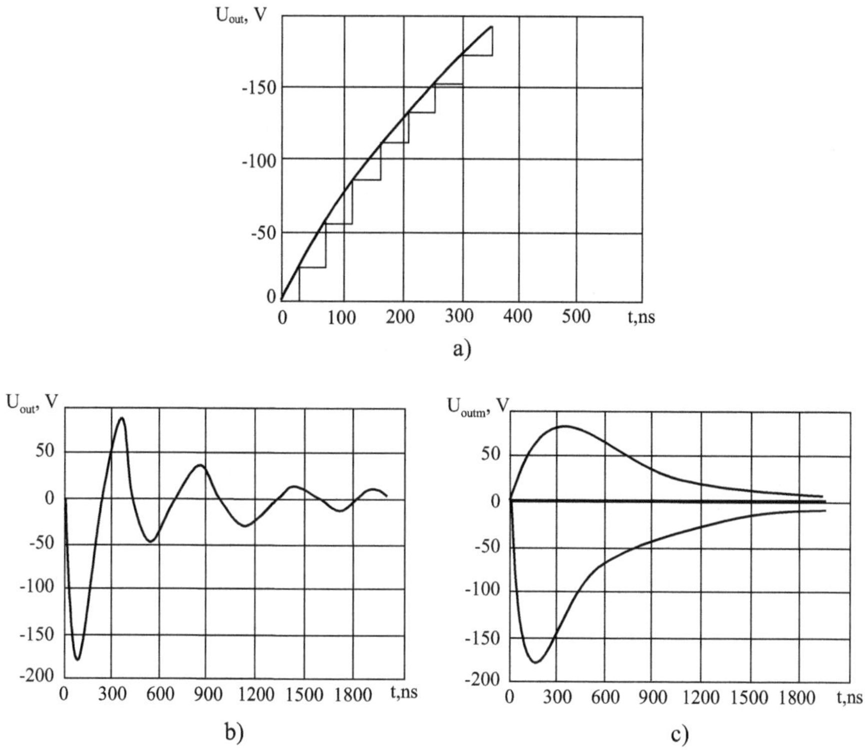

Fig. 3.84 Filter simulation results

Fig. 3.85 Amplitude–frequency characteristic of the filters

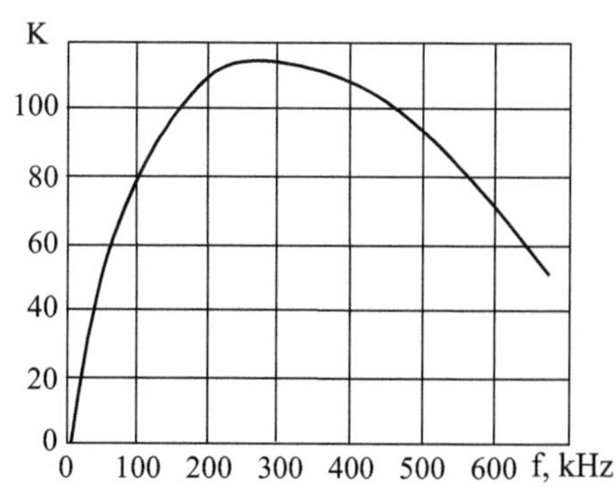

The calculation time for 300 clock cycles was 160 s. Figure 3.85 shows amplitude–frequency characteristic of the same circuit. The calculation time for the amplitude–frequency characteristic points was 240 s.

3.4.2 Improving the Accuracy of Integrated Circuit Models on a SC

To increase the accuracy of the results of calculating circuits on a SC, it is necessary to take into account the real parameters of OA, SC, and keys in integrated models given in Sect. 3.4. Consider, for example, how the equations of models change when finite values of the gain (k) and the bias voltage (U_{bias}) of the OA are taken into account. Assume that the SC was disconnected from the input of the OA at the beginning of the clock, and in the middle of the clock it switched to the input of the OA. If one assumes that the charge time of the capacitance is small in comparison with T, then for the time $t \gg t_{\text{switch}}$ the equivalent circuit of the OA with the capacitance C_{feed} in the feedback takes the form shown in Fig. 3.86. The voltage on the capacitance C_{feed} at the beginning of this clock, when C is still disconnected from the OA, is set equal to $U_{Cf_{\text{eed}}n-1} = U_{\text{in}}(t - T) - U_{\text{out}}(t \cdot T)$

This voltage is maintained until the capacitance C switches to the input of the OA, after which it changes to the value $U_{Cf_{\text{eed}}n}$ and remains so until the end of the clock cycle. Proceeding from this, the following can be written: $U_{Cf_{\text{eed}}n} = U_{\text{in}}(t) - U_{\text{out}}(t)$. The increment of the voltage on the capacitance C_{feed} will be determined as follows:

$$\Delta U_{C_{\text{feed}}} = U_{C_{\text{feed}}n} - U_{C_{\text{feed}}n-1} = U_{\text{in}}(t) - U_{\text{out}}(t) - U_{\text{in}}(t-T) + U_{\text{out}}(t-T) \quad (3.110)$$

From Fig. 3.86 it can be seen that:

$$U_{\text{in}}(t) = U_{\text{bias}} - \frac{U_{\text{out}}(t)}{k} \quad (3.111)$$

In the same way:

$$U_{\text{in}}(t - T) = U_{\text{bias}} - \frac{U_{\text{out}}(t - T)}{k} \quad (3.112)$$

Proceeding from (3.111) and (3.112), the expression (3.110) can be written as follows:

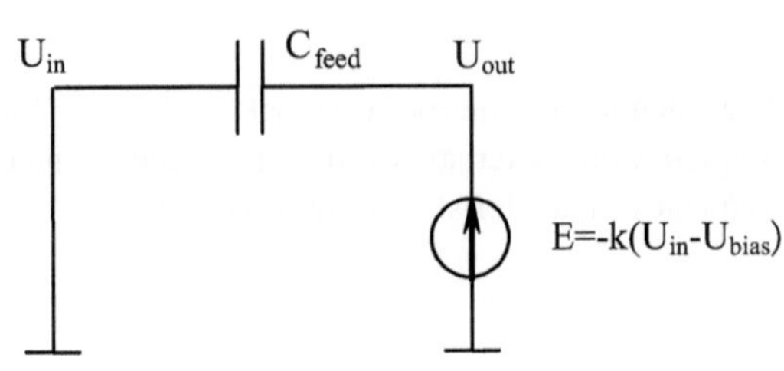

Fig. 3.86 OA model with capacitance

$$\Delta U_{C_{\text{feed}}} = -\left(\frac{1}{k} - 1\right) \cdot (U_{\text{out}}(t) - U_{\text{out}}(t - T)). \tag{3.113}$$

If, like (3.102), to assume that the voltage $U_{\text{out}}(t)$ is a continuous function of t, then:

$$U_{\text{out}}(t) = U_{\text{out}}(t - T) + T \cdot U'_{\text{out}} \tag{3.114}$$

Then (3.113) takes the following form:

$$\Delta U_{C_{\text{feed}}} = -\left(\frac{1}{\text{k}} - 1\right) \cdot T \cdot U'_{\text{out}} \tag{3.115}$$

The voltage increment on the capacitance C_{feed} per clock cycle $(\Delta U_{C_{\text{feed}}})$ is related to the change in charge on this capacitance:

$$\Delta Q_{C_{\text{feed}}} = C_{\text{feed}} \cdot \Delta U_{C_{\text{feed}}} = -C_{\text{feed}} \cdot T \cdot \left(\frac{1}{k} - 1\right) \cdot U'_{\text{out}} \tag{3.116}$$

The average current for the period flowing into the C_{feed}, if one assumes that the input resistance of the OA is sufficiently large, will be equal to its input current and will be determined as follows:

$$I_{\text{aver.in}} = -C_{\text{feed}} \cdot \left(1 - \frac{1}{k}\right) \cdot U_{\text{out}} \tag{3.117}$$

Equation (3.117), in contrast to (3.108), takes into account the decrease in current $I_{\text{aver. in}}$ at a finite value of k. By analogy, integrated models that take into account other real parameters of the elements on the SC can be obtained.

Chapter 4
Algorithmic Implementation of the Automated System of Gate-Level Simulation of Digital Circuits with Consideration of DF

4.1 Algorithm of Gate-Level Simulation of Digital Circuits with Consideration of DF

In the existing systems of gate-level simulation [150, 152, 155, 156, 160–200], various simulation algorithms are used. They take into account the spatial (only gates participating in the switching of digital circuits are considered) and timing (only the moments of time when the events occur are considered, i.e. changes in the state of the nodes) events of the circuit. Since, on average, only a small number of digital cells participate in the switching of digital circuits, then due to the event, a significant gain in simulation time is provided. However, the known algorithms of event-driven simulation [163, 165, 181, 182, 188, 189, 191, 194–196, 200] are not suitable for gate-level simulation of digital circuits with consideration of DF for the following reasons. Existing algorithms:

1. Are oriented to simpler gate-level models in comparison with the MCE, presented in Sects. 1.4 and 2.2. The use of the MCE requires construction of a new algorithm that takes into account all its specific aspects: switching stages, rules for summing excitations, methods for determining the parameters of the output signal, etc.
2. Do not take into account the dependence of the MCE parameters $\vec{P} = \{p_1, p_2, \ldots, p_u\}$ (Sect. 1.4) on the DF $\vec{V} = v_1, v_2, \ldots$, i.e. MDE (Sect. 2.1).
3. Consider that the timing parameters of gate are known before the beginning of the simulation and they do not change during the simulation. This assumption greatly simplifies the process of gate-level simulation, but does not satisfy the requirements of gate-level simulation of digital circuits with consideration of DF. The point is that in the latter case, pulses of arbitrary duration and amplitude (Sect. 2.2) may appear at the nodes of a digital circuit, whose values cannot be foreseen in advance.

V. Melikyan, *Simulation and Optimization of Digital Circuits*,
https://doi.org/10.1007/978-3-319-71637-4_4

4. Operate with quantized states of gate, which simplifies the decomposition procedures for simulated digital circuits.

Proceeding from this, there was a need to develop a new algorithm for gate-level simulation, having the advantages of known algorithms (the ability to take into account spatial and timing events) and satisfying the requirements of gate-level simulation of digital circuits with consideration of DF.

In Sect. 1.4 it was noted that when simulating a digital circuit with consideration of DF, the spectrum of possible simulation modes is wider than that of general gate-level simulation. Therefore, consider the essence of the developed algorithm [60, 166, 167, 282] for the most general and complex case—simulation of transient processes with consideration of DF effects, when $\vec{I}(t) \neq \text{const}$, $\vec{V}(t) \neq \text{const}$, $\vec{W}(t) \neq \text{const}$. Then consider the possibility of extending the proposed algorithm to other cases (Sect. 1.4). To simplify the essence of the developed algorithm, suppose that the static mode is already calculated and the steady state of the circuit is obtained. For clarity, the description of the algorithm is accompanied by parallel comments using the example of a digital circuit divider into 5 (Fig. 4.1) using a synchronous JK flip-flop (Fig. 4.2) [122]. Figure 4.3 shows the input signals of a digital circuit. For simplicity, it is assumed in the example that only the following DFs affect the digital circuit: the ambient temperature V_2 (varying in the way shown

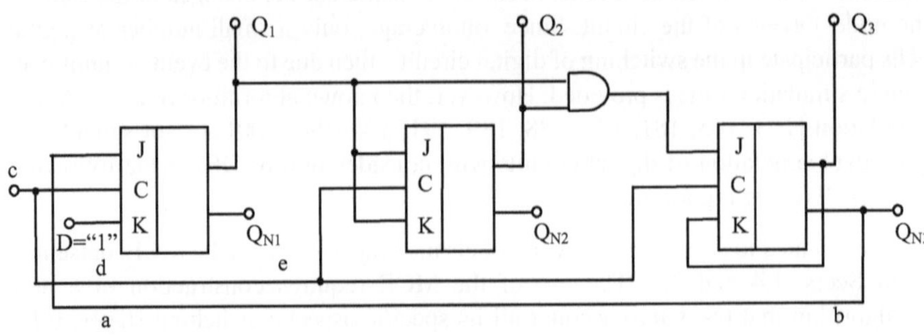

Fig. 4.1 The digital circuit divider into 5

Fig. 4.2 Synchronous JK flip-flop

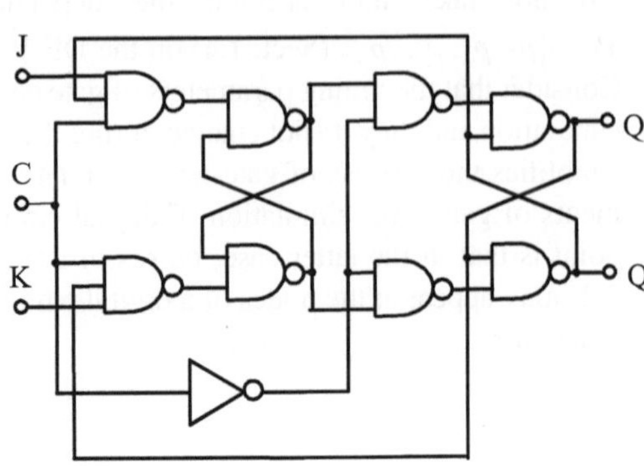

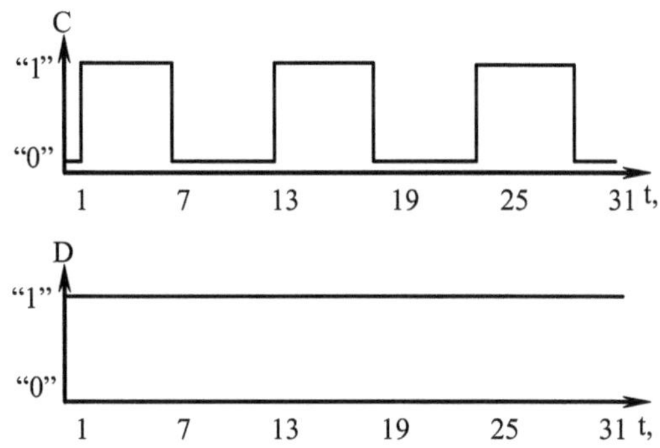

Fig. 4.3 Input signals of the digital circuit divider into 5

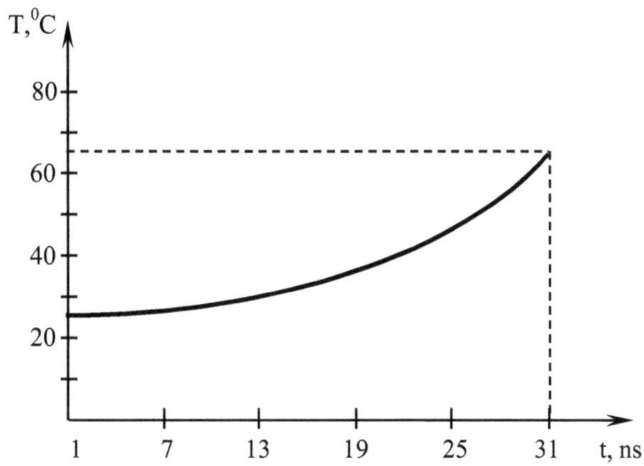

Fig. 4.4 Change in ambient temperature

in Fig. 4.4), the number of loads N (V_5), the nonideal interconnects (W_1) and power rails (W_2).

The following transformations of the source digital circuits are carried out by preprocessing algorithms immediately before simulation (Fig. 4.5):

1. Based on the considerations given in Sect. 3.1, and using topological data on interconnects, the number of sequentially connected LBs that replace each interconnect interval is determined. For example, in the circuit in Fig. 4.4, the interconnect *ab* is replaced by a sequential chain of three LBs, de—of two LBs, etc. With the help of the MDE from Sect. 3.1.4, the timing parameters of these LBs are determined. There are also interconnects, which are considered ideal.
2. Separately, the nodes and the gates of digital circuits are numbered, taking into account the added LB (Fig. 4.5).
3. The distances between the nodes connecting to the power rails of all possible gate pairs and the interconnect lengths are also determined from the digital circuit topology.

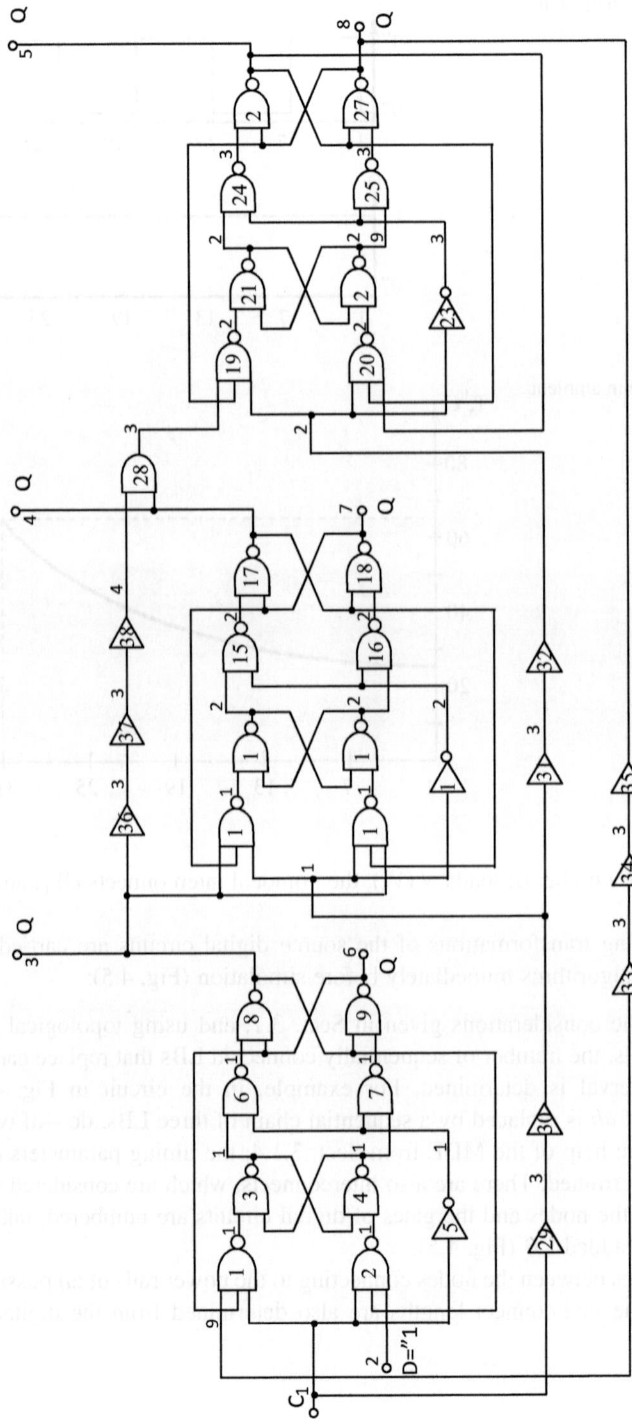

Fig. 4.5 Digital circuit

- The input information for the algorithm is:
  the vector of the nodes of the circuit (the nodes are the inputs and outputs that make up the digital circuit of the gate): ND = $\{nd_1, nd_2, \ldots, nd_i, \ldots, nd_m\}$, where m is the number of nodes of a digital circuit; nd_i is the number of the ith node of a digital circuit. For the considered example: ND = {1, 2, 3, …, 39, 40}.
- The vector of the inputs of a digital circuit: $I = \{i_1, i_2, \ldots, i_i, \ldots, i_p\}$, where p is the number of inputs of a digital circuit; i_i—number of the ith input of a digital circuit. For an example from Fig. 4.5: $I = \{1, 2\}$.
- The output vector of the circuit: $O = \{o_1, o_2, \ldots, o_i, \ldots, o_q\}$, where q is the number of outputs of a digital circuit; O_i is the number of the ith output of a digital circuit. For the example: $O = \{3, 4, 5, 6, 7, 8\}$.
- List of sets of timing parameters: $P = \{p_1, p_2, \ldots, p_i, \ldots, p_r\}$, where p_i is the ith set of timing parameters; r is the number of such sets; P_i contains the set number and the values of all eight MCE parameters for the nominal values of the considered DF, in this example—$T = 25$ °C and $N = 1$ (Sect. 2.1). One of these sets corresponds to each gate of a digital circuit. It is assumed that all gates of the considered digital circuits are characterized by one of the following sets of timing parameters (Table 4.1).

- List of logical functions: $\vartheta = \{\varphi_1, \varphi_2, \ldots, \varphi_i, \ldots, \varphi_o\}$, where φ_i is the ith function; o—number of functions. Each gate circuit implements one of the functions of this list. For this example, there is the following list (Table 4.2):

- Structure of circuit elements: $E = \{e_1, e_2, \ldots, e_i, \ldots, e_g\}$, where g is the number of gates. e_i contains the following information related to the ith gate of a digital circuit: p_i (number of the set of timing parameters of the gate), φ_i (number of the logical function realized by the ith gate), $i_{1i}, i_{2i}, \ldots, i_{ji}, \ldots, i_{pi}, o_i$. Here i_{ji} is the node number of the jth input, p is the number of inputs, and o_i is the output number of the ith gate. Naturally, $i_{ji} \in$ ND and $o_i \in$ ND. For this example, there is Table 4.3:

For example, the third column describes the gate with the number 10, in which the timing parameters are indicated in the set 1 (Table 4.1), the gate implements the function $y = a + b - \min(x_1, x_2)$, its inputs are nodes numbered 3.7, 17, and the output is node 18.

- List of followers of nodes FW_{ij}, $i = 1, 2, \ldots, m; j = 1, 2, \ldots, q_i$, where FW_{ij} is the jth gate of a digital circuit, with at least one input connected to the ith node of the circuit; q_i—the number of gate, at least one input connected to the ith node of the circuit. Naturally, $\mathrm{FW}_{ij} \in E$.
  For this example, there is Table 4.4:

- The list of distances between the connection points of the gate to the power rail: $H = \{l_{r1}, l_{r2}, \ldots, l_{ri}, \ldots, l_{rs}\}$, where l_{ri} is the distance between the connection points to the power supply of the gate pair, s is the number of all possible pairs between gates from the set E. This list is formed by preprocessing algorithms of

Table 4.1 Sets of timing parameters

Set number	$\Delta t_{\mathrm{min}i}^{01}$ (ns)	Δt_{bi}^{01} (ns)	$\Delta t_{\mathrm{max}i}^{01}$ (ns)	Δt_{si}^{01} (ns)	$\Delta t_{\mathrm{min}i}^{10}$ (ns)	Δt_{bi}^{10} (ns)	$\Delta t_{\mathrm{max}i}^{10}$ (ns)	Δt_{si}^{10} (ns)
1	0.3	0.4	0.7	0.8	0.1	0.2	0.5	0.7
2	0.3	0.5	0.8	1.0	0.2	0.3	0.6	0.8
3	0.1	0.2	0.4	0.5	0.1	0.2	0.4	0.5

Table 4.2 List of functions

Function number	Boolean function
1	$y = a + b - \min(x_1, x_2)$
2	$y = \max(x_1, x_2)$
3	$y = a + b - x$
4	$y = x$

the topology of digital circuits, and its elements are arguments for the MDE to detect noise (3.1.5). For this example, the list is shown in Table 4.5.

- List of interconnect lengths: $L_m = \{l_{m1}, l_{m2}, \ldots, l_{mi}, \ldots, l_{mt}\}$, where l_{mi} is the length of the ith interconnect between the output of the xth gate and the input of the jth gate, t is the number of interconnects. This list is also formed by preprocessing algorithms of the topology of digital circuits, and its elements are arguments for the MDE from Sect. 3.1.4. For this example, the list is shown in Table 4.6.

- List of input signals: $S = \{s_1, s_2, \ldots, s_i, \ldots, s_n\}$, where s_i is the input signal at the ith input of a digital circuit, containing information on the initial state of the input i $s_i(0)$, as well as sequential moments of signal switching at the ith input of a digital circuit: $t_{i1}, t_{i2}, \ldots, t_{ij}, \ldots, t_{iu}$, where t_{ij} is the time of the jth switching at the ith input of a digital circuit, and u is the number of switching operations in the ith input signal. For example, for signal C from Fig. 4.3: 0.1.7, 13, 19, 25, 31.
- The simulation time, after which the simulation process is suspended (T_{sim}). In this example: $T_{\text{sim}} = 80$.
- In the process of simulation of digital circuits, the following working information is also used:

  - Excitation list: $\text{EXC} = \{\text{exc}_1, \text{exc}_2, \ldots, \text{exc}_i, \ldots, \text{exc}_d\}$, where d is the number of excitations in this list; exc_i contains information about switching the ith node of a digital circuit nd_i, which will happen in the future, and has the following components:

    (a) Node number nd_i
    (b) The state of Q_i into which node nd_i should switch
    (c) The moments of completion of certain switching stages of the node nd_i: $t_{\text{min}i}, t_{\text{b}i}, t_{\text{max}i}, t_{\text{s}i}$
    (d) Switching stage ST_i, which can take the following values: 1 when $0 \le t_{\text{sim}} \le t_{\text{min}i}$; $2 - t_{\text{min}i} \le t_{\text{sim}} \le t_{\text{b}i}$; $3 - t_{\text{b}i} \le t_{\text{sim}} \le t_{\text{max}i}$; $4 - t_{\text{max}i} \le t_{\text{sim}} \le t_{\text{s}i}$ (when the state of the node Q increases, respectively, 5, 6, 7, 8—when decreases) $0 - t_{\text{sim}} > t_{\text{s}i}$

The latter case corresponds to the completion of excitation. At the same time, the excitation is not physically erased, and the value $\text{ST}_i = 0$ indicates to the excitation processing unit that a new excitation can be written in place of the ith excitation. This way of organizing the list of excitations allows making the number of excitations in

Table 4.3 Structure of digital circuit elements

	e_1	e_2	e_3	e_4	e_5	e_6	e_7	e_8	e_9	e_{10}	e_{11}	e_{12}	e_{13}	e_{14}	e_{15}	e_{16}	e_{17}	e_{18}	e_{19}
P	1	1	1	1	3	1	1	1	1	1	1	1	1	3	1	1	1	1	1
φ	1	1	1	1	3	1	1	1	1	1	1	1	1	3	1	1	1	1	1
i_1	9	1	10	12	1	12	14	15	3	3	17	18	20	17	20	22	23	4	33
i_2	6	3	13	11	–	14	13	6	15	7	4	21	19	7	22	21	7	24	8
i_3	1	2	–	–	–	–	–	–	–	17	3	–	–	–	–	–	–	–	25
o	10	11	12	13	14	15	16	3	6	18	19	20	21	22	23	24	4	7	26

	e_{20}	e_{21}	e_{22}	e_{23}	e_{24}	e_{25}	e_{26}	e_{27}	e_{28}	e_{29}	e_{30}	e_{31}	e_{32}	e_{33}	e_{34}	e_{35}	e_{36}	e_{37}	e_{38}
P	1	1	1	1	3	1	1	1	2	3	3	3	3	3	3	3	3	3	3
φ	1	1	1	1	3	1	1	1	2	4	4	4	4	4	4	4	4	4	4
i_1	25	26	28	25	28	30	31	5	40	1	34	15	35	9	36	37	3	38	39
i_2	5	29	27	–	30	29	8	32	4	–	–	–	–	–	–	–	–	–	–
i_3	5	–	–	–	–	–	–	–	–	–	–	–	–	–	–	–	–	–	–
o	27	28	29	30	31	32	5	8	33	34	15	35	25	36	37	8	38	39	40

Table 4.4 List of followers of nodes

ND	1	2	3	4	5	6	7	8	9	10	11	12	13	14	15	16	17	18	19	20
e_1	1	2	36	28	20	1	10	19	1	3	4	4	3	6	8	9	10	12	13	13
e_2	2	–	–	–	27	8	17	26	–	–	–	6	7	7	–	–	11	–	–	15
e_3	5	–	–	–	–	–	–	35	–	–	–	–	–	–	–	–	14	–	–	–
e_4	–	–	–	–	–	–	–	–	–	–	–	–	–	–	–	–	31	–	–	–
ND	21	22	23	24	25	26	27	28	29	30	31	32	33	34	35	36	37	38	39	40
e_1	12	15	17	18	19	21	22	22	21	24	26	27	19	30	32	33	34	37	38	33
e_2	16	16	–	–	20	–	–	24	25	25	–	–	–	–	–	–	–	–	–	–
e_3	–	–	–	–	23	–	–	–	–	–	–	–	–	–	–	–	–	–	–	–
e_4	–	–	–	–	–	–	–	–	–	–	–	–	–	–	–	–	–	–	–	–

Table 4.5 Distances between gates via power rails

e_i	e_j	L_r, mkm
1	2	10.2
1	3	14.4
1	4	7.8
...	...	...
1	38	18.2
2	3	8.8
2	4	10.6
...	...	...
38	38	11.4

Table 4.6 Interconnect length

e_i	e_j	L_m, mkm
1	3	1.2
3	4	4.8
3	6	5.4
5	6	8.4
4	7	5.3
...	...	...
27	19	14.6
26	20	14.8
38	28	6.6

the list variable. D_{max} is equal to the maximum number of simultaneously excited nodes of a digital circuit. Since this number is on average 3% ... 5% [163] of the total number of nodes of a digital circuit, a significant gain in memory is obtained;

– List of states of nodes of a digital circuit: $C = \{c_1, c_2, \ldots, c_i, \ldots, c_m\}$, where c_i is the state of the ith node of a digital circuit, taking arbitrary values from the range $[a, b]$ (Sect. 2.2). c_i is the function of the current simulation time t_{sim}. It should be noted that the list of states of digital circuit nodes is also the result of the algorithm work, since the values of the function c_i (t_{sim}) for $0 \leq t_{sim} \leq T_{sim}$ are the time diagram of the state change of the ith node of a digital circuit.

The initial value of the C list is formed as a result of calculating the static mode. For this example, it is assumed that the initial state of the circuit has this form (Table 4.7):

In order to facilitate understanding of the essence of the proposed gate-level simulation algorithm, with consideration of DF effects in Sect. 4.3, the list of states of the nodes of circuit C and the list of excitations EXC for the considered example to simulation moment $t_{sim} = 2.35$ are given. Below is a brief description of the developed algorithm. At the beginning of simulation (steps 1 ... 13), the EXC list is filled with the exc_i elements, each of which corresponds to one switching of some input signal of a digital circuit (moment $t_{sim} = 0$). Therefore, the initial number of elements of the EXC d_{max} list is equal to the number of all switching operations on all inputs of the circuit—6. For each switching of the input signal in the EXC list, an

Table 4.7 List of node states

ND	1	2	3	4	5	6	7	8	9	10	11	12	13	14	15	16	17	18	19	20
C	0	1	0	0	0	1	1	1	1	1	1	0	1	1	1	0	0	1	1	0
ND	21	22	23	24	25	26	27	28	29	30	31	32	33	34	35	36	37	38	39	40
C	1	1	1	0	0	1	1	0	1	1	1	0	0	0	0	1	1	0	0	0

excitation with parameters is written: $t_{\text{min}i} = t_{\text{b}i} = t_{\text{max}i} = t_{\text{s}i} = t_{ja}$, where t_{ja} is the moment of the ath switching on jth input of a digital circuit. The further sequence of the algorithm ensures that each of these excitations is erased immediately at the time the input state changes.

At the beginning of each stage of the simulation, the current analysis time, at a minimum, of all t_{min}, t_{b}, t_{max}, t_{s} from the EXC list is determined first. Immediately after this (steps 17 … 38), the states of the nodes of a digital circuit of the GME are calculated (Sect. 1.4), and it is determined whether the gate switches as a result of noise. If switching is to occur, the consequences are determined (steps 39 … 77) and the corresponding excitations are entered in the EXC list.

t_{sim} can simultaneously be equal to several t_{min}, t_{b}, t_{max}, and t_{s} from the EXC list. If $t_{\text{sim}} = t_{\text{s}}$, then the corresponding excitations are quenched (steps 91 … 93 and 158 … 159). In this example, such a case arose at the moment $t = 1.9$ for the node numbered 12. In case where $t_{\text{sim}} = t_{\text{min}}$ or $t_{\text{sim}} = t_{\text{max}}$, the excitation is transferred to the next switching stage (steps 79 … 82, 94 … 95 and, accordingly, 87 … 90, 156 … 157). In this example, this situation arose at the moment of 1.1 for the node with the number 10. If the relation $t_{\text{sim}} = t_{\text{b}}$ holds, then the corresponding excitations from the EXC list are realized. In this case, the state of the corresponding node changes in the list C (steps 98 … 99). Then the excitation stage changes (step 98). After that, those gates of the circuit are considered which could be affected by the change in the state of this node. Depending on whether the change in the state of the node leads to the switching of the output of the followers or not, as well as the state of the gate at the current time of the simulation (whether excited or not, at what stage of switching, etc.), different switching rules are implemented, described in Sect. 2.2 (steps 106 … 114). The implementation of switching in this example occurred at times $t_{\text{sim}} = 1, 2$ for node 10, 1.36 for node 12, etc.

After processing all the excitations, the definition of t_{sim} is repeated. Simulation ends in one of two cases:

1. When the EXC list becomes empty, i.e. transient processes in the circuit are set before the expiration of the time T_{sim} ($\text{ST}_i = 0$ for all $i = 1, 2, \ldots, d_{\text{max}}$).
2. When T_{sim} time expires, although the transient processes in the circuit did not end.

The above algorithm provides an event-driven calculation that considers only those moments in time when events occur, and only those MCEs that respond to these events. The order of processing the events occurring simultaneously on different inputs of the MCE does not affect the result due to the rules of superimposing excitations adopted in the MCE (Sect. 2.2). Therefore, no special measures are required when processing multiple events [172]. For nonzero timing parameters of the MCE, the convergence of the algorithm in sense of the absence of a loop at some moment of time is obvious. With zero timing parameters of the MCE, the algorithm is equivalent to Seidel iterations with preliminary ranking of a digital circuit, which is achieved by adding new excitations after the others, referring to the same moment of t_{sim}, and by choosing the first of multiple excitations. Thus, the wave of events is monitored. In this case, the preliminary ranking of a digital circuit is not required—it is done automatically during the simulation process. Another possible way, which leads to the same results, but somewhat facilitates the maintenance of the EXC list, is the replacement of all zero timing parameters of the MCE with very small, but still finite values.

The considered algorithm provides a calculation of transient processes in digital circuits, which was initially in a stable static state, from which it was derived by the action of input signals. In this case, at the start of the time, the EXC list contains only the lines relating to the input signals. However, the same algorithm can be used to calculate the static state of a digital circuit ($\vec{I}(t) = \text{const}$, $\vec{V}(t) = \text{const}$, $\vec{W}(t) = \text{const}$). In this case, at the beginning of the calculation, the signals at the inputs of digital circuits are given, and signals can be specified by the user in some internal nodes of digital circuits. The signals in the remaining nodes are not defined. They can be taken equal to x, if such a state is provided by models, or 0 or 1 is given.

The algorithm for calculating the static mode, storing as much user-defined signal values as possible, can be as follows:

Procedure A. MCE, at the outputs of which the signals are set by the user, are temporarily excluded from the digital circuit. Exceptions related to their outputs and inputs are recorded in the EXC. As $t_{\text{min}i}$, $t_{\text{b}i}$, $t_{\text{max}i}$, $t_{\text{s}i}$, zeros are indicated, and as Q—given signal values. Event-driven simulation is performed according to the algorithm considered above, until the EXC is empty or generation is fixed (the number of calculation steps exceeds the specified maximum value).

Procedure B. The conditions for the excitation of all MCE are checked. If in some MCE they are executed, then the corresponding excitation is recorded in the EXC. If in the end the EXC turns out to be non-empty, event-driven simulation is repeated.

Procedure C. MCEs, excluded during the operation of procedure A from the description of a digital circuit, return to the circuit and the excitation conditions are checked therein. If excitation occurs, they are recorded in the EXC. If the result of the EXC is non-empty, then an event-driven simulation is executed.

The procedure C checks whether the detected state of a digital circuit corresponds to the specified signal values. If there are no excitations, then the desired state is found. The appearance of excitations indicates that the required state was not obtained. In this case, the automated gate-level simulation system should give the user a corresponding message and calculate the digital circuit, changing the values of specified signals. The reasons for the appearance of excitations during the operation of procedure C can be two. First, a user can mistakenly specify a combination of signals that is not implemented in this circuit. Secondly, the given signals are contradictory. In both cases, the user must fix the task or agree with the correction that is made by the algorithm.

The case of calculating the transient processes of a digital circuit without consideration of DF ($\vec{I}(t) \neq \text{const}$, $\vec{V}(t) = \text{const}$, $\vec{W}(t) = \text{const}$) differs from the algorithm given in Sect. 4.2 only in that the blocks of computation by MDE and GME are actually switched off. If the static mode of operation of a digital circuit is simulated with consideration of DF ($\vec{I}(t) = \text{const}$, $\vec{V}(t) \neq \text{const}$, $\vec{W}(t) \neq \text{const}$), then the above algorithm can be applied. This simulation mode can only be conditionally called static, because as a result of the changes $\vec{W}(t)$ and $\vec{V}(t)$ by the MDE, such values of Q in digital circuit nodes can be obtained which can lead to switching of some gates and transient processes will actually start in digital circuits. In this case, the processing will pass through the corresponding branches of the algorithm. Thus, the developed gate-level simulation algorithm with consideration of DF provides all simulation modes described in Sect. 1.4.

4.2 Description of the Algorithm

(Start of filling in the list of excitations EXC on the input signals of a digital circuit: i—index of the EXC list; j—current number of the input of a digital circuit; a—sequence number of the switching at the jth input)

step 1:	$i = 0$;
step 2:	$j = 1$;
step 3:	if $j > n$, go to step 14;
step 4:	$Q = 1 - s_j(0)$;
step 5:	$a = 1$;
step 6:	If $a > u_j$, go to step 12;
step 7:	$i = i + 1$;
step 8:	input excitation exc_i for the ath switching of the jth input signal: $\mathrm{nd}_i = \mathrm{in}_j$, Q, $t_{\mathrm{min}i} = t_{ja}$, $t_{fi} = t_{ja}$, $t_{\mathrm{max}i} = t_{ja}$, $t_{si} = t_{ja}$, $\mathrm{ST}_i = 1$;
step 9:	$a = a + 1$;
step 10:	$Q = 1 - Q$;
step 11:	go to step 6;
step 12:	$j = j + 1$;
step 13:	go to step 3;

(the end of filling out the EXC list, the beginning of simulation process)

step 14:	$d_{\mathrm{max}} = i$;
step 15:	if the EXC list is empty ($\mathrm{ST}_i = 0$ for all $i = 1, 2, \ldots, d_{\mathrm{max}}$), go to step 16;
step 16:	finding from the list EXC $t_{\mathrm{sim}} = \min\{t_{\mathrm{min}i}, t_{\mathrm{b}i}, t_{\mathrm{max}i}, t_{\mathrm{bd}i}\}$, $i = 1, 2, \ldots, d_{\mathrm{max}}$

(the beginning of the block for calculating the states of the nodes of a digital circuit by GME)

step 17:	$i = n + 1$; (i is the number of nodes of a digital circuit)
step 18:	$f_i = 0$; (1.23)
step 19:	$j = n + 1$;
step 20:	if $j = i$, go to step 24;
step 21:	if $j > m$, go to step 26;
step 22:	determination of the values of the MCE parameters by the MDE from Sect. 2.4.
step 23:	application of PDE (1.23)
step 24:	$j = j + 1$;
step 25:	go to step 20;

(continued)

step 26:	application of RSE (1.20)
step 27:	if the ith node of the gate is excited (there is an excitation with the number q in the EXC list), go to step 39;
step 28:	if $Q_i = c_i(t_{\text{sim}})$, go to step 36 (the current state is confirmed)
Entering a new excitement, node i must switch as a result of GME calculation)	
step 29:	$d = 1$;
step 30:	if $d > d_{\max}$, go to step 34;
step 31:	if $\text{ST}_d = 0$, go to step 35;
step 32:	$d = d + 1$;
step 33:	go to step 30;
step 34:	$d_{\max} = d_{\max} + 1$, $d = d_{\max}$;
step 35:	Entering of a new excitation: $\text{nd}_d = i$, $t_{\text{min}d} = t_{\text{sim}} + \Delta t_{\min}$, $t_{\text{bd}} = t_{\text{sim}} + \Delta t_{\text{b}}$, $t_{\text{max}d} = t_{\text{sim}} + \Delta t_{\max}$, $t_{sd} = t_{\text{sim}} + \Delta t_{\text{s}}$; ($\Delta t_{\min}$, Δt_{b}, $\Delta t_{\max}$, Δt_{s} are determined from the list P, and the switching direction: $s_i(t_{\text{sim}}) \rightarrow Q_i$)
step 36:	$i = i + 1$;
step 37:	if $i > m$, go to step 78;
step 38:	go to step 18;
step 39:	if $Q_i > Q_q$, go to step 36;
step 40:	if $t_{\text{sim}} < t_{\text{min}q}$, go to step 69;
step 41:	if $t_{\text{sim}} < t_{\text{b}q}$, go to step 60;
step 42:	if $t_{\text{sim}} < t_{\text{max}q}$, go to step 51;
step 43:	$d = 1$;
step 44:	if $d > d_{\max}$, go to step 48;
step 45:	if $\text{ST}_d = 0$, go to step 49;
step 46:	$d = d + 1$;
step 47:	go to step 44;
step 48:	$d_{\max} = d_{\max} + 1$, $d = d_{\max}$;
step 49:	$\text{nd}_d = \text{nd}_q$, $Q_d = a + b - Q_q$, $t_{\text{mind}} = t_{\text{s}q}$, $t_{\ell d} = \Delta t_{\text{min}d} + \Delta t_{\ell} - \Delta t_{\min}$, $t_{\max} = t_{\text{min}d} + \Delta t_{\max} - \Delta t_{\min}$, $t_{\text{s}d} = \Delta t_{\text{min}d} + \Delta t_{\text{s}} - \Delta t_{\min}$, $\text{ST}_d = 5$;
step 50:	go to step 36;
step 51:	the definition of t_2 (Fig. 2.29) according to formula (2.60);
step 52:	$d = 1$;
step 53:	if $d > d_{\max}$, go to step 57;
step 54:	if $\text{ST}_d = 0$, go to step 58;
step 55:	$d = d + 1$;
step 56:	go to step 53;
step 57:	$d_{\max} = d_{\max} + 1$, $d = d_{\max}$;
step 58:	$\text{nd}_d = \text{nd}_q$, $Q_d = a + b - Q_q$, $t_{\text{bd}} = t_2 + \Delta t_{\text{b}} - \Delta t_{\min}$, $t_{\max} = t_2 + \Delta t_{\max} - \Delta t_{\min}$, $t_{\text{sd}} = t_2 + \Delta t_{\text{s}} - \Delta t_{\min}$, $\text{ST}_d = 6$;
step 59:	go to step 36;
step 60:	the definition of t_2 (Fig. 2.27) according to formula (2.55);
step 61:	$d = 1$;
step 62:	if $d > d_{\max}$, go to step 66;
step 63:	if $\text{ST}_d = 0$, go to step 67;
step 64:	$d = d + 1$;

(continued)

step 65:	go to step 62;
step 66:	$d_{\max} = d_{\max} + 1, d = d_{\max}$;
step 67:	$\text{nd}_d = \text{nd}_q$, $Q_d = a + b - Q_q$, $t_{\max} = t_2 + \Delta t_{\max} - \Delta t_{\min}$, $t_{\text{sd}} = t_2 + \Delta t_{\text{s}} - \Delta t_{\min}$, $\text{ST}_d = 7$;
step 68:	go to step 36;
step 69:	the definition of t_2 (Fig. 2.24) according to formula (2.39);
step 70:	$d = 1$;
step 70:	if $d > d_{\max}$, go to step 75;
step 72:	if $\text{ST}_d = 0$, go to step 76;
step 73:	$d = d + 1$;
step 74:	go to step 71;
step 75:	$d_{\max} = d_{\max} + 1, d = d_{\max}$;
step 76:	$\text{ND}_d = \text{ND}_q$, $t_{sd} = t_2 + \Delta t_{\text{s}} - \Delta t_{\min}$, $\text{ST}_d = 8$;
step 77:	go to step 36;
(the end of the block for calculating the states of a digital circuit nodes by GME)	
step 78:	if $t_{\text{sim}} > T_{\text{sim}}$, go to step 160;
step 79:	if $t_{\text{sim}} \neq t_{\min}$, at least for one $i = 1, 2, \ldots, d_{\max}$, go to step 83;
(t_{sim} is found from such exc_i for which $t_{\text{sim}} = t_{\min i}$)	
step 80:	search for the next i, for which $t_{\text{sim}} = t_{\min i}$;
step 81:	if there exists i, for which $t_{\text{sim}} = t_{\min i}$, go to step 94;
step 82:	go to step 15;
step 83:	if $t_{\text{sim}} \neq t_{fi}$, at least for one $i = 1, 2, \ldots, d_{\max}$, go to step 87;
(t_{sim} is found from such exc_i for which $t_{\text{sim}} = t_{\text{b}i}$)	
step 84:	search for the next i, for which $t_{\text{sim}} = t_{\text{b}i}$;
step 85:	if there exists i, for which $t_{\text{sim}} = t_{\text{b}i}$, go to step 96;
step 86:	go to step 15;
step 87:	if $t_{\text{sim}} \neq t_{\max i}$, at least for one $i = 1, 2, \ldots, d_{\max}$, go to step 91;
(t_{sim} is found from such exc_i for which $t_{\text{sim}} = t_{\max i}$)	
step 88:	search for the next i, for which $t_{\text{sim}} = t_{\max i}$;
step 89:	if there exists i, for which $t_{\text{sim}} = t_{\max i}$, go to step 156;
step 90:	go to step 15; (t_{sim} is found for such v_i for which $t_{\text{sim}} = t_{si}$)
step 91:	search for the next i, for which $t_{\text{sim}} = t_{si}$;
step 92:	if there exists i, for which $t_{\text{sim}} = t_{si}$, go to step 158;
step 93:	go to step 15;
step 94:	$\text{ST}_i = 2$;
step 95:	go to step 80;
(the beginning of implementing excitation)	
step 96:	$b = \text{nd}_i$;
step 97:	$c_{\text{b}}(t_{\text{sim}}) = Q_i$;
step 98:	if $t_{\min i} = t_{fi} = t_{\max i} = t_{si}$, then $\text{ST}_i = 0$; otherwise $\text{ST}_i = 3$;
step 99:	Search for all gates with at least one input connected to node b by FW list and definition of the number of such *gates*—q_b;
step 100:	if $q_b = 0$, ($b \in O$), go to step 84;
step 101:	$j = 1$; (j—the sequence number of the gate, the input of the connections to the node b)
step 102:	$e = \text{FW}_{bj}$; (e is the number of the jth element-follower in the list E)

(continued)

step 103:	definition of P_e, φ_c, i_{1e}, i_{2e}, …, i_{pe}, o_e from the structure E;
step 104:	calculation of a new state Q_n of the output of the o_j *j*th gate in case of the current values of input stats i_{1e}, i_{2e}, …, i_{pe} by the function φ_c.
step 105:	if the *j*th gate is excited (there is an excitation with the number q in the EXC list), go to step 117;
step 106:	if $Q_n = c_{\text{ine}}(t_{\text{sim}})$, go to step114; (the current state of the gate is confirmed)
(entering a new excitation, since the gate must switch to another state)	
step 107:	$d = 1$; (d—the excitation number)
step 108:	if $d > d_{\max}$, go to step 112;
step 109:	if $\text{ST}_d = 0$, go to step 113;
step 110:	$d = d + 1$;
step 111:	go to step 108;
step 112:	$d_{\max} = d_{\max} + 1, d = d_{\max}$;
step 113:	Entering a new excitation: $\text{bd}_d = \text{in}_e$, $t_{\text{min}d} = t_{\text{sim}} + \Delta t_{\min}$, $t_{\text{bd}} = t_{\text{sim}} + \Delta t_{\text{b}}$, $t_{\text{max}d} = t_{\text{sim}} + \Delta t_{\max}$, $t_{\text{s}d} = t_{\text{sim}} + \Delta t_{\text{s}}$, $\text{ST}_d = 1$; ($\Delta t_{\min}$, Δt_{b}, $\Delta t_{\max}$, Δt_{s} are determined from the list P, and the switching direction: $c_{\text{ine}}(t_{\text{sim}}) \to Q_n$)
step 114:	$j = j + 1$;
step 115:	if $j \le q_b$, go to step 102;
step 116:	go to step 84;
step 117:	If the new excitation confirms the old one ($Q_n = Q_q$), go to step 114;
step 118:	if $t_{\text{sim}} < t_{\text{min}q}$, go to step 147;
step 119:	if $t_{\text{sim}} < t_{\text{b}q}$, go to step 138;
step 120:	if $t_{sim} < t_{\text{maxb}q}$, go to step 129;
step 121:	$d = 1$;
step 122:	if $d > d_{max}$, go to step 126;
step 123:	if $\text{ST}_d = 0$, go to step 127;
step 124:	$d = d + 1$;
step 125:	go to step 122;
step 126:	$d_{\max} = d_{\max} + 1, d = d_{\max}$;
step 127:	$\text{bd}_d = \text{bd}_q$, $Q_d = a + b - Q_q$, $t_{\text{mind}} = t_{\text{bd}q}$, $t_{\text{bd}} = t_{\text{mind}} + \Delta t_{\text{b}} - \Delta t_{\min}$, $t_{\max} = t_{\text{mind}} + \Delta t_{\max} - \Delta t_{\min,}$ $t_{\text{bd}d} = t_{\text{mind}} + \Delta t_{\text{s}} - \Delta t_{\min}$, $ST_{\text{d}} = 5$; (timing parameters are defined like step 113)
step 128:	go to step 114;
step 129:	the definition of t_2 (Fig. 2.29) according to formula (2.60);
step 130:	$d = 1$;
step 131:	if $d > d_{\max}$, go to step 135;
step 132:	if $\text{ST}_d = 0$, go to step 136;
step 133:	$d = d + 1$;
step 134:	go to step 131;
step 135:	$d_{\max} = d_{\max} + 1, d = d_{\max}$;
step 136:	$\text{bd}_d = \text{bd}_q$, $\text{Q}_d = a + b - Q_q$, $t_{\text{bd}} = t_2 + \Delta t_{\text{b}} - \Delta t_{\min}$, $t_{\max} = t_2 + \Delta t_{\max} - \Delta t_{\min}$, $t_{\text{bd}d} = t_2 + \Delta t_{\text{bd}} - \Delta t_{\min}$, $\text{ST}_d = 6$; (*timing parameters are defined like step 113*)
step 137:	*go to step 114*;
step 138:	the definition of t_2 (Fig. 2.27) according to formula (2.55)
step 139:	$d = 1$;
step 140:	if $d > d_{\max}$, go to step *144*;

(continued)

step 141:	if $\mathrm{ST}_d = 0$, *go to step 145*;
step 142:	$d = d + 1$;
step 143:	*go to step 140*;
step 144:	$d_{\max} = d_{\max} + 1$, $d = d_{\max}$;
step 145:	$\mathrm{bd}_d = \mathrm{bd}_q$, $Q_d = a + b - Q_q$, $t_{\mathrm{iax}d} = t_2 + (t_{\max} - t_{\min}$, $t_{\mathrm{bd}d} = t_2 + \Delta t_{\mathrm{bd}} - \Delta t_{\min}$, $\mathrm{ST}_d = 7$; (timing parameters are defined like step 113)
step 146:	go to step 114;
step 147:	the definition of t_2 (Fig. 2.24) according to formula (2.39);
step 148:	$d = 1$;
step 149:	if $d > d_{\max}$, go to step 153;
step 150:	if $\mathrm{ST}_d = 0$, go to step 154;
step 151:	$d = d + 1$;
step 152:	Go to step 149;
step 153:	$d_{\max} = d_{\max} + 1$, $d = d_{\max}$;
step 154:	$\mathrm{bd}_d = \mathrm{bd}_q$, $t_{\mathrm{s}d} = t_2 + \Delta t_{\mathrm{s}} - \Delta t_{\min}$, $\mathrm{ST}_d = 8$; (timing parameters are defined like step 113)
step 155:	go to step114;
(end of implementing excitation)	
step 156:	$\mathrm{ST}_i = 4$;
step 157:	go to step 88;
step 158:	$\mathrm{ST}_i = 0$;
step 159:	go to step 91;
step 160:	fixing the fact of noncompletion of transient processes in time T_{sim};
step 161:	the end of simulation.

4.3 Simulation Example

Tables 4.8 and 4.9.

4.4 Mixed-Mode Simulation Algorithms with Consideration of DF

The efficiency of mixed-mode simulation of analog-digital circuits is greatly influenced by the solution of the problem of timing (algorithmic) matching of circuit-level and gate-level simulation (Sect. 1.2). In the known algorithms of mixed-mode simulation, the effect of DF was not taken into account in ML. Therefore, mixed-mode simulation algorithms do not take into account the features of the developed MCE (continuity of states of gate output, the possibility of changing the gate parameters during the ML calculation, etc. (Sect. 2.2)).

Table 4.8 List of node states

nd	t_{sim} 0	1	1.1	1.2	1.27	1.36	1.43	1.44	1.5	1.51	1.54	1.7	1.81	1.82	1.83	1.9	1.92	2.02	2.03	2.13	2.16	2.24	2.25	2.35
1	0	1	1	1	1	1	1	1	1	1	1	1	1	1	1	1	1	1	1	1	1	1	1	1
2	1	1	1	1	1	1	1	1	1	1	1	1	1	1	1	1	1	1	1	1	1	1	1	1
3	0	0	0	0	0	0	0	0	0	0	0	0	0	0	0	0	0	0	0	0	0	0.97	0.97	0.97
4	0	0	0	0	0	0	0	0	0	0	0	0	0	0	0	0	0	0	0	0	0	0	0	0
5	0	0	0	0	0	0	0	0	0	0	0	0	0	0	0	0	0	0	0	0	0	0	0	0
6	1	1	1	1	1	1	1	1	1	1	1	1	1	1	1	1	1	1	1	1	1	1	0.94	0.94
7	1	1	1	1	1	1	1	1	1	1	1	1	1	1	1	1	1	1	1	1	1	1	1	1
8	1	1	1	1	1	1	1	1	1	1	1	1	1	1	1	1	1	1	1	1	1	1	1	1
9	1	1	1	1	1	1	1	1	1	1	1	1	1	1	1	1	1	1	1	1	1	1	1	1
10	1	1	1	0	0	0	0	0	0	0	0	0	0	0	0	0	0	0	0	0	0	0	0	0
11	1	1	1	1	1	1	1	1	1	1	1	1	1	1	1	1	1	1	1	1	1	1	1	1
12	0	0	0	0	0	0.98	0.98	0.98	0.98	0.98	0.98	0.97	0.97	0.97	0.97	0.97	0.97	0.97	0.97	0.97	0.97	0.97	0.97	0.97
13	1	1	1	1	1	1	1	1	1	0.05	0.05	0.05	0.06	0.06	0.06	0.06	0.06	0.06	0.06	0.06	0.06	0.06	0.06	0.06
14	1	1	1	1	1	1	1	1	1	1	1	1	1	1	1	1	1	1	1	1	1	1	1	1
15	1	1	1	1	1	1	1	1	1	1	1	1	1	1	0.04	0.04	0.04	0.04	0.04	0.04	0.04	0.04	0.04	0.04
16	0	0	0	0	0	0	0	0	0	0	0	0	0	0.95	0.95	0.95	0.95	0.95	0.95	0.95	0.95	0.95	0.95	0.95
17	0	0	0	0	0	0	0	0	0	0	0	0	0	0	0	0	0	0	0	0	0	0	0	0
18	1	1	1	1	1	1	1	1	1	1	1	1	1	1	1	1	1	1	1	1	1	1	1	1
19	1	1	1	1	1	1	1	1	1	1	1	1	1	1	1	1	1	1	1	1	1	1	1	1
20	0	0	0	0	0	0	0	0	0	0	0	0	0	0	0	0	0	0	0	0	0	0	0	0
21	1	1	1	1	1	1	1	1	1	1	1	1	1	1	1	1	1	1	1	1	1	1	1	1
22	1	1	1	1	1	1	1	1	1	1	1	1	1	1	1	1	1	1	1	1	1	1	1	1
23	1	1	1	1	1	1	1	1	1	1	1	1	1	1	1	1	1	1	1	1	1	1	1	1
24	0	0	0	0	0	0	0	0	0	0	0	0	0	0	0	0	0	0	0	0	0	0	0	0
25	0	0	0	0	0	0	0	0	0	0	0	0	0	0	0	0	0	0	0	0	0	0	0	0

(continued)

Table 4.8 (continued)

nd	t_{sim} 0	1	1.1	1.2	1.27	1.36	1.43	1.44	1.5	1.51	1.54	1.7	1.81	1.82	1.83	1.9	1.92	2.02	2.03	2.13	2.16	2.24	2.25	2.35
26	1	1	1	1	1	1	1	1	1	1	1	1	1	1	1	1	1	1	1	1	1	1	1	1
27	1	1	1	1	1	1	1	1	1	1	1	1	1	1	1	1	1	1	1	1	1	1	1	1
28	0	0	0	0	0	0	0	0	0	0	0	0	0	0	0	0	0	0	0	0	0	0	0	0
29	1	1	1	1	1	1	1	1	1	1	1	1	1	1	1	1	1	1	1	1	1	1	1	1
30	1	1	1	1	1	1	1	1	1	1	1	1	1	1	1	1	1	1	1	1	1	1	1	1
31	1	1	1	1	1	1	1	1	1	1	1	1	1	1	1	1	1	1	1	1	1	1	1	1
32	0	0	0	0	0	0	0	0	0	0	0	0	0	0	0	0	0	0	0	0	0	0	0	0
33	0	0	0	0	0	0	0	0	0	0	0	0	0	0	0	0	0	0	0	0	0	0	0	0
34	0	0	0	0	0	0	0	0	0	0	0	0	0	0	0	0	0	0	0	0	0	0	0	0
35	0	0	0	0	0	0	0	0	0	0	0	0	0	0	0	0	0	0	0	0	0	0	0	0
36	1	1	1	1	1	1	1	1	1	1	1	1	1	1	1	1	1	1	1	1	1	1	1	1
37	1	1	1	1	1	1	1	1	1	1	1	1	1	1	1	1	1	1	1	1	1	1	1	1
38	0	0	0	0	0	0	0	0	0	0	0	0	0	0	0	0	0	0	0	0	0	0	0	0
39	0	0	0	0	0	0	0	0	0	0	0	0	0	0	0	0	0	0	0	0	0	0	0	0
40	0	0	0	0	0	0	0	0	0	0	0	0	0	0	0	0	0	0	0	0	0	0	0	0

Table 4.9 Excitation list EX

t_{sim}	bd	Q	t_{min}	t_b	t_{max}	t_s	ST
1	2	3	4	5	6	7	8
0	1	1	1	1	1	1	1
	1	0	7	7	7	7	1
	1	1	13	13	13	13	1
	1	0	19	19	19	19	1
	1	1	25	25	25	25	1
	1	0	31	31	31	31	1
1	10	0	1.1	1.2	1.5	1.7	1
	1	0	7	7	7	7	1
	1	1	13	13	13	13	1
	1	0	19	19	19	19	1
	1	1	25	25	25	25	1
	1	0	31	31	31	31	1
1.1	10	0	1.1	1.2	1.5	1.7	2
	1	0	7	7	7	7	1
	1	1	13	13	13	13	1
	1	0	19	19	19	19	1
	1	1	25	25	25	25	1
	1	0	31	31	31	31	1
1.2	10	0	1.1	1.2	1.5	1.7	3
	1	0	7	7	7	7	1
	1	1	13	13	13	13	1
	1	0	19	19	19	19	1
	1	1	25	25	25	25	1
	1	0	31	31	31	31	1
	12	0.98	1.27	1.36	1.69	1.88	1
1.27	10	0	1.1	1.2	1.5	1.7	3
	1	0	7	7	7	7	1
	1	1	13	13	13	13	1
	1	0	19	19	19	19	1
	1	1	25	25	25	25	1
	1	0	31	31	31	31	1
	12	0.98	1.27	1.36	1.69	1.88	2
1.36	10	0	1.1	1.2	1.5	1.7	3
	1	0	7	7	7	7	1
	1	1	13	13	13	13	1
	1	0	19	19	19	19	1
	1	1	25	25	25	25	1
1.366	1	0	31	31	31	31	1
	12	0.98	1.27	1.36	1.69	1.88	3
	15	0.03	1.44	1.54	1.83	2.03	1
	13	0.05	1.43	1.51	1.81	2.02	1

(continued)

Table 4.9 (continued)

t_{sim}	bd	Q	t_{min}	t_b	t_{max}	t_s	ST
1	2	3	4	5	6	7	8
1.43	10	0	1.1	1.2	1.5	1.7	3
	1	0	7	7	7	7	1
	1	1	13	13	13	13	1
	1	0	19	19	19	19	1
	1	1	25	25	25	25	1
	1	0	31	31	31	31	1
	12	0.98	1.27	1.36	1.69	1.88	3
	15	0.03	1.44	1.54	1.83	2.03	1
	13	0.05	1.43	1.51	1.81	2.02	2
1.44	10	0	1.1	1.2	1.5	1.7	3
	1	0	7	7	7	7	1
	1	1	13	13	13	13	1
	1	0	19	19	19	19	1
	1	1	25	25	25	25	1
	1	0	31	31	31	31	1
	12	0.98	1.27	1.36	1.69	1.88	3
	15	0.03	1.44	1.54	1.83	2.03	2
	13	0.05	1.43	1.51	1.81	2.02	2
1.5	10	0	1.1	1.2	1.5	1.7	4
	1	0	7	7	7	7	1
	1	1	13	13	13	13	1
	1	0	19	19	19	19	1
	1	1	25	25	25	25	1
	1	0	31	31	31	31	1
	12	0.98	1.27	1.36	1.69	1.88	3
	15	0.03	1.44	1.54	1.83	2.03	2
	13	0.05	1.43	1.51	1.81	2.02	2
1.51	10	0	1.1	1.2	1.5	1.7	4
	1	0	7	7	7	7	1
	1	1	13	13	13	13	1
	1	0	19	19	19	19	1
	1	1	25	25	25	25	1
	1	0	31	31	31	31	1
	12	0.97	1.28	1.38	1.7	1.9	3
	15	0.03	1.44	1.54	1.83	2.03	2
	13	0.05	1.43	1.51	1.81	2.02	3
	16	0.97	1.82	1.92	2.24	2.35	1

(continued)

Table 4.9 (continued)

t_{sim}	bd	Q	t_{min}	t_b	t_{max}	t_s	ST
1	2	3	4	5	6	7	8
1.54	10	0	1.1	1.2	1.5	1.7	4
	1	0	7	7	7	7	1
	1	1	13	13	13	13	1
	1	0	19	19	19	19	1
	1	1	25	25	25	25	1
	1	0	31	31	31	31	1
	12	0.97	1.28	1.38	1.7	1.9	3
	15	0.03	1.44	1.54	1.83	2.03	3
	13	0.05	1.43	1.51	1.81	2.02	3
	16	0.97	1.82	1.92	2.24	2.35	1
1.7	10	0	1.1	1.2	1.5	1.7	0
	1	0	7	7	7	7	1
	1	1	13	13	13	13	1
	1	0	19	19	19	19	1
	1	1	25	25	25	25	1
	1	0	31	31	31	31	1
	12	0.97	1.28	1.38	1.7	1.9	4
	15	0.04	1.44	1.54	1.83	2.03	3
	13	0.06	1.43	1.51	1.81	2.02	3
	16	0.97	1.82	1.92	2.24	2.35	1
1.81	10	0	1.1	1.2	1.5	1.7	0
	1	0	7	7	7	7	1
	1	1	13	13	13	13	1
	1	0	19	19	19	19	1
	1	1	25	25	25	25	1
	1	0	31	31	31	31	1
	12	0.96	1.28	1.38	1.7	1.9	4
	15	0.04	1.44	1.54	1.83	2.03	3
	13	0.06	1.43	1.51	1.81	2.02	4
	16	0.95	1.82	1.92	2.24	2.35	1
1.82	6	0.94	2.13	2.25	2.52	2.62	1
	1	0	7	7	7	7	1
	1	1	13	13	13	13	1
	1	0	19	19	19	19	1
	1	1	25	25	25	25	1
	1	0	31	31	31	31	1
	12	0.96	1.28	1.38	1.7	1.9	4
	15	0.04	1.44	1.54	1.83	2.03	3
	13	0.06	1.43	1.51	1.81	2.02	4
	16	0.95	1.82	1.92	2.24	2.35	2

(continued)

Table 4.9 (continued)

t_{sim}	bd	Q	t_{min}	t_b	t_{max}	t_s	ST
1	2	3	4	5	6	7	8
1.83	6	0.94	2.13	2.25	2.52	2.62	1
	1	0	7	7	7	7	1
	1	1	13	13	13	13	1
	1	0	19	19	19	19	1
	1	1	25	25	25	25	1
	1	0	31	31	31	31	1
	12	0.96	1.28	1.38	1.7	1.9	4
	15	0.04	1.44	1.54	1.83	2.03	4
	13	0.06	1.43	1.51	1.81	2.02	4
	16	0.95	1.82	1.92	2.24	2.35	2
	3	0.97	2.16	2.24	2.48	2.58	1
1.9	6	0.94	2.13	2.25	2.52	2.62	1
	1	0	7	7	7	7	1
	1	1	13	13	13	13	1
	1	0	19	19	19	19	1
	1	1	25	25	25	25	1
	1	0	31	31	31	31	1
	12	0.96	1.28	1.38	1.7	1.9	0
	15	0.04	1.44	1.54	1.83	2.03	4
	13	0.06	1.43	1.51	1.81	2.02	4
	16	0.95	1.82	1.92	2.24	2.35	2
	3	0.97	2.16	2.24	2.48	2.58	1
1.92	6	0.94	2.13	2.25	2.52	2.62	1
	1	0	7	7	7	7	1
	1	1	13	13	13	13	1
	1	0	19	19	19	19	1
	1	1	25	25	25	25	1
	1	0	31	31	31	31	1
	12	0.96	1.28	1.38	1.7	1.9	0
	15	0.04	1.44	1.54	1.83	2.03	4
	13	0.06	1.43	1.51	1.81	2.02	4
	16	0.95	1.82	1.92	2.24	2.35	3
	3	0.97	2.16	2.24	2.48	2.58	1
2.02	6	0.94	2.13	2.25	2.52	2.62	1
	1	0	7	7	7	7	1
	1	1	13	13	13	13	1
	1	0	19	19	19	19	1
	1	1	25	25	25	25	1
	1	0	31	31	31	31	1
	12	0.96	1.28	1.38	1.7	1.9	0
	15	0.04	1.44	1.54	1.83	2.03	4
	13	0.06	1.43	1.51	1.81	2.02	0
	16	0.95	1.82	1.92	2.24	2.35	3
	3	0.97	2.16	2.24	2.48	2.58	1

(continued)

Table 4.9 (continued)

t_{sim}	bd	Q	t_{min}	t_b	t_{max}	t_s	ST
1	2	3	4	5	6	7	8
2.03	6	0.94	2.13	2.25	2.52	2.62	1
	1	0	7	7	7	7	1
	1	1	13	13	13	13	1
	1	0	19	19	19	19	1
	1	1	25	25	25	25	1
	1	0	31	31	31	31	1
	12	0.96	1.28	1.38	1.7	1.9	0
	15	0.04	1.44	1.54	1.83	2.03	0
	13	0.06	1.43	1.51	1.81	2.02	0
	16	0.95	1.82	1.92	2.24	2.35	3
	3	0.97	2.16	2.24	2.48	2.58	1
2.13	6	0.94	2.13	2.25	2.52	2.62	2
	1	0	7	7	7	7	1
	1	1	13	13	13	13	1
	1	0	19	19	19	19	1
	1	1	25	25	25	25	1
	1	0	31	31	31	31	1
	12	0.96	1.28	1.38	1.7	1.9	0
	15	0.04	1.44	1.54	1.83	2.03	0
	13	0.06	1.43	1.51	1.81	2.02	0
	16	0.95	1.82	1.92	2.24	2.35	3
	3	0.97	2.16	2.24	2.48	2.58	1
2.16	6	0.94	2.13	2.25	2.52	2.62	2
	1	0	7	7	7	7	1
	1	1	13	13	13	13	1
	1	0	19	19	19	19	1
	1	1	25	25	25	25	1
	1	0	31	31	31	31	1
	12	0.96	1.28	1.38	1.7	1.9	0
	15	0.04	1.44	1.54	1.83	2.03	0
	13	0.06	1.43	1.51	1.81	2.02	0
	16	0.95	1.82	1.92	2.24	2.35	3
	3	0.97	2.16	2.24	2.48	2.58	2
2.24	6	0.94	2.13	2.25	2.52	2.62	2
	1	0	7	7	7	7	1
	1	1	13	13	13	13	1
	1	0	19	19	19	19	1
	1	1	25	25	25	25	1
	1	0	31	31	31	31	1
	38	0.93	2.33	2.53	2.73	2.83	1
	15	0.04	1.44	1.54	1.83	2.03	0
	13	0.06	1.43	1.51	1.81	2.02	0
	16	0.95	1.82	1.92	2.24	2.35	4
	3	0.97	2.16	2.24	2.48	2.58	2

(continued)

Table 4.9 (continued)

t_{sim}	bd	Q	t_{min}	t_b	t_{max}	t_s	ST
1	2	3	4	5	6	7	8
2.25	6	0.94	2.13	2.25	2.52	2.62	3
	1	0	7	7	7	7	1
	1	1	13	13	13	13	1
	1	0	19	19	19	19	1
	1	1	25	25	25	25	1
	1	0	31	31	31	31	1
	38	0.93	2.33	2.53	2.73	2.83	1
	15	0.04	1.44	1.54	1.83	2.03	0
	13	0.06	1.43	1.51	1.81	2.02	0
	16	0.95	1.82	1.92	2.24	2.35	3
	3	0.97	2.16	2.24	2.48	2.58	2
2.35	6	0.94	2.13	2.25	2.52	2.62	3
	1	0	7	7	7	7	1
	1	1	13	13	13	13	1
	1	0	19	19	19	19	1
	1	1	25	25	25	25	1
	1	0	31	31	31	31	1
	38	0.93	2.33	2.53	2.73	2.83	2
	15	0.04	1.44	1.54	1.83	2.03	0
	13	0.06	1.43	1.51	1.81	2.02	0
	16	0.95	1.82	1.92	2.24	2.35	4
	3	0.97	2.16	2.24	2.48	2.58	2

In addition, the spatial and timing sparseness of the joint functioning of digital and analog blocks is not fully taken into account at this level.

In [233], for example, the maximum step h_c is bounded from above by the largest common multiple of all gate delays. In this case, ML is calculated once in the beginning of the next step of the circuit-level simulation. This is due to the fact that the change in Q_{in} due to the change of U_{in} at a given step does not affect the value of Q_{out} during h_c. Of course, this algorithm is quite simple, but the number of MC calculations is large, and this requires considerable computer time. In addition, in all cases when Q_{out} does not change during step h_c, an extra calculation of ML is made, which also leads to an increase in the expenditure of computer time.

In [231], the step h_c may be greater than the minimum delay of the gate (t_{dmin}), but should not exceed the interval between events in the ML. This limitation of h_c is not justified, either, since it can lead to unnecessary expenditure of computer time for circuit-level simulation in all cases when internal nodes of the digital part that are not connected with the MC, switch. Consideration of spatial and timing sparseness within digital and analog blocks of analog-digital circuits (event-driven gate-level algorithms, algorithms that take into account the latency of analog circuits), of course, reduces the costs of computer resources for gate-level and circuit-level simulation. However, an incomplete consideration of the sparse interference of these blocks can lead to unnecessary expenditure of computer time.

Consider, at what moments of time it is necessary to carry out of timing matching of MC and ML and by what methods it can be made.

In MC and ML, the simulation is performed with different timing steps and at different times: in the MC, the time step h_c is determined by the predetermined accuracy of the calculation and depends on the rate of change of the variables, and in the ML by the time interval between the nearest current events. To obtain the maximum calculation speed, it is necessary to ensure the maximum possible independence of h_c and h_l. In general case, the step h_c can turn out to be both smaller and larger than h_l.

The step h_c may be less than h_l in cases where the potentials of nodes and branch currents are rapidly changing in the analog part of the circuit. To ensure the required accuracy of calculations it is necessary to perform circuit-level simulation with a small step h_c. The necessity of ML limiting the step h_c occurs only in two cases:

1. If during the given step h_c the input voltages ML cross the threshold voltage level (Q_{in} changes) and this change leads to excitation of the input gate.

   The determination of the excitation moment of the input gate can be carried out in two ways:

   (a) Using the interpolation U_{in} (Fig. 4.6). Since the values of the voltage U_{in} at the beginning of Q_{inn} and at the end of the Q_{inn+1} nth simulation step are known, as well as the value of the threshold voltage U_{th}, then, for example, in the first-order integration method, a linear approximation of U_{in} can be carried out:

$$t_{ein} = t_n + \frac{U_{th} - U_{inn}}{U_{inn+1} - U_{inn}} \cdot h_c. \tag{4.1}$$

The undisputed advantage of this method is the speed of determination of t_{ein}, since in this case there is no need to decrease h_c. However, the use of the interpolation U_{in} for the refinement of the time t_{ein} in general case does not guarantee a reliable result. The point is that the time instant t_{ein} must be tracked with a sufficiently small error Δt. Otherwise, an error in the determination of t_{ein} can lead to a qualitatively incorrect result in the ML (if, for example, critical contests are possible

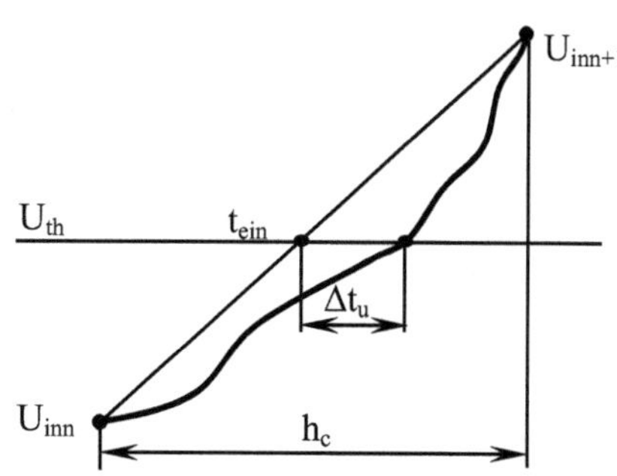

Fig. 4.6 Interpolation of input voltage

in it) or to large timing errors in the simulation of the circuit. Hence it follows that an increase in the time Δt may lead to inadequate simulation. On the other hand, a decrease in the value of Δt will lead to an increase in the expenditure of computer time for the determination of t_{ein}. Therefore, the value of Δt should be chosen from the following condition:

$$\Delta t \leq \Delta t_{bmin}, \tag{4.2}$$

where t_{bmin}—the minimum value of Δt_b for all gates, whose inputs are analog nodes.

However, the error in determining the time instant t_{ein} by interpolation U_{in} (Fig. 4.6—Δt_u) in general case may not satisfy the condition (4.2), and in these cases a violation of the adequacy of the simulation is possible. In addition, a restriction on the order of the integration method is imposed, since it is possible to determine t_{ein} in explicit form, as in (4.1), for the method of integration not higher than the second order.

(b) By the dichotomy method, i.e. the decrease of h_c is always twofold. With the use of the dichotomy method, the number of decrements produced h_c for the determination of t_{ein} with an accuracy of Δt is equal to $\log_2(h_c/\Delta t)$, which in general is greater than one. However, in this case, unlike the previous method, the determination of the moment t_{ein} by the given accuracy Δt is guaranteed. In addition, with decreasing h_c, only verification of the presence of a time t_{ein} can sometimes be done carried out, and this does not require a complete calculation of ML.

In cases where the first or second order integration method is used, or the specified value of Δt is greater than the interpolation error, it is more convenient to use the first method, and in the remaining cases—the second method.

2. If during this step h_c U_{out} can change more than once. For the adequacy of the simulation, it is necessary to monitor the switching at the outputs ML (U_{out}) and not allow multiple changes of U_{out} during h_c, since these switches may remain invisible to the MC (Fig. 4.7). In this case, h_c must decrease so that the following condition is met:

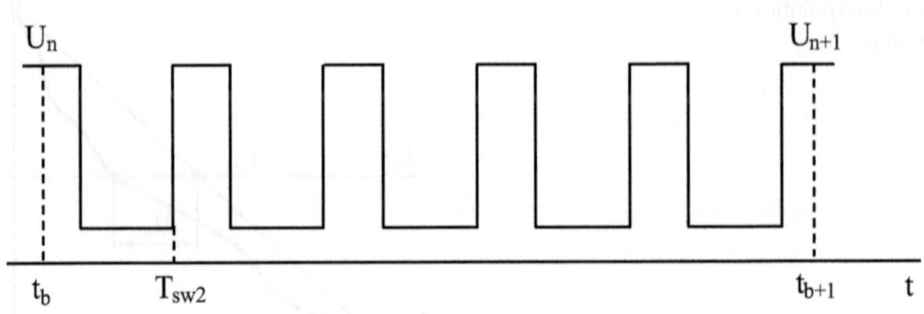

Fig. 4.7 Multiple signal changes during the simulation step

$$t_b + h_c < t_{sw2}, \tag{4.3}$$

where t_b is the time instant corresponding to the beginning of the current step; t_{sw2}—time of second switching of U_{out}.

In the mentioned situations, the decrease in h_c is necessary, since otherwise qualitatively inaccurate simulation results can be obtained. In all other cases, there is no need to reduce the step h_c.

In the simplest case, ML can be considered as the usual model of one component of the MC. However, when constructing the mixed-mode simulation algorithm, one must take into account that in any nth step, in general, a multiple calculation of the MC is required. First, if the MC is nonlinear, then a repeated inversion to it occurs in the iteration process using the Newton method. Secondly, the accuracy simulation may require decreasing h_c and calculating the MC at the same value of t_n, but with a reduced value of t_{n+1}. However, by virtue of the above considerations, the ML calculation can be performed less frequently than the MC calculation. In the process of iterations by the Newton method, h_c, t_n and t_{n+1} are constant. Consequently, the values of Q_{out} and Q^h_{out} are also constant, and ML calculation is not required. References to ML must be made only at the beginning of each step and with decreasing h_c at a given step. If the moments t_{ein} are determined by the dichotomy method, then ML will be required at the end of each step. In the latter case, the ML approach does not aim at calculating Q_{out}, but only checking whether the instant t_{ein} has arisen at this step, has it disappeared due to the reduction of the step and whether it is not necessary to reduce h_c in connection with this.

The developed algorithms of mixed-mode gate-level circuit-level simulation [239], which fully considers the sparse interaction of MC and ML are outlined, and their effectiveness is evaluated. The block diagram of the first algorithm is shown in Fig. 4.8. The determination of the time t_{ein} is performed by the dichotomy method. Assume that the value h_c is reduced to such an extent that $h_c < \Delta t$. Then the time t_{ein} can be attributed to any point of the interval $t_n \ldots t_{n+1}$, not exceeding the given error Δt. The simplest algorithm is obtained by assigning this unknown moment t_{ein} to the end of the interval, i.e. assume that $t_{ein} = t_{n+1}$. In this case, the result of the calculation of the ML state and, in particular, of the Q_{out} at the instant t_{n+1} at any step is independent of $U_{in}(t_{n+1})$, but is determined only by the value of h_c, the state ML, U_{in} at the beginning of the step (at time t_n) and internal processes in ML. If the moment of excitation t_{ein} does not arise at this step, then the independence of $Q_{out}(t_{n+1})$ from $U_{in}(t_{n+1})$ is obvious for any value of h_c, but if the time t_{ein} appears at a given step, then the above procedure for finding t_{ein} with accuracy Δt by the dichotomy method ensures that equating t_{in} to t_{n+1}, which ensures the independence of $Q_{out}(t_{n+1})$ from $U_{in}(t_{n+1})$, leads to a time error less than a given value of Δt.

Now the circuit of those operations are outlined that must be carried out each time the ML is accessed. Denote by C (EXC, Q) the state of ML including the list of excitations EXC, and Q is the state vector of all nodes of ML, with the exception of Q_{in}. Denote by C^t the state at the current time t, C^n is the state characteristic for the beginning of this step, i.e. at $t = t_n$. Introduce three vectors Q_{in}: Q^{n+1}_{in}, where current

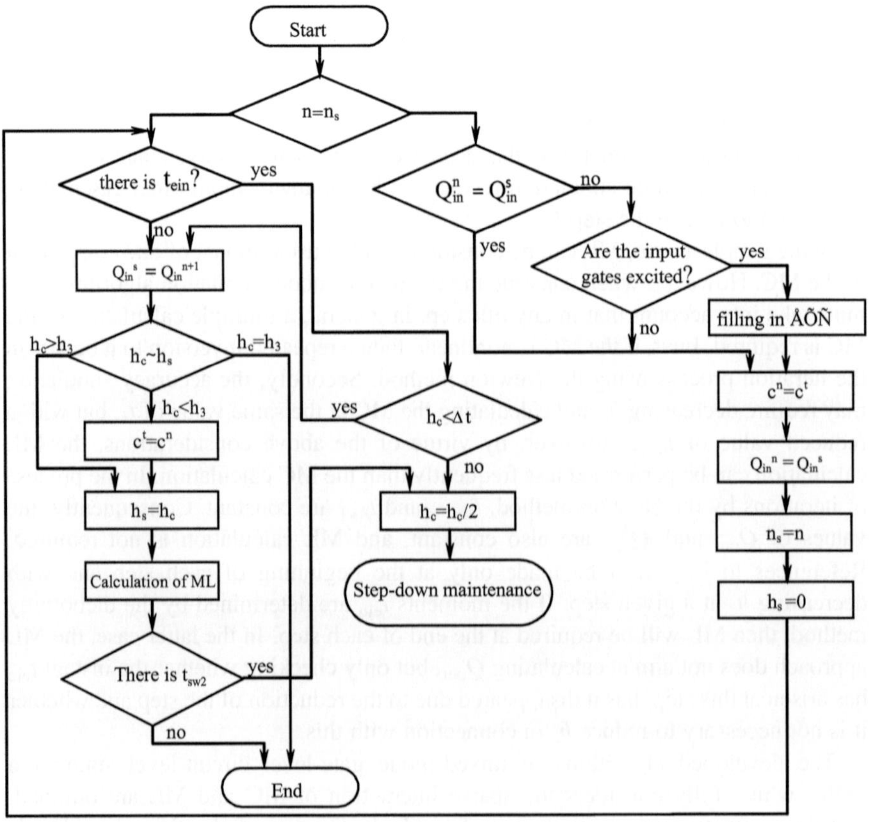

Fig. 4.8 The first algorithm of mixed-mode gate-level circuit-level simulation

values of $Q_{\text{in}}(t_{n+1})$ are entered at each reference to ML, recalculated from $U_{\text{in}}(t_{n+1})$; Q^n_{in}—the value of the vector $Q_{\text{in}}(t_n)$, corresponding to the beginning of this step; Q^s_{in}—the value of the vector stored from the previous step Q^{n+1}_{in}, which should be taken as Q^n_{in} for this step. Also these notations are introduced: $n_{s'}$ and $h_{s'}$—the number and magnitude of the step for the previous reference to ML. The values $n_{s'}$, h_c, Q^n_{in}, $Q^{s'}_{\text{in}}$, $C^{s'}$, C^n, $Q^{s'}_{\text{out}}$ are stored in memory as they were the result of the previous reference. At the moment of reference, the values n, h_c, t_{n+1}, Q^{n+1}_{in} are also known. Immediately after referencing to ML, it is checked whether this is a new step or repeated reference in the previous step. If the step is new ($n \neq n_ç$), then it is determined whether the vector Q_{in} has changed during the previous step. This can be done by comparing the vector Q^n_{in}, which in this case is equal to $Q_{\text{in}}(t_{n+1})$, with the vector $Q_{\text{in}} = Q_{\text{in}}(t_n)$. If the vector Q_{in} has changed and this causes excitation of the input gate, the list of excitations B (correspondingly, that the excitation moment $t_{\text{ein}} = t_n$) is updated accordingly. Then, as a new vector Q^n_{in} is taken as Q^s_{in}, the current state C^t is stored as a new step number, n is stored and the information about the value of the previous step h_3 is erased.

Further, it is checked whether during this step, there is an excitation moment t_{ein} of input gate due to the change of Q_{in}. To do this, compare the vectors Q_{in}^{n} and Q_{in}^{n+1}.If they are not equal and, consequently, the input signals have changed in the interval $t_n \ldots t_{n+1}$, it is estimated whether this change leads to excitation of the input gate. If this change leads to excitation (i.e., there is a moment t_{ein} at the interval $t_n \ldots t_{n+1}$), the value h_c is checked, and if it exceeds Δt, a step reduction is performed.

In case there is no need to reduce h_c, Q_{in}^{n+1} is remembered. Further operations depend on the ratio of h_c and h_s. If $h_c = h_s$ (this corresponds to a reference at the end of the step), then no calculations are made, since the state C^t and the vector Q_{out}^{n} have already been calculated for previous references. If $h_c < h_s$ (this corresponds to a repeated reference at a given step with a reduced value of h_c), then before the calculation begins, the initial state ML ($C^t = C^n$) is restored. When $h_c > h_s$ (this situation occurs during the first reference at this step), the recovery of C^t is not required, since it already corresponds to the beginning of the step. Then the value of h_c is memorized and the actual ML calculation is carried out. The calculation (that is, the definition of C^t and Q_{out}^{n}) is carried out at $Q_{in} = Q_{out}^{n}$and $t \le t_{n+1}$. Maintenance to reduce the step is also made if it is found out that Q_{out} should switch more than once. If the determination of the presence of a moment t_{ein} can be carried out without simulating ML, in this case it cannot be done.

The algorithm, in contrast to [231, 233], allows selecting the steps h_c independently of the ML simulation steps (with the exception of the refining time t_{ein} and the multiple switching of Q_{out}). Thus, it is possible to calculate the MC in large steps exceeding the intervals between events in the ML. In addition, the ML calculation is less frequent than the MC calculation. These advantages are achieved due to additional memory costs in relation to the mentioned algorithms: it is necessary to double the storage space for the ML state and triple for the storage of Q_{in}.

As it was mentioned in the introduction, the main limiting factor in simulation of analog-digital circuits is computer time, so the developed algorithm is more effective than in [231, 233]. The main advantage of the above algorithm is its reliability, since the dichotomy method guarantees the determination of t_{ein} with a predetermined error Δt. The disadvantages of the algorithm include the following:

(a) The number of step-downs to determine t_{ein} is large.
(b) To ensure a given accuracy in the calculation of voltage or current for the output nodes of ML in which switching occurred during h_c, the simulation steps in the MC are reduced until the errors in the voltages (or currents) in these nodes during h_c become less than the set value.

To reduce the number of steps for determining t_{ein} in cases where this is allowed, the interpolation U_{in} can be used. In order to eliminate the last drawback, the following can be proposed: if at least on one output ML, connected to the MC, the signal should switch, then the ML gives a signal for decreasing h_c. But since the moment of this switching t_n and setting the signal t_u to the corresponding output are known in the ML, the reduced value of the current step ($h_{cn} = t_s - t_b$) and the next step ($h_{cn+1} = \alpha \cdot (t_s - t_b)$) are transferred to the MC. Here α is a constant value less

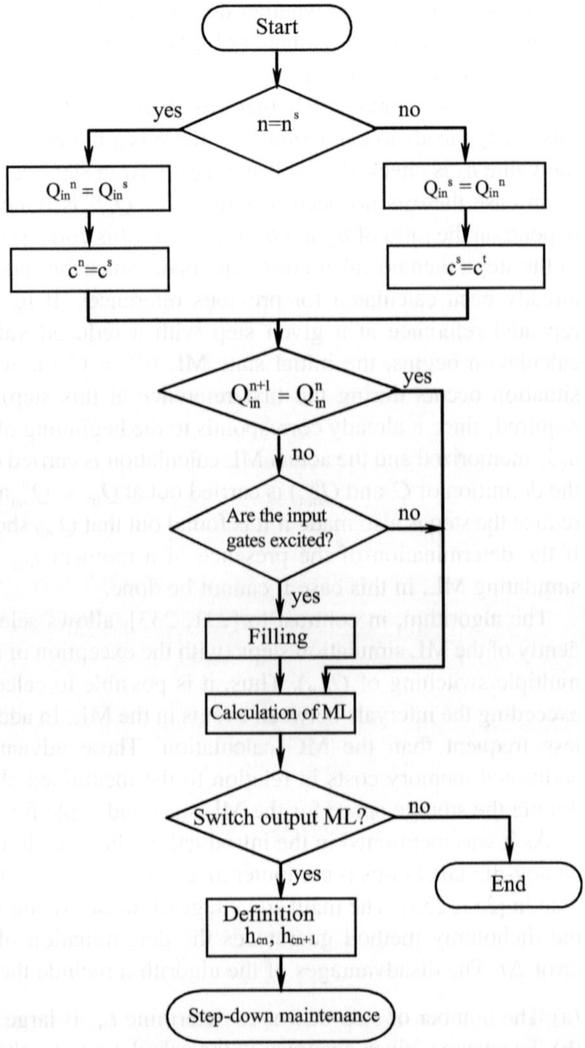

Fig. 4.9 The second algorithm of mixed-mode gate-level circuit-level simulation

than 1. The value should be chosen from the interval from 0.1 to 0.5. Thus, unnecessary reductions of the steps h_c are eliminated to track the change in the signal at the ML outputs. In addition, the condition (4.3) is automatically satisfied.

Figure 4.9 illustrates a block diagram of the second mixed-mode simulation algorithm, which eliminates the drawbacks of the previous one (Fig. 4.8). The notation is the same as in the description of the previous algorithm.

If the step is new ($n \neq n_ç$), then the states of all ML nodes are remembered. Further, it is checked whether there are changes in Q_{in} during the h_c that excite the input gate. If such changes exist in Q_{in}, then the corresponding excitations are recorded. The excitation moment is determined from the interpolation Q_{in}. Further, ML is calculated. If during this calculation it turns out that for at least one ML output connected to the MC, Q_{out} should change, then the signal to reduce the h_c and the magnitude of the current and future steps is transmitted to the MC. If several ML outputs are switched during h_c, then h_{cn+1} is determined by the minimum switching time.

If the step is old, which can happen, for example, when referencing at the same step with a reduced h_c, then the ML state is restored at the initial instant of the step, and then the algorithm is similar.

The last algorithm (Fig. 4.9) has the highest possible performance, since the ML restricts the value of h_c only when switching Q_{out}, and as much as necessary. However, because of the interpolation of U_{in}, the reliability of the algorithm is low. Memory costs are the same as for the previous algorithm. Practical implementation of algorithms has confirmed the above. When simulating various analog-to-digital circuits, the second algorithm requires an average of 20% … 50% less computer time compared to the first one.

Chapter 5
Optimization of Digital Circuits with Consideration of DF

5.1 Optimization of Critical Timing Paths of Digital Circuits with Consideration of DF

It is known [32, 163, 290, 310] that verification of the timing characteristics of a digital circuit is applied at almost all stages of digital IC design (behavioral description, logical synthesis, physical design, post-layout verification, etc.). One of the most important tasks of verification of digital IC design is the optimization of critical timing paths (CTP) of digital circuits. Critical is the path between the *i*th input and the *j*th output of a digital circuit, which has a delay of $t_{\mathrm{d.CTP}}$, greater than a predetermined permissible value T_{per} (Fig. 5.1) [311].

At present, there is a great variety of [128, 129, 134, 135, 210, 253, 255, 257, 312–316] algorithms for optimization of the CTP (AOCTP) of digital circuits. The most successful algorithms are based on the principles of changing the dimensions of gate, introduction of logical buffers (LB), separation of interconnects and their sorting into groups, etc.

The AOCTP of a digital circuit described in [317] uses the idea of excluding existing connections and adding additional connections (Fig. 5.2). It is supposed to exclude separate connections ($G_{\mathrm{p}} - G_{\mathrm{q}}$) from the CTP and replace them with alternative connections ($G_{\mathrm{r}} - G_{\mathrm{s}}$) or gate not on the CTP. The $G_{\mathrm{p}} - G_{\mathrm{q}}$ interconnect is called a possible line if it is on the CTP of a digital circuit. The logic D is considered to be alternative for a possible line δ, if the values of all digital circuit outputs are not changed when the line δ is deleted from the CTP and logic D is added. A possible pair is (D, δ), if δ is a possible line, and D is an alternative logic for the line δ. Each possible line δ_i can have more than one alternative logic (D_{ij}, δ_i), $_{i=1,2,...,y,\ j=1,2,...,x}$. Therefore, the goal is to find the best pairs (D_{ij}, δ_i), $_{i=1,2,...,y,\ j=1,2,...,x}$ (y is the number of possible CTP lines, x-the number of possible alternative logics for *i*th possible line) at which the total delay of the signal on the CTP is minimal. In Fig. 5.2 the line $\delta_{\mathrm{t}} = G_{\mathrm{p}} - G_{\mathrm{q}}$ is possible, and $\delta_{\mathrm{w}} = G_{\mathrm{r}} - G_{\mathrm{s}}$—alternative. Since, when the possible line δ_{t} is excluded, the G_{p}

V. Melikyan, *Simulation and Optimization of Digital Circuits*,
https://doi.org/10.1007/978-3-319-71637-4_5

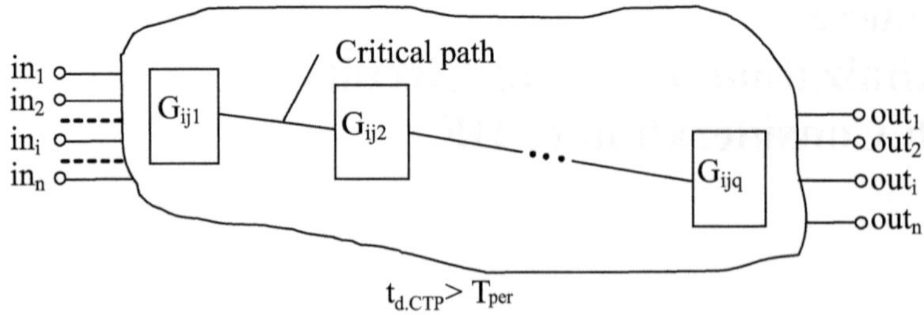

Fig. 5.1 CTP of a digital circuit

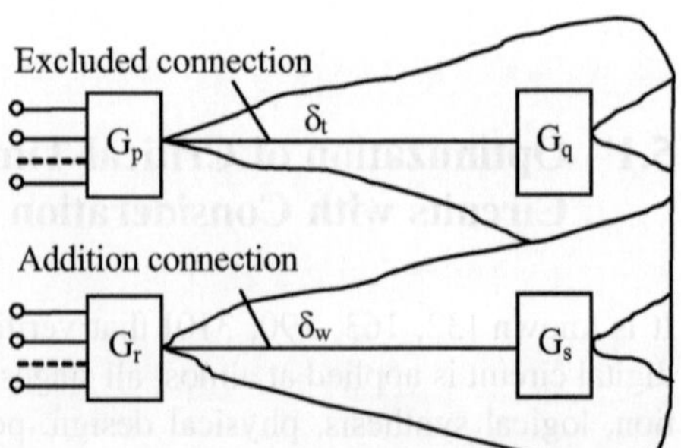

Fig. 5.2 Directions for signal propagation in digital circuits when adding and removing connections

load decreases, hence, the delays of all paths including the G_p are reduced. On the other hand, the addition of an alternative line $G_r - G_s$ leads to the formation of some new connections in digital circuits. The load G_r increases, which can lead to an increase in delays on paths including G_s. Changes in digital circuits can affect other CTP of the same digital circuits, which must be taken into account when choosing a possible pair (D_{ij}, δ_i). For each possible pair, according to (5.1), the value of the function $E(i, j)$ ($F(i, j)$ is calculated, the decrease in the delay of digital circuits, $G(i, j)$ is the decrease in the delay on the longest CTP, β—weighting factor). Those transformations of digital circuits are selected for which the function $E(i, j)$ has a maximum value. The application of the AOCTP of digital circuits described in a number of test digital circuits, resulted in a decrease in the CTP delays by an average of 7.6%.

$$E(i,j) = F(i,j) + b \times G(i,j) \tag{5.1}$$

In [318], the developed AOCTP of digital circuits combines the addition and elimination of connections with the introduction of LB and the rearrangement of the gate. In this case, the maximum value $E(i, j)$ (5.1) is again used as the objective function. The strategy, shown in Fig. 5.3 is used to exclude a number of alternative transformations in this AOCTP of a digital circuit.

If for a possible line δ_u there are two alternative connections δ_1 and δ_2 and both are inputs of the same G_q, and if after the addition of an alternative connection, the changes in the G_{p1} and G_{p2} delays are insignificant compared to the inverter and the

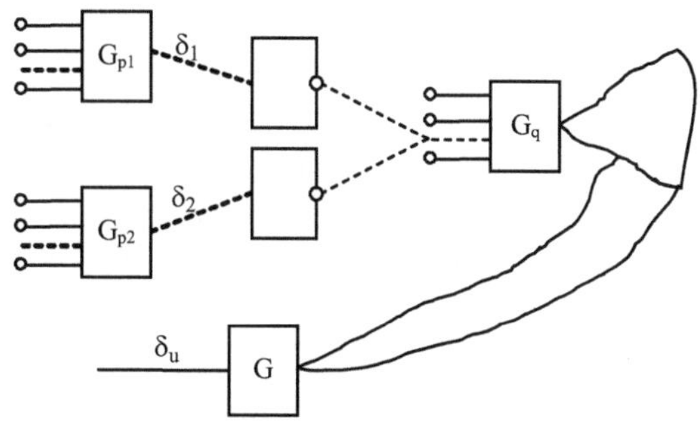

Fig. 5.3 The process of selecting changing interconnects

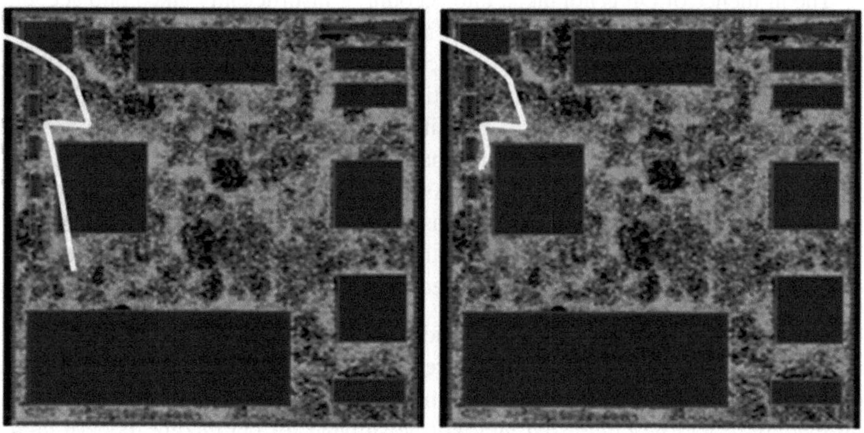

Fig. 5.4 Obtained CTP using AOCTP. (**a**) Without preliminary decomposition of a digital circuit; (**b**) with preliminary decomposition of a digital circuit

introduced alternative delay line, then those gates for which, as a result of adding an alternative connection, the delay of the input signal changes, they are identical for δ_1 and δ_2. Therefore, it is possible to determine in advance which of the transformations is more conducive to improving the time characteristics of the CTP.

The application of this AOCTP in digital circuits on a number of test digital circuits has reduced the delay of the CTP by an average of 17.9%.

In [319], prior to the beginning of the operation of the AOCTP, the digital circuit is decomposed. Gates connected to the same clock signal are combined in one group. If some gate groups contain a large number of elements, they are divided into smaller groups. The exclusion and the addition of connections are made first of all within the obtained groups. Experiments have shown that the application of preliminary decomposition of a digital circuit greatly improves the results of optimization of the CTP of digital circuits (Fig. 5.4).

The experimental results show that the AOCTP of digital circuits described in [319] can provide a reduction in the delay of the CTP to 22%.

However, simulation of the existing AOCTP of digital circuits shows that they have the following important limitations:

1. Only timing characteristics and area of digital circuits are optimized. In the meantime, the power consumed by digital circuits is not minimized. Moreover, the improvement of timing parameters of digital circuits is realized precisely due to the consumed power of digital IC. This is natural, but solutions to the problem of optimizing the CTP of digital circuits with lower power consumption of the same digital circuit are possible. However, these solutions are not considered. Such algorithms are especially unsuitable for the design of currently widespread low-power digital circuits [67–70, 147].
2. The influence of DF is not taken into account or insufficiently taken into account (Sect. 1.1). In the best case, only the influence of parasitic parameters of interconnects is taken into account [129, 253, 256, 317, 322, 323], and rather rough models are used (at best-RC). In particular, RLC interconnect models are not used, interconnect interference, parasitic parameters of power circuits are not taken into account, only one timing parameter is taken into account (averaged delay of interconnect and gate), etc. As a result, with the CMOS IC 0.13 μm or below manufacturing technologies as well as operating frequencies above 0.5 GHz, the existing AOCTP of digital circuits do not provide results acceptable from the point of view of practical needs.

Therefore, there was an extreme need to develop AOCTP of digital circuits, which will take into account the influence of real parameters of physical design of digital ICs, simultaneously with the optimization of the CTP, the power consumed by the circuit will be minimized and will be suitable for optimizing modern high-speed and low-power digital circuits, regardless of their circuit implementation, etc.

Below the developed AOCTP of a digital circuit with consideration of DF [132, 320, 321] is described.

With a view to a clearer presentation of the essence of the developed AOCTP of a digital circuit, the following definitions are introduced.

Gate e_i is considered to belong to a CTP if the output (o_{e_i}) and one of the inputs $\left(i_{j_{e_i}}\right)$ of the gate e_i are on the CTP, i.e. $e_i \in$ CTP, if the $o_{e_i} \in$CTP and $i_{j_{e_i}} \in$CTP$,_{j=1\vee 2\vee \ldots \vee m}$, where m is the number of inputs of gate e_i. For example, gate e_4, e_6, e_8, and e_9 in a digital circuit in Fig. 5.5 are owned by the CTP.

The gate e_j is considered to be adjacent to the belonging CTP of gate e_i, if e_j is not a CTP belonging, but its output (o_{e_j}) is the input of the gate e_i, i.e. $e_j \notin$ CTP and $e_i \in$ CTP and $o_{e_j} = i_{k_{e_i}},_{k=1\vee 2\vee \ldots \vee m}$, where m is the number of inputs of gate e_i. For example, the gate e_3 is adjacent to e_6, and e_5 to e_8 (Fig. 5.5).

A logical group (LG) is a set of gates connected to each other, containing at least one gate belonging to the CTP and adjacent gate, i.e. $=\{e_1, e_2, \ldots, e_x\}$, $e_1 \vee e_2 \vee \ldots \vee e_x \in$ CTP, x is the number of gate in LG. In a digital circuit in Fig. 5.5 LGs are: LG_1—$\{e_3, e_4, e_6\}$, LG_2—$\{e_5, e_8\}$, and LG_3—$\{e_9\}$.

Gate e_i is an input of order M for gate e_j if there exists a shortest path between e_i and e_j containing at most M branches.

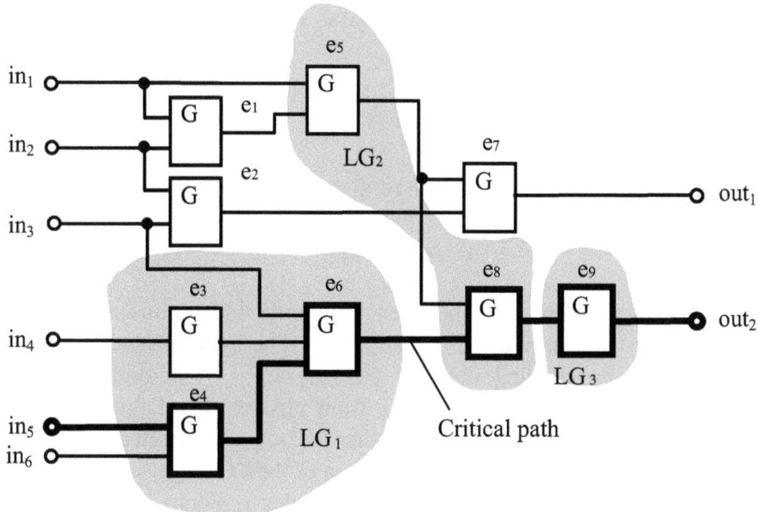

Fig. 5.5 Formation of LG

Fig. 5.6 Delay of LG

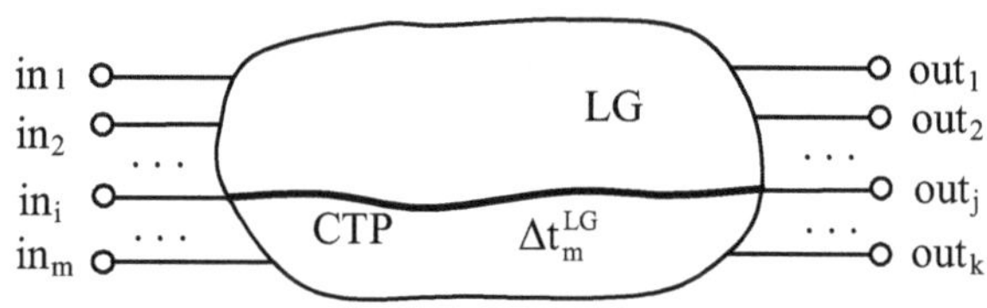

The algorithm of forming LG is quite simple [317]. At first, the $LG_{j,\ j=1,2,...,y}$ are formed, each of which includes one gate of each e_i belonging to the CTP. In fact, every gate e_i belonging to a *CTP* can enter only one of the LG_j. Therefore, the number of LG cannot exceed the number of gates belonging to the *CTP*. Then, the expansion of LG_j, including input LGs of the first order occurs under the condition that these gates are not included in other LGs. In case the same gate e_i is a candidate for including two different LGs, the gate e_i is included in the LG with a longer delay. At the end of the operation of the algorithm, if there is LG_j with one output and LG_1 connected to this single output LG_j, then LG_j and LG_1 are combined.

The described algorithm for allocating LG in a digital circuit allows concluding the following: LG_m has only one input (in_i) and one output (out_j), located on the CTP (Fig. 5.6). The delay between $\mathrm{in_i}$ and $\mathrm{out_j}$ is the delay of this LG: Δt_m^{LG}

Δt_m^{LG} includes the timing parameters of the MCE on the path between the in_i and out_j, as well as the delays in interconnects between them. For example, Δt_1^{LG} for the digital circuit in Fig. 5.5 includes the timing parameters of the MCE for gate e_4 and e_6, as well as interconnects between e_4 and e_6. These timing parameters, in turn, are DF functions and are determined by means of the MDE from Sect. 3.1.

The goal of the AOCTP of a digital circuit is the simultaneous implementation of the following conditions:

1. Providing a delay of a CTP of digital circuits ($t_{\mathrm{d.CTP}}$), smaller than a predetermined value T_{per}:

$$T_{\mathrm{d.CTP}} < T_{\mathrm{per}}. \tag{5.2}$$

2. Minimizing the power consumption of digital circuits P_{con}:

$$P_{\mathrm{con}} \to \min. \tag{5.3}$$

This provides a compromise between the conflicting requirements for digital circuits on performance and power consumption. The implementation of condition (5.2) is achieved by singling out the sets $\mathrm{LG}_{j,\ j=1,2,\ldots,y}$ in digital circuits and optimizing the timing delay of each LG_j from this set. The latter is achieved by replacing $\mathrm{LG}_{jk,\ k=1,2,\ldots}$ inside the LG_j by other $\mathrm{LG'}_{jk,\ k=1,2,\ldots}$ from digital cell library with a higher load capacitance, the arrangement of the $\mathrm{LG'}_{jk,\ k=1,2,\ldots}$ within one LG_j more closely related to their physical design in order to reduce interconnect delays. As a result of such changes, the delay of the $\mathrm{LG}_{j,\ j=1,2,\ldots,y}$ decreases, thereby—$t_{\mathrm{d.CTP}}$. However, if the changes are carried out within each $\mathrm{LG}_{j,\ j=1,2,\ldots,y}$ in the presence of several possible solutions, the algorithm preference is given to those solutions that satisfy the condition (5.3).

It is known [67] that the delay LG_i can be determined as follows:

$$\Delta t_i^{\mathrm{LG}} = \frac{C_{\mathrm{L}i}^{\mathrm{LG}} \cdot U_{\mathrm{L}}^2}{P_{\mathrm{con}\cdot i}^{\mathrm{LG}}}, \tag{5.4}$$

where $C_{\mathrm{L}i}^{\mathrm{LG}}$ is the capacitive load of LG_i; U is the supply voltage; $P_{\mathrm{con}\cdot i}^{\mathrm{LG}}$—power consumption of LG_i. In its turn,

$$C_{\mathrm{L}i}^{\mathrm{LG}} = C_{i+1}^{\mathrm{LG}} + \sum_{i=1}^{k}\left(\frac{\varepsilon w_{Mi} l_{Mi}}{t_{\mathrm{ox}}} + C_i^{G_i,\mathrm{LG}_j}\right), \tag{5.5}$$

where C_{i+1}^{LG} is the input capacitance LG_{i+1}, connected to the output LG_i; $\frac{\varepsilon w_{\rangle li} l_{\rangle li}}{t_{\mathrm{ox}}}$ capacitance of the jth interconnect, connected to the output LG_i with length $l_{\rangle i}$ and width $w_{\rangle i}$; $C_i^{G_j,\mathrm{LG}_i}$—input capacitance G_j, connected to the output LG_i, but not belonging to LG_{i+1}.

The power consumed by the $G_i(P_{\mathrm{con}\cdot i}^{\mathrm{LG}})$ is defined as the sum of the consumed powers of all the LG_i belonging to the G_i.

$$P_{\mathrm{con}\cdot i}^{\mathrm{LG}} = \sum_{j=1}^{n} p_{ij}^{G_j,\mathrm{LG}_i}. \tag{5.6}$$

The LG_i delay includes the timing parameters of the G_1 that make up the LG_i $\left(\Delta t_l^{G_j,\mathrm{LG}_i}\right)$) and belonging to CTP, as well as interconnects $\left(\Delta t_{Ms,s+1}^{\mathrm{LG}_i}\right)$ between them:

$$\Delta t_i^{\mathrm{LG}} = \sum_{l=1}^{z} \Delta t_l^{G_l,\mathrm{LG}_i} + \sum_{s=1}^{m} \Delta t_{Ms,s+1}^{\mathrm{LG}_i}, \tag{5.7}$$

where z is the number of gates on the CTP in LG_i, and m is the number of interconnects.

Taking into account the parameters of the developed MCE (Sect. 2.2) as well as the relations (2.72), the following is obtained:

$$\Delta t_i^{\mathrm{LG}} = \sum_{l=1}^{z} \Delta t_{bl}^{G_l,\mathrm{LG}_i} + \Delta t_s^{G_l,\mathrm{LG}_i} + \sum_{s=1}^{m} \Delta t_{Ms,s+1}^{\mathrm{LG}_i}. \tag{5.8}$$

Here the quantities $\Delta t_{bl}^{G_l,\mathrm{LG}_i}$, $\Delta t_s^{G_l,\mathrm{LG}_i}$ and $\Delta t_{Ms,s+1}^{\mathrm{LG}_i}$ are functions of the DF and are determined with the use of the MDE (Sect. 3.1) and the PDE (Sect. 1.4), i.e.

$$\begin{aligned}\Delta t_i^{\mathrm{LG}} = &\sum_{l=1}^{z} \left\{\Delta t_{bl}^{G_l,\mathrm{LG}_i} = \mathrm{PDE}_l\left[\mathrm{MDE}_{l\gamma}(w_\gamma), \mathrm{MDE}_{l\eta}(v_\eta)\right]\right\} + \\ &+ \left\{\Delta t_s^{G_l,\mathrm{LG}_i} = \mathrm{PDE}_o\left[\mathrm{MDE}_{o\gamma}(w_\gamma), \mathrm{MDE}_{o\eta}(v_\eta)\right]\right\} \\ &+ \sum_{s=1}^{m}\left[\Delta t_{Ms,s+1}^{\mathrm{LG}_i} = \mathrm{MDE}_q(l_{Ms,s+1})\right],\end{aligned} \tag{5.9}$$

where γ is the number of external, and η is the number of internal DFs.

Substituting (5.5)–(5.9) into (5.4) and taking into account that $U_n = \mathrm{const}$, the following objective function is obtained:

$$\sum_{i=1}^{n}\left(\frac{C_{i+1}^{\mathrm{LG}} + \sum_{i=1}^{k}\left(\frac{\varepsilon w_{Mi} l_{Mi}}{t_{ox}} + C_i^{G_i,\mathrm{LG}_j}\right)}{\begin{array}{l}\sum_{l=1}^{z}\left\{\Delta t_{bl}^{G_l,\mathrm{LG}_i} = \mathrm{PDE}_l\left[\mathrm{MDE}_{l\gamma}(w_\gamma), \mathrm{MDE}_{l\eta}(v_\eta)\right]\right\} + \\ +\left\{\Delta t_s^{G_l,\mathrm{LG}_i} = \mathrm{PDE}_o\left[\mathrm{MDE}_{o\gamma}(w_\gamma), \mathrm{MDE}_{o\eta}(v_\eta)\right]\right\} \\ +\sum_{s=1}^{m}\left[\Delta t_{Ms,s+1}^{\mathrm{LG}_i} = \mathrm{MDE}_q(l_{Ms,s+1})\right]\end{array}}\right) \rightarrow \min \tag{5.10}$$

Although the aim of the AOCTP is to optimize the delay of the CTP of a digital circuit, the objective function is (5.10), and the CTP delay is specified as the constraint (5.2), which takes the following form after substituting the values from (5.9):

$$\begin{aligned}\Delta t_i^{\mathrm{LG}} = & \sum_{l=1}^{z}\left\{\Delta t_{bl}^{G_l,\mathrm{LG}_i} = \mathrm{PDE}_l\left[\mathrm{MDE}_{l\gamma}(w_\gamma), \mathrm{MDE}_{l\eta}(v_\eta)\right]\right\}+ \\ & + \left\{\Delta t_s^{G_l,\mathrm{LG}_i} = \mathrm{PDE}_o\left[\mathrm{MDE}_{o\gamma}(w_\gamma), \mathrm{MDE}_{o\eta}(v_\eta)\right]\right\} \\ & + \sum_{s=1}^{m}\left[\Delta t_{Ms,s+1}^{G_i} = \mathrm{MDE}_q(l_{Ms,s+1})\right] \le T_{per}.\end{aligned} \tag{5.11}$$

Prior to minimizing the objective function, different solutions are formed for the same $\mathrm{LG}_{j,\ j=1,2,\ldots,y}$. Each of the solutions SOL_{jm} for LG_j is characterized by the power consumed by $\mathrm{LG}_j P_m^{\mathrm{LG}_j}$, the capacitance $\mathrm{LG}_j C_m^{\mathrm{LG}_j}$ and the delay $\Delta t_m^{\mathrm{LG}_j}$, i.e. $\mathrm{SOL}_{jm} = \left\{P_m^{\mathrm{LG}_j}, C_m^{\mathrm{LG}_j}, \Delta t_m^{\mathrm{LG}_j}\right\}$.

The decision SOL_{jm1} is considered to be inferior to the solution of SOL_{jm2} if the following conditions are simultaneously satisfied: $P_{m1}^{\mathrm{LG}_j} > P_{m2}^{\mathrm{LG}_j}$, $C_{m1}^{\mathrm{LG}_j} > C_{m2}^{\mathrm{LG}_j}$ and $\Delta t_{m1}^{\mathrm{LG}_j} > \Delta t_{m2}^{\mathrm{LG}_j}$. Out of the entire set of solutions LG_j, inferior solutions are eliminated and the following restrictions are determined:

$$\begin{cases}\varphi_1 \le C_1^{\mathrm{LG}} \le \psi_1, \\ \cdots \\ \varphi_i \le C_i^{\mathrm{LG}} \le \psi_i, \\ \cdots \\ \varphi_n \le C_n^{\mathrm{LG}} \le \psi_n,\end{cases} \quad \begin{cases}\alpha_1 \le \Delta t_1^{\mathrm{LG}} \le \beta_1, \\ \cdots \\ \alpha_i \le \Delta t_i^{\mathrm{LG}} \le \beta_i, \\ \cdots \\ \alpha_n \le \Delta t_n^{\mathrm{LG}} \le \beta_n\end{cases} \tag{5.12}$$

to the possible values of the input capacitances and timing delays of the LG—followers for $\mathrm{LG}_{j,\ j=1,2,\ldots,y}$.

In fact, when solving the problem of optimizing a CTP of a digital circuit, the objective function $f(x)$ is minimized for m constraints of the type $g_i(x) \le \varepsilon_{i,\ i=1,2,\ldots,m}$. The most convenient [297] for solving the optimization problem of the mentioned type is the Lagrange minimization method with verification of Kuhn Tucker conditions.

Constraints in the form of parameters can be transformed into equalities if each of them is supplemented with a nonnegative value σ_i^2:

$$g_i(x) + \sigma_i^2 - \varepsilon_i = 0. \tag{5.13}$$

To determine the minimum conditions, the Lagrange function [297] is applied:

$$F(x, y, \lambda) = f(x, y) + \lambda \cdot g(x, y). \tag{5.14}$$

For this case, there is:

$$F(x, y, \lambda) = f(x) + \sum_{i=1}^{m} \lambda_i \left[g_i(x) + \sigma_i^2 - \varepsilon_i \right]. \tag{5.15}$$

The minimum conditions are as follows:

$$\frac{\partial F}{\partial x_j} = 0 = \frac{\partial f}{\partial x_j} + \sum_{i=1}^{m} \lambda_i \frac{\partial g_i}{\partial x_j}, \quad j=1,2,\ldots,n, \tag{5.16}$$

$$\frac{\partial F}{\partial \lambda_j} = 0 = g_i(x) + \sigma_i^2 - \varepsilon_i, \quad i=1,2,\ldots,m, \tag{5.17}$$

$$\frac{\partial F}{\partial \sigma_i} = 0 = 2\lambda_i \sigma_i, \quad i=1,2,\ldots,m. \tag{5.18}$$

Multiplying (5.18) by $\frac{\sigma_i}{2}$, this is obtained $\lambda_i \sigma_i^2 = 0$, i.e.

$$\lambda_i [\varepsilon_i - g_i(x)] = 0, \quad i=1,2,\ldots,m. \tag{5.19}$$

The expressions (5.16), (5.17) and (5.19) are necessary conditions of the minimum at the point x^* under the restrictions (5.16) and (5.17) and repeat the restriction $g_i(x) \leq 0$. It follows from (5.19) that either $\lambda_i = 0$ or $\varepsilon_i - g_i(x^*) = 0$. If $\lambda_i \neq 0$, then $g_i(x^*) = \varepsilon_i$. In this case, the restriction is considered active and constitutes the restriction in the form of equality. On the other hand, if the restriction is a strict inequality $g_i(x^*) < \varepsilon_i$, then $\lambda_i = 0$. There is also an additional condition $\lambda_i \geq 0$, which is satisfied at the point of minimum.

Thus, the restrictions in the form of equalities are as follows:

$$\sum_{l=1}^{h} \left(\Delta t_i^{\mathrm{LG}} + \Delta t_{Ms,s+1}^{\mathrm{LG}} \right) + \sigma_i^2 - T_{\mathrm{per}} = 0, \tag{5.20}$$

$$\left\{ \begin{array}{l} \varphi_2 - C_2^{\mathrm{LG}} + \sigma_{2a}^2 = 0, \\ \quad \cdots \\ \varphi_i - C_i^{\mathrm{LG}} + \sigma_{ia}^2 = 0, \\ \quad \cdots \\ \varphi_n - C_n^{\mathrm{LG}} + \sigma_{na}^2 = 0, \\ C_2^{\mathrm{LG}} - \psi_2 + \sigma_{2b}^2 = 0, \\ \quad \cdots \\ C_i^{\mathrm{LG}} - \psi_i + \sigma_{ib}^2 = 0, \\ \quad \cdots \\ C_n^{\mathrm{LG}} - \psi_n + \sigma_{nb}^2 = 0, \end{array} \right. \quad \left\{ \begin{array}{l} x_1 - \Delta t_1^{\mathrm{LG}} + \sigma_{1x}^2 = 0, \\ \quad \cdots \\ x_i - \Delta t_i^{\mathrm{LG}} + \sigma_{ix}^2 = 0, \\ \quad \cdots \\ x_m - \Delta t_m^{\mathrm{LG}} + \sigma_{mx}^2 = 0, \\ \Delta t_1^{\mathrm{LG}} - y_1 + \sigma_{1y}^2 = 0, \\ \quad \cdots \\ \Delta t_i^{\mathrm{LG}} - y_i + \sigma_{iy}^2 = 0, \\ \quad \cdots \\ \Delta t_m^{\mathrm{LG}} - \psi_m + \sigma_{my}^2 = 0. \end{array} \right. \tag{5.21}$$

The Lagrange function is as follows:

$$F\left(C^{\mathrm{LG}},\Delta t^{\mathrm{LG}},\lambda,\sigma\right)=$$

$$=u_n^2\sum_{i=1}^{n}\left(\frac{C_{i+1}^{\mathrm{LG}}+\sum_{i=1}^{k}\left(\frac{\varepsilon w_{\grave{l}i}l_{\grave{l}i}}{t_{\mathrm{ox}}}+C_i^{G_i,\mathrm{LG}_j}\right)}{\begin{array}{c}\sum_{l=1}^{z}\left\{\Delta t_{bl}^{G_l,\mathrm{LG}_i}=\mathrm{PDE}_l\left[\mathrm{MDE}_{l\gamma}(w_\gamma),\mathrm{MDE}_{l\eta}(v_\eta)\right]\right\}+\\+\left\{\Delta t_s^{G_l,\mathrm{LG}_i}=\mathrm{PDE}_o\left[\mathrm{MDE}_{o\gamma}(w_\gamma),\mathrm{MDE}_{o\eta}(v_\eta)\right]\right\}\\+\sum_{s=1}^{m}\left[\Delta t_{Ms,s+1}^{,\mathrm{LG}_i}=\mathrm{MDE}_q(l_{Ms,s+1})\right]\end{array}}\right)+$$
$$+\lambda_1\sum_{l=1}^{z}\left\{\Delta t_{bl}^{G_l,\mathrm{LG}_i}=\mathrm{PDE}_1\left[\mathrm{MDE}_{l\gamma}(w_\gamma),\mathrm{MDE}_{l\eta}(v_\eta)\right]\right\}+$$
$$+\left\{\Delta t_s^{G_l,\mathrm{LG}_i}=\mathrm{PDE}_o\left[\mathrm{MDE}_{o\gamma}(w_\gamma),\mathrm{MDE}_{o\eta}(v_\eta)\right]\right\}$$
$$+\sum_{s=1}^{m}\left[\Delta t_{Ms,s+1}^{,\mathrm{LG}_i}=\mathrm{MDE}_q(l_{Ms,s+1})\right]++\sigma_1^2-T_{\mathrm{per}}$$
$$+\sum_{i=1}^{n}\lambda_{i+1}\left(\varphi_{i+1}-C_{i+1}^{\mathrm{LG}}+\sigma_{ia}^2\right)+\sum_{i=1}^{n}\lambda_{n+i+1}\left(C_{i+1}^{\mathrm{LG}}-\psi_{i+1}+\sigma_{ib}^2\right)+$$
$$+\sum_{i=1}^{n}\lambda_{2n+i+1}\left(x_i-\Delta t_i^{LG}+\sigma_{ix}^2\right)+\sum_{i=1}^{n}\lambda_{2n+m+i+1}\left(\Delta t_i^{LG}-y_i+\sigma_{ib}^2\right) \tag{5.22}$$

Calculating the partial derivatives (5.22), the final conditions of Kuhn Tucker are obtained:

$$\begin{cases}\frac{\partial F}{\partial C_2{}^{\mathrm{LG}}}=\frac{1}{\left\{\Delta t_{bl}^{\mathrm{LG}}=\mathrm{PDE}_l\left[\mathrm{MDE}_{l\gamma}(w_\gamma),\mathrm{MDE}_{l\eta}(v_\eta)\right]\right\}+\left[\Delta t_{M1,2}^{,\mathrm{LG}_i}=\mathrm{MDE}_q(l_{M1,2})\right]}-\lambda_2+\lambda_{n+2}=0\\ \cdots\cdots\\ \frac{\partial F}{\partial C_i{}^{\mathrm{LG}}}=\frac{1}{\left\{\Delta t_{bi}^{\mathrm{LG}}=\mathrm{PDE}_i\left[\mathrm{MDE}_{i\gamma}(w_\gamma),\mathrm{MDE}_{i\eta}(v_\eta)\right]\right\}+\left[\Delta t_{Mi,i+1}^{,\mathrm{LG}_i}=\mathrm{MDE}_q(l_{Mi,i+1})\right]}-\lambda_3+\lambda_{n+3}=0\\ \cdots\cdots\\ \frac{\partial F}{\partial C_k{}^{\mathrm{LG}}}=\frac{1}{\left\{\Delta t_{i(k-1)}^{\mathrm{LG}}=\mathrm{PDE}_{k-1}\left[\mathrm{MDE}_{(k-1)\gamma}(w_\gamma),\mathrm{MDE}_{(k-1)\eta}(v_\eta)\right]\right\}+\left[\Delta t_{Mk-1,k}^{,\mathrm{LG}_i}=\mathrm{MDE}_q(l_{Mk-1,k})\right]}-\\ -\lambda_k+\lambda_{n+k}=0\\ \frac{\partial F}{\partial \Delta t_1{}^{\mathrm{LG}}}=-\frac{\left(C_2^{\mathrm{LG}}+C^{\mathrm{LG}_1}\right)}{\left(\left\{\Delta t_{bl}^{\mathrm{LG}}=\mathrm{PDE}_l\left[\mathrm{MDE}_{l\gamma}(w_\gamma),\mathrm{MDE}_{l\eta}(v_\eta)\right]\right\}+\left[\Delta t_{M1,2}^{,\mathrm{LG}_i}=\mathrm{MDE}_q(l_{M1,2})\right]\right)^2}+\\ +\lambda_1-\lambda_{2n+2}+\lambda_{2n+m+2}=0\\ \cdots\cdots\\ \frac{\partial F}{\partial \Delta t_i{}^{\mathrm{LG}}}=-\frac{\left(C_i^{\mathrm{LG}}+C^{\mathrm{LG}_i}\right)}{\left(\left\{\Delta t_{bi}^{\mathrm{LG}}=\mathrm{PDE}_i\left[\mathrm{MDE}_{i\gamma}(w_\gamma),\mathrm{MDE}_{i\eta}(v_\eta)\right]\right\}+\left[\Delta t_{Mi,i+1}^{,\mathrm{LG}_i}=\mathrm{MDE}_q(l_{Mi,i+1})\right]\right)^2}+\\ +\lambda_1-\lambda_{2n+i+1}+\lambda_{2n+m+i+1}=0\\ \cdots\cdots\\ \frac{\partial F}{\partial \Delta t_q{}^{\mathrm{LG}}}=-\frac{\left(C_{q+1}^{\mathrm{LG}}+C^{\mathrm{LG}_q}\right)}{\left(\left\{\Delta t_{iq}^{\mathrm{LG}}=\mathrm{PDE}_q\left[\mathrm{MDE}_{q\gamma}(w_\gamma),\mathrm{MDE}_{q\eta}(v_\eta)\right]\right\}+\left[\Delta t_{Mq,q+1}^{,\mathrm{LG}_i}=\mathrm{MDE}_q(l_{Mq,q+1})\right]\right)^2}+\\ +\lambda_1-\lambda_{2n+q+1}+\lambda_{2n+m+q+1}=0\end{cases} \tag{5.23}$$

$$\begin{cases} -C_2^{\mathrm{LG}} \le -\varphi_2 \\ \dots \\ -C_i^{\mathrm{LG}} \le -\varphi_i \\ \dots \\ -C_n^{\mathrm{LG}} \le -\varphi_n \\ -C_2^{\mathrm{LG}} \le -\psi_2 \\ \dots \\ -C_i^{\mathrm{LG}} \le -\psi_i \\ \dots \\ -C_n^{\mathrm{LG}} \le -\psi_n \\ -\left\{\Delta t_{bl}^{\mathrm{LG}} = \mathrm{PDE}_l\left[\mathrm{MDE}_{l\gamma}(w_\gamma), \mathrm{MDE}_{l\eta}(v_\eta)\right]\right\} \le -x_1 \\ \dots \\ -\left\{\Delta t_{bi}^{\mathrm{LG}} = \mathrm{PDE}_i\left[\mathrm{MDE}_{i\gamma}(w_\gamma), \mathrm{MDE}_{i\eta}(v_\eta)\right]\right\} \le -x_i \\ \dots \\ -\left\{\Delta t_{bn}^{\mathrm{LG}} = \mathrm{PDE}_n\left[\mathrm{MDE}_{n\gamma}(w_\gamma), \mathrm{MDE}_{n\eta}(v_\eta)\right]\right\} \le -x_n \\ \sum_{i=1}^{n}\left(\left\{\Delta t_{bi}^{\mathrm{LG}} = \mathrm{PDE}_i\left[\mathrm{MDE}_{i\gamma}(w_\gamma), \mathrm{MDE}_{i\eta}(v_\eta)\right]\right\} + \left[\Delta t_{Mi,i+1}^{\mathrm{LG}_i} = \mathrm{MDE}_q(l_{Mi,i+1})\right]\right) \le T_{per} \\ \lambda_1, \lambda_2, \dots, \lambda_{2n+2m+1} \ge 0 \end{cases} \tag{5.24}$$

$$\begin{cases} \lambda_1 \sum_{i=1}^{n}\left(\left\{\Delta t_{bi}^{\mathrm{LG}} = \mathrm{PDE}_i\left[\mathrm{MDE}_{i\gamma}(w_\gamma), \mathrm{MDE}_{i\eta}(v_\eta)\right]\right\} + \left[\Delta t_{Mi,i+1}^{\mathrm{LG}_i} = \mathrm{MDE}_q(l_{Mi,i+1})\right]\right) - T_{\widehat{\ddot{a}\ddot{i}}} = 0 \\ \lambda_2\left(\varphi_2 - C_2^{\mathrm{LG}}\right) = 0 \\ \dots \\ \lambda_i\left(\varphi_i - C_i^{\mathrm{LG}}\right) = 0 \\ \dots \\ \lambda_n\left(\varphi_n - C_n^{\mathrm{LG}}\right) = 0 \\ \lambda_{n+1}\left(C_2^{\mathrm{LG}} - \psi_2\right) = 0 \\ \dots \\ \lambda_{2n+1}\left(C_n^{\mathrm{LG}} - \psi_n\right) = 0 \\ \lambda_{2n+i}\left(x_1 - \left\{\Delta t_{b1}^{\mathrm{LG}} = \mathrm{PDE}_1\left[\mathrm{MDE}_{1\gamma}(w_\gamma), \mathrm{MDE}_{1\eta}(v_\eta)\right]\right\}\right) = 0 \\ \dots \\ \lambda_{2n+m+1}\left(x_m - \left\{\Delta t_{im}^{\mathrm{LG}} = \mathrm{PDE}_m\left[\mathrm{MDE}_{m\gamma}(w_\gamma), \mathrm{MDE}_{m\eta}(v_\eta)\right]\right\}\right) = 0 \\ \dots \\ \lambda_{2n+m+i}\left(\left\{\Delta t_{i(m+i)}^{\mathrm{LG}} = \mathrm{PDE}_{m+i}\left[\mathrm{MDE}_{(m+i)\gamma}(w_\gamma), \mathrm{MDE}_{(m+i)\eta}(v_\eta)\right]\right\} - y_1\right) = 0 \\ \dots \\ \lambda_{2n+2m+1}\left(\left\{\Delta t_{im}^{\mathrm{LG}} = PDE_m\left[\mathrm{MDE}_{m\gamma}(w_\gamma), \mathrm{MDE}_{m\eta}(v_\eta)\right]\right\} - y_m\right) = 0 \end{cases} \tag{5.25}$$

Thus, the objective function of the problem of optimizing a digital circuit with consideration of DF is (5.10) with restrictions (5.24), (5.25). According to the classification [297], it is a nonlinear, convex optimization problem with n variables and with restrictions. The developed AOCTP in its general form is shown in Fig. 5.7.

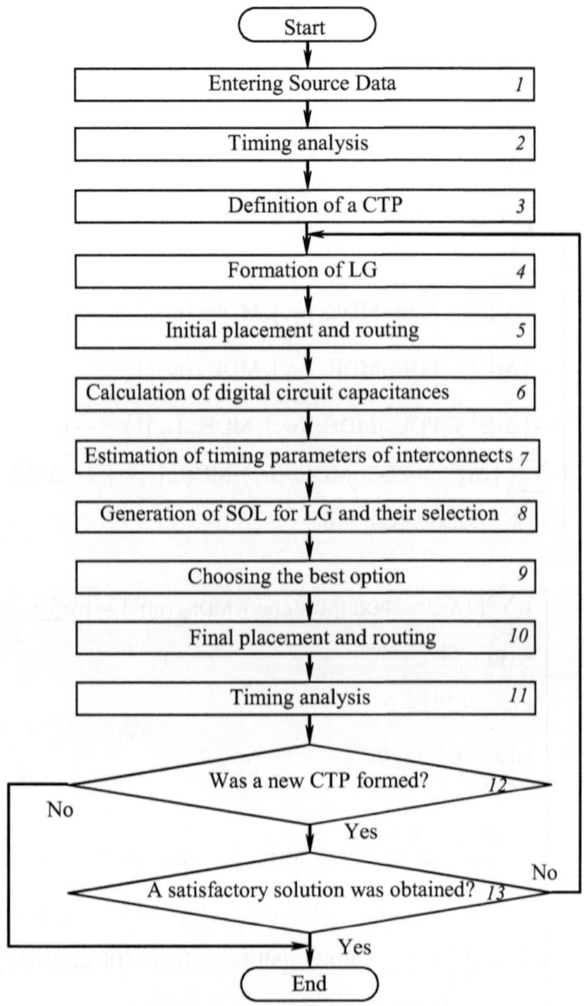

Fig. 5.7 AOCTP in general form

step 1:	Digital Cell Libraries (LEF, LIB, GDSII, Verilog, SP), the optimized digital circuit models (DEF, Verilog, CIR, GDSII), T_{per}, operating frequency f, and technological models (TECH, TF) are introduced.
step 2:	As a result of timing analysis, delays between all inputs and outputs of digital circuits are estimated.
step 3:	Using the results of the previous step, the CTP of the CA, for which $t_{\text{d.CTP}} > T_{\text{per}}$ are determined.
step 4:	LG are formed according to the algorithm described above.
step 5:	As a result of the initial placement and routing of the LG, their components are as close to each other as possible (Fig. 5.8) [132], which leads to a reduction in the LG delay and interconnect.

(continued)

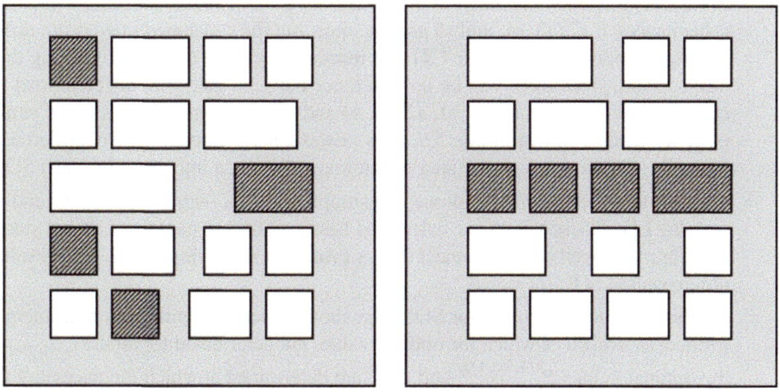

Fig. 5.8 Changes in the initial placement and routing stage

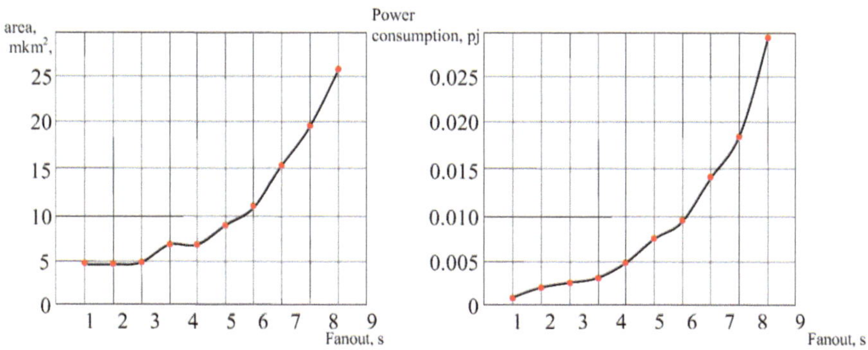

Fig. 5.9 Dependences of the area and power consumption of the inverter on the load capacitance

step 6:	The computation of capacitances of digital circuits is carried out by SPICE [74, 75]. These values are subsequently used to estimate the gate loads.
step 7:	The timing of interconnects is estimated using RLC models.
step 8:	The sets of solutions SOL_{ij} for $_{i=1,2,\ldots,y},\ _{j=1,2,\ldots,x}$ of each $LG_j,\ _{j=1,2,\ldots,x}$ are generated. The number of SOL for each $LG_j,\ _{j=1,2,\ldots,y}$ depends on the number of variants with different load capacitances. It is known that the input capacitance of the gate is directly proportional to its size. Therefore, the use of digital cells with a higher load capacitance from the library leads to an increase in the load of the previous gate, resulting in an increase in timing delay. The noted capacitance C_L can also be reduced by the correct choice of gate and their matching. Naturally, with the increase in the performance of digital circuits, the power consumed by the digital circuit and the occupied area also increase. Obviously, with a decrease in digital circuit delay, these quantities must also be taken into account. It is known that these quantities are directly proportional. But, unlike the power consumption, gate areas with different load capacitances can be the same. Figure 5.9 as an example, shows the dependences of the areas and the power consumed on the load capacitance for inverters from the same digital cell library obtained using SPICE [74, 75]. As can be seen from Fig. 5.9, the areas of inverters with the load

(continued)

	capacitances 1, s2, s3, s4, and s5 are the same, but the consumed powers are different. During the optimization of the CTP, an inaccuracy can be obtained by taking the area into account, since there will be no difference between gate with different load capacitances (in the example, s1, s2, s3, s4 and s5). Taking into account the similarity of dependencies shown in Fig. 5.9, only power consumption can be considered, assuming that in this case the area is automatically taken into account. Each SOL_{ij} contains three values: $P_i^{\mathrm{LG}_j}$—power consumption, $C_i^{\mathrm{LG}_j}$—input capacitance and $\Delta t_i^{\mathrm{LG}_j}$—delay LG_j. These values are calculated based on timing models of digital standard cell library, as well as interconnect delays evaluated in previous steps. As a result, only nonexpanded SOL_{ij} is stored.
	On the basis of the noninferior SOL_{ij} the above described optimization problem is formed, as a result of which the optimal values are determined for all the $\mathrm{LG}_{j},\, _{j=1,2,\ldots,y}$, the optimal values $P_i^{\mathrm{LG}_j}$, $\tilde{N}_i^{\mathrm{LG}_j}$ and $\Delta t_i^{\mathrm{LG}_j}$ are determined at which the necessary CTP delay is provided with the minimum power consumed by digital circuits.
step 9:	From the set SOL_{ij}, those are chosen, the parameters of which are closest to the optimal values. Hence it follows that the optimization process always converges, since there is always a SOL_{ij}, whose parameters are closest to the obtained values.
step 10:	At the stage of final placement and routing, minimal changes are possible, since the placement remains the same as in the preliminary stage, and changes are possible only if the gate area differs from the area in the preliminary stage.
step 11:	Timing analysis is performed based on the RLC description of layout.
step 12:	The presence of a new CTP is verified. If no new CTP is formed during optimization, the timing delay of which exceeds the CTP, then the AOCTP of digital circuits terminates. Otherwise, at step 13, the case of the formation of a new CTP is checked, for which $t_{\mathrm{d.stp}} > T_{\mathrm{per}}$. If this condition is violated, the AOCTP is repeated for the new CTP. Practical application of the developed AOCTP shows that the number of cases of forming new CTP usually does not exceed 2 … 3. A detailed AOCTP is given in Sect. 5.2. An example of the application of the developed AOCTP for a digital circuit s289 from the ISCAS 89 test series is also shown. The optimization results of digital circuit radar from this series are given in Table 5.1.

As seen from Table 5.1, the application of the developed AOCTP results in a reduction in the delay of the CTP by 25% … 30% with an increase in the power consumption of digital circuits by only 4% … 5%. These results exceed the performance of other AOCTP, as they provide a maximum reduction in the delay of the CTP by 20% with an increase in power consumption by 10%.

However, such an advantage developed by the AOCTP was achieved due to the large expenditures of computer time. Figure 5.10 shows the computer time costs for the developed AOCTP and the algorithm from [319] when optimizing a digital circuit from the test series ISCAS 89.

After the approximation shown in Fig. 5.10, the following relation is obtained:

$$t_{\mathrm{M}} = -0.5002n^2 + 32.65n - 209.8, \tag{5.26}$$

which indicates a relation that is close to linear. Here n is the number on the CTP.

The developed AOCTP on average requires 22% … 24% more computer time compared with [319]. However, due to this, an average 5% reduction in the delay of

Table 5.1 Results of optimization of a digital circuit from the ISCAS 89 series

Before the AOCTP			After the AOCTP			
Digital circuit name	CTP delay (ns)	Power consumption (mcWt)	CTP delay (ns)	Gain in %	Power consumption (mcWt)	Loss in %
s27	0.67	84.61	0.4891	27	90.533	7
s298	0.84	865.62	0.6888	18	908.901	5
s344	0.97	1138.98	0.5820	40	1230.098	8
s349	1.42	1145.48	0.9230	35	1168.390	2
s382	1.12	1165.01	0.5712	49	1258.211	8
s386	1.21	1073.89	0.7502	38	1116.846	4
s400	1.17	1204.06	0.8658	26	1167.938	−3
s420	1.54	1522.98	0.9394	39	1553.440	2
s444	1.04	1314.71	0.8424	19	1288.416	−2
s510	0.87	1421.34	0.4785	45	1549.261	9
s526	1.14	1392.81	0.8094	29	1462.451	5
s641	2.15	2590.37	1.4620	32	2823.503	9
s713	2.07	2681.49	1.6353	21	2869.194	7
s820	1.45	1913.57	0.8845	39	1990.113	4
s832	0.98	1908.87	0.7056	28	1889.781	−1
s953	2.24	2759.58	1.8816	16	2952.751	7

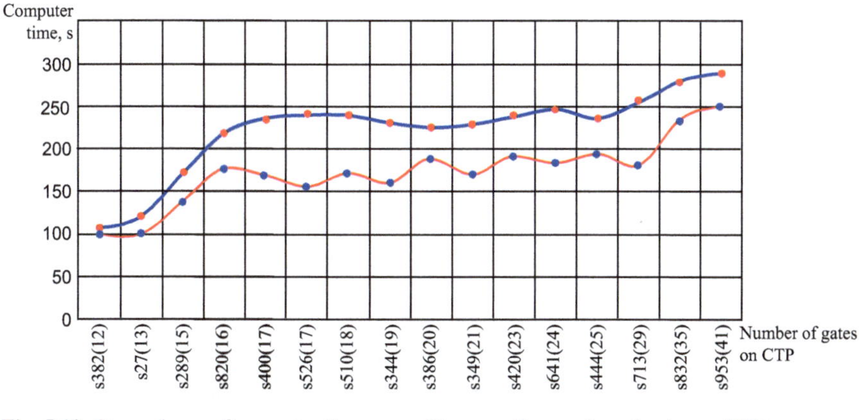

Fig. 5.10 Dependence of computer time expenditures on the number of gates on CTP

the CTP with simultaneous gain in the power consumption by 4% ... 5%, is obtained which indicates the efficiency of the AOCTP of a digital circuit. In addition, the accuracy of calculations carried out with the help of the developed AOCTP of digital circuits on average is 20% ... 30% higher due to DF consideration.

5.2 Algorithm for Optimization of Critical Timing Paths of Digital Circuits with Consideration of DF

step 1:	Input of models (LEF, LIB, GDSII, Verilog, SP, DEF, CIR) of digital cell library.
step 2:	Input of technological models (TF, TECH) and T_{per}.
step 3:	Calculation of delays $\{\Delta t_{ij}\}$ between inputs ($i = 1, 2, \ldots, n$) and outputs ($j = 1, 2, \ldots, m$) of digital circuits.
step 4:	Definition of CTP with $\Delta t_{ij} > \text{T}_{\text{per}}$.
step 5:	Performing step 6 for each G_{ij}, belonging to the CTP_i
step 6:	The formation of $\text{LG}_i = \{G_{ij}\}$, $_{i=1,2,\ldots,y;\ j=1,2,\ldots,x}$
step 7:	Performing steps 8–12 for each of the G_i, $_{i=1,2,\ldots,y}$
step 8:	Performing step 9–12 for each input G_m of Mth order G_{ij}, $_{i=1,2,\ldots,y,\ j=1,2,\ldots,x}$
step 9:	If the G_{ij} is not included in some LG_i, then go to step 10, otherwise—to step 11
step 10:	The inclusion of G_{ij} in LG_i
step 11:	Performing step 12 for each output branch from G_{ij} to LG_i
step 12:	The inclusion of LG_i in LG_i
step 13:	Initial placement and routing, based on the input models with minimization of the interconnect delays $\Delta t_{l,n,n+1}^{\text{LG}_i} = \text{MDE}_q(l_{l,n,n+1}), \quad n = 1, 2, \ldots, m$ between CTP gates.
step 14:	Calculation of capacitances $C^{\text{LG}} = \{C_1^{\text{LG}}, \ldots, C_n^{\text{LG}}\}$ and $C^{G,\text{LG}} = \{C^{G_1,\text{LG}_1}, \ldots, C^{G_{\delta},\text{LG}_{\delta}}\}$ based on the initial layout of digital circuits.
step 15:	Estimating the timing delays $\Delta t_{l,n,n+1}^{\text{LG}_i} = \text{MDE}_q(l_{l,n,n+1}), \quad n = 1, 2, \ldots, m$ of interconnects of digital circuits.
step 16:	Generation of the solution set $\text{SOL}^{\text{LG}} = \{\text{SOL}_1, \ldots, \text{SOL}_n\}$ for each LG
step 17:	Calculation of parameters $\text{SOL}_n(Đ_n^{\text{LG}}, \tilde{N}_n^{\text{LG}}, \Delta t_n^{\text{LG}})$ for each solution of each $\text{LG}_{j,\ j=1,2,\ldots,y}$
step 18:	Comparison of solutions of each SOL^{LG_j}, $_{j=1,2,\ldots,y}$
step 19:	If for each of the solutions x and y the following occurs $P_x^{\text{LG}} \geq P_y^{\text{LG}}, C_x^{\text{LG}} \geq C_y^{\text{LG}}$ and $\Delta t_x^{\text{LG}} \geq \Delta t_y^{\text{LG}}$, then classification of the solution x is as inferior and its ignoring.
step 20:	Taking into account the previously calculated $C^{G,\text{LG}} = \{C^{G_1,\text{LG}_1}, \ldots, C^{G_{\delta},\text{LG}_{\delta}}\}$ and $\Delta t_{\text{M},n,n+1}^{\text{LG}_i} = \text{MDE}_q(l_{\text{M},n,n+1}), \quad n = 1, 2, \ldots, m$, obtaining the objective function (5.10).
step 21:	Taking into account the previously calculated sets $C^{\text{LG}} = \{C_1^{\text{LG}}, \ldots, C_m^{\text{LG}}\}$ and $\Delta t^{\text{LG}} = \{\Delta t_1^{\text{LG}}, \ldots, \Delta t_q^{\text{LG}}, \ldots, \Delta t_m^{\text{LG}}\}$, obtaining the constraints (5.12) of the optimization problem
step 22:	Reduction of restrictions to the form (5.13)
step 23:	The derivation of the Lagrange functions (5.22), taking into account the objective function (5.10) and the constraints (5.21)
step 24:	Calculation of partial derivatives of the obtained Lagrange function (5.23) and equating it to 0.
step 25:	Checking the minimum conditions (5.19) and obtaining optimal values for C_i^{LG} and Δt_i^{LG}, $i = 1, 2, \ldots, n$, at which the LG power consumption is minimal under the given constraints

(continued)

step 26:	The choice of solutions from the set SOL_i^{LG}, the difference of parameters between them and the ones calculated at step 25, is minimal.
step 27:	Final placement and routing based on the selected solutions and introduced in steps 1 and 2 of the models.
step 28:	Calculation of delays $\{\Delta t_{ij}\}$ between inputs $i = 1, 2, \ldots, n$ and outputs $j = 1, 2, \ldots, m$ of digital circuits
step 29:	If none of the values of Δt_{ij} is more than T_{per}, go to step 31
step 30:	If the generated CTP belongs to the set of CTP, the confirmation of the composition of gates and the topology of optimized CTP and the transition to step 5.
step 31:	Completion of digital circuit optimization process.

5.3 An example of Optimization of Critical Timing Paths of Digital Circuits with Consideration of DF

Below is an example of optimization of critical timing paths of digital circuits with consideration of DF in order to evaluate the effectiveness of the developed algorithm. The optimization was carried out using the example of a digital circuit s289 (Fig. 5.11) from a number of ISCAS89 test series. The digital circuit contains 131 gates: 31 "AND", 9 "NAND", 16 "OR", 19 "NOR" and 56 "NOT".

The following initial data are given:

(a) Digital circuit (Fig. 5.11), the Verilog description of which is as follows:

```
wire  G10,G29,G11,G30,G12,G34,G13,G39,G14,G44,G15,G56,G16,G86,
      G17,G92,G18,G98,
      G19,G102,G20,G107,G21,G113,G22,G119,G23,G125,G28,G130,G38,
      G40,G45,G46,G50,
      G51,G54,G55,G59,G60,G64,II155,II158,G76,G82,G87,G91,G93,
      G96,G99,G103,G108,
      G112,G114,II210,II213,G120,G124,G121,II221,G126,G131,G127,
      II229,II232,
      II235,II238,G26,G27,G31,G32,G33,G35,G36,G37,G42,G41,G48,
      G47,G49,G52,G57,
      G61,G58,G65,G62,G63,G74,G75,G88,G89,G90,G94,G95,G100,G105,
      G104,G110,G109,
      G111,G115,G122,G123,G128,G129,G24,G25,G68,G69,G70,G71,G72,
      G73,G77,G78,G79,
      G80,G81,G83,G84,G85,G43,G97,G101,G106,G116,G53;
```

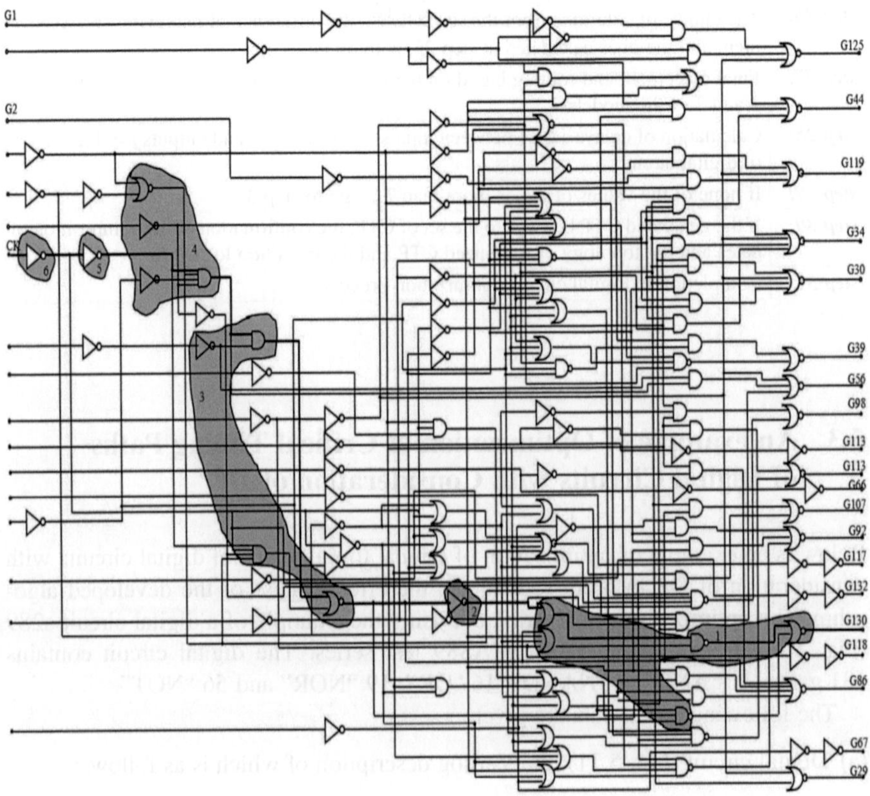

Fig. 5.11 Optimized digital circuit s289

```
dff DFF_0(CK,G10,G29);
dff DFF_1(CK,G11,G30);
dff DFF_2(CK,G12,G34);
dff DFF_3(CK,G13,G39);
dff DFF_4(CK,G14,G44);
dff DFF_5(CK,G15,G56);
dff DFF_6(CK,G16,G86);
dff DFF_7(CK,G17,G92);
dff DFF_8(CK,G18,G98);
dff DFF_9(CK,G19,G102);
dff DFF_10(CK,G20,G107);
dff DFF_11(CK,G21,G113);
dff DFF_12(CK,G22,G119);
dff DFF_13(CK,G23,G125);
not NOT_0(G28,G130);
```

```
not NOT_1(G38,G10);
not NOT_2(G40,G13);
not NOT_3(G45,G12);
not NOT_4(G46,G11);
not NOT_5(G50,G14);
not NOT_6(G51,G23);
not NOT_7(G54,G11);
not NOT_8(G55,G13);
not NOT_9(G59,G12);
not NOT_10(G60,G22);
not NOT_11(G64,G15);
not NOT_12(II155,G16);
not NOT_13(G66,II155);
not NOT_14(II158,G17);
not NOT_15(G67,II158);
not NOT_16(G76,G10);
not NOT_17(G82,G11);
not NOT_18(G87,G16);
not NOT_19(G91,G12);
not NOT_20(G93,G17);
not NOT_21(G96,G14);
not NOT_22(G99,G18);
not NOT_23(G103,G13);
not NOT_24(G108,G112);
not NOT_25(G114,G21);
not NOT_26(II210,G18);
not NOT_27(G117,II210);
not NOT_28(II213,G19);
not NOT_29(G118,II213);
not NOT_30(G120,G124);
not NOT_31(G121,G22);
not NOT_32(II221,G2);
not NOT_33(G124,II221);
not NOT_34(G126,G131);
not NOT_35(G127,G23);
not NOT_36(II229,G0);
not NOT_37(G130,II229);
not NOT_38(II232,G1);
not NOT_39(G131,II232);
not NOT_40(II235,G20);
not NOT_41(G132,II235);
not NOT_42(II238,G21);
not NOT_43(G133,II238);
```

```
and AND2_0(G26,G28,G50);
and AND2_1(G27,G51,G28);
and AND3_0(G31,G10,G45,G13);
and AND2_2(G32,G10,G11);
and AND2_3(G33,G38,G46);
and AND3_1(G35,G10,G11,G12);
and AND2_4(G36,G38,G45);
and AND2_5(G37,G46,G45);
and AND2_6(G42,G40,G41);
and AND4_0(G48,G45,G46,G10,G47);
and AND3_2(G49,G50,G51,G52);
and AND4_1(G57,G59,G11,G60,G61);
and AND2_7(G58,G64,G65);
and AND4_2(G62,G59,G11,G60,G61);
and AND2_8(G63,G64,G65);
and AND3_3(G74,G12,G14,G19);
and AND3_4(G75,G82,G91,G14);
and AND2_9(G88,G14,G87);
and AND2_10(G89,G103,G96);
and AND2_11(G90,G91,G103);
and AND2_12(G94,G93,G13);
and AND2_13(G95,G96,G13);
and AND3_5(G100,G99,G14,G12);
and AND3_6(G105,G103,G108,G104);
and AND2_14(G110,G108,G109);
and AND2_15(G111,G10,G112);
and AND2_16(G115,G114,G14);
and AND2_17(G122,G120,G121);
and AND2_18(G123,G124,G22);
and AND2_19(G128,G126,G127);
and AND2_20(G129,G131,G23);
or OR4_0(G24,G38,G46,G45,G40);
or OR3_0(G25,G38,G11,G12);
or OR4_1(G68,G11,G12,G13,G96);
or OR2_0(G69,G103,G18);
or OR2_1(G70,G103,G14);
or OR3_1(G71,G82,G12,G13);
or OR2_2(G72,G91,G20);
or OR2_3(G73,G103,G20);
or OR4_2(G77,G112,G103,G96,G19);
or OR2_4(G78,G108,G76);
or OR2_5(G79,G103,G14);
or OR2_6(G80,G11,G14);
```

```
  or OR2_7(G81,G12,G13);
  or OR4_3(G83,G11,G12,G13,G96);
  or OR3_2(G84,G82,G91,G14);
  or OR3_3(G85,G91,G96,G17);
  nand NAND3_0(G41,G12,G11,G10);
  nand NAND3_1(G43,G24,G25,G28);
  nand NAND4_0(G52,G13,G45,G46,G10);
  nand NAND4_1(G65,G59,G54,G22,G61);
  nand NAND4_2(G97,G83,G84,G85,G108);
  nand NAND4_3(G101,G68,G69,G70,G108);
  nand NAND2_0(G106,G77,G78);
  nand NAND4_4(G109,G71,G72,G73,G14);
  nand NAND4_5(G116,G79,G80,G81,G108);
  nor NOR2_0(G29,G10,G130);
  nor NOR4_0(G30,G31,G32,G33,G130);
  nor NOR4_1(G34,G35,G36,G37,G130);
  nor NOR2_1(G39,G42,G43);
  nor NOR3_0(G44,G48,G49,G53);
  nor NOR2_2(G47,G50,G40);
  nor NOR2_3(G53,G26,G27);
  nor NOR3_1(G56,G57,G58,G130);
  nor NOR2_4(G61,G14,G55);
  nor NOR4_2(G86,G88,G89,G90,G112);
  nor NOR3_2(G92,G94,G95,G97);
  nor NOR2_5(G98,G100,G101);
  nor NOR2_6(G102,G105,G106);
  nor NOR2_7(G104,G74,G75);
  nor NOR2_8(G107,G110,G111);
  nor NOR2_9(G112,G62,G63);
  nor NOR2_10(G113,G115,G116);
  nor NOR3_3(G119,G122,G123,G130);
  nor NOR3_4(G125,G128,G129,G130);
endmodule
```

(b) Digital cell library for 28 nm technology, operating voltages—1 V ± 10%, temperature range −40 to 125 °C. The composition of the library is shown in Table 5.2.

Behavioral, physical, timing, and topological models are given for each library gate. Here, to reduce the scope of the example, these models are given only for the "AND" element with the load capacitance s1.

Table 5.2 Digital standard cell library structure

Fanout of a digital cell	Logic function "NOT"	"NAND2"	"AND2"	"OR2"	"NOR2"
s1	•	•	•	•	•
s2	•	•	•	•	•
s3	•	•	•	•	•
s4	•	•	•		
s5	•				
s6	•				
s7	•				
s8	•				
s9	•				
s10	•				

—Physical model

```
MACRO and2s1
CLASS  CORE ;
 FOREIGN and2s1 0.000 0.000 ;
 ORIGIN 0.000 0.000 ;
 SIZE 2.240 BY 3.920 ;
  SYMMETRY X Y  ;
  SITE CoreSite ;
  PIN Q
    DIRECTION OUTPUT ;
    PORT
      LAYER METAL1 ;
        RECT 1.740 1.090 1.930 1.350 ;
        RECT 1.770 1.090 1.930 3.000 ;
        RECT 1.770 2.090 2.110 2.390 ;
    END
  END Q
  PIN SVSS12
    DIRECTION INOUT ;
    USE GROUND ;
    SHAPE ABUTMENT ;
    PORT
      LAYER METAL1 ;
        RECT 1.170 0.330 1.330 1.300 ;
        RECT 1.120 1.140 1.380 1.300 ;
        RECT 0.000 0.330 2.240 0.790 ;
    END
  END SVSS12
  PIN DIN2
```

```
    DIRECTION INPUT ;
    PORT
      LAYER METAL1 ;
        RECT 1.060 1.480 1.550 1.830 ;
    END
  END DIN2
  PIN DIN1
    DIRECTION INPUT ;
    PORT
      LAYER METAL1 ;
        RECT 0.130 1.530 0.540 1.830 ;
    END
  END DIN1
  PIN BVSS12
    DIRECTION INOUT ;
    USE GROUND ;
    SHAPE ABUTMENT ;
    PORT
      LAYER METAL1 ;
        RECT 0.000 -0.110 2.240 0.110 ;
    END
  END BVSS12
  PIN VDD12
    DIRECTION INOUT ;
    USE POWER ;
    SHAPE ABUTMENT ;
    PORT
      LAYER METAL1 ;
        RECT 0.210 2.360 0.370 4.150 ;
        RECT 1.260 2.360 1.420 4.150 ;
        RECT 0.000 3.690 2.240 4.150 ;
    END
  END VDD12
  OBS
      LAYER METAL1 ;
        RECT 0.720 2.020 1.590 2.180 ;
        RECT 0.720 1.140 0.880 2.620 ;
        RECT 0.240 1.140 0.880 1.300 ;
  END
END and2s1
```

—Timing model

```
cell (and2s1){ /* 2-input AND, 1x */
  area : 8.7808;
/* using cmos4 delaymodel */
  pin (Q) {
    max_transition : 0.499;
    max_capacitance : 0.04737;
    function : "DIN1 & DIN2";
    direction : output;
    timing (){
      timing_sense : positive_unate;
      related_pin  : "DIN2";
      cell_rise(li8X6){
         index_1("0.00000, 0.00395, 0.00791, 0.01582, 0.02373, 0.03164, 0.03955,
0.04746");
        index_2("0.030, 0.067, 0.115, 0.213, 0.312, 0.511");
        values("0.071, 0.075, 0.078, 0.083, 0.086, 0.090",\
            "0.097, 0.101, 0.105, 0.110, 0.114, 0.119",\
            "0.117, 0.121, 0.125, 0.131, 0.136, 0.141",\
            "0.153, 0.157, 0.162, 0.168, 0.173, 0.179",\
            "0.188, 0.192, 0.197, 0.203, 0.208, 0.215",\
            "0.223, 0.227, 0.232, 0.238, 0.243, 0.250",\
            "0.258, 0.262, 0.266, 0.273, 0.278, 0.285",\
            "0.292, 0.296, 0.301, 0.308, 0.313, 0.320");
      }
      rise_transition(li8X6){
         index_1("0.00000, 0.00395, 0.00791, 0.01582, 0.02373, 0.03164, 0.03955,
0.04746");
        index_2("0.030, 0.067, 0.115, 0.213, 0.312, 0.511");
        values("0.036, 0.036, 0.038, 0.042, 0.043, 0.050",\
            "0.075, 0.076, 0.078, 0.080, 0.084, 0.091",\
            "0.113, 0.113, 0.115, 0.116, 0.119, 0.126",\
            "0.189, 0.189, 0.190, 0.193, 0.195, 0.198",\
            "0.263, 0.266, 0.265, 0.265, 0.270, 0.277",\
            "0.339, 0.340, 0.341, 0.340, 0.344, 0.350",\
            "0.414, 0.418, 0.419, 0.417, 0.421, 0.424",\
            "0.493, 0.496, 0.497, 0.494, 0.498, 0.502");
      }

cell_fall(li8X6){
         index_1("0.00000, 0.00399, 0.00798, 0.01595, 0.02393, 0.03190, 0.03988,
0.04785");
        index_2("0.026, 0.064, 0.113, 0.210, 0.309, 0.506");
        values("0.069, 0.080, 0.092, 0.113, 0.130, 0.160",\
            "0.097, 0.108, 0.120, 0.142, 0.160, 0.191",\
            "0.119, 0.130, 0.143, 0.164, 0.183, 0.215",\
            "0.160, 0.171, 0.184, 0.206, 0.224, 0.257",\
            "0.200, 0.211, 0.224, 0.245, 0.264, 0.297",\
            "0.240, 0.250, 0.263, 0.285, 0.304, 0.337",\
            "0.279, 0.290, 0.303, 0.325, 0.343, 0.377",\
            "0.319, 0.330, 0.343, 0.364, 0.383, 0.416");
      }

       fall_transition(li8X6){
         index_1("0.00000, 0.00399, 0.00798, 0.01595, 0.02393, 0.03190, 0.03988,
0.04785");
        index_2("0.026, 0.064, 0.113, 0.210, 0.309, 0.506");
        values("0.029, 0.030, 0.032, 0.036, 0.040, 0.049",\
            "0.069, 0.069, 0.071, 0.077, 0.079, 0.086",\
            "0.107, 0.106, 0.108, 0.112, 0.115, 0.122",\
            "0.183, 0.183, 0.185, 0.185, 0.187, 0.194",\
            "0.260, 0.259, 0.259, 0.263, 0.266, 0.268",\
            "0.340, 0.336, 0.338, 0.341, 0.344, 0.343",\
            "0.419, 0.418, 0.420, 0.421, 0.422, 0.423",\
            "0.497, 0.495, 0.496, 0.498, 0.499, 0.499");
      }
    }

       internal_power (){
      related_pin  : "DIN2";
```

```
      rise_power(li8X6){
         index_1("0.00000, 0.00395, 0.00791, 0.01582, 0.02373, 0.03164, 0.03955,
0.04746");
        index_2("0.030, 0.067, 0.115, 0.213, 0.312, 0.511");
        values("0.0040, 0.0039, 0.0040, 0.0042, 0.0045, 0.0052",\
            "0.0039, 0.0038, 0.0038, 0.0040, 0.0042, 0.0049",\
            "0.0039, 0.0038, 0.0038, 0.0039, 0.0042, 0.0048",\
            "0.0039, 0.0038, 0.0038, 0.0039, 0.0042, 0.0047",\
            "0.0039, 0.0039, 0.0039, 0.0040, 0.0042, 0.0047",\
            "0.0040, 0.0039, 0.0039, 0.0040, 0.0042, 0.0048",\
            "0.0040, 0.0039, 0.0039, 0.0040, 0.0042, 0.0048",\
            "0.0040, 0.0040, 0.0039, 0.0040, 0.0042, 0.0048");
      }

       fall_power(li8X6){
         index_1("0.00000, 0.00399, 0.00798, 0.01595, 0.02393, 0.03190, 0.03988,
0.04785");
        index_2("0.026, 0.064, 0.113, 0.210, 0.309, 0.506");
        values("0.0036, 0.0036, 0.0037, 0.0040, 0.0044, 0.0054",\
            "0.0036, 0.0035, 0.0036, 0.0038, 0.0042, 0.0050",\
            "0.0036, 0.0036, 0.0036, 0.0038, 0.0041, 0.0049",\
            "0.0036, 0.0036, 0.0036, 0.0039, 0.0042, 0.0049",\
            "0.0037, 0.0037, 0.0037, 0.0039, 0.0042, 0.0049",\
            "0.0037, 0.0037, 0.0037, 0.0039, 0.0042, 0.0049",\
            "0.0038, 0.0038, 0.0038, 0.0040, 0.0043, 0.0049",\
            "0.0038, 0.0037, 0.0038, 0.0040, 0.0042, 0.0049");
      }
    }
    timing (){
      timing_sense : positive_unate;
      related_pin  : "DIN1";

      cell_rise(li8X6){
         index_1("0.00000, 0.00395, 0.00790, 0.01579,
                    0.02369, 0.03158, 0.03948, 0.04737");
        index_2("0.030, 0.067, 0.115, 0.213, 0.312, 0.512");
        values("0.067, 0.074, 0.080, 0.089, 0.096, 0.104",\
            "0.093, 0.100, 0.107, 0.116, 0.123, 0.133",\
            "0.113, 0.120, 0.127, 0.137, 0.144, 0.155",\
            "0.150, 0.157, 0.164, 0.174, 0.181, 0.193",\
            "0.185, 0.191, 0.199, 0.209, 0.217, 0.228",\
            "0.219, 0.226, 0.233, 0.243, 0.251, 0.263",\
            "0.254, 0.261, 0.268, 0.278, 0.286, 0.297",\
            "0.289, 0.295, 0.303, 0.313, 0.321, 0.332");
      }

      rise_transition(li8X6){
         index_1("0.00000, 0.00395, 0.00790, 0.01579,
                    0.02369, 0.03158, 0.03948, 0.04737");
        index_2("0.030, 0.067, 0.115, 0.213, 0.312, 0.512");
        values("0.035, 0.036, 0.038, 0.042, 0.045, 0.053",\
            "0.075, 0.076, 0.078, 0.082, 0.085, 0.092",\
            "0.112, 0.113, 0.114, 0.117, 0.120, 0.127",\
            "0.187, 0.189, 0.190, 0.192, 0.193, 0.201",\
            "0.263, 0.265, 0.264, 0.268, 0.272, 0.277",\
            "0.343, 0.339, 0.341, 0.345, 0.347, 0.351",\
            "0.418, 0.417, 0.416, 0.420, 0.422, 0.425",\
            "0.495, 0.495, 0.494, 0.497, 0.499, 0.500");
      }

      cell_fall(li8X6){
         index_1("0.00000, 0.00398, 0.00797, 0.01594,
                    0.02391, 0.03188, 0.03985, 0.04782");
        index_2("0.026, 0.064, 0.113, 0.210, 0.308, 0.505");
        values("0.064, 0.075, 0.086, 0.105, 0.121, 0.148",\
            "0.091, 0.102, 0.114, 0.134, 0.150, 0.179",\
            "0.113, 0.124, 0.136, 0.156, 0.173, 0.202",\
            "0.154, 0.165, 0.177, 0.197, 0.215, 0.245",\
```

```
                "0.194, 0.205, 0.217, 0.238, 0.255, 0.285",\
                "0.233, 0.244, 0.257, 0.277, 0.294, 0.324",\
                "0.273, 0.284, 0.296, 0.317, 0.334, 0.364",\
                "0.313, 0.324, 0.336, 0.357, 0.373, 0.403");
        }

   fall_transition(li8X6){
             index_1("0.00000, 0.00398, 0.00797,
                    0.01594, 0.02391, 0.03188, 0.03985, 0.04782");
            index_2("0.026, 0.064, 0.113, 0.210, 0.308, 0.505");
            values("0.028, 0.029, 0.031, 0.035, 0.039, 0.047",\
                "0.068, 0.068, 0.070, 0.073, 0.078, 0.086",\
                "0.106, 0.105, 0.107, 0.111, 0.116, 0.121",\
                "0.181, 0.182, 0.184, 0.186, 0.189, 0.194",\
                "0.260, 0.258, 0.259, 0.261, 0.264, 0.270",\
                "0.339, 0.335, 0.337, 0.338, 0.340, 0.344",\
                "0.414, 0.417, 0.419, 0.416, 0.420, 0.424",\
                "0.494, 0.494, 0.496, 0.497, 0.497, 0.499");
        }
      }
      internal_power (){
        related_pin  : "DIN1";

        rise_power(li8X6){
          index_1("0.00000, 0.00395, 0.00790,
                 0.01579, 0.02369, 0.03158, 0.03948, 0.04737");
          index_2("0.030, 0.067, 0.115, 0.213, 0.312, 0.512");
          values("0.0037, 0.0038, 0.0038, 0.0041, 0.0044, 0.0052",\
              "0.0036, 0.0036, 0.0037, 0.0039, 0.0041, 0.0048",\
              "0.0036, 0.0036, 0.0036, 0.0038, 0.0041, 0.0047",\
              "0.0037, 0.0036, 0.0037, 0.0038, 0.0040, 0.0046",\
              "0.0037, 0.0037, 0.0037, 0.0038, 0.0041, 0.0046",\
              "0.0038, 0.0037, 0.0037, 0.0039, 0.0041, 0.0046",\
              "0.0038, 0.0038, 0.0038, 0.0039, 0.0041, 0.0046",\
              "0.0038, 0.0038, 0.0038, 0.0039, 0.0041, 0.0046");
        }

        fall_power(li8X6){
          index_1("0.00000, 0.00398, 0.00797,
              0.01594, 0.02391, 0.03188, 0.03985, 0.04782");
          index_2("0.026, 0.064, 0.113, 0.210, 0.308, 0.505");
          values("0.0033, 0.0033, 0.0034, 0.0038, 0.0041, 0.0050",\
              "0.0033, 0.0033, 0.0034, 0.0036, 0.0039, 0.0047",\
              "0.0034, 0.0033, 0.0034, 0.0036, 0.0039, 0.0046",\
              "0.0034, 0.0034, 0.0034, 0.0036, 0.0039, 0.0046",\
              "0.0035, 0.0035, 0.0035, 0.0037, 0.0039, 0.0046",\
              "0.0035, 0.0035, 0.0036, 0.0037, 0.0040, 0.0047",\
              "0.0036, 0.0036, 0.0036, 0.0038, 0.0040, 0.0047",\
              "0.0036, 0.0035, 0.0036, 0.0038, 0.0040, 0.0047");
        }
      }
    }
    pin (DIN1) {
      max_transition : 0.506;
      direction : input;
      capacitance : 0.00130;

      internal_power(){
        rise_power(i6){
          index_1("0.030,0.067,0.114,0.211,0.314,0.510");
          values("0.0000,0.0000,0.0000,0.0000,0.0000,0.0000");
        }
        fall_power(i6){
          index_1("0.026,0.064,0.112,0.211,0.309,0.506");
          values("0.0003,0.0003,0.0003,0.0003,0.0003,0.0003");
        }
      }
    }

    pin (DIN2) {
      max_transition : 0.506;
      direction : input;
      capacitance : 0.00140;
```

```
        internal_power(){
          rise_power(i6){
            index_1("0.030,0.067,0.115,0.214,0.312,0.511");
            values("-0.0000,-0.0001,-0.0001,-0.0001,-0.0001,-0.0001");
          }
          fall_power(i6){
            index_1("0.026,0.064,0.112,0.211,0.310,0.506");
            values("0.0002,0.0002,0.0002,0.0002,0.0002,0.0002");
          }
        }
      }
      leakage_power () {
        when :"DIN1&!DIN2&!Q";
        value : 23250.10000;
      }
      leakage_power () {
        when :"!DIN1&DIN2&!Q";
        value : 23894.50000;
      }
      leakage_power () {
        when :"!DIN1&!DIN2&!Q";
        value : 16244.10000;
      }
      leakage_power () {
        when :"DIN1&DIN2&Q";
        value : 23597.05000;
      }
      cell_leakage_power : 23894.50000;
    }
```

—Behavioral model

```
`celldefine
// 2-input AND, 2x
// Q = DIN1 & DIN2
module and2s1 (Q, DIN1, DIN2);
    output Q;
    input  DIN1;
    input  DIN2;
`protect
    specify
        // Pin-to-pin timing.
    (DIN1 => Q) = (0,0);
    (DIN2 => Q) = (0,0);
    endspecify
    // Gate-level description.
    and _i0 (Q,DIN1,DIN2);
`endprotect
endmodule
`endcelldefine
```

(c) The boundary value of the timing delay is $T = 0.5$ ns
(d) Operating frequency—1 GHz

First, timing simulation of digital circuits is performed based on the use of parasitic RLC parameters. As a result, a critical timing path is determined, as well as timing delays of all other digital circuit paths. The results of timing simulation of digital circuits are as follows:

```
+-----------------------------------------+
| Report            | report_timing       |
|------------------+----------------------|
| Options           | -from CK -to G117   |
+------------------+----------------------+
| Module            | s298                |
| Timing            | LATE                |
| Slew Propagation  | WORST               |
| PVT Mode          | max                 |
| Tree Type         | worst_case          |
| Process           | 1.00                |
| Voltage           | 1.08                |
| Temperature       | 125.00              |
| time unit         | 1.00 ns             |
| capacitance unit  | 1.00 pF             |
| resistance unit   | 1.00 kOhm           |
+-----------------------------------------+
Path 1: MET External Delay Assertion
Other End Arrival Time       0.00
- External Delay             0.00
+ Path Delay                 0.50
= Required Time              0.90
- Arrival Time               0.77
= Slack Time                 0.13
+-----------------------------------------+
| Report            | report_timing       |
|------------------+----------------------|
| Options           | -from CK -to G132   |
+------------------+----------------------+
| Module            | s298                |
| Timing            | LATE                |
| Slew Propagation  | WORST               |
| PVT Mode          | max                 |
| Tree Type         | worst_case          |
| Process           | 1.00                |
| Voltage           | 1.08                |
| Temperature       | 125.00              |
| time unit         | 1.00 ns             |
| capacitance unit  | 1.00 pF             |
| resistance unit   | 1.00 kOhm           |
+-----------------------------------------+
Path 1: MET External Delay Assertion
Other End Arrival Time       0.00
- External Delay             0.00
+ Path Delay                 0.50
= Required Time              0.90
- Arrival Time               0.82
= Slack Time                 0.08
+-----------------------------------------+
| Report            | report_timing       |
|------------------+----------------------|
| Options           | -from CK -to G66    |
+------------------+----------------------+
| Module            | s298                |
| Timing            | LATE                |
| Slew Propagation  | WORST               |
| PVT Mode          | max                 |
| Tree Type         | worst_case          |
| Process           | 1.00                |
| Voltage           | 1.08                |
| Temperature       | 125.00              |
| time unit         | 1.00 ns             |
| capacitance unit  | 1.00 pF             |
```

```
| resistance unit   | 1.00 kOhm              |
+-------------------------------------------+
Path 1: MET External Delay Assertion
Other End Arrival Time       0.00
- External Delay             0.00
+ Path Delay                 0.50
= Required Time              0.85
- Arrival Time               0.71
= Slack Time                 0.14
+-------------------------------------------+
| Report            | report_timing          |
|-------------------+-----------------------|
| Options           | -from CK -to G118      |
+-------------------+-----------------------+
| Module            | s298                   |
| Timing            | LATE                   |
| Slew Propagation  | WORST                  |
| PVT Mode          | max                    |
| Tree Type         | worst_case             |
| Process           | 1.00                   |
| Voltage           | 1.08                   |
| Temperature       | 125.00                 |
| time unit         | 1.00 ns                |
| capacitance unit  | 1.00 pF                |
| resistance unit   | 1.00 kOhm              |
+-------------------------------------------+
Path 1: MET External Delay Assertion
Other End Arrival Time       0.00
- External Delay             0.00
+ Path Delay                 0.50
= Required Time              0.85
- Arrival Time               0.75
= Slack Time                 0.10
+-------------------------------------------+
| Report            | report_timing          |
|-------------------+-----------------------|
| Options           | -from CK -to G133      |
+-------------------+-----------------------+
| Module            | s298                   |
| Timing            | LATE                   |
| Slew Propagation  | WORST                  |
| PVT Mode          | max                    |
| Tree Type         | worst_case             |
| Process           | 1.00                   |
| Voltage           | 1.08                   |
| Temperature       | 125.00                 |
| time unit         | 1.00 ns                |
| capacitance unit  | 1.00 pF                |
| resistance unit   | 1.00 kOhm              |
+-------------------------------------------+
Path 1: VIOLATED External Delay Assertion
Other End Arrival Time       0.00
- External Delay             0.00
+ Path Delay                 0.50
= Required Time              0.72
- Arrival Time               0.95
= Slack Time                -0.23
+-------------------------------------------+
| Report            | report_timing          |
|-------------------+-----------------------|
| Options           | -from CK -to G67       |
+-------------------+-----------------------+
| Module            | s298                   |
| Timing            | LATE                   |
| Slew Propagation  | WORST                  |
```

```
| PVT Mode          | max                  |
| Tree Type         | worst_case           |
| Process           | 1.00                 |
| Voltage           | 1.08                 |
| Temperature       | 125.00               |
| time unit         | 1.00 ns              |
| capacitance unit  | 1.00 pF              |
| resistance unit   | 1.00 kOhm            |
+----------------------------------------+
Path 1: MET External Delay Assertion
Other End Arrival Time      0.00
- External Delay            0.00
+ Path Delay                0.50
= Required Time             0.90
- Arrival Time              0.88
= Slack Time                0.02
+----------------------------------------+
| Report            | report_timing      |
|-------------------+--------------------|
| Options           | -from G0 -to G117  |
+-------------------+--------------------+
| Module            | s298               |
| Timing            | LATE               |
| Slew Propagation  | WORST              |
| PVT Mode          | max                |
| Tree Type         | worst_case         |
| Process           | 1.00               |
| Voltage           | 1.08               |
| Temperature       | 125.00             |
| time unit         | 1.00 ns            |
| capacitance unit  | 1.00 pF            |
| resistance unit   | 1.00 kOhm          |
+----------------------------------------+
Path 1: MET External Delay Assertion
Other End Arrival Time      0.00
- External Delay            0.00
+ Path Delay                0.50
= Required Time             1.00
- Arrival Time              0.92
= Slack Time                0.08
+----------------------------------------+
| Report            | report_timing      |
|-------------------+--------------------|
| Options           | -from G0 -to G132  |
+-------------------+--------------------+
| Module            | s298               |
| Timing            | LATE               |
| Slew Propagation  | WORST              |
| PVT Mode          | max                |
| Tree Type         | worst_case         |
| Process           | 1.00               |
| Voltage           | 1.08               |
| Temperature       | 125.00             |
| time unit         | 1.00 ns            |
| capacitance unit  | 1.00 pF            |
| resistance unit   | 1.00 kOhm          |
+----------------------------------------+
Path 1: MET External Delay Assertion
Other End Arrival Time      0.00
- External Delay            0.00
+ Path Delay                0.50
= Required Time             1.00
- Arrival Time              0.82
= Slack Time                0.18
```

```
+--------------------------------------------+
| Report            | report_timing          |
|-------------------+------------------------|
| Options           | -from G0 -to G66       |
+-------------------+------------------------+
| Module            | s298                   |
| Timing            | LATE                   |
| Slew Propagation  | WORST                  |
| PVT Mode          | max                    |
| Tree Type         | worst_case             |
| Process           | 1.00                   |
| Voltage           | 1.08                   |
| Temperature       | 125.00                 |
| time unit         | 1.00 ns                |
| capacitance unit  | 1.00 pF                |
| resistance unit   | 1.00 kOhm              |
+--------------------------------------------+
No constrained timing paths found.
Paths may be unconstrained or may not exist.

+--------------------------------------------+
| Report            | report_timing          |
|-------------------+------------------------|
| Options           | -from G0 -to G118      |
+-------------------+------------------------+
| Module            | s298                   |
| Timing            | LATE                   |
| Slew Propagation  | WORST                  |
| PVT Mode          | max                    |
| Tree Type         | worst_case             |
| Process           | 1.00                   |
| Voltage           | 1.08                   |
| Temperature       | 125.00                 |
| time unit         | 1.00 ns                |
| capacitance unit  | 1.00 pF                |
| resistance unit   | 1.00 kOhm              |
+--------------------------------------------+
No constrained timing paths found.
Paths may be unconstrained or may not exist.

+--------------------------------------------+
| Report            | report_timing          |
|-------------------+------------------------|
| Options           | -from G0 -to G133      |
+-------------------+------------------------+
| Module            | s298                   |
| Timing            | LATE                   |
| Slew Propagation  | WORST                  |
| PVT Mode          | max                    |
| Tree Type         | worst_case             |
| Process           | 1.00                   |
| Voltage           | 1.08                   |
| Temperature       | 125.00                 |
| time unit         | 1.00 ns                |
| capacitance unit  | 1.00 pF                |
| resistance unit   | 1.00 kOhm              |
+--------------------------------------------+
Path 1: MET External Delay Assertion
Other End Arrival Time        0.00
- External Delay              0.00
+ Path Delay                  0.50
= Required Time               1.00
- Arrival Time                0.88
= Slack Time                  0.12
```

```
+---------------------------------------+
| Report            | report_timing     |
|-------------------+-------------------|
| Options           | -from G0 -to G67  |
+-------------------+-------------------+
| Module            | s298              |
| Timing            | LATE              |
| Slew Propagation  | WORST             |
| PVT Mode          | max               |
| Tree Type         | worst_case        |
| Process           | 1.00              |
| Voltage           | 1.08              |
| Temperature       | 125.00            |
| time unit         | 1.00 ns           |
| capacitance unit  | 1.00 pF           |
| resistance unit   | 1.00 kOhm         |
+---------------------------------------+
No constrained timing paths found.
Paths may be unconstrained or may not exist.

+---------------------------------------+
| Report            | report_timing     |
|-------------------+-------------------|
| Options           | -from G1 -to G117 |
+-------------------+-------------------+
| Module            | s298              |
| Timing            | LATE              |
| Slew Propagation  | WORST             |
| PVT Mode          | max               |
| Tree Type         | worst_case        |
| Process           | 1.00              |
| Voltage           | 1.08              |
| Temperature       | 125.00            |
| time unit         | 1.00 ns           |
| capacitance unit  | 1.00 pF           |
| resistance unit   | 1.00 kOhm         |
+---------------------------------------+
Path 1: MET External Delay Assertion
Other End Arrival Time        0.00
- External Delay              0.00
+ Path Delay                  0.50
= Required Time               1.00
- Arrival Time                0.97
= Slack Time                  0.03

+---------------------------------------+
| Report            | report_timing     |
|-------------------+-------------------|
| Options           | -from G1 -to G132 |
+-------------------+-------------------+
| Module            | s298              |
| Timing            | LATE              |
| Slew Propagation  | WORST             |
| PVT Mode          | max               |
| Tree Type         | worst_case        |
| Process           | 1.00              |
| Voltage           | 1.08              |
| Temperature       | 125.00            |
| time unit         | 1.00 ns           |
| capacitance unit  | 1.00 pF           |
| resistance unit   | 1.00 kOhm         |
+---------------------------------------+
Path 1: MET External Delay Assertion
Other End Arrival Time        0.00
- External Delay              0.00
```

```
+ Path Delay              0.50
= Required Time           1.00
- Arrival Time            0.93
= Slack Time              0.07

+-----------------------------------------+
| Report            | report_timing       |
|-------------------+---------------------|
| Options           | -from G1 -to G66    |
+-------------------+---------------------+
| Module            | s298                |
| Timing            | LATE                |
| Slew Propagation  | WORST               |
| PVT Mode          | max                 |
| Tree Type         | worst_case          |
| Process           | 1.00                |
| Voltage           | 1.08                |
| Temperature       | 125.00              |
| time unit         | 1.00 ns             |
| capacitance unit  | 1.00 pF             |
| resistance unit   | 1.00 kOhm           |
+-----------------------------------------+
No constrained timing paths found.
Paths may be unconstrained or may not exist.

+-----------------------------------------+
| Report            | report_timing       |
|-------------------+---------------------|
| Options           | -from G1 -to G118   |
+-------------------+---------------------+
| Module            | s298                |
| Timing            | LATE                |
| Slew Propagation  | WORST               |
| PVT Mode          | max                 |
| Tree Type         | worst_case          |
| Process           | 1.00                |
| Voltage           | 1.08                |
| Temperature       | 125.00              |
| time unit         | 1.00 ns             |
| capacitance unit  | 1.00 pF             |
| resistance unit   | 1.00 kOhm           |
+-----------------------------------------+
No constrained timing paths found.
Paths may be unconstrained or may not exist.
+-----------------------------------------+
| Report            | report_timing       |
|-------------------+---------------------|
| Options           | -from G1 -to G133   |
+-------------------+---------------------+
| Module            | s298                |
| Timing            | LATE                |
| Slew Propagation  | WORST               |
| PVT Mode          | max                 |
| Tree Type         | worst_case          |
| Process           | 1.00                |
| Voltage           | 1.08                |
| Temperature       | 125.00              |
| time unit         | 1.00 ns             |
| capacitance unit  | 1.00 pF             |
| resistance unit   | 1.00 kOhm           |
+-----------------------------------------+
Path 1: MET External Delay Assertion
Other End Arrival Time    0.00
- External Delay          0.00
+ Path Delay              0.50
```

```
= Required Time              1.00
- Arrival Time               0.96
= Slack Time                 0.04

+-----------------------------------------+
| Report            | report_timing       |
|-------------------+---------------------|
| Options           | -from G1 -to G67    |
+-------------------+---------------------+
| Module            | s298                |
| Timing            | LATE                |
| Slew Propagation  | WORST               |
| PVT Mode          | max                 |
| Tree Type         | worst_case          |
| Process           | 1.00                |
| Voltage           | 1.08                |
| Temperature       | 125.00              |
| time unit         | 1.00 ns             |
| capacitance unit  | 1.00 pF             |
| resistance unit   | 1.00 kOhm           |
+-----------------------------------------+
No constrained timing paths found.
Paths may be unconstrained or may not exist.

+-----------------------------------------+
| Report            | report_timing       |
|-------------------+---------------------|
| Options           | -from G2 -to G117   |
+-------------------+---------------------+
| Module            | s298                |
| Timing            | LATE                |
| Slew Propagation  | WORST               |
| PVT Mode          | max                 |
| Tree Type         | worst_case          |
| Process           | 1.00                |
| Voltage           | 1.08                |
| Temperature       | 125.00              |
| time unit         | 1.00 ns             |
| capacitance unit  | 1.00 pF             |
| resistance unit   | 1.00 kOhm           |
+-----------------------------------------+
Path 1: MET External Delay Assertion
Other End Arrival Time       0.00
- External Delay             0.00
+ Path Delay                 0.50
= Required Time              1.00
- Arrival Time               0.90
= Slack Time                 0.10

+-----------------------------------------+
| Report            | report_timing       |
|-------------------+---------------------|
| Options           | -from G2 -to G132   |
+-------------------+---------------------+
| Module            | s298                |
| Timing            | LATE                |
| Slew Propagation  | WORST               |
| PVT Mode          | max                 |
| Tree Type         | worst_case          |
| Process           | 1.00                |
| Voltage           | 1.08                |
| Temperature       | 125.00              |
| time unit         | 1.00 ns             |
| capacitance unit  | 1.00 pF             |
| resistance unit   | 1.00 kOhm           |
+-----------------------------------------+
```

```
Path 1: MET External Delay Assertion
Other End Arrival Time       0.00
- External Delay             0.00
+ Path Delay                 0.50
= Required Time              1.00
- Arrival Time               0.89
= Slack Time                 0.11

+-------------------------------------------+
| Report             | report_timing        |
|--------------------+----------------------|
| Options            | -from G2 -to G66     |
+--------------------+----------------------+
| Module             | s298                 |
| Timing             | LATE                 |
| Slew Propagation   | WORST                |
| PVT Mode           | max                  |
| Tree Type          | worst_case           |
| Process            | 1.00                 |
| Voltage            | 1.08                 |
| Temperature        | 125.00               |
| time unit          | 1.00 ns              |
| capacitance unit   | 1.00 pF              |
| resistance unit    | 1.00 kOhm            |
+-------------------------------------------+
No constrained timing paths found.
Paths may be unconstrained or may not exist.

+-------------------------------------------+
| Report             | report_timing        |
|--------------------+----------------------|
| Options            | -from G2 -to G118    |
+--------------------+----------------------+
| Module             | s298                 |
| Timing             | LATE                 |
| Slew Propagation   | WORST                |
| PVT Mode           | max                  |
| Tree Type          | worst_case           |
| Process            | 1.00                 |
| Voltage            | 1.08                 |
| Temperature        | 125.00               |
| time unit          | 1.00 ns              |
| capacitance unit   | 1.00 pF              |
| resistance unit    | 1.00 kOhm            |
+-------------------------------------------+
No constrained timing paths found.
Paths may be unconstrained or may not exist.
+-------------------------------------------+
| Report             | report_timing        |
|--------------------+----------------------|
| Options            | -from G2 -to G133    |
+--------------------+----------------------+
| Module             | s298                 |
| Timing             | LATE                 |
| Slew Propagation   | WORST                |
| PVT Mode           | max                  |
| Tree Type          | worst_case           |
| Process            | 1.00                 |
| Voltage            | 1.08                 |
| Temperature        | 125.00               |
| time unit          | 1.00 ns              |
| capacitance unit   | 1.00 pF              |
| resistance unit    | 1.00 kOhm            |
+-------------------------------------------+
Path 1: MET External Delay Assertion
```

```
Other End Arrival Time      0.00
- External Delay            0.00
+ Path Delay                0.50
= Required Time             1.00
- Arrival Time              0.91
= Slack Time                0.09

+---------------------------------------+
| Report            | report_timing     |
|-------------------+-------------------|
| Options           | -from G2 -to G67  |
+-------------------+-------------------+
| Module            | s298              |
| Timing            | LATE              |
| Slew Propagation  | WORST             |
| PVT Mode          | max               |
| Tree Type         | worst_case        |
| Process           | 1.00              |
| Voltage           | 1.08              |
| Temperature       | 125.00            |
| time unit         | 1.00 ns           |
| capacitance unit  | 1.00 pF           |
| resistance unit   | 1.00 kOhm         |
+---------------------------------------+
No constrained timing paths found.
Paths may be unconstrained or may not exist.
```

The timing simulation is performed at a supply voltage of 1.08V and an operating temperature of 125 °C.

As seen from the above results, the critical path is between the input CK and the output G130. Then, logical groups are formed (Fig. 5.11). After that, initial placement and routing of an optimized digital circuit are performed. As a result of this procedure, logical groups and their elements are located close to each other. Therefore, the interconnect lengths, i.e. their parasitic parameters are shortened. Thus interconnect delays are reduced.

After the formation of logical groups, the digital circuit is simulated to assess interconnect delays. An example of a simulation task is shown below:

```
***Spice Model
*.prot
 .lib 'gl018.l' tt
*.unprot
**********************************************

***Netlist
 .inc 's298_cx.spf'

x1 G23 N76 N73 N113 N106 N69 G0 N87 N92 N108 G12 N96 N95 G13 G67 CK
N102 G117 N83 N65 G14 N107 G39 G11 G10 G132 N85 N109 N110 N68 G113
N104 N84 N103 N64 N112 N105 N70 G22 G125 N98 N97 G118 G86 G44 N77
N99 G1 G102 VDD GND G92 N66 N71 N101 N90 G98 N86 N79 N78 N91 N81 N82
N111 N94 N93 N74 N75 G107 G133 G2 G119 G56 G15 N100 N88 N72 N89 N80
G34 G30 G66 N67 s298
```

```
vG23 G23 0 0
vG23N76 23N76 0 0
vN73 N73 0 0
vN113 N113 0 0
vN106 N106 0 0
vN69 N69 0 0
vG0 G0 0 0
vN87 N87 0 0
vN92 N92 0 0
vN108 N108 0 0
vG12 G12 0 0
vN96 N96 0 0
vN95 N95 0 0
vG13 G13 0 0
vG67 G67 0 0
vCK CK 0 0
vN102 N102 0 0
vG117 G1170 0
vN83 N83 0 0
vN65 N65 0 0
vG14 G14 0 0
vN107 N107 0 0
vG39 G39 0 0
vG11 G11 0 0
vG10 G10 0 0
vG132 G132 0 0
vN85 N85 0 0
vN109 N109 0 0
vN110 N110 0 0
vN68 N68 0 0
vG113 G113 0 0
vN104 N104 0 0
vN84 N84 0 0
vN103 N103 0 0
vN64 N64 0 0
vN112 N112 0 0
vN105 N105 0 0
vN70 N70 0 0
vG22 G22 0 0
vG125 G125 0 0
vN98 N98 0 0
vN97 N97 0 0
vG118 G118 0 0
vG86 G86 0 0
vG44 G44 0 0
vN77 N77 0 0
vN99 N99 0 0
```

```
vG1 G1 0 0
vG102 G102 0 0
vGND GND 0 0
vG92 G92 0 0
vN66 N66 0 0
vN71 N71 0 0
vN101 N101 0 0
vN90 N90 0 0
vG98 G98 0 0
vN86 N86 0 0
vN79 N79 0 0
vN78 N78 0 0
vN91 N91 0 0
vN81 N81 0 0
vN82 N82 0 0
vN111 N111 0 0
vN94 N94 0 0
vN93 N93 0 0
vN74 N74 0 0
vN75 N75 0 0
vG107 G107 0 0
vG133 G133 0 0
vG2 G2 0 0
vG119 G119 0 0
vG56 G56 0 0
vG15 G15 0 0
vN100 N100 0 0
vN88 N88 0 0
vN72 N72 0 0
vN89 N89 0 0
vN80 N80 0 0
vG30 G30 0 0
vG34 G34 0 0
v66 66 0 0
vN67 N67 0 0

***Operating Voltage
.param vdd = 1.8

***Operating Temperature
.temp 25

***Supply Voltage
 vvdd vdd 0 dc vdd
```

```
 vvss vss 0 dc 0.0
*********************************************

.probe v(*)
*********************************************

***Options
 .option captab
 .option post probe
*********************************************

***Analysis
.tran 0.01n '10n'
*********************************************
.end
```

Calculated values of parasitic parameters of capacitances are shown below.

```
maximum nodal capacitance=   1.330E-12         on node      0:g92

    nodal  capacitance table

    node      =     cap       node      =     cap       node      =     cap
 +0:23        =   0.         0:66       =   0.         0:ck       =  23.7010f
 +0:g0        =  13.7235f 0:g1          =  15.5921f 0:g10         =  12.2435f
 +0:g102      =   5.0613f 0:g107        =   7.1256f 0:g11         =  10.6369f
 +0:g113      =   8.7903f 0:g117        =  25.7831f 0:g1170       =   0.
 +0:g118      =  11.9535f 0:g119        =   4.3587f 0:g12         =  17.6676f
 +0:g125      =   7.5478f 0:g13         =  41.1509f 0:g132        =  17.4351f
 +0:g133      =   6.7991f 0:g14         =  14.9227f 0:g15         =   7.2365f
 +0:g2        =   9.5621f 0:g22         =  11.5652f 0:g23         =   6.4724f
 +0:g30       =   7.2850f 0:g34         =   7.7360f 0:g39         =  23.8615f
 +0:g44       =   9.9309f 0:g56         =   8.1022f 0:g66         =   9.4053f
 +0:g67       =  20.2354f 0:g86         =  10.1459f 0:g92         =   1.3298p
 +0:g98       =   7.1811f 0:n100        =   6.0688f 0:n101        =  17.0512f
 +0:n102      =  20.0951f 0:n103        =   9.9542f 0:n104        =  16.5013f
 +0:n105      =   8.2695f 0:n106        =   6.1970f 0:n107        =  29.3109f
 +0:n108      =  28.9325f 0:n109        =   7.4307f 0:n110        =   6.6265f
 +0:n111      =   8.7692f 0:n112        =  12.1189f 0:n113        =   8.0761f
 +0:n64       =   8.0756f 0:n65         =  17.2165f 0:n66         =   1.0902p
 +0:n67       =   6.0994f 0:n68         =   8.0070f 0:n69         =  43.5871f
 +0:n70       =   6.2415f 0:n71         =   7.6807f 0:n72         =   6.4748f
 +0:n73       =   7.4135f 0:n74         =  14.1086f 0:n75         =   9.2806f
 +0:n76       =   6.1614f 0:n77         =   9.1492f 0:n78         =   4.9648f
 +0:n79       =   7.4991f 0:n80         =   8.0068f 0:n81         =   9.0756f
```

```
+0:n82      =    5.5445f 0:n83      =  47.3782f 0:n84      =  15.4917f
+0:n85      =    8.6848f 0:n86      =   6.1887f 0:n87      =  29.3073f
+0:n88      =    7.4984f 0:n89      =   6.0489f 0:n90      =   6.3180f
+0:n91      =    9.0002f 0:n92      =  38.6111f 0:n93      =   8.3806f
+0:n94      =   12.1799f 0:n95      =  45.3875f 0:n96      =  19.8638f
+0:n97      =   11.6616f 0:n98      =  20.2178f 0:n99      =  15.2218f
+0:vdd      =   12.8328f 0:vss      =   0.       1:xdff_13_= 444.3073a
+1:xdff_13_= 476.1981a 1:xdff_13_=   1.3812f 1:xdff_13_= 734.1624a
+1:xdff_13_= 728.2341a 1:xdff_13_=   3.5901f 1:xdff_12_=   6.5509f
+1:xdff_12_=   5.6447f 1:xdff_12_=   7.5417f 1:xdff_12_=   6.6327f
+1:xdff_12_=   7.7135f 1:xdff_12_=   7.6641f 1:xdff_12_= 574.8212a
+1:xdff_12_= 406.8594a 1:xdff_12_= 425.4619a 1:xdff_12_=   1.3707f
+1:xdff_12_= 742.6424a 1:xdff_12_= 731.0467a 1:xdff_12_=   3.4875f
+1:xdff_5_x=   6.3286f 1:xdff_5_x=   5.6258f 1:xdff_5_x=   7.4778f
+1:xdff_5_x=   6.5472f 1:xdff_5_x=   7.5011f 1:xdff_5_x=   7.6814f
+1:xdff_5_x= 576.7433a 1:xdff_5_x= 423.2218a 1:xdff_5_x= 455.0632a
+1:xdff_5_x=   1.4017f 1:xdff_5_x= 731.0431a 1:xdff_5_x= 754.8868a
+1:xdff_5_n=   3.4207f 1:xdff_4_x=   6.3078f 1:xdff_4_x=   5.7055f
+1:xdff_4_x=   7.7521f 1:xdff_4_x=   6.7316f 1:xdff_4_x=   7.3608f
+1:xdff_4_x=   7.5765f 1:xdff_4_x= 572.4132a 1:xdff_4_x= 413.6206a
+1:xdff_4_x= 430.7915a 1:xdff_4_x=   1.3890f 1:xdff_4_x= 747.5271a
+1:xdff_4_x= 730.4520a 1:xdff_4_n=   3.3966f 1:xdff_2_x=   6.2145f
+1:xdff_2_x=   5.4462f 1:xdff_2_x=   7.8449f 1:xdff_2_x=   6.6239f
+1:xdff_2_x=   6.9876f 1:xdff_2_x=   7.2526f 1:xdff_2_x= 567.6043a
+1:xdff_2_x= 413.2511a 1:xdff_2_x= 456.7109a 1:xdff_2_x=   1.3833f
+1:xdff_2_x= 747.7400a 1:xdff_2_x= 737.3520a 1:xdff_2_n=   3.4728f
+1:xdff_3_x=   6.2879f 1:xdff_3_x=   5.7705f 1:xdff_3_x=   7.6022f
+1:xdff_3_x=   6.4957f 1:xdff_3_x=   7.3841f 1:xdff_3_x=   7.5475f
+1:xdff_3_x= 574.8385a 1:xdff_3_x= 410.4162a 1:xdff_3_x= 441.8111a
+1:xdff_3_x=   1.3762f 1:xdff_3_x= 745.3833a 1:xdff_3_x= 738.5094a
+1:xdff_3_n=   3.3604f 1:xdff_1_x=   6.3283f 1:xdff_1_x=   5.7057f
+1:xdff_1_x=   7.4961f 1:xdff_1_x=   6.3883f 1:xdff_1_x=   7.3677f
+1:xdff_1_x=   7.4128f 1:xdff_1_x= 568.4790a 1:xdff_1_x= 413.6787a
+1:xdff_1_x= 424.5887a 1:xdff_1_x=   1.3815f 1:xdff_1_x= 745.7053a
+1:xdff_1_x= 736.1916a 1:xdff_1_n=   3.4567f 1:xdff_0_x=   6.4907f
+1:xdff_0_x=   5.6791f 1:xdff_0_x=   7.5781f 1:xdff_0_x=   6.5626f
+1:xdff_0_x=   7.4874f 1:xdff_0_x=   7.4659f 1:xdff_0_x= 564.9599a
+1:xdff_0_x= 408.8507a 1:xdff_0_x= 420.2224a 1:xdff_0_x=   1.3697f
+1:xdff_0_x= 749.4944a 1:xdff_0_x= 746.0386a 1:xdff_0_n=   3.3385f
+1:xdff_8_x=   9.4648f 1:xdff_8_x=   9.6312f 1:xdff_8_x=   9.4414f
+1:xdff_8_x=   9.6016f 1:xdff_8_x=  17.2355f 1:xdff_8_x=  15.9433f
```

```
+1:xdff_8_x= 405.8525a 1:xdff_8_x=   1.0581f 1:xdff_8_x= 447.4954a
+1:xdff_8_x= 733.1420a 1:xdff_8_x=   2.1086f 1:xdff_8_x= 745.2135a
+1:xdff_8_n=   9.0809f 1:xdff_11_=   9.5319f 1:xdff_11_=   9.6051f
+1:xdff_11_=   9.2683f 1:xdff_11_=   9.7017f 1:xdff_11_=  16.9004f
+1:xdff_11_=  15.9455f 1:xdff_11_= 420.4059a 1:xdff_11_=   1.0787f
+1:xdff_11_= 446.6612a 1:xdff_11_= 728.2970a 1:xdff_11_=   2.1108f
+1:xdff_11_= 731.2647a 1:xdff_11_=   8.9465f 1:xdff_7_x=   9.5440f
+1:xdff_7_x=   9.8699f 1:xdff_7_x=   9.1838f 1:xdff_7_x=   9.7602f
+1:xdff_7_x=  17.2792f 1:xdff_7_x=  15.6597f 1:xdff_7_x= 408.7364a
+1:xdff_7_x=   1.0816f 1:xdff_7_x= 458.0605a 1:xdff_7_x= 743.2035a
+1:xdff_7_x=   2.1282f 1:xdff_7_x= 744.2543a 1:xdff_7_n=   8.9688f
+1:xdff_9_x=   9.6623f 1:xdff_9_x=   9.8239f 1:xdff_9_x=   9.5390f
+1:xdff_9_x=  10.0281f 1:xdff_9_x=  16.8302f 1:xdff_9_x=  15.8328f
+1:xdff_9_x= 406.9811a 1:xdff_9_x=   1.9247f 1:xdff_9_x= 448.7892a
+1:xdff_9_x= 728.8787a 1:xdff_9_x=   2.1331f 1:xdff_9_x= 729.6178a
+1:xdff_9_n=   8.9719f 1:xdff_6_x=   9.7672f 1:xdff_6_x=  10.0501f
+1:xdff_6_x=   9.2635f 1:xdff_6_x=   9.9972f 1:xdff_6_x=  16.8728f
+1:xdff_6_x=  15.5688f 1:xdff_6_x= 403.8139a 1:xdff_6_x=   1.0883f
+1:xdff_6_x= 488.7444a 1:xdff_6_x= 736.3448a 1:xdff_6_x=   2.1327f
+1:xdff_6_x= 728.3847a 1:xdff_6_n=   8.8974f 1:xdff_10_=   9.5236f
+1:xdff_10_=  10.0520f 1:xdff_10_=   9.3683f 1:xdff_10_=   9.9071f
+1:xdff_10_=  17.0161f 1:xdff_10_=  15.8827f 1:xdff_10_= 402.0408a
+1:xdff_10_=   1.0644f 1:xdff_10_= 445.2030a 1:xdff_10_= 751.6647a
+1:xdff_10_=   2.1117f 1:xdff_10_= 736.4096a 1:xdff_10_=   9.0535f
+1:xdff_13_=   6.6458f 1:xdff_13_=   5.5007f 1:xdff_13_=   7.5692f
+1:xdff_13_=   7.0386f 1:xdff_13_=   7.1403f 1:xdff_13_=   7.7832f
+1:xdff_13_= 579.5513a 1:xu129_ne=   6.3512f 1:xu129_ne=   3.8539f
+1:xu130_ne=   6.5356f 1:xu130_ne=   1.4650f 1:xu130_ne= 851.5045a
+1:xu131_ne=   6.3505f 1:xu131_ne=   3.8403f 1:xu133_ne=   5.9142f
+1:xu134_ne=   1.4081f 1:xu134_ne=   5.7237f 1:xu135_ne=   6.4838f
+1:xu135_ne=   1.4898f 1:xu135_ne= 873.7689a 1:xu136_ne=   6.4832f
+1:xu136_ne=   1.4821f 1:xu136_ne= 845.1670a 1:xu137_ne= 989.6650a
+1:xu137_ne=   8.0192f 1:xu137_ne=   1.0162f 1:xu137_ne=   1.0619f
+1:xu139_ne=   3.2289f 1:xu139_ne=   3.2351f 1:xu139_ne=   3.8345f
+1:xu140_ne=   6.7174f 1:xu140_ne=   1.4839f 1:xu140_ne= 861.8236a
+1:xu141_ne=   6.6766f 1:xu141_ne=   1.4879f 1:xu141_ne= 862.4577a
+1:xu142_ne=   6.4655f 1:xu142_ne=   3.6692f 1:xu143_ne=   1.2481f
+1:xu143_ne=   3.0626f 1:xu143_ne=   6.1041f 1:xu144_ne=   1.2480f
+1:xu144_ne=   3.0592f 1:xu144_ne=   5.9112f 1:xu147_ne= 994.2204a
+1:xu147_ne=   8.1267f 1:xu147_ne=   1.0112f 1:xu147_ne=   1.0528f
+1:xu148_ne=   6.7722f 1:xu148_ne=   1.4662f 1:xu148_ne= 846.9596a
+1:xu149_ne=   1.7578f 1:xu149_ne=   5.9573f 1:xu149_ne=   5.7809f
+1:xu149_ne=   1.4933f 1:xu151_ne=   2.6939f 1:xu152_ne=   6.4712f
+1:xu152_ne=   3.7534f 1:xu153_ne=   2.9071f 1:xu153_ne=   3.1439f
+1:xu155_ne=   6.2266f 1:xu155_ne=   3.7803f 1:xu156_ne=   2.6693f
+1:xu158_ne=   2.9219f 1:xu158_ne=   3.1206f 1:xu159_ne=   4.7093f
```

```
+1:xu159_ne=   8.1244f 1:xu159_ne=   3.1164f 1:xu159_ne=   6.0395f
+1:xu160_ne=   2.6843f 1:xu161_ne=   4.4332f 1:xu161_ne=   7.9104f
+1:xu161_ne=   3.0967f 1:xu161_ne=   5.9481f 1:xu168_ne=   1.0051f
+1:xu168_ne=   7.9042f 1:xu168_ne=   1.0253f 1:xu168_ne=   1.0346f
+1:xu171_ne=   6.3257f 1:xu171_ne=   3.5428f 1:xu172_ne=   2.6928f
+1:xu173_ne=   4.5670f 1:xu173_ne=   8.3460f 1:xu173_ne=   3.1849f
+1:xu173_ne=   6.0244f 1:xu174_ne=   2.6905f 1:xu176_ne=   2.9323f
+1:xu176_ne=   3.1325f 1:xu177_ne=   1.4457f 1:xu177_ne=   5.0099f
+1:xu178_ne=   4.8384f 1:xu178_ne=   1.4892f 1:xu179_ne=   4.8953f
+1:xu179_ne=   1.5009f 1:xu180_ne=   2.0974f 1:xu180_ne=   6.5605f
+1:xu180_ne=   2.0923f 1:xu180_ne=   2.1503f 1:xu181_ne=   4.7763f
+1:xu181_ne=   1.4990f 1:xu182_ne=   2.2448f 1:xu182_ne=   5.6208f
+1:xu182_ne=   2.1847f 1:xu183_ne=   4.9011f 1:xu183_ne=   1.5059f
+1:xu184_ne=   4.9222f 1:xu184_ne=   1.4945f 1:xu185_ne=   2.2174f
+1:xu185_ne=   5.5735f 1:xu185_ne=   2.1653f 1:xu186_ne=   4.7110f
+1:xu186_ne=   1.4908f 1:xu187_ne=   2.1975f 1:xu187_ne=   5.5371f
+1:xu187_ne=   2.1888f 1:xu188_ne=   4.8504f 1:xu188_ne=   1.4915f
+1:xu189_ne=   4.8466f 1:xu189_ne=   1.4952f 1:xu190_ne=   4.9710f
+1:xu190_ne=   1.4878f 1:xu191_ne=   4.8841f 1:xu191_ne=   1.4990f
```

Then, for each logical group, a lot of possible solutions are generated. For each solution, the timing delay and the power consumption are calculated using the timing parameters of the library cells, as well as the calculated interconnect capacitances.

The critical path to (a) and after (b) of the optimization step is given in Fig. 5.12. After the optimization stage, the final placement and routing of digital circuits is performed (Fig. 5.13.)

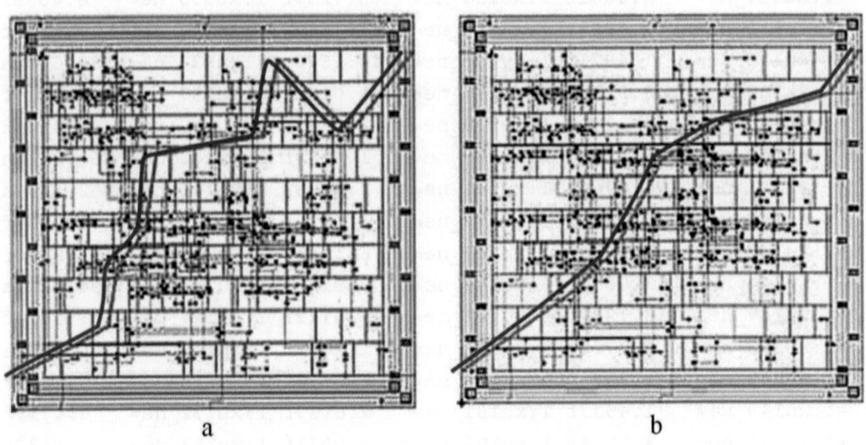

Fig. 5.12 Critical path of a digital circuit to (**a**) and after (**b**) optimization stage

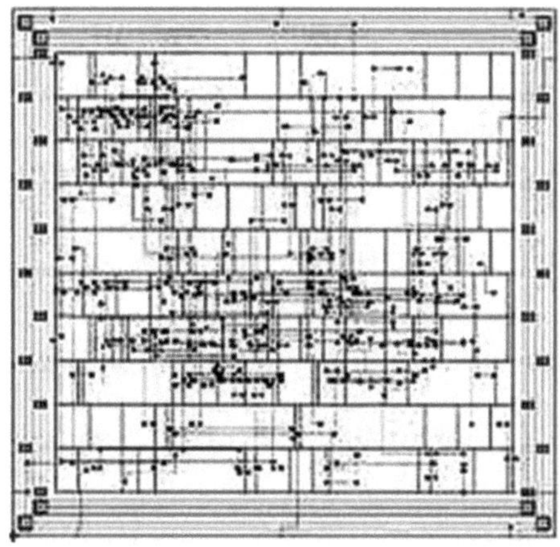

Fig. 5.13 Topology of a digital circuit s289 after optimization

Prior to optimization, the power consumption of a digital circuit s289 was 94.87 mcWt, and after optimization—102.45 mcWt.

5.4 Optimization of Power Consumption of Digital Circuits with Consideration of DF

At present, the problem of optimizing the power consumption (OP) of digital circuits has become particularly relevant in connection with the expansion of the market for portable electronic devices, increasing the cost of special cooling subsystems for IC and their heat-resistant packages, etc.

The power consumption of digital circuit technology produced by CMOS is determined as follows [67]:

$$P_n = 0.5\ CU_n^2 fN + Q_ç U_n fN + I_y U_n, \tag{5.27}$$

where P_n is the total power consumed by digital circuits; U_n is the value of the supply voltage; f—value of the clock frequency of digital circuit operation. The first term in (5.27) is the power of charge and discharge of capacitances C of digital circuits. Switching activity N is the number of switching outputs of digital circuits per clock cycle. The second term is the power of a short circuit. $Q_ç$ is the magnitude of the electric charge carried by the short-circuit current for one switching. The third term is the static power consumed by the leakage current I_c. The first term, called the switching power P_{con}, makes more than 90% of the total power consumed by digital circuits. For this reason, in most papers [324, 325] devoted to the OP of digital circuits, the question of optimizing P_n is considered.

At the logical level of digital IC design, the following OP methods of digital circuits are known [326]: the predicted calculation of logical values at the output of digital circuits per clock before they are required, and the subsequent use of these

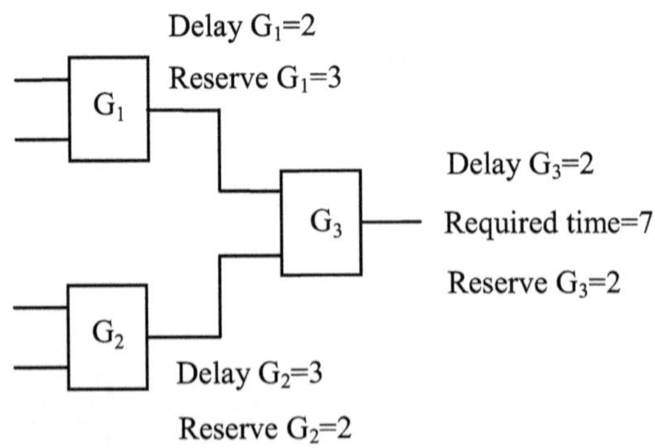

Fig. 5.14 Timing reserve of the gate

values during the next clock; inclusion of additional flip-flops at the outputs of nodes with a high level of useless switching and a large load capacitance in digital circuits; balancing of the values of delays of all true paths in digital circuits to exclude useless switching of gates, etc.

The calibration method is attractive as it does not change the topology of optimized digital circuit and at the same time allows to get a digital circuit of lower power consumption and size than before optimization.

In [324], an algorithm is proposed for solving the problem of an OP of digital circuits using the calibration method. The application of this method allowed to reduce the power consumption of digital circuit ISCAS'89 test series by an average of 15.52%. The algorithm consists of two stages. At the first stage, the time reserve of the gate delay in digital circuits is determined, i.e. the value of the permissible increase in the delay of each separate gate, at which the total delay of digital circuits does not exceed the permissible limit (Fig. 5.14).

From Fig. 5.1 it follows that the values of the time reserves of G_1G_2 and G_3 are equal to 3, 2 and 2, respectively, at the required time of appearance of the signal at the output of G_3, equal to 7.

At the second stage, the algorithm replaces certain gate, whose time reserve is greater than zero, for equivalent but less fast gate of smaller size and therefore consuming less power. At the same time, the decision on which gate with a positive time reserve are subject to calibration is important, since the maximum gain in the power consumption of the entire digital circuit depends on this.

In the heuristic algorithm proposed in [325], the digital circuit is optimized, beginning with the gates, which are closer to the main outputs of digital circuits, and further moves deeper into the digital circuit inputs. Once the gate meets the positive value of the time reserve, it is replaced by an equivalent gate from the library of digital cells of a smaller size. After the gate is replaced, information on the amount of its delay is updated, and the value of the time reserve of those gates whose outputs are connected to the inputs of this one is recalculated. The optimization process is completed if there are no more gates left in digital circuits, the calibration of which would not violate the maximum total delay limit of digital circuits.

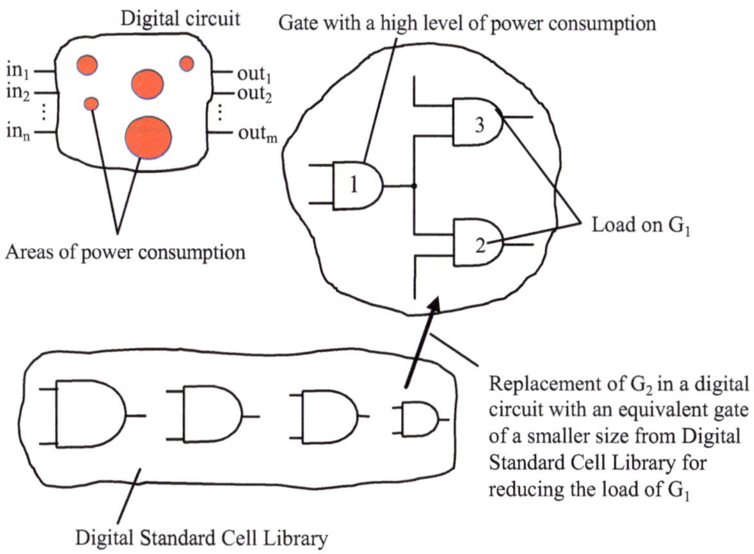

Fig. 5.15 Principle of optimizing power consumption by gate calibration method

An analysis of numerous works devoted to the OP of digital circuits shows that at present the most effective is the calibration of the gate. Due to this, it is possible to reduce the power consumption level of digital circuits by an average of 15% . . . 20% in comparison with the initial one. The essence of the method of optimization of power consumption by calibration of the gate is shown in Fig. 5.15. In digital circuit technology produced by the CMOS, depending on the intensity of switching and the value of the load capacitance, different gate units consume different amount of power. Fig. 5.15 shows the "areas" of power consumption, indicated by dark circles and represent the locations of gate, consuming significant power. In this case, the larger the diameter of the circle, the more power is consumed by the gate.

As seen from Fig. 5.15, G_1 consumes the greatest power. Reduction of the level of power consumption of digital circuits can be carried out by "dampening" this area by reducing the value of the load capacitance at the output of G_1. This can be achieved by replacing the gate, connected to the output of G_1 (for example, G_2), by equivalent gate of smaller physical size from the digital cell library. This process is also called gate calibration.

However, the gain in the power consumption obtained from the calibration is accompanied by a decrease in performance, because as a result of the replacement of the G_2 with its equivalent of a smaller size, the delay of the G_2 and, consequently, the delay of all signal paths in digital cells that pass through the G_2 increase. It is therefore important, after calibrating the gate, to simulate the operation of a digital circuit to verify the implementation of two criteria for maintaining the accuracy of its operation:

1. Preservation of timing balance of signaling paths in digital circuits. This means that the reference mutual switching sequence of the digital circuit outputs,

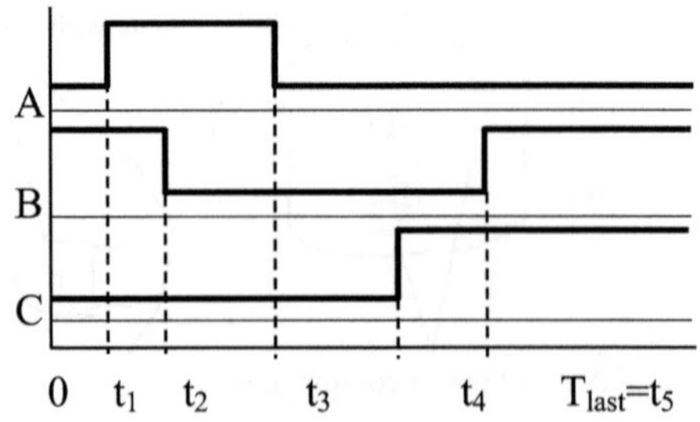

Fig. 5.16 The order of switching outputs A, B, C of digital circuits before the calibration of the gate

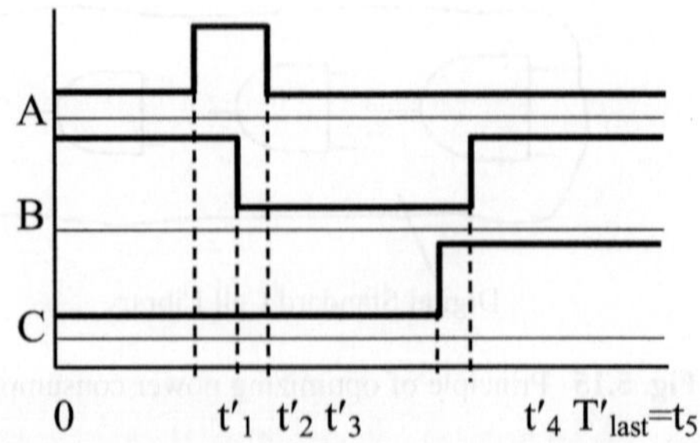

Fig. 5.17 The order of switching outputs A, B, C of digital circuits after the calibration of the gate

observed before the start of the optimization process, should remain unchanged after the calibration of the gate.

2. The time interval during which the digital circuit manages to process the input signals, should not increase. This interval is equal to the time interval from the moment of the arrival of the very first switching of the signal to the digital circuit inputs up to the moment of the last switching of the signal at its outputs.

The above criteria are explained in Figs. 5.16 and 5.17. They show, as an example, the diagrams of the output signals A, B and C of digital circuits before and after the calibration of the gate. As it can be seen, the first criterion for the correct functioning of a digital circuit is fulfilled, since the mutual order of switching of outputs A, B and C remained unchanged after the calibration of the gate. A second criterion is also fulfilled, since the time T_{last} from the last switching of the signal before calibration is equal to the moment T'_{last} after the last switching of the signal at digital circuit outputs after calibration.

The power consumption of digital circuits is reduced by consecutive "muting" of power consumption centers. The optimization process ends if the calibration of any gate leads to a violation of at least one of the specified criteria for the correct functioning of a digital circuit.

Obtaining a positive result of the OP of digital circuits is real, since there is reason to believe that a digital circuit, as a rule, has a certain amount of performance. It is due to the use of this stock that it is possible to reduce the power consumption of digital circuits while maintaining its correct functioning.

However, in all works, the calibration of the gate is carried out without taking into account the influence of the DF. This means that the real parasitic parameters of interconnects, power circuits, etc., are not taken into account. Since the necessity to take into account the influence of DF does not raise any doubts (Sect. 1.1), it became necessary to develop a calibration method to take into account the influence of DF.

Below the developed OP algorithm (OPA) of a digital circuit with consideration of DF [131, 250, 251] is described.

The goal is to minimize the value of power consumption of digital circuits for given reference diagrams of input and output signals:

$$P_{\text{con}} \to \min. \tag{5.28}$$

Power consumption P_{con} is defined as the sum of the consumed powers of all G_i belonging to a digital circuit:

$$P_{\text{con}} = \sum_{i=1}^{g} P_i, \tag{5.29}$$

where g –the number of gates in digital circuits.

In turn, the power consumption P_i of each G_i is determined as follows:

$$P_i = 0.5C_i \cdot U_n^2 \cdot F_i, \tag{5.30}$$

where U_n is the value of the supply voltage; F_i—the average frequency of switching outputs G_i; C_i—load capacitance of G_i:

$$F_i = \frac{q_i}{T_{\text{con}}}, \tag{5.31}$$

where T_{con} is the duration of transient processes of digital circuits, and q_i is the number of switching operations at the output of the G_i in the time T_{con}.

C_i is a set of interconnect capacitances and input capacitances G_j, $_{j=1,2,\ldots,k}$, connected to the output G_i. Similar to (5.5):

$$C_i = \sum_{j=1}^{k} \left(\frac{\varepsilon w_{Mj} l_{Mj}}{t_{\text{ox}}} + C_{\text{in}j} \right), \tag{5.32}$$

where $\varepsilon w_{Mj} l_{Mj}/t_{\text{ox}}$—capacitance of the jth interconnect connected to the output G_i with length l_{Mj} and width w_{Mj}; $C_{\text{in}j}$—input capacitance of G_j connected to the output of the G_i; k—the number of gates connected to the output of G_i.

Taking into account (5.4), (5.8), the constants U_n, T_{con} and applying the corresponding transformations, the following objective function of the OP of digital circuits problem is obtained:

$$\sum_{i=1}^{g} q_i \sum_{j=1}^{k} \left(\frac{\varepsilon w_{Mj} l_{Mj}}{t_{\text{ox}}} + \{\Delta t_{bj} = \text{PDE}_j[\text{MDE}_{j\gamma}(w_\gamma), \text{MDE}_{j\eta}(v_\eta)]\} \cdot P_j \right) \to \min. \tag{5.33}$$

Restrictions are:

1. $$T'_{\text{con}}\left(\Delta t'_{\min_i}, \Delta t'_{b_i}, \Delta t'_{\max_i}, \Delta t'_{s_i}\right) \le T_{\text{con}}(\Delta t_{\min_i}, \Delta t_{b_i}, \Delta t_{\max_i}, \Delta t_{s_i})_{i=1,2,\ldots,g}. \tag{5.34}$$

2. $$t'_x\left(\Delta t'_{\min_i}, \Delta t'_{b_i}, \Delta t'_{\max_i}, \Delta t'_{s_i}\right) < t'_{x+1}\left(\Delta t'_{\min_i}, \Delta t'_{b_i}, \Delta t'_{\max_i}, \Delta t'_{s_i}\right),$$
$$\text{if } t_x(\Delta t_{\min_i}, \Delta t_{b_i}, \Delta t_{\max_i}, \Delta t_{s_i}) < t_{x+1}(\Delta t_{\min_i}, \Delta t_{b_i}, \Delta t_{\max_i}, \Delta t_{s_i}) \tag{5.35}$$

for all $x = 1, 2, \ldots, r$, where $T'_{\text{con}}\left(\Delta t'_{\min_i}, \Delta t'_{b_i}, \Delta t'_{\max_i}, \Delta t'_{s_i}\right)$—the last switching time at digital circuit outputs after calibration, i.e. replacing some G_i of digital circuits with equivalent G_i from digital cell library; $T_{\text{con}}(\Delta t_{\min_i}, \Delta t_{b_i}, \Delta t_{\max_i}, \Delta t_{s_i})$—the same value before calibration; $t'_x\left(\Delta t'_{\min_i}, \Delta t'_{b_i}, \Delta t'_{\max_i}, \Delta t'_{s_i}\right)$—time of the xth switching at digital circuit outputs after calibration, and $t_x(\Delta t_{\min_i}, \Delta t_{b_i}, \Delta t_{\max_i}, \Delta t_{s_i})$—before calibration; r—number of switchings on the outputs of digital circuits. It is clear that since the timing parameters of the MCE (Sect. 2.2) are functions of the DF, then, T'_{last}, T_{last}, t'_x and t_x also are functions of the DF. Values, T'_{last}, T_{last}, t'_x and t_x are determined by gate-level simulation with consideration of DF (Chap. 2 and Sect. 4.1).

A key feature of the developed OPA of digital circuits is the method of selecting the "best" gate in terms of achieving reduction in the level of power consumption of the digital circuits as a result of its calibration in each particular situation. To this end, the reserves G_i, $_{i=1,2,\ldots,g}$, are determined, which are arranged in descending order and the calibration is performed in this order, as long as the constraints (5.34) and (5.35) are satisfied. The determination of the switching activity of each gate of digital circuits F_i (5.30) is performed as a result of digital circuit simulation taking into account the DF.

For each switching of the outputs of digital circuits, the corresponding path of the input signal that triggers this switching is constructed, which starts from the main input of a digital circuit and ends with the main output of a digital circuit where the switching occurred. This path, called the gate net, consists of gates and interconnects connecting these gates (Fig. 5.18). For each net, the time at which the corresponding input signal is input to its input and the time at which the given input signal reaches its output is fixed. Information about the direction of its switching ("1" "0" or "0" "1") is fixed for each net of the gate.

Below is a detailed description of the developed OPA of digital circuits considering DF.

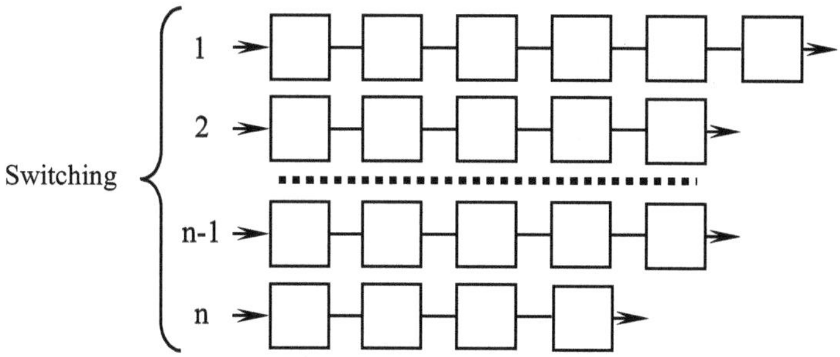

Fig. 5.18 Gate nets

The following notations are introduced:

- g—total number of gates in digital circuits;
- e_i—gates in digital circuits with the index i, $1 \le i \le g$;
- e_i^s—gate e_i, where s is the physical size identifier;
- $E^{\text{in}}(e_i)$—the set of gates in digital circuits whose outputs are connected to the inputs of LE e_i;
- $E^{\text{out}}(e_i)$—the set of gates in digital circuits, to the inputs of which the gate output e_i is connected;
- ΔC_i^{in}—value of decreasing the total value of the switched capacitance at the inputs of the gate e_i;
- ΔC_i^{out}—load capacitance at the output of the gate e_i;
- $\Delta t_{\min}^{01}\left(e_i^s, C_i^{\text{out}}\right)$, $\Delta t_{\text{b}}^{01}\left(e_i^s, C_i^{\text{out}}\right)$, $\Delta t_{\max}^{01}\left(e_i^s, C_i^{\text{out}}\right)$, $\Delta t_{\text{s}}^{01}\left(e_i^s, C_i^{\text{out}}\right)$, $\Delta t_{\min}^{10}\left(e_i^s, C_i^{\text{out}}\right)$, $\Delta t_{\text{b}}^{10}\left(e_i^s, C_i^{\text{out}}\right)$, $\Delta t_{\max}^{10}\left(e_i^s, C_i^{\text{out}}\right)$, $\Delta t_{\text{s}}^{10}\left(e_i^s, C_i^{\text{out}}\right)$—parameters of the MCE e_i^s from digital circuits with an output load capacitance equal to C_i^{out};
- SN_q—the gate net corresponding to switching q on the outputs of digital circuits, $1 \le q \le x$, x—the total number of switches on the outputs of digital circuits;
- $T_{\text{in}}^{\text{SN},q}$—the moment of getting input signal, generating switching q at the output of digital circuits;
- $T_{\text{out}}^{\text{SN},q}$—switching time q at the output of digital circuits;
- $\Delta t_{\text{SN},\, q}$—the value of the total delay of the chain net SN_q;
- λ—the reference mutual switching order at the outputs of digital circuits (5.35);
- λ'—the mutual sequence of switching at digital circuit outputs, obtained as a result of simulation of the operation of digital circuits after the calibration of the gate;
- T_{last}—the moment of the last switching on the outputs of digital circuits before the optimization process begins;
- T'_{last}—the moment of the last switching on the outputs of digital circuits, obtained as a result of simulation of the operation of digital circuits after the calibration of the gate;

– X—is the set of gate, the calibration attempt of which led to violation of constraints (5.34) or (5.35).

step 1:	The choice of gate e_j^s in a digital circuit ($1 \le j \le g$), for which $\Delta C_j^{\text{in}} \ge \Delta C_i^{\text{in}}$ for $\forall i \in 1 \ldots g$, $e_j \notin X$ and $\Delta C_j^{\text{in}} > 0$.
step 2:	Calculation C_j^{out} for e_j^s.
step 3:	Calculation
	$\Delta t_{\text{min}}^{01}\left(e_j^{s-1}, C_j^{\text{out}}\right) = \text{MDE}_{\text{min01}}\left(s-1, C_j^{\text{out}}\right)$, $\Delta t_{\text{b}}^{01}\left(e_j^{s-1}, C_j^{\text{out}}\right) = \text{MDE}_{n01}\left(s-1, C_j^{\text{out}}\right)$, $\Delta t_{\text{max}}^{01}\left(e_j^{s-1}, C_j^{\text{out}}\right) = \text{MDE}_{\text{max01}}\left(s-1, C_j^{\text{out}}\right)$, $\Delta t_{\text{s}}^{01}\left(e_j^{s-1}, C_j^{\text{out}}\right) = \text{MDE}_{c01}\left(s-1, C_j^{\text{out}}\right)$, $\Delta t_{\text{min}}^{10}\left(e_j^{s-1}, C_j^{\text{out}}\right) = \text{MDE}_{\text{min01}}\left(s-1, C_j^{\text{out}}\right)$, $\Delta t_{\text{b}}^{10}\left(e_j^{s-1}, C_j^{\text{out}}\right) = \text{MDE}_{n01}\left(s-1, C_j^{\text{out}}\right)$, $\Delta t_{\text{max}}^{10}\left(e_j^{s-1}, C_j^{\text{out}}\right) = \text{MDE}_{\text{max01}}\left(s-1, C_j^{\text{out}}\right)$, $\Delta t_{\text{s}}^{10}\left(e_j^{s-1}, C_j^{\text{out}}\right) = \text{MDE}_{c01}\left(s-1, C_j^{\text{out}}\right)$.
step 4:	Calculation C_j^{out} for $e_i \in E^{\text{out}}(e_j)$.
step 5:	Calculation $\Delta t_{\text{min}}^{01}$, Δt_{b}^{01}, $\Delta t_{\text{max}}^{01}$, Δt_{s}^{01}, $\Delta t_{\text{min}}^{10}$, Δt_{b}^{10}, $\Delta t_{\text{max}}^{10}$, Δt_{s}^{10} for all $e_i \in E^{\text{in}}(e_j)$.
step 6:	Allocation of a subset U of all nets of SN_q containing any of the following gates: e_j, $e_i \in E^{\text{in}}(e_j)$.
step 7:	Substitution of the information on the delay of the gate e_j^{s-1} and all $e_i \in E^{\text{in}}(e_j)$ into the corresponding chains $\text{SN}_q \in U$.
step 8:	Calculation of the value of $\Delta t_{\text{SN},q}$, for all $\text{SN}_q \in U$.
step 9:	For all $\text{SN}_q \in U$ calculation of the value $T_{\text{out}}^{\text{SN},q} = T_{\text{in}}^{\text{SN},q} + \Delta t_{\text{SN},q}$.
step 10:	The arrangement of all the instants of time, $T_{\text{out}}^{\text{SN},q}$, $q \in 1, \ldots, x$ in ascending order
step 11:	The designation of the obtained mutual order of switching outputs of digital circuits λ'.
step 12:	Verification of the condition $\lambda = \lambda'$.
step 13:	On coincidence, go to step 17.
step 14:	The inclusion of the gate e_j in the set X: $e_j \subset X$.
step 15:	Choice of the gate e_b in a digital circuit ($1 \le b \le g$), for which $\Delta C_b^{\text{in}} \ge \Delta C_i^{\text{in}}$ for $\forall$ $i \in 1 \ldots g$, $e_b \notin X$ an $\Delta C_b^{\text{in}} > 0$.
step 16:	Go to step 2.
step 17:	Check the condition $T_{\text{out}}^{\text{SN},q} \le T_{\text{last}}$ for $\forall$ $q \in 1 \ldots x$.
step 18:	If the condition is not fulfilled, go to step 14.
step 19:	Performing the calibration of the gate in a digital circuit: replacing e_j^s with e_j^{s-1}.
step 20:	Gate-level simulation of a digital circuit with consideration of DF.
step 21:	Checking the condition $\lambda' = \lambda$
step 22:	If the condition is met, go to step 25.

(continued)

step 23:	Performing the reverse calibration of the gate in a digital circuit: replacing e_j^{s-1} back to the e_j^s.
step 24:	Go to step 14.
step 25:	Checking the restrictions (5.34), (5.35).
step 26:	If the condition is not met, go to step 23.
step 27:	Fixing the calibration in a digital circuit.
step 28:	Verification of the presence of gate e_b in a digital circuit ($1 \leq b \leq g$), for which $e_b \notin X$.
step 29:	If the condition is not fulfilled, go to step 1.
step 30:	Completion of the optimization process.

To assess the effectiveness of the proposed OPA of digital circuits as criteria, the following are used: (a) a relative decrease in the level of power consumption; (b) the computer time spent on obtaining the result of optimizing the power consumption of the IC.

The total operating time of the OPA of a digital circuit is represented in the following form:

$$T_{\text{tot}} = T_1 + T_2 + T_3 = T_{\text{prep}} + n^* \left(T_{\text{ext}} + T_{\text{par}} + T_{\text{net}} \right) + m^* T_{\text{sim}}, \tag{5.36}$$

where T_{prep}—the value of time interval of the "preparatory" simulation of digital circuit operation before the optimization process begins.

The goal of simulation is to obtain information on the reference mutual order of digital circuit outputs (5.35), to determine the value of T_{last} and the switching activity of the gate F_i, to construct the gate nets (Fig. 5.18), to locate all the gates in the descending order of the values of the total value of the switched capacitance at their inputs. The number of iterations of the "preparatory" simulation during the OPA of digital circuit operation is equal to 1. T_{ext}—the amount of time required to extract information about the load capacitance at the output of the gate; n is the number of iterations of this action during the algorithm operation. T_{par} is the value of the time interval required to calculate new values of timing parameters of the MCE after calibration; n is the number of iterations of this action during the OPA of digital circuit operation. T_{net} is the amount of time required for preliminary verification of compliance with constraints using gate nets. This is necessary when considering the calibration of gates; n is the number of iterations during the OPA of digital circuit operation. T_{sim}—the amount of time required to test the work of a digital circuit after calibrating one gate; m is the number of iterations of this action during the OPA of digital circuit operation ($m < n$). The value of m is equal to the number of gates, the verification of the possibility of calibrating them with the help of gate nets gave a positive result.

Note that the maximum weight in the expression (5.36) falls on the term T_3. However, since the verification of compliance with constraints (5.34) and (5.35) is carried out in simulation process of digital circuit operation, and not after its

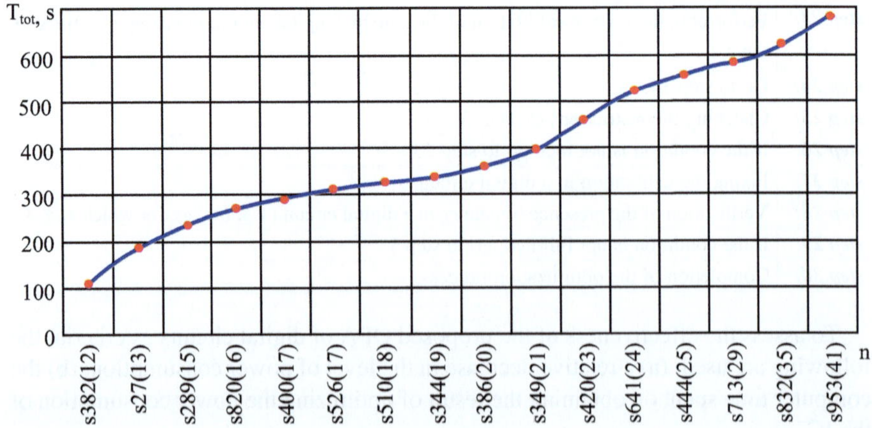

Fig. 5.19 Dependence of the total operating time of the OPA on the number of gates in a digital circuit

Table 5.3 Results of the OP of a digital circuit from the ISCAS89 series

Circuit	Number of gates	ΔP_1 (%)	ΔP_2 (%)	ΔP_3 (%)	ΔP_4 (%)
s27	10	15.3	11.3	7.8	22.9
s298	119	12.0	17.1	14.6	12.5
s344	160	16.2	18.9	20.4	24.1
s349	161	15.8	5.6	10.4	9.6
s382	158	20.3	27.0	21.6	19.4
s386	159	5.3	3.1	8.0	6.0
s400	162	7.9	7.5	18.1	16.0
s420	196	24.4	28.3	28.1	25.2
s444	181	19.1	15.4	3.9	22.6
s510	211	21.5	16.3	15.4	12.2
s526	193	22.4	16.5	15.1	12.4
s641	379	19.5	17.3	15.2	24.3
s713	393	24.2	16.8	16.2	26.1
s820	289	22.1	19.3	18.2	28.8
s832	287	18.9	26.4	25.1	29.4
s953	395	23.7	18.3	18.4	29.5

termination, the real weight of this term is usually much smaller than the value $m \cdot T_{\text{sim}}$. This fact becomes more important as the number of gates in a digital circuit increases.

Experimentally, the dependence of T_{tot}, on the number of gates (m) was obtained (Fig. 5.19).

After the approximation shown in Fig. 5.19 the following dependence is obtained:

$$t_{\mathrm{M}} = -0.6346n^2 + 53.529n - 448. \quad (5.37)$$

Comparing the costs of computer time with [324], an average loss of 7% ... 9% is obtained, but the gain in power consumption exceeds 12% ... 15%. This is confirmed by the results of the OP of digital circuits from the ISCAS89 test series, given in Table 5.3.

The OP of each digital circuit was produced with three different sets of input signals. The percentage decrease in the level of power consumption of a digital circuit, denoted by ΔP_i ($i \in 1 \ldots 3$) for each of the sets of input signals, is given in Table 5.3.

Chapter 6
Linguistic and Software Development of the Automated System of Gate-Level Simulation of Digital Circuits with Consideration of DF

6.1 Software Structure

The automated system of gate-level simulation and optimization (ASGSO) of digital circuits with consideration of DF is an important EDA subsystem of digital circuits. In turn, ASGSO of digital circuits is a composite complex, consisting of many subsystems for solving various tasks of digital circuit automated design. Separate subsystems of ASGSO of digital circuits, with consideration of DF can be used both in an autonomous mode, and together with other subsystems.

Based on the analysis of the tasks of gate-level simulation and optimization of digital circuits with consideration of DF as well as the requirements given in Sect. 1.4, the following general structure of the ASGSO of digital circuit software is proposed (Fig. 6.1).

The information link between the ASGSO of digital circuit subsystems, with consideration of DF, is organized in the form of a database (DB). The concept of subsystem is conditional. For example, in ASGSO of digital circuits as an independent subsystem, it is possible to single out the program of gate-level simulation of digital circuits with consideration of DF if it is not algorithmically connected with other subsystems. Consider the purpose and interrelations of various subsystems that are part of ASGSO of digital circuits.

The subsystem of gate-level simulation of digital circuits with consideration of DF [60, 166–170, 275–279, 310] is the main one for simulating digital circuits, containing up to tens of millions of gates. The mathematical apparatus is the equations of the functioning of the MCE (Sects. 1.4 and 2.2). The purpose of the calculation is to determine the sequence of signals in different nodes of digital circuits in the modes described in Sects. 1.4 and 4.1. The subsystem is oriented to combine with the subsystem of circuit simulation for functioning within the framework of mixed-mode simulation system [234–239] as well as with the physical design subsystem [100–103, 137, 175]—to obtain information on the topology of digital circuits with the purpose of its use in the MDE, GME (Sects. 1.4 and 3.1), and

V. Melikyan, *Simulation and Optimization of Digital Circuits*,
https://doi.org/10.1007/978-3-319-71637-4_6

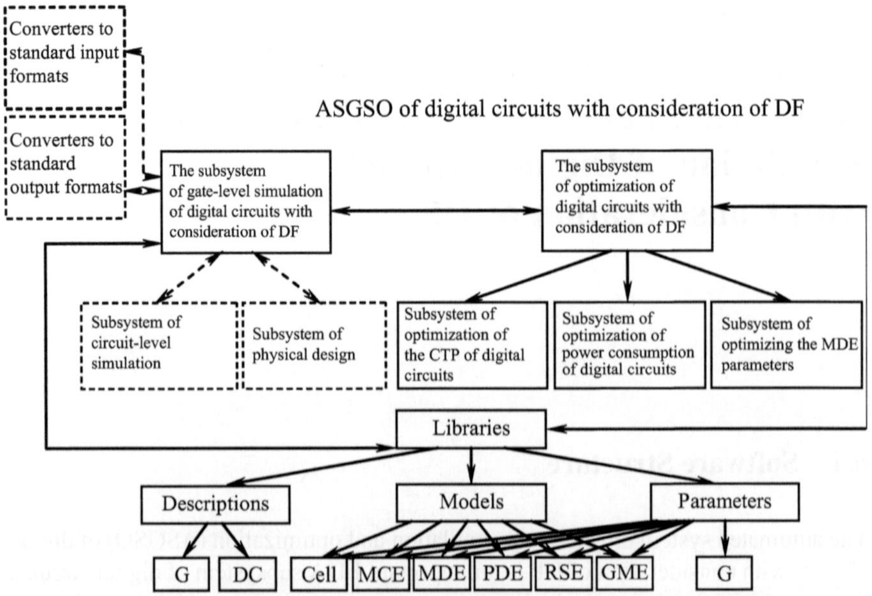

Fig. 6.1 The structure of the ASGSO of digital circuit software, with consideration of DF

so on. The input information comes from the input data stream or graphical user interface, as well as through the database from the subsystem of optimizing the MDE parameters, from the description libraries of gates, digital circuits, models and parameters, as well as from the physical design subsystem. The output information can be transferred to the same subsystems.

The subsystem of optimization of the CTP of digital circuits [132, 320, 321], with consideration of DF, is required primarily in EDA of digital circuits to improve the timing characteristics of a digital circuit in order to meet the requirements of the design specification of digital circuit design. Usually such a need arises at the stage of post-layout design of digital circuits. The used mathematical apparatus is described in Sect. 5.1. For the work of the subsystem, multiple repeated reference to the subsystem of gate-level simulation of digital circuits with consideration of DF is needed, but it can also function separately from power optimization subsystem of digital circuits, although simultaneous use of the marked subsystems can significantly improve the quality of digital circuit design.

The subsystem of optimization of power consumption of digital circuits [131, 250, 251] is also one of the main subsystems of the ASGSO of digital circuits, taking into account the DF, and is designed to improve the power consumption of digital circuits with a view to reducing the speed requirements for separate gates of digital circuits. This subsystem is also used at the stage of post-layout design, although it can be applied to initial stages of digital circuits design, if an MDE (Sect. 3.1) is used with predicted [22] topological data values. For the work of the subsystem, multiple repeated reference to the subsystem of gate-level simulation of digital circuits with

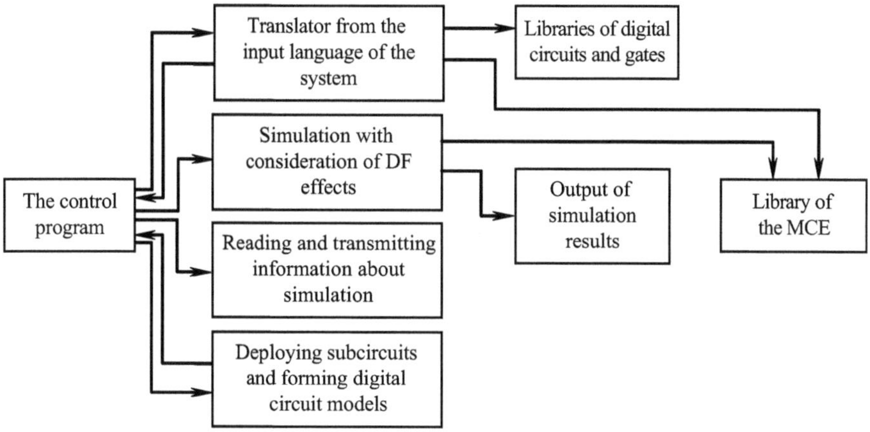

Fig. 6.2 Structural diagram of the subsystem of gate-level simulation of digital circuits with consideration of DF

consideration of DF is needed, but it can also function as an independent system for optimizing the power consumption of digital circuits. At the same time, as it was already noted, the joint use of this subsystem with the subsystem of optimization of the CTP of digital circuits will only contribute to improving the quality of digital circuit design.

The subsystem of optimizing the MDE parameters [291, 300, 301] is designed to provide all other components of the ASGSO of digital circuits with information about the MDE parameters (Sect. 3.1). The received results are transferred to the remaining program subsystems via the database, i.e. this subsystem is connected to others only by information. Its purpose is to optimize the MDE parameters on model information, as well as to calculate the coefficients of the MDE equations. All the coefficients of the MDE equations are computed once before the simulation of digital circuits and are stored in the database so that during the simulation of digital circuits with consideration of DF, computer time is not wasted on it repeatedly.

As it was already said, the main subsystems of ASGSO of digital circuits, taking into account DF, in turn, are complex systems and have their own structure. Figure 6.2 shows the structure of the subsystem of gate-level simulation of digital circuits with consideration of DF.

A number of functional blocks are allocated in the program of gate-level simulation of digital circuits with consideration of DF (Fig. 6.2).

Initially, the dictionary of the translator from the input language of the system is initiated. The dictionary contains abbreviations for recognizing the operators of the input language of the system. Then the descriptions of digital circuits are read. The procedures of the translator from the input language of the system perform syntactic control of the input information and the formation of internal information arrays. Then the whole text of the description of digital circuits is translated, even if errors are found during translation. If there are errors in the description of digital circuits,

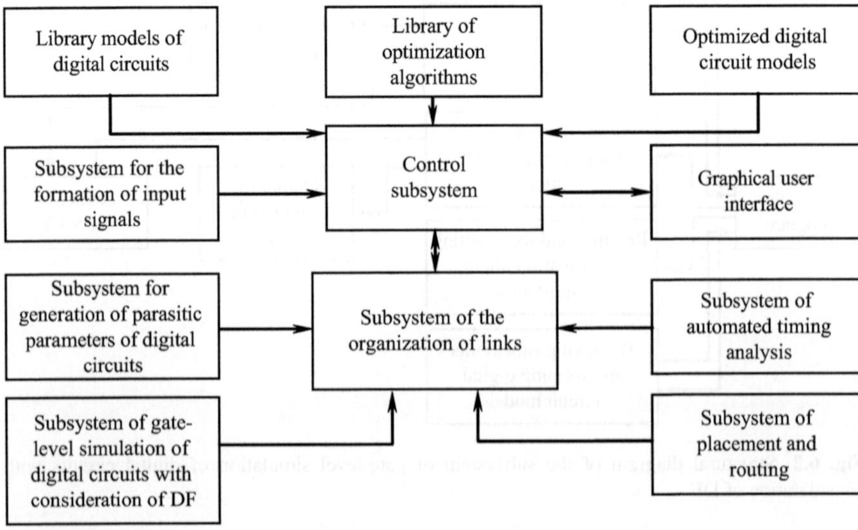

Fig. 6.3 Structure of the subsystem of optimization of the CTP of digital circuits with consideration of DF

the subsystem issues the corresponding diagnostic messages. After translation, the description of digital circuits is deployed, the main information arrays are formed for simulation programs. At the same time, a semantic control of the description of digital circuits is performed. If there are semantic errors in the description of digital circuits, the program issues the corresponding diagnostic messages. After the digital circuit description is deployed, information about the simulation, its syntactic and semantic control, simulation, and output of the analysis results are read.

The structure of the subsystem for optimizing the CTP of digital circuits with consideration of DF is shown in Fig. 6.3.

Models of digital standard cell libraries are nodal in the design process of digital circuits. The correct choice of a digital standard cell library is due to the optimal design of digital circuits and the provision of the required characteristics of digital circuits. The digital standard cell library is represented by several files of different formats, which are gate models. These are physical (LEF), timing (LIB, TLF, ALF, DB), behavioral (Verilog, VHDL), topological (GDSII), and schematic (SP) models. The technological files (TF, TECH) containing the characteristic parameters of the used technological process are also introduced into this subsystem.

Since the proposed (Sect. 5.1) AOCTP of digital circuits belongs to the class of post-layout optimization algorithms, it also requires existence of a subsystem of optimized digital circuit models. The digital circuit that is being designed, is specified after the stages of placement and routing. Therefore, in digital circuit model subsystem, a transfer (DEF) model that provides the ability to enter the digital circuit into the placement and routing subsystem, as well as the behavioral (Verilog, VHDL) and topological (GDSII) models are used.

The graphical user interface of the subsystem of optimizing CTP of digital circuits provides the ability to visualize the operation of some subsystems. The order of the menu items is adapted to the steps of solving the problem of optimizing the CTP of digital circuits and facilitates the use and development of the subsystem.

The input signal generation subsystem provides an estimate of timing delays between digital circuit nodes. With its help, signals are described on all inputs of digital circuits, and all the measurements are determined by means of the corresponding description language. As a result, a text file is obtained, which is the input for the subsystem of gate-level simulation of digital circuits with consideration of DF.

The input data for the subsystem of placement and routing are the models of digital standard cell libraries and optimized digital circuits. Output data are changed topological and physical models of optimized digital circuits.

The topology of optimized digital circuits serves as the input information of the subsystem for generating parasitic parameters of digital circuits. On the basis of this, parasitic RLC parameters are calculated. The output information is a text description of optimized digital circuits, including parasitic parameters. This subsystem also provides options for verifying the compliance of optimized digital circuits with technological requirements and electrical connections.

The subsystem of the automated timing analysis is intended for detection of CTP of digital circuits. It allows to quickly calculate the timing delays of all digital circuit paths. The input data are the models of digital standard cell libraries and optimized digital circuits, excluding topological models. Output information is a file of timing delays of digital circuits, used for decision making.

The subsystem of gate-level simulation of digital circuits with consideration of DF allows to accurately calculate the delays introduced by interconnects due to their parasitic parameters. The input data of the subsystem is a text file formed by the input signal generation subsystem, in which all the input signals and the measurements are described. In addition, a text description and a corresponding technological file generated from optimized digital circuit layout are also used. The output information is a file containing timing delays distributed among the nodes of digital circuits.

The structure of the subsystem for optimizing the power consumption of digital circuits with consideration of DF is shown in Fig. 6.4.

The description libraries of digital circuits, gates, digital cell models, and optimization algorithms as well as the automated subsystem for generating the parasitic parameters of digital circuits have a similar purpose, as in the optimization subsystem of the CTP of digital circuits with consideration of DF.

An expandable library of input signals is a collection of text files with descriptions of input signals in a language specially designed for this purpose (Sect. 6.4.1). Using the input signal library, the user can quickly select input signals required for digital circuit simulation, add descriptions of new input signals to the library, remove input signals from the library, and so on.

The language processor is a subsystem whose functions include reading from specialized languages and translating the digital circuit descriptions into internal

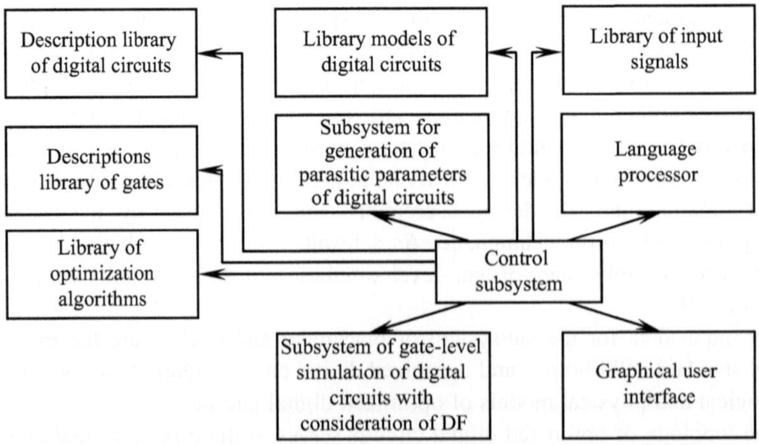

Fig. 6.4 Structure of the power consumption optimization subsystem with consideration of DF

format, input signals when the user starts the process of optimizing the power consumption of digital circuits with consideration of DF, or the process of conventional gate-level simulation of digital circuits.

The control subsystem for processing sends all events generated through the graphical user interface. It is a "dispatcher," whose functions include coordinating all actions (for example, calling subsystems) necessary to perform the task requested by the user.

6.2 Linguistic Development

Linguistic support is an important part of ASGSO of digital circuits with consideration of DF [150, 327]. This is means of the user's description of digital circuits, tasks for calculation, etc. Gate-level simulation and optimization of digital circuits with consideration of DF, in addition to the usual requirements for naturalness, simplicity of the language, uniqueness of syntactic and semantic rules, also impose specific requirements on the input language of the ASGSO of digital circuits: the hierarchy of the description of various digital circuits with a large number of digital cells, the ease of editing the description of digital circuits, systematization and transformation of models (MCE, MDE, RGE, RSE, GME), the convenience of working with the libraries of the system, adaptation to the classes of designed digital circuits [291, 293], etc.

Most of these requirements are met by the language developed for the ASGSO of digital circuits with consideration of DF of ELAIS-L [167, 328]. The description of the input language of ELAIS-L is given in Sect. 6.4.1—a description in the form of Backus-Naur, Sect. 6.4.2—an informal description. Using the example of this language, it is convenient to consider ways of solving the problem of linguistic support of ASGSO of digital circuits taking into account DF.

Description of the circuits is carried out with the help of subcircuits with an unlimited degree of their nesting. The limitation is related only to the size of the RAM of the used computer. The description of each subcircuit can be contained in the input information stream or stored in the circuits description library. In the latter case, only the library name of the subcircuit is specified in the input information and, if required, changes in the subcircuit. As a subcircuit, a description can also be used of any previously designed digital circuit stored in the library. It is enough to indicate only which of its nodes should be considered external. In addition, gate models and digital circuits can be quickly created in the input language, made out in the form of subcircuits and be entered into the library. However, this applies only to the gates, the digital circuits, and the MCE. As for the MDE, RGE, GME, RSE, they are library models and have advantages over quickly determined ones from the point of view of effectiveness of determining their parameters. In case of using a library MDE, it is sufficient to indicate only its library number in the input information. If there are no parameters in the library, they are specified in the input information. Widely used and the known principle of setting the default parameters [150, 226]: for all parameters whose values are not specified by the user, their typical values are taken. The input language constructions for describing the MCE are oriented to the default mode.

However, the ASGSO of digital circuit input language with consideration of DF is only one of the possibilities of connecting with the system. The ELAIS-L system also has a modern, user-friendly interface.

The main window of the ASGSO of digital circuits ELAIS-L has the form shown in Fig. 6.5.

This window shows the initial state of the system before the opening of the graphical editor of a digital circuit. Initially, only the digital standard cell library is available. The library of digital circuits is divided into combinational, sequential, and other digital cells. The library has a hierarchical structure (Fig. 6.6)

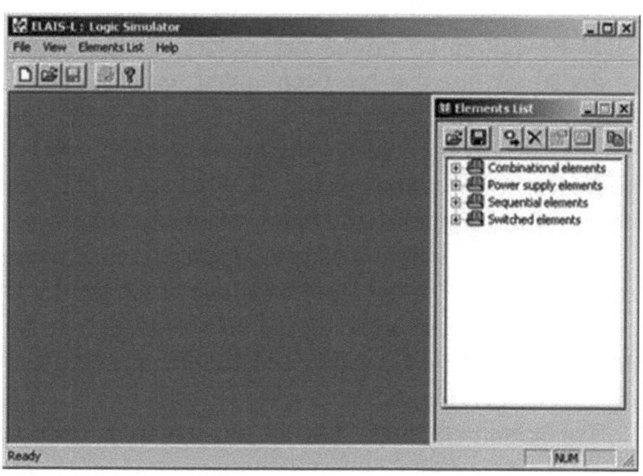

Fig. 6.5 Main window of ASGSO of a digital circuit ELAIS-L

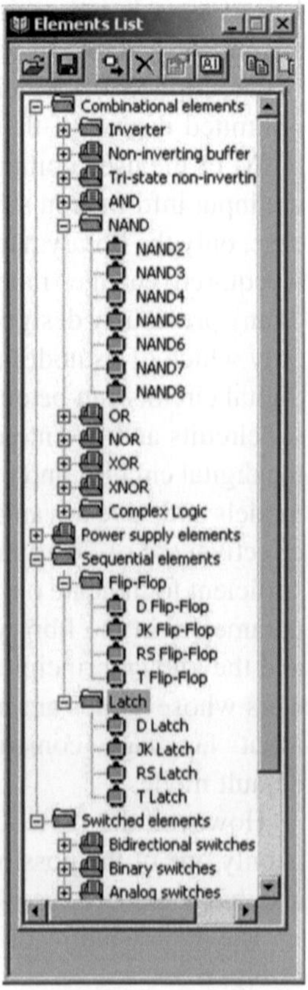

Fig. 6.6 Hierarchic structure of digital standard cell library

The main menu has the following items: FILE (work with digital circuit files), VIEW (display), ELEMENTS LIST (tools for working with standard or additional libraries), HELP (help, additional information).

The FILE mode (Fig. 6.7) has standard sub-modes: NEW—work with a new digital circuit; OPEN—opening the existing file of a digital circuit; PRINT SETUP—setting up printer parameters; EXIT—shutting down the system. A list of digital circuits that has been edited recently is also used. Hot keys can also be used.

The window for editing a digital circuit (Fig. 6.8) is divided into two parts. When you select NEW command, the NEW window (Fig. 6.9) appears, giving the user the choice of the type of used editor: schematic, graphical, and language.

The language editor is a standard text editor for typing and correcting descriptions of digital circuits in the input language of ELAIS-L. The graphical editor allows to create images of separate digital cells used during the operation of the schematic editor. The schematic editor allows creating and correcting images of arbitrary digital circuits (Fig. 6.10).

The schematic editor has all the standard functions of such programs. When creating a new digital circuit, first the corresponding digital cell is selected from the

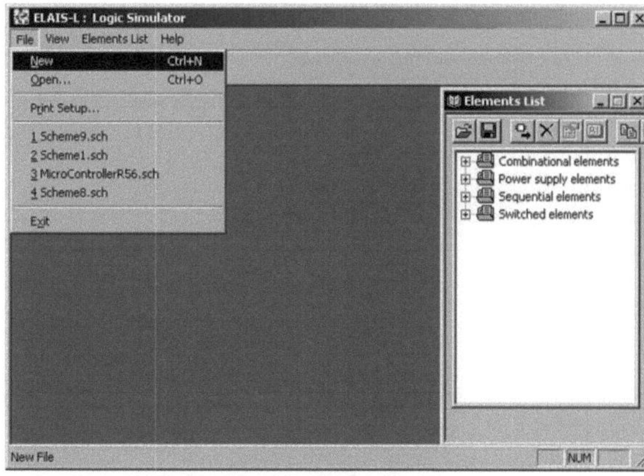

Fig. 6.7 FILE mode

Fig. 6.8 Window of editing a digital circuit

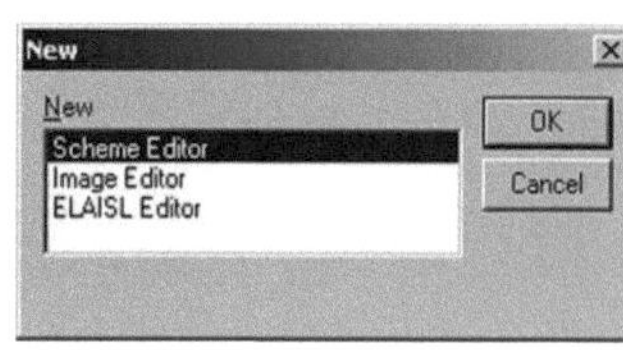

Fig. 6.9 NEW window

standard digital cell library (Fig. 6.11), then the selected digital cell is assigned a unique name for the design (Fig. 6.12), after which the digital cell appears in the schematic editor window (Fig. 6.13), then connections of digital cells (Fig. 6.14) with other digital cells are given, as well as other properties (timing parameters, names of inputs, outputs, etc.) (Fig. 6.15), including the DF (Fig. 6.16) (Fig. 6.17).

In the schematic editor, two-sided envelopes of digital circuit images are also built in the standard formats Verilog, VHDL, and also in the own format of the ELAIS-L program (Sect. 6.4) (Fig. 6.18).

The graphical interface of ASGSO of digital circuits ELAIS-L also has the capability to enter assignment parameters for calculation: descriptions of input signals (Fig. 6.19), setting the list of output signals (Fig. 6.20), etc.

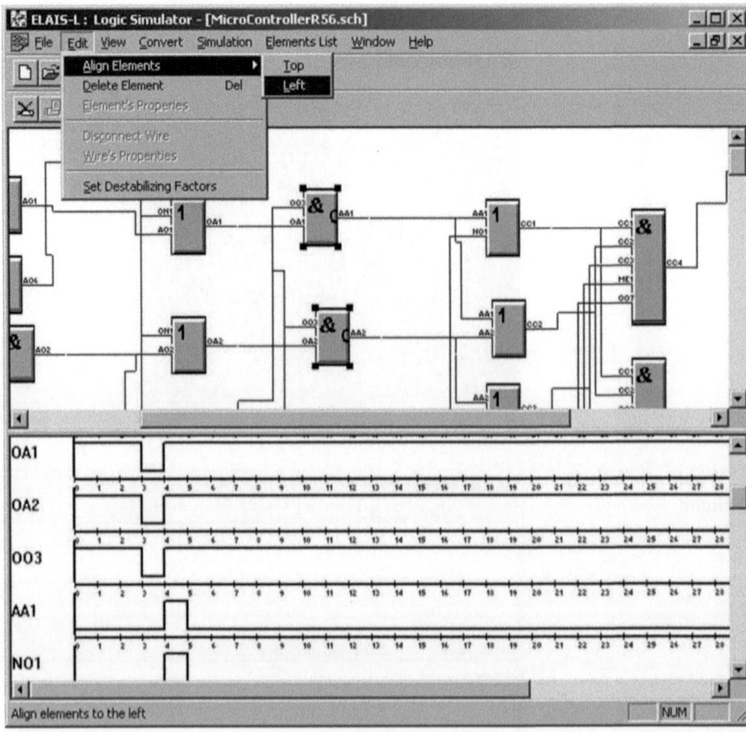

Fig. 6.10 Schematic editor

Fig. 6.11 Selecting a digital cell from the library

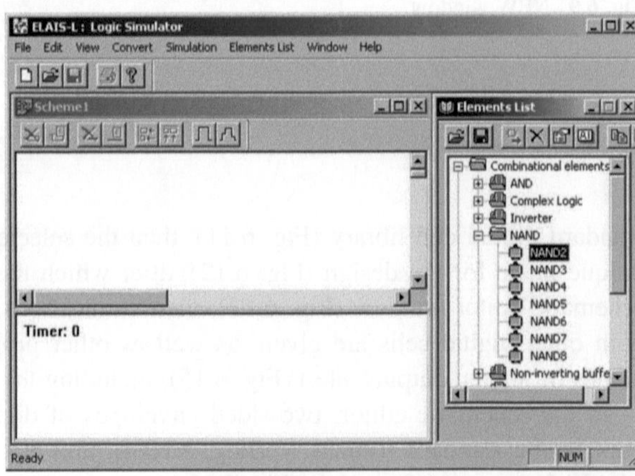

Fig. 6.12 Giving a unique digital cell name

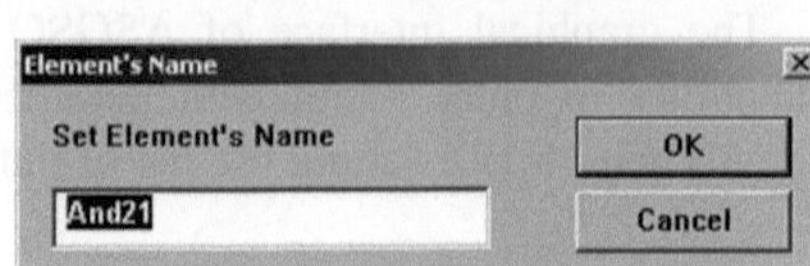

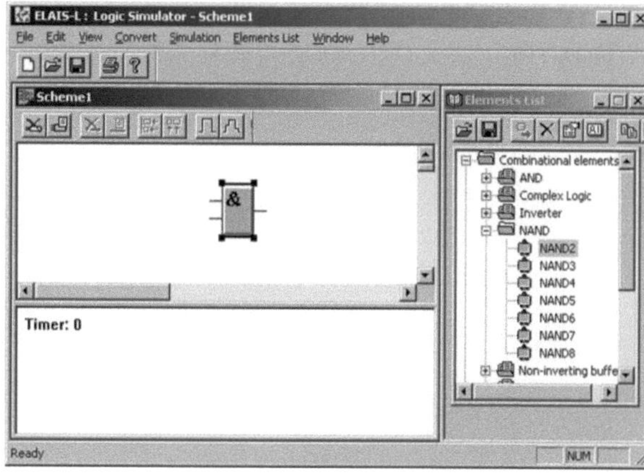

Fig. 6.13 Location of selected digital cell

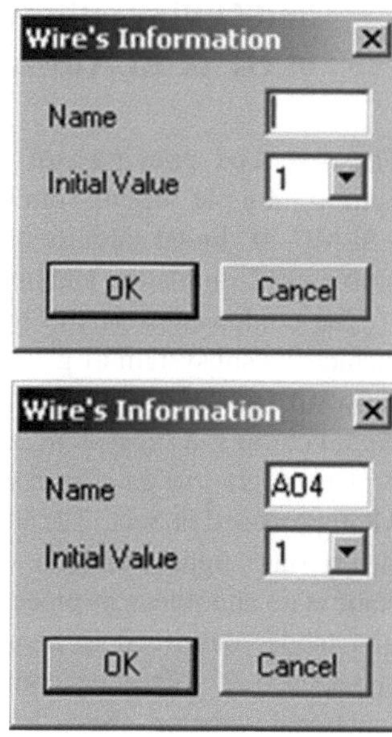

Fig. 6.14 Specifying digital cell connections

The ELAIS-L program has the capability (Fig. 6.21) to simulate digital circuits both without consideration of DF (in order to save the computer time in the initial stages of digital circuit design), and with consideration of DF.

In simulating with consideration of DF, not rectangular pulses, but waveforms close to circuit-level simulation are produced (Fig. 6.22).

Other ASLO digital circuit subsystems, with consideration of DF ELAIS-L, also have a modern user-friendly interface. Figures 6.23 and 6.24 show examples of windows of the subsystems of optimization of the CTP "ICO Ultra" and optimization of power consumption "ECOS" of digital circuits respectively.

The ELAIS-L system also has a number of training programs for users working with a set of programs based on the principles outlined in [329].

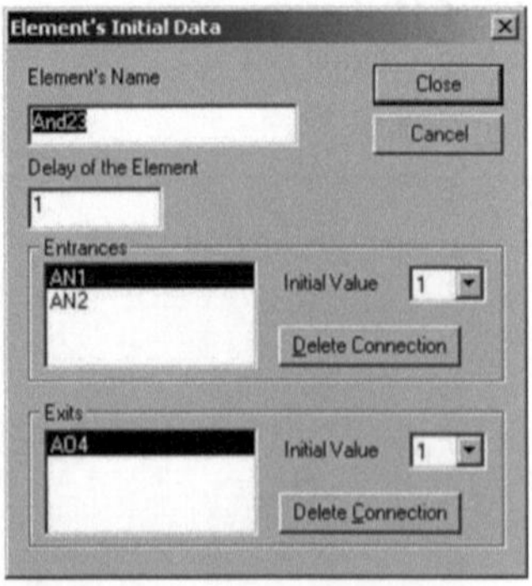

Fig. 6.15 Setting properties of digital cells

6.3 System of Programs of Gate-Level Simulation and Optimization of Digital Circuits with Consideration of DF of ELAIS-L

The described principles of constructing a system of gate-level simulation and optimization of digital circuits, algorithms and models are implemented in ASLMO of digital circuits of ELAIS-L. ELAIS-L allows to solve a complex of problems of simulation and optimization of digital circuits with consideration of DF.

The composition of the basic software of ELAIS-L is shown in Fig. 6.25. It includes: a subsystem of gate-level simulation of ELAIS-L, a subsystem of optimization of CTP of digital circuits, a subsystem of optimizing power consumption of digital circuits, a subsystem of parametric optimization.

The program of gate-level simulation, with consideration of DF of ELAIS-L, uses MCE proposed in Sect. 2.2, and their modifications [176–178, 201]. The alphabet of steady-state signal states is virtually infinite-valued. The program calculates the static state and transient processes in digital circuits. In addition, additional calculation modes described in Sect. 1.4 are implemented. The program can detect the appearance of forbidden combinations of signals, the occurrence of failures and functional violations (Sect. 2.1). There is a mode for calculating the limiting speed of digital circuits, which reduces to finding a scale factor that shows how many times the time intervals of the input signal sequence can be shortened without causing failures in digital circuits. The volume of calculated circuits is up to tens of millions of gates and is limited solely to the memory of the used computer. MCE are described in the input language (Sect. 6.4) by logical equations and references to other models that can be specified in the input information or stored in the library of digital circuits in the form of subcircuits. The library includes standard digital circuits of various complexity levels, including a number of ISCAS89 test circuits, as well as a large set of MDE, RSE, PDE. The library can be filled by the user using the input language ELAIS-L. Any digital circuit written in it can later be used as a subcircuit. When using library subcircuits, their timing parameters can be adjusted.

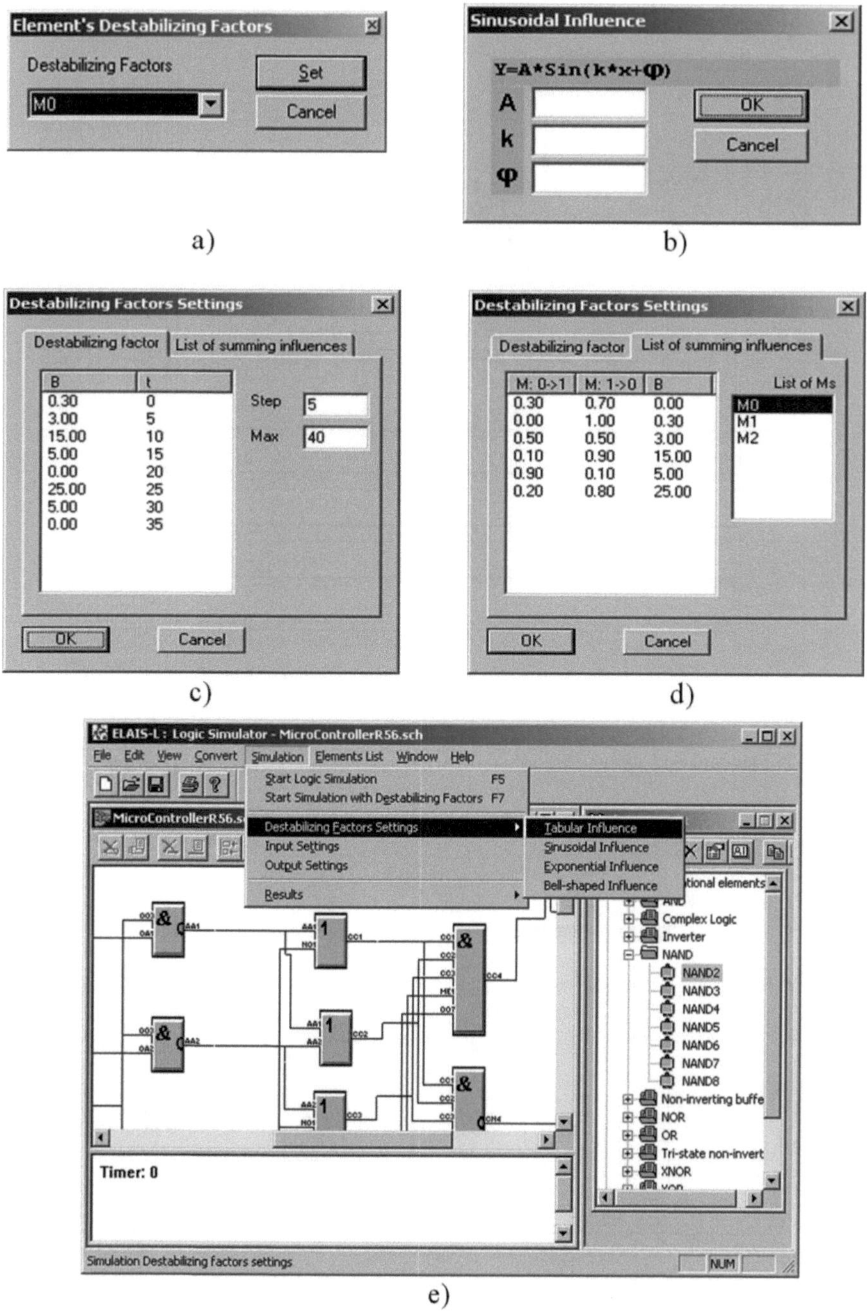

Fig. 6.16 Defining DF parameters

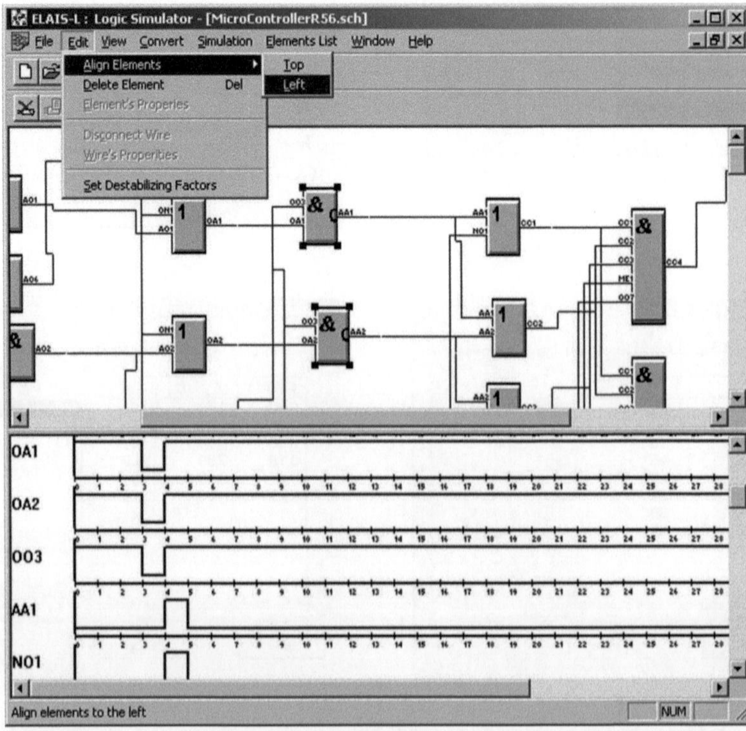

Fig. 6.17 Standard operations with digital cells

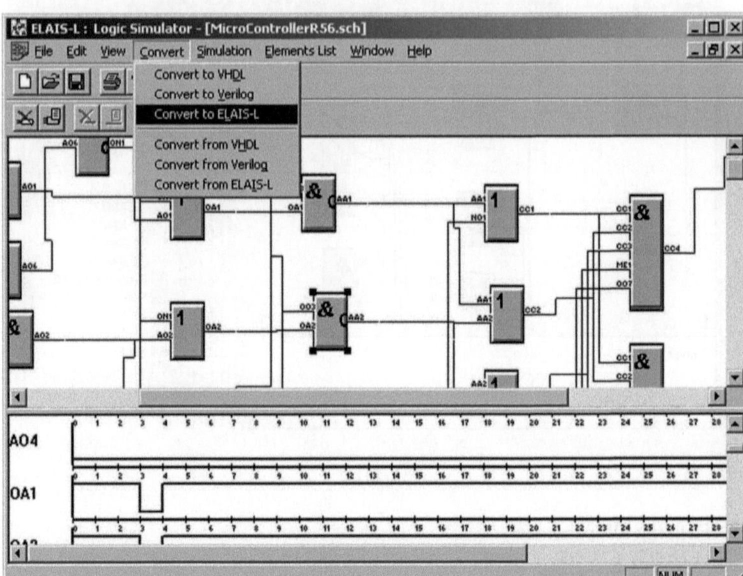

Fig. 6.18 Conversion to standard formats

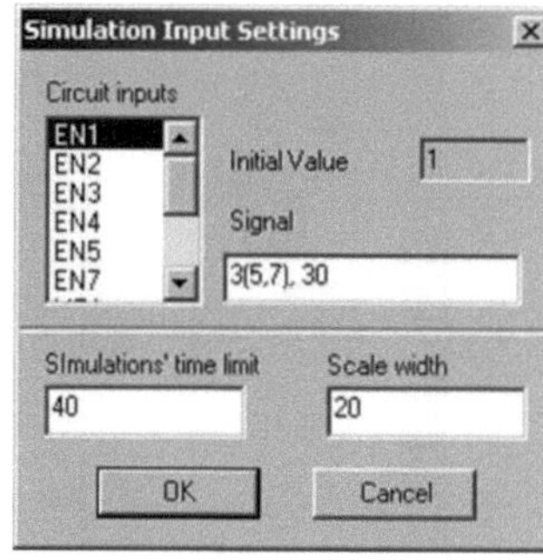

Fig. 6.19 Setting input signals

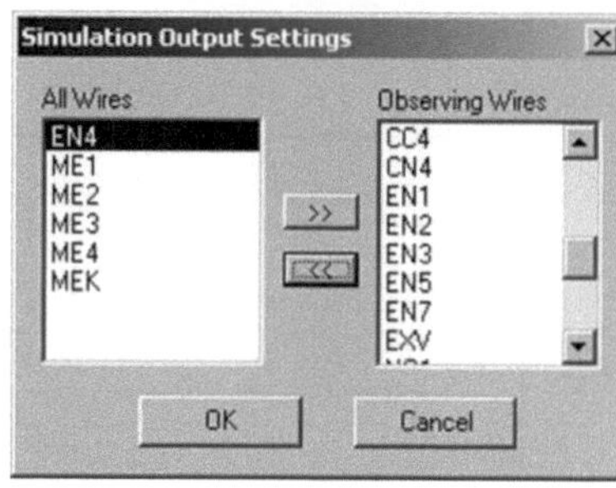

Fig. 6.20 Setting the list of output signals

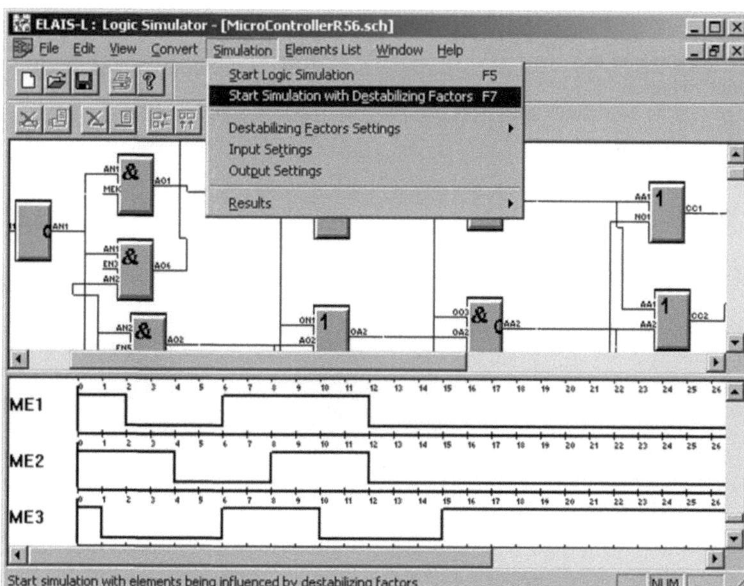

Fig. 6.21 Selecting the type of digital circuit simulation

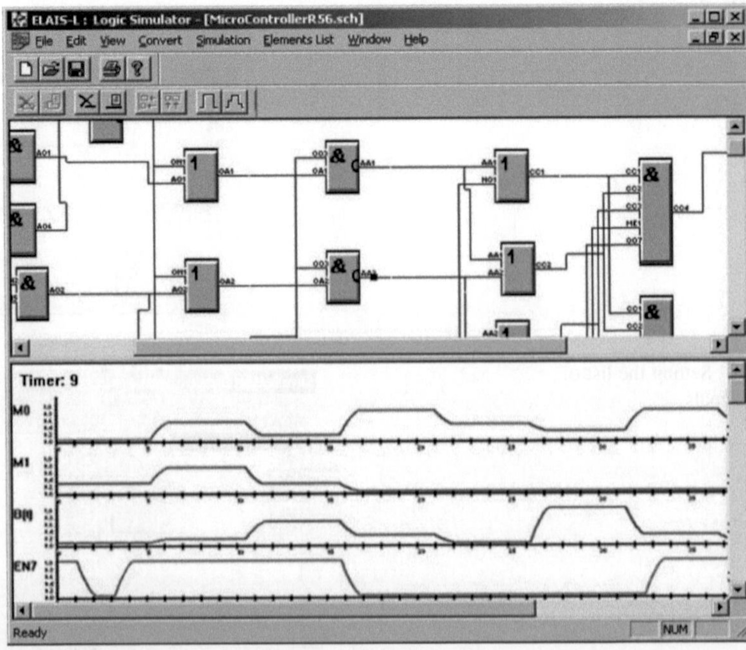

Fig. 6.22 Simulation of digital circuits with consideration of DF

Fig. 6.23 Example of the "ICO Ultra" window

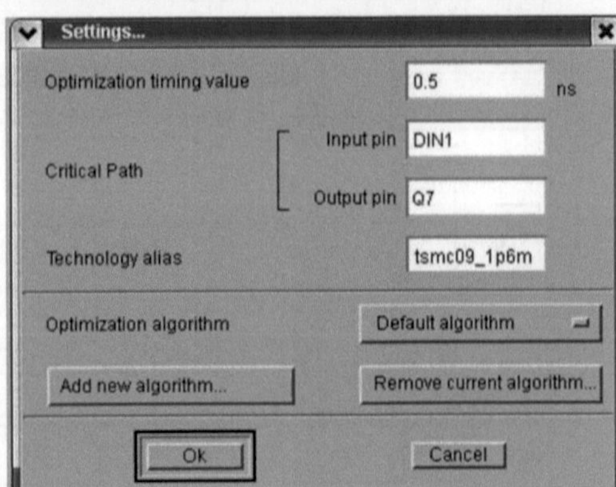

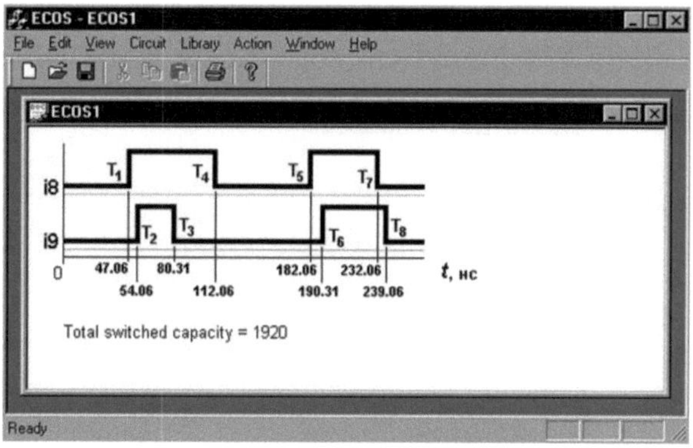

Fig. 6.24 Example of the "ECOS" window

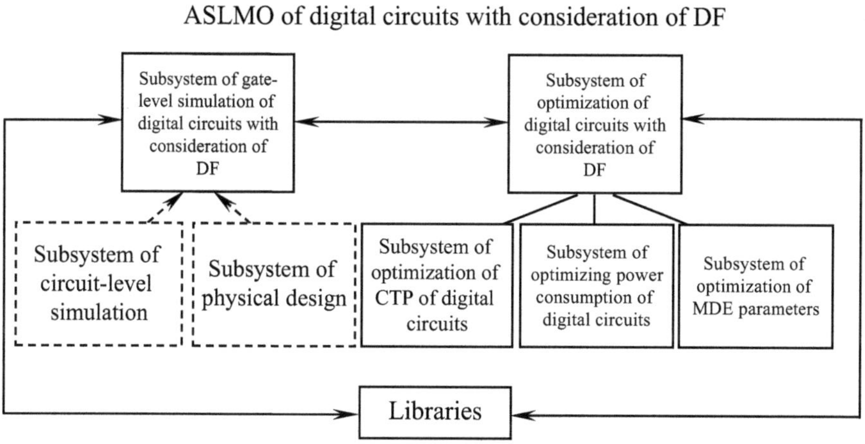

Fig. 6.25 The composition of the basic software of ELAIS-L

6.4 Description of the Input Language of ELAIS-L

6.4.1 Description According to the Backus-Naur Form

Basic and elementary constructions

```
<letter> :: = A|B|C|D|E|F|G|H|I|J|K|L|M|N|O|P|Q|R|S|T|U|V|W|X|Y|Z
             |a|b|c|d|e|f|g|h|i|j|k|l|m|n|o|p|q|r|s|t|u|v|w|x|y|z
<number> :: = 0|1|2|3|4|5|6|7|8|9
<underlining> ::= _
<space> ::= {/ /}
<key_symbols> ::= .|,|;|:|(|)|=|+|-|*
```

basic identifiers

```
<identifier>::=
      <identifier><letter>
      |<identifier><number>

<identifier>::=
      <first_symbol>{<next_symbol>}

<first_symbol>::=
      <letter>
      |<number>
      |<underlining>

<next_symbol>::=
      <letter>
      |<number>
      |<underlining>
      |<space>

<next_symbols>::=
      {<next_symbol>}

<comments>::=
      [<next_symbols>]

<end_operator>::=
      ;<comments>

<identifier_4>::=
      <first_symbol><next_symbol><next_symbol><next_symbol>

<identifier_8>::=
      <identifier_4> <next_symbols_4>

<next_symbols_4>::=
      <next_symbol><next_symbol><next_symbol><next_symbol >
```

Identifiers, used in the description of a digital circuit

```
<name_circuit>::= <identifier_4>

<name_ subcircuit>::= <identifier_8>
```

values, used in the language

```
<value_parameter>::=
      <numeric_value>
      |<integer_value>

<logical_value>::= 0 |1
```

literals

```
<numeric_value>::=
      <number>{<number>} [.{<number>}]
      |.<number>{<number>}

<integer_value>::=
      <number>{<number>}

<sign>::=
      - | +
```

Expressions

```
<expression>::=
      [<sign>] <element_ expression>
      {<operator_ expression > <expressions _in_ brackets>}
      {<operator_ expression >[<sign>]<element_ expression>}

<element_ expression>::=
      <constant>
      |<name_variable>
      | (<expression >)

<expressions _in_ brackets>::=
      (<operator_ expression>[<sign>]<element_ expression> )
      |( <expression> )

<operator_ expression>::=
      + | * |-

<constant> ::=
      <value_parameter>

<name_ variable>::=
      <identifier>
```

```
<logical_ expression>::=
     <element_logical_ expression>
   {<logical_operator_ expression> <logical_ expressions_in_ brackets>}
  {<logical_operator_ expression> <element_logical_ expression>}

<logical_ expressions_in_ brackets>::=
     (<logical_operator_ expression> <element_logical_ expression>)
     |(<logical_ expression>)

<element_logical_ expression>::=
     <constant>
     |<identifier>
     |<logical_ expression>
     |<logical_ expressions_in_ brackets >

<logical_operator_ expression>::=
     + | * |-

<model_ expression> ::=
     [<sign>] <element_model_ expression>
  {<operator_model_ expression> <model_ expressions_in_ brackets>}
  {<operator_model_ expression>[<sign>]<element_model_ expression>}

<element_model_ expression>::=
     <constant>
     |<name_library set_parameters>
     |(<model_ expressions>)

<model_ expressions_in_ brackets>::=
     (<operator_model_ brackets>[<sign>]<element_model_ brackets>)
     |( <model_ brackets> )

<operator_model_ brackets>::=
     + | * |-

<name_ library set_parameters>::=
     <identifier_parameter>
     |<name_index_in_place_value>

<identifier_parameter>::=
     <identifier>

<name_index_in_place_value>::=
     "<identifier>"
```

Values of pointers and parameters are stored in libraries used in the automated system of gate-level simulation and optimization of digital circuits of ELAIS-L. In the language, spaces are ignored, or the compiler neglects them.

Each description statement starts at any position in the new line and ends with a semicolon. Each statement can contain several lines. The operator can be transferred after delimiters or key symbols (parentheses, colons, commas, signs of logical operations, etc.). Any comment can be written after the semicolon.

Circuit description

```
<description_circuit> ::=
      <operator_name_circuit>
      <declaration_used_circuits><description_ subcircuit>
      {<description _ subcircuit>} <description_main_circuit>
      {<operator_ ANALYSIS> <parameters_ ANALYSIS>
<operator_ next_actions>}
      <operator_ END>

<operator_name_circuit> ::=
      <name_circuit> <next_symbols> <end_operator>

<operator_name_ subcircuit> ::=
      <name_ subcircuit> <next_symbols> <end_operator>

<declaration_used_circuits> ::=
      {<type_circuit> <end_operator>}

<type_circuit> ::=
      <name_ subcircuit>, L
      |<name_ subcircuit>, I
      |<name_ subcircui>, E
```

When working with a library, the operator looks like this:

L—if the library circuit is calculated;
I—if the circuit is included in the library;
E—if the circuit is excluded from the library;

```
< description_subcircuit> ::=
        <operator_name_ subcircuit> <list_inputs> <list_outputs>
        [<list_names_sets_ parameters>]
        [<description_sets_ parameters>] < description_logic> <operator_END>

< description_main_circuit>
        <operator_name_circuit> <list_inputs> <list_outputs>
        [<list_names_sets_ parameters>]
        [<description_sets_ parameters>] < description_logic> <operator_END>

<list_inputs> ::=
        IN<next_symbols>:<incoming_identifiers> <end_operator>
        | INPUTS:< identifiers_ inputs> <end_operator>

< identifiers_ inputs> ::=
        < identifier _inputs >
        | < identifier_input> {,< identifier_input>}
< identifier_ input> ::=
        [<name_ identifier_ input>]
<name_ identifier_ input> ::=
        < identifier>

<list_outputs> ::=
        OUT<next_symbols>:< identifiers_output> <end_operator>
        | OUTPUTS:< identifiers_output> <end_operator>
< identifiers_output> ::=
        < identifier_output>
        | < identifier_output> {,< identifier_output>}
< identifier_output> ::=
        <name_ identifier_output>
<name_ identifier_output> ::=
        < identifier>

<list_names_sets_ parameters> ::=
        PA<next_symbols>:< timing_identifiers>
<end_operator>
        | PARAMETERS:< timing_identifiers> < end_operator>

< timing_ identifiers> ::=
        <name_parameter>
        | <name_parameter> {,< name_parameter>}

< name_parameter> ::=
        < identifier>

< description_sets_ parameters> ::=
        <name_parameter>:< timing_parameters> <end_operator>

< timing_ parameters> ::=
        <first_ timing_ parameter>
        {,< timing_parameter_front_edge>}
        {,< timing_ parameter_falling_edge>}
<first_ timing_parameter> ::=
        < timing_parameter_front_edge>
        | < timing_parameter_falling_edge>
```

< timing_parameter_front_edge> ::=
[TF=<value_ timing_parameter>]
| [TF01=< value_ timing_parameter>]
| [TF10=< value_ timing_parameter>]

< timing_parameter_falling_ edge> ::=
[TP=< value_ timing_parameter>]
| [TP01=< value_ timing_parameter>]
| [TP10=< value_ timing_parameter>

< value_ timing_parameter> ::=
< value_parameter>
| < identifier _ timing_parameter >
| <model_ timing_parameters>
| < expression_parameter>

< identifier_ timing_parameter> ::=
<element_model_ expression>

< expression_parameter> ::=
<model_ expression>

< assignment_value_ timing_parameters> ::=
< assignment_ value> <end_operator>
| < assignment_ value> {,< assignment_ value> } <end_ operator>

< assignment_ value> ::=
< identifier_parameter>=< value_parameter>

< model_ timing_parameter> ::=
<name_ model>,< quantity_parameters>=< value_parameters><end_ operator>

< name_ model>::=
< identifier>

< quantity_parameters>::=
< integer_ value>

< value_parameters>::=
< value_parameter>{,< value_parameter>}

Quantity of < value_parameter > should be by value (<quantity_parameters >).

< description_ logic> ::=
< operator_ LOGIC>
{<call_ library_ subcircuit>}
{< logic_ subcircuit>}

< call_ library_ subcircuit> ::=
<name_ subcircuit>(< arguments _ subcircuit>)
[,(<timing_ identifier>)] <end_operator>

< arguments _ subcircuit> ::=
< arguments _ subcircuit>{,< argument_ subcircuit>}

< argument _ subcircuits> ::=
< input_ identifier>
| <output_ identifier>
<logic_ subcircuits> ::=
< function_ subcircuits> = <logical_ expression> <end_operator>
< function_ subcircuit> ::=
<name_ subcircuit>(< timing_ identifiers>)

< operator_LOGIC> ::=
LO<next_symbols> <end_operator>
| LOGIC <end_ operator>

< operator _END> ::=
END<next_symbols> <end_operator>
| END<end_operator>

< operator_ ANALYSIS> ::=
AN<next_symbols> < end_operator>
| ANALYSIS< end_operator>

<parameters_ ANALYSIS> ::=
{<operators_ ANALYSIS>}
{< identifiers_ ANALYSIS>}
{< functions_ ANALYSIS>}

< operators_ ANALYSIS> ::=
< operator_MODEL>
| < operator_INIT_STATE>
| < operator_KGEN>
| < operator_TIME>
| < operator_STAT>
| < operator_NOSTAT>
| < operator_FAILURES>
| < operator_SCALE>
| < operator_ LIMIT_PERFORMANCE>

< operator _MODEL> ::=
MO<next_symbols> = <type_model> <end_operator>
| MODEL = <type_model> <end_operator>

<type_model> ::= 3

<operator_INIT_STATE> ::=
IN<next_symbols> (<initial_state>) =
< chain_input_parameters>
| INIT STATE (<initial_state>) = < chain_input_ parameters>
< chain_input_ parameter> ::=
< identifiers_input> <end_operator>

< initial_ state> ::= < numeric_value>

<operator_KGEN> ::=
KG<next_symbols>=< integer_value> <end_operator>
| KGEN=< integer_ value> < end_operator>

<operator_TIME> ::=
TI< next_symbols>=< duration_ transition_ process> < end_operator>
| TIME=< duration_ transition_ process> < end_operator>

< duration_ transition_ process>::=
<value_time>

<operator_STAT> ::=
ST<next_symbols> <end_operator>
| STAT <end_operator>

<operator_NOSTAT> ::=
NO< next_symbols> < end_operator>
| NOSTAT < end_operator>

<operator_FAILURES> ::=
FA<next_symbols> <end_operator>
| FAILURES < end_operator>

<operator_ SCALE> ::=
SC<next_symbols> = <value_Km> < end_operator>
SCALE= <value_Km> < end_operator>

<value_Km>::=
< numerical_ value>
| < integer_value>

<operator_ LIMIT_PERFORMANCE> ::=
LIM<next_symbols> < end_operator>
| LIMIT PEROMORMANCE< end_operator>

< identifiers_ ANALYSIS> ::=
< identifiers_ signal>

< identifiers_ signal> ::=
< identifier_ signal>(< primary_value_signal>)=< set_values_time>
< set_ values_time> ::=
<moments_ switching> < end_operator>

< identifier_signal> ::=
< identifier>
< primary_value_signal> ::=
<logical_value>
<moments_switching> ::=
<moment_ switching>{,< moment_ switching>}

```
< moment _ switching> ::=
        <value_time>

< value_time> ::=
        <numerical_value>
        | < numerical_value>

< function_ ANALYSIS> ::=
        < function_ PRINTING>

< function_ PRINTING> ::=
        <operator_ PRINTING> < identifiers_for_printing>
        [<operator_ INTERVAL>]
        [<operator_ STEP_ PRINTING>]

< operator_ PRINTING> ::=
        PR<next_symbols> <end_operator>
        | PRINTING < end_operator>

< operator_ INTERVAL> ::=
        IN< next_symbols> = <interval> <end_operator>
        | INTERVAL = <interval> <end_operator>
<interval> ::=
        <primary_value_interval>,<final_value_interval>
<final_value_interval> ::=
        <value_time>
< final_ value_interval> ::=
        < value_time>

<operator_ STEP_ PRINTING > ::=
        PR<next_symbols>  = <value_step> < end_operator>
        | STEP PRINTING = < value_step> < end_operator>
< value_step> ::=
        < numerical_ value>
        | < integer_ value>

< identifiers_for_printing> ::=
        <list_ nodes>
< list_ nodes> ::=
        < argument_ subcircuit>
        | <others_nodes>
<other_nodes> ::=
        < identifier>

<operator_ subsequent_actions> ::=
        < operator_ ANALYSIS>
        | < operator_END>
        | <end_operator>
```

6.4.2 Informal Description

Digital circuits usually consist of a large number of repeating fragments of different complexity degrees: gates, flip-flop systems, register bits, counters, adders, etc. Therefore, before the description of the circuit, it is necessary to allocate such fragments—subcircuits.

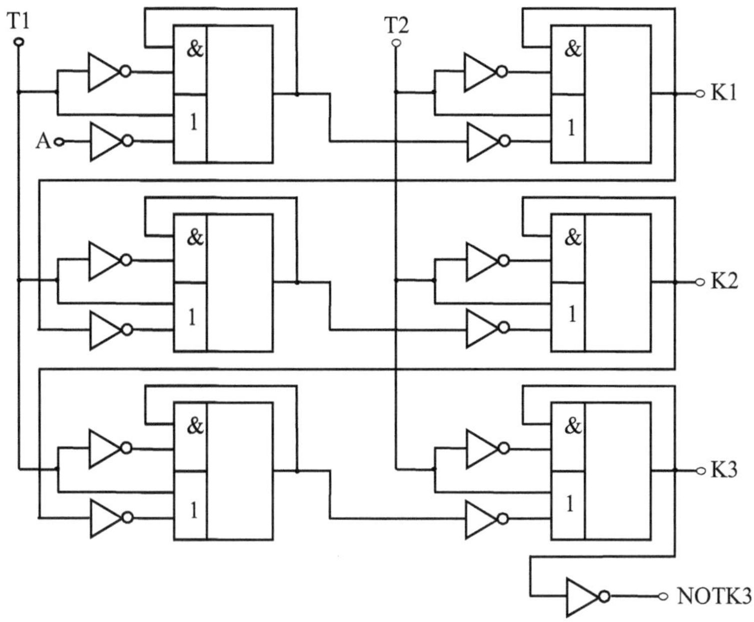

Fig. 6.26 Circuit of a shift register

Subcircuits can be of different levels. First-level subcircuits are the fragments of digital circuits, which are represented as a "black box" having several inputs and several outputs. For a first-level subcicuit, logical functions are assigned that connect the values of signals at its outputs with the values of signals at its inputs and outputs. First-level subcircuits can be either simple logic gates or more complex digital circuits (flip-flops, registers, etc.) if the user is not interested in changes in the states of internal nodes. A second-level subcircuit consists of subcircuits of the first level. A third-level subcircuit can contain subcircuits of the first and second levels. The subcircuit of the K-th level can contain subcircuits of the first, second, … $(K - 1)$th levels. The maximum level is limited only by the amount of memory.

Each type of subcircuit is assigned an identifier name containing up to ten nonblank characters. Each node in the subcircuit is associated with a certain designation—an identifier containing up to four significant characters. The designations in various subcircuits can be repeated.

As an example, consider the circuit of a shift register shown in Fig. 6.26. The register has three discharges with outputs K1, K2, K3 and inverted output of the third digit NOTK3: T1 and T2—clock inputs; *A* is the input of the register.

The partitioning of the calculated digital circuit into subcircuits is ambiguous task. It is possible, for example, to allocate a register bit as a subcircuit and name this type a subcircuit DISCHARGE. Then the circuit will take the form shown in Fig. 6.27. Each discharge can be represented as consisting of two subcircuits of half-discharge PR (Fig. 6.28). And each PR can be represented as consisting of

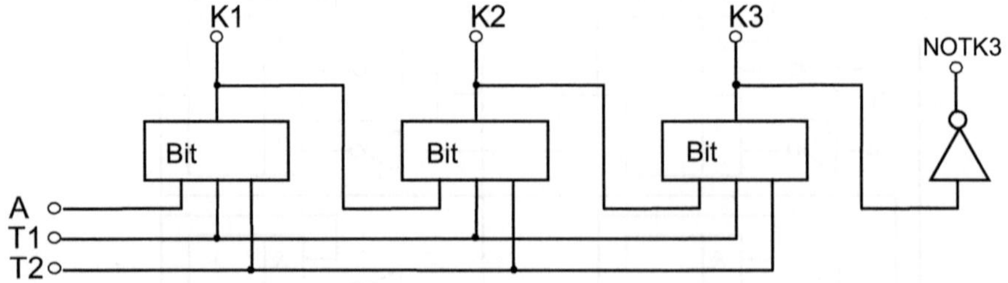

Fig. 6.27 Example of subcircuit allocation

Fig. 6.28 Example of subcircuit allocation

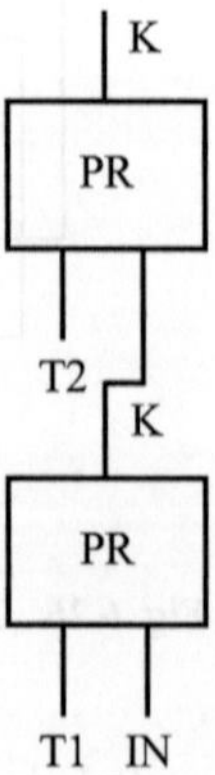

Fig. 6.29 Example of subcircuit allocation

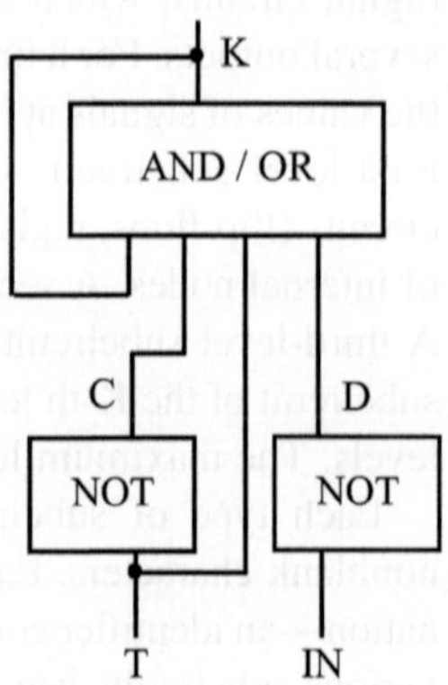

subcircuits AND/OR and NOT (Fig. 6.29). The feedback can be included inside the digital circuit AND/OR and present a PR, consisting of the subcircuits TR and NO (Fig. 6.30). In contrast, AND/OR (Fig. 6.31) can be detailed. Finally, it is possible not to break PR into subcircuits, but rather to describe the logical functions of the PR directly.

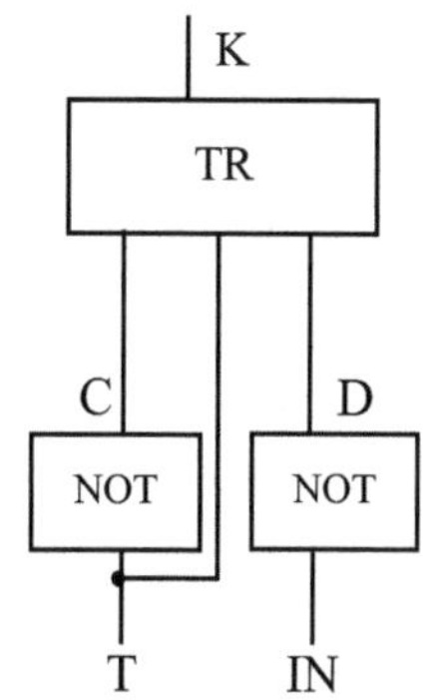

Fig. 6.30 Example of subcircuit allocation

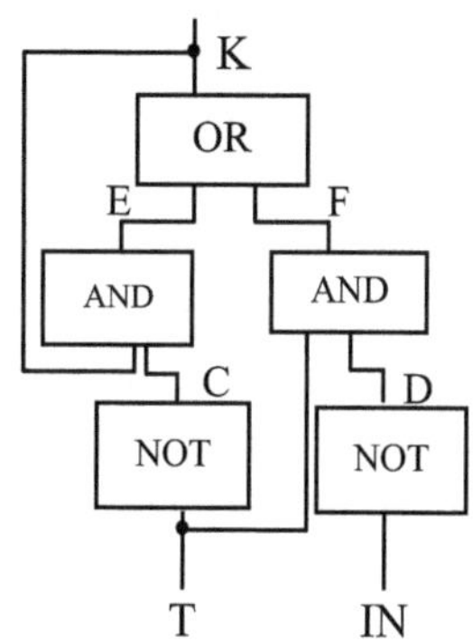

Fig. 6.31 Example of subcircuit allocation

Some of the required subcircuits can be contained in the system library. Then the catalog of the library should set the name of the subcircuit, the order of enumerating the inputs and outputs in it.

Each description operator starts at any position in the new line and ends with a semicolon. Each statement can contain several lines. The operator can be transferred to another line after delimiters (parentheses, colons, commas, signs of logical operations, etc.). Any comment can be written after the semicolon.

The description of the circuit has this form:

```
Operator with circuit name
Description of the subcircuit
Description of the subcircuit
....................................
```

After the description of the circuit, information about the simulation is given.

In the simplest case, the operator with the circuit name consists only of circuit name, for example, REGISTER. When working with a library, the operator looks like this:

- NAME, I;—if the circuit is included in the library. For example, FLIPFLOP, I;. In this case, the description of the circuit must be specified. The circuit is enabled only if there are no errors in the description;
- NAME, I;—the library circuit is calculated. For example, REGISTER, I;. After this operator, information about simulation must follow;
- NAME, E;—the library circuit is excluded. After this operator, either a new operator with a circuit name or an operator indicating the end of the entire assignment can be specified.
- INMDE;—if inclusion of the MDE (Sect. 3.1) is required in the library. After this operator, MDE must be listed. The list of these models must be terminated by the operator END;. Inclusion of the MDE is carried out only in the absence of errors in the description of this MDE. After the operator END; either a new operator with a circuit name or an operator indicating the end of the entire assignment can be specified.
- LIB;—outputs the contents of the library. After this operator, either a new operator with a circuit name or an operator indicating the end of the entire assignment can be specified.

The description of the circuit consists of descriptions of subcircuits, and, as one of subcircuits, the circuit itself is described. Thus, the circuit name is repeated twice: in the operator with the circuit name and in the description of one of subcircuits. The description of the circuit itself in the same form in which the subcircuits are described makes it possible to easily increase the complexity of the tasks to be solved, and, in the future, to calculate the devices in which this circuit is actually one of the subcircuits.

Descriptions of subcircuits can be arranged in any order. Subcircuits are identified by their names. When describing the subcircuits with the same names, the last one is perceived (the previous ones are ignored). Each subcircuit can be represented in two forms. If the given subcircuit is present in the library, its description is reduced to one NAME, I: operator. For example, NOT, I;. If the subcircuit is introduced only in the given task, its description has this form:

name

List of inputs
List of outputs
Description of the MDE parameters (not mandatory)
Description of timing parameters
Timing parameters
Description of logic
Operator END.

As seen from the following, some keywords are used in describing the circuit (LOGIC, END, INPUTS, OUTPUTS, etc.). Keywords can be shortened to the first two letters.

For example, instead of the word INPUTS, you can write IN or INPUT, etc.

6.4.2.1 Lists of Inputs and Outputs

The list of inputs has the form IN[PUTS]: I_1, I_2, . . ., where I_i—the designations of input nodes. For example,

IN: T1, T2, IN;

The list of outputs has the form OUT[PUTS]: I_1, I_2, . . ., where I_i—designation of output nodes. For example,

OUTPUTS: K1, K2, K3;

6.4.2.2 Description of the MDE Parameters

This operator is intended for setting the MDE number, as well as the parameter values necessary for this model.

The operator has the form:

MDE [<identifier>]; NM [parameter list], where NM is the model number.

A list of parameters can be specified either by enumerating parameter values, or by setting parameter values together with their names (depending on the model number). For example,

MDE1:3, A = 13, B = 25;
MDEA:1;
MDEDC:2, 3.5, 4.9;

This operator should be used only when describing those subcircuits of the first level for which the MDE <identifier> is used in the logic part.

In addition, it is possible to use MDE models, prerecorded in the library of the system (with the help of the INMDE operator;) by the operator

MDE [identifier], I
For example:
MDE1, I;
MDEDC, I;

6.4.2.3 Lists and Description of Parameters

The description of timing parameters of the subcircuit is carried out as follows:

PA [RAMETERS]: I_1, I_2, . . ., I_i;
I_1: a set of timing parameters 1;
I_2: a set of timing parameters 2;
. . . .
I_i: a set of timing parameters i;

I_i—identifier of the *i*th set of timing parameters of the given subcircuit. To set the timing parameters, the following keywords are used: TP01 (t_{pd01}), TP10(t_{pd10}), TF01 (t_r), TF10(t_f), TP (t_{pd} if $t_{pd01} = t_{pd10}$), TF (t_f, if $t_r = t_f$).

The timing parameters can be specified in any order, and any of the parameters can be omitted if they are 0.

Example of setting timing parameters:

PA: A1, A2, ABC;
A1: TP01 = 10, TP10 = 22, TF01 = 5;
A2: TP = 20, TF10 = 2.5;
ABC: TR = 20, TF = 2.5;

6.4.2.4 Description of Logic

The description of logic of the subcircuit operation starts with the LOGIC operator :. After it the operators that specify the logic either using logical expressions or links to subcircuits follow.

In logical expressions, parentheses and some operation signs (+, *, ∧), which in different logical bases may have different interpretations, can be used. The choice of the logical base is carried out by the corresponding operator of information about simulation. For example, using minimax logic (Sect. 2.1) * is a minimum operation, + is a maximum operation, and ^ is a negation operation. If for the given MDE [<identifier>] in the description of the subcircuit, the model number is not given, then it is assumed by default that the model number is 1. For example, the OR cell is described by this logical expression:

$$Y = \hat{}(\hat{}x1 * \hat{}x2 + \mathrm{MDE1} + \mathrm{MDE2}) * (\hat{}x2 + \hat{}x2 + \mathrm{MDE3}) + \mathrm{MDE1} \qquad (6.1)$$

where *x*1, *x*2 are the inputs of the cell, MDE1, MDE2, MDE3—identifiers, with the help of which DF effect is considered.

The description of the logic of operation with a reference to the subcircuit is made by the operator: ID (I_{in1}, I_{in2}, ... I_{out1}, I_{out2} ...), where ID is the name of the subcircuit, I_{in} is the designation of input nodes in the order in which the inputs were listed in INPUTS operator of the subcircuit. I_{out}—is the designation of output nodes in the order in which they were specified in the OUTPUTS operator when describing the subcircuit. For example, if the INPUTS operator is indicated in the description of the subcircuit PR, then logic is described by the following expressions when describing the subcircuit BIT (Fig. 6.28):

PR (T2, KB, K):
PR (T1, IN, KB):

6.4.2.5 Description of the Analysis Information

The complete task when working with ELAIS-L is:

Circuit description
Operator ANALYSIS
Information about simulation
Follow-up operator.

The follow-up operator can be either END—the end of the entire simulation, or ANALYSIS—transition to the simulation of the next version of the same circuit, or any comment (empty statement)—transition to the simulation of the next circuit.

Now consider information about the simulation, which indicates the types of calculation and the form of the output of the results.

The initial states of all nodes except those that are connected to the signal sources are defaulted to 1. But before calculation, the values of external nodes can be changed by the operator:

INIT STATE (0) = LIST OF NODES;

For example, INIT STATE(0) = K1, K2, K3;. If the given initial states are inconsistent, they can change during the calculation. But not for any circuit there is a static mode, because generation can occur in the circuit. In this case, the calculation of the static mode could continue indefinitely. To prevent this from happening, the K_{gen} generation factor is introduced. Calculation of the static mode continues only as long as in the process of searching for it the number of switchings of the circuit does not exceed $K_{gen} * K_n$, where K_n is the number of nodes of the circuit. If the static mode is not established for such a number of switching operations, a corresponding message is printed and the calculation of the static mode ends. By default $K_{gen} = 1$. If it needs to be changed, it can be done by the operator KGEN = K_{gen}.

To set the interval for changing the value of the DF, the operator

EF[FECT]: <number 1>, <number 2> is used

where <number 1> is the initial value of the DF; <number 2> is the final value of the DF. By default, these values are 1 and 60, respectively.

The number of considered values of DF in this interval are given by operator:

HBO3: <integer>;

where <integer> is the number of considered DF values. By default, <number> is equal to 30.

VGEN: <integer>;

If <integer> = 1, additional information about the produced generation is not output.

If <integer> = 2, then when generation occurs, the names of the nodes are output, where there is generation.

If <integer> = 3, then when generation occurs, the names and values of the nodes are output, where there is generation.

To set the logarithmic scale of changing the DF values, use the operator

LOG;

The magnitude of the impact is determined as follows:

$$F_i = F\,(i-1) * dF,$$

where

$$dF = \frac{\mathrm{F2}}{\mathrm{F1}} ** \frac{1}{\mathrm{NT}-1},$$

where F_i—new value of DF; $F\,(i-1)$—the previous value of DF; dF—increment of DF; F2—finite value of DF; F1—initial value of DF; NT—the number of considered values of DF.

If this operator does not exist, this redefined DF value is calculated according to the formula:

$$F_i = F(i-1) + \frac{\mathrm{F2}-\mathrm{F1}}{\mathrm{NT}-1}. \tag{6.2}$$

i.e. the linear scale is used.

When calculating the transient process, the DF is specified using models from the library.

FEFF=NM;

NM—specifies the function number by which the DF changes.

In this case, the NEFF operator has a different meaning:

HEFF = T_d;

T_d is the time variation step for calculating the DF.

To determine the logical calculation basis, use the operator

LOGB: <integer>;

where <integer> is the number of the logical database. By default, the logical database number is 1.

Print results can be carried out in two forms: in the form of tables and in the form of graphs. To output simulation results in the form of graphs, use the operator

GRAPH: <graph name>: <integer>;
$I_1, I_2, \ldots, I_i$;,

where <graph name> is the identifier of this graph. The presence of this name leads to the creation of a library of enumerated node values, <integer> is the number of pages on which graphics are to be placed. By default, it is 1. And I_i is the name of the node for which the graph should be displayed. The maximum value is $i = 5$. If necessary, several GRAPH operators can be used.

To output the simulation results in the form of tables, use the operator:

TABLE:

$I_1, I_2, \ldots, I_i;,$

where I_i is the name of the node whose states should be output in the form of a string.

The maximum value is $i = 5$. If necessary, several TABLE operators can be used.

For the interval of changing the DF values when issuing output information (in the form of a graph or table), use the operator:

VINT=<number >. <number >;

If this operator does not exist, then the print interval is taken equal to the DF interval.

To specify input digital circuit sets, use the operator:

INSG:<I1>,<I2>,. . .,<In>; <Q1>,<Q2>,. . .,<Qn>;

where <I_i> is the identifier of the input, set to the Q_i state ("0" or "1"). If any input needs to be permanently assigned a value of 1, then the corresponding identifier can be omitted from the list of input identifiers of this operator. After setting the table header (the list of inputs), the corresponding sets of combinations of states at the inputs of the circuit are given. The length of this string is $N \times M$, where M is the number of combinations of input signals. For example, if it is necessary to set the following combinations of input signals to the inputs of T1, T2, IN of the subcircuit in Fig. 6.28:

1 0 1
1 1 1
1 1 0

then it can be done as follows:

INSG: T2, IN;

0 1 1 1 1 0;

If it is necessary to simulate a digital circuit with all possible combinations of input signals, it is sufficient to specify only the INSG; Often there is a need to obtain not only the states of different nodes of a digital circuit, but also various information calculated on the basis of these states for certain functions. The program provides several such functions. To select one of these functions, which has the number <integer>, can use the operator:

OUTF = <integer>;

The RESB operator is used to store the simulation results.

RESB;
[File name];

In this file, the simulation results will be stored. The file extension is defined by default [.RAB].

When calculating transient processes, some operators lose meaning or are replaced by another operator. For example, EFF → FEFF, INSG → [source of signals], VINT → INTERVAL.

Sources of signals are given by operators of the form:

I(*C*)=T1,T2,. . .,*NM*,N(T1,. . .,TK);

where *I* is the name of the signal, *C* is the initial value of the signal (0 and 1), and T_i is the time of the signal change. The operator part, placed in parentheses, records the periodic part of the signal.

The calculation of the transient process is given by the operator TIME = T_d, where T_d is the duration of the transient process. The calculation is carried out on a time interval from 0 to T_d. Printing of transient processes is carried out in the time interval 0–T_d. If one wants to print in the interval T1–T2, this can be done by the operator

INTERVAL = T1, T2;

It is convenient to carry out the research of the performance of a digital circuit by changing the time scale with the help of the operator

SCALE = Km;
By default Km = 1.

When calculating the transient process, the operators OUTF, RESB, LOG are ignored.

6.4.2.6 Examples of the Description of a Circuit in the ELAIS-L language

Two digital circuits shown in Figs. 6.32 and 6.33 are described below.

The description of the circuit in Fig. 6.32:

```
EX1;
AND;
IN:X1, X2;
OUT:X3;
```

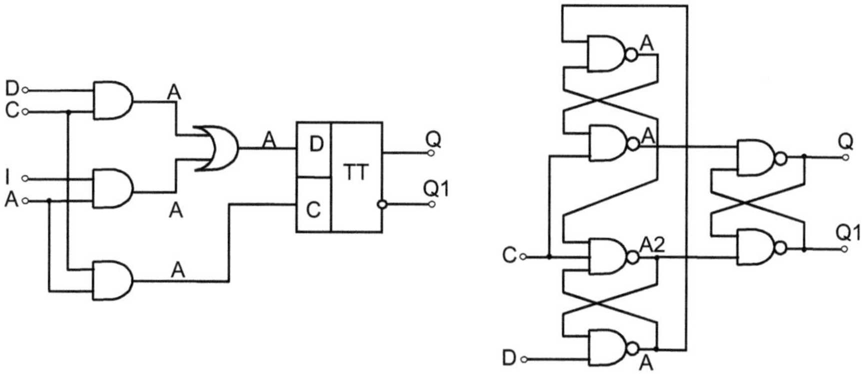

Fig. 6.32 An example of a circuit

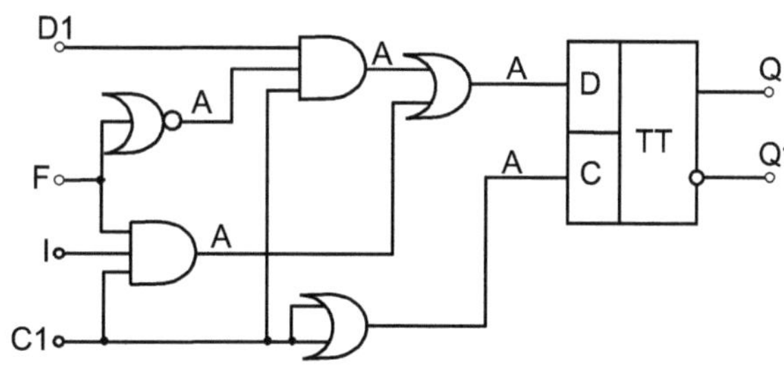

Fig. 6.33 An example of a circuit

```
LOGIC;
X3=X1*X2;
END;
OR;
IN:Y1, Y2;
OUT:Y3;
MDE:3, 15, 100;
MDE:3, 45, 80;
MDE:1;
LOGIC;
Y3=^(^Y1*^Y2)+(MDE1+MDE2)*(^Y1+^Y2+MDE3)+MDE1;
END;
OR;
IN:Z1, Z2;
OUT:Z3;
MDEK:3, 28, 59;
LOGIC;
Z3=^(Z1*Z2)+MDEK;
END;
XNAND;
IN:A1, A2, A3;
```

```
OUT:A4;
MDEK1:3, 18, 49;
LOGIC;
A4=^(A1*A2*A3)+MDEK1;
END;
FLIPFLOP;
IN:C, D;
OUT:Q, Q1;
LOGIC;
NAND(D, A2, A1);
NAND3(A1, A3, C, A2);
NAND(C, A4, A3);
NAND(A3, A1, A4);
NAND(A3, Q1, Q);
NAND(A2, Q, Q1);
END;
EX1;
IN:D1, C1, I, A;
OUT:Q, Q1;
LOGIC;
AND(D1, C1, A1);
AND(I, A, A2);
AND(C1, A, A3);
OR(A1, A2, A4);
FLIPFLOP(A3, A4, Q, Q1);
END;
ANALYSIS;
EFF:13, 109;
NEFF:22;
LOG;
LOGB:1;
GRAPH:12;
A3, A4, Q, Q1;
VINT:15, 100;
INSG:D1, C1, 1, A;
0, 0, 0, 0, 1, 0, 1, 0;
0, 1, 1, 0, 1, 1, 1, 1;
OUTF=1;
END;
END;
```

In the description of subcircuit AND, the DF effect is not taken into account. MDE1, MDE2, MDE3 are inserted in the logical function of the OR element, by the help of which the change in the output state of the element as a function of the DF is simulated. The meanings of the values with the names MDE1 and MDE2 are determined by the model with the number 3, the values of the model parameters are respectively 15, 100 and 45, 80. The MDE3 is determined by the model with the number 1. The value of the parameters is taken by default. The DF value varies from 13 to 109, and simulation is performed for 22 DF values in this range. The diagrams will show the state changes of the nodes A3, A4, Q, Q1, and the graphs will show the states of these nodes only at effect values from 15 to 100. Simulation will be performed with the following combinations of input signals:

```
D1, C1, 1, A;
0,  0,  0, 0,
1,  0,  1, 0,
0,  1,  1, 0,
1,  1,  1, 1,
```

In addition to the graphs, the information calculated from the function with the number 1 will be given.

The description of the circuit in Fig. 6.33:

```
EX2;
NOT;
IN:A;
OUT:A1;
PA:B;
B:TF=10, TR=15;
MDE1:3, 17, 39;
LOGIC;
A1(B)=^A+MDE1;
END;
AND3;
IN:A1, A2, A3;
OUT:A4;
PA:C;
C:TF=10, TR=15;
LOGIC;
A4(C)=A1*A2*A3;
END;
OR;
IN:A8, A9;
OUT:A10;
MDE3:1;
PA:C;
C:TF=10, TR=15;
LOGIC;
A10=^(^A8*^A9+(MDE3+MDEC)*(^A8+^A9+MDEC))+MDE3;
END;
```

```
NAND;
IN:A1, A3;
OUT:A3;
MDE3:3, 15, 48;
LOGIC;
A3=^(A1+A2)+MDE3;
END;
FLIP;
IN:D, C;
OUT:Q, Q1;
LOGIC;
NAND(A1, A3, A4);
NAND(A4, C, A3);
NAND3(C, A1, A3, A2);
NAND(D, A2, A1);
NAND(A2, Q, Q1);
NAND(A3, Q1, Q);
END;
EX2;
IN: D1, F, I, C1;
OUT:Q, Q1;
LOGIC;
NOT(F, A1);
XNAND(C1, I, F, A4);
OR(A2, A4, A5);
OR(C1, C1, A6);
FLIP(A5, A6, Q, Q1);
END;
ANALYSIS;
EFF:15, 118;
NEFF:12;
(FEFF=1;)
(NEFF=5;)
LOGB:1;
GRAPH:10;
A5, A6, Q, Q1;
INSG;
D1(0)=4, 8, 14;
F(0)=3, 8, 10, 17;
C1(0)=5, 9, 14;
D(0)=4, 8, 14, 20;
END;
END;
```

The values of the MDET and MDEC variables, present in the logical function of the subcircuit OR, are calculated from the model number 1, since they do not specifically indicate the model number. The logical database number is 1, i.-e. minimax logic should be used. Simulation will be performed for possible combinations of input signals.

References

1. Kang S., Leblebici Y., Kim Ch. CMOS Digital Integrated Circuits Analysis & Design. McGraw-Hill Education; 4 edition, 2014. 736p.
2. Mehler R.W. Digital Integrated Circuit Design Using Verilog and Systemverilog. Newnes; 1 edition, 2014. 448p.
3. Kirgizova A.V., Nikiforov A.Y., Grigor'ev N.G., Poljakov I.V., Skorobogatov P.K. Dominant mechanisms of transient-radiation upset in CMOS RAM VLSI circuits realized in SOS technology // Mikroelektronika, Vol. 35, No. 3, 2006. P. 191–208.
4. Messomo E.A. Radiation and Temperature Effects on the APV25 Readout Chip for the CMS Tracker. PhD Dissertation, London University, 2002. 132p.
5. Ashok K. Goel. High-Speed VLSI Interconnections. Wiley India; 2 edition. 2015. 432p.
6. Moiseev K., Kolodny A., Wimer S. Multi-Net Optimization of VLSI Interconnect. Springer. 2015. 233p.
7. Lemieux G., Lewis D. Design of Interconnection Networks for Programmable Logic. Springer; Softcover reprint of the original 1st ed. 2004. 206 p.
8. Duan C., LaMeres B.J. On and Off-Chip Crosstalk Avoidance in VLSI Design. Springer. 2010. 240p.
9. Sahoo M., H. Rahaman. Impact of mutual inductance on the crosstalk induced effects in single-walled carbon nanotube bundle interconnects // 31st Symposium on Microelectronics Technology and Devices (SBMicro). 2016. P. 286–290.
10. Zhang X., Jiang W. Impact of crosstalk on signal integrity of high density ceramic package for IC // 17th International Conference on Electronic Packaging Technology (ICEPT). Paris, 2016. Vol. 21, No. 12. P. 429–433.
11. Li Ch., Rakhra P., Norman P. Practical computation of di/dt for high-speed protection of DC microgrids // IEEE DC Microgrids (ICDCM). 2017. Vol. 64, P. 153–159.
12. Zhao S., Roy K., Koh C.K. Estimation of Inductive and Resistive Switching Noise on Power Supply Network in Deep Sub-micron CMOS Circuits // ACM/IEEE International Conference on Computer Aided Design (ICCAD). San Jose, 2000. P. 65–72.
13. John H. Lau. 3D IC Integration and Packaging // McGraw-Hill Education; 1 edition. 2015. P. 480
14. Vaisband I., Jakushokas R., Popovich M., Mezhiba A., Köse S., Friedman E. On-Chip Power Delivery and Management. Springer. 2016. P. 742.
15. Tramel R.W., Turowski M., Przekwas A., Schultz J., Frey R.G. Modeling of Electromagnetic Fields in High Speed Electronic Interconnects and Flex Circuit Boards Using a Least Squares FD-TD Algorithm // Fourth International Conference on Modeling and Simulation of Microsystems (MSM). Hilton Head Island, South Carolina, 2001. P. 602–605.

V. Melikyan, *Simulation and Optimization of Digital Circuits*,
https://doi.org/10.1007/978-3-319-71637-4

16. Chumakov A.I., Yanenko A.V., Kalashnikov O.A. RAM radiation functional upsets//Third Workshop on Electronics for LHC Experiments. London, 1997. P. 419–425.
17. Liden P. On Latching Probability of Particle Induced Transients in Combinational Networks // IEEE International Symposium on Fault Tolerant Computing (FTCS). Madison, 1994. P. 340–349.
18. Metra C., Favalli M., Ricco B. On-Line Detection of Logic Errors due to Crosstalk, Delay and Transient Faults // IEEE International Test conference. Washington, 1998. Vol. 8, No. 4. P. 844–944.
19. Brayton R.K., Sangiovanni-Vincentelli A.L. Cross-Talk Noise Immune VLSI Design Using Regular Layout Fabrics. Springer; 2001 edition. 2012. P. 112.
20. Agakhanyan T.M. Mathematical Modeling of Ionizing-Radiation Effects in ICs: A Review // Russian Microelectronics. 2004. Vol. 33, No. 2. P. 64–67.
21. Lourenco N.E., Fleetwood Z.E., Ildefonso A. The Impact of Technology Scaling on the Single-Event Transient Response of SiGe HBTs // IEEE Transactions on Nuclear Science. 2017. Vol. 64, Issue: 1. P. 406–414.
22. Celik M., Pileggi L., Odabasioglu A. IC Interconnect Analysis. Kluwer Academic Publishers, 2002. 320p.
23. Hall S.H., Heck H.L. Advanced Signal Integrity for High-Speed Digital Designs. Wiley-IEEE Press; 1 edition, 2009. 680p.
24. Pasricha S., Dutt N. On-Chip Communication Architectures: System on Chip Interconnect (Systems on Silicon). Morgan Kaufmann, 2008. 544p.
25. Nurmi J., Tenhunen H., Isoaho J., Jantcsh A. Interconnect-Centric Design for Advanced SOC and NOC. Springer, 2010. 460p.
26. Theis T.N. The future of interconnection technology // IBM Journal Research and Development, 2000. Vol. 44, No. 3, P. 379–390.
27. Zarkesh-Ha P., Meindl J.D. Optimum On-Chip Power Distribution Networks for Gigascale Integration (GSI) // International Interconnect Technology Conference (IITC). Austin, 2001. P. 125–127.
28. Wang X., Zhang D., Su D. A Novel Peak Power Supply Noise Measurement and Adaptation System for Integrated Circuits // IEEE Transactions on Very Large Scale Integration (VLSI) Systems. Anaheim, California, 2016. Vol. 24, Issue: 5. P. 1715–1727.
29. Wu X., Hong X., Cai Y., Cheng C.K., Gu J., Dai W.M. Area Minimization of Power Distribution Network Using Efficient Nonlinear Programming Techniques // ACM/IEEE International Conference on Computer Aided Design (ICCAD). San Jose, 2001. P. 153–157.
30. Wing-Hung Ki. Power Management Integrated Circuit Analysis and Design. Wiley-IEEE Press; 1 edition, 2017. 448p.
31. Bagad V.S. VLSI Design. Technical Publications; 1 edition, 2011. 288p.
32. Ashok B. Mehta. ASIC/SoC Functional Design Verification: A Comprehensive Guide to Technologies and Methodologies. Springer, 2017. 328p.
33. Boylestad R., Nashelsky L. Electronic Devices and Circuit Theory. Prentice Hall; 10 edition, 2008. 912p.
34. Zhang H., Krooswyk S., Ou J. High Speed Digital Design: Design of High Speed Interconnects and Signaling. Morgan Kaufmann; 1 edition, 2015. 272p.
35. Jespers P., Murmann B. Systematic Design of Analog CMOS Circuits: Using Pre-Computed Lookup Tables. Cambridge University Press, 2017. 342p.
36. Chen W. Analog and VLSI Circuits. CRC Press; 3 edition, 2009. 702p.
37. Dobkin B., Williams D. Analog Circuit Design: A Tutorial Guide to Applications and Solutions. Newnes; 1 edition, 2011. 960p.
38. Ott H.W. Electromagnetic Compatibility Engineering. Wiley; 1st edition, 2009. 872p.
39. Guo F., Feng X., Wang Z. Research on Time Domain Characteristics and Mathematical Model of Electromagnetic Radiation Noise Produced by Single Arc // IEEE Transactions on Components, Packaging and Manufacturing Technology. 2017. Vol. PP, Issue: 99-P. 1–10.

40. Chumakov A.I. The effect of cosmic radiation on IC. Ì.: Radio and communication, 2004. 320p. (in Russian)
41. Cherniak M.E., Smolin A.A., Ulanova A.V., Nikiforov A.Y. Investigation of Nonuniform Degradation of CMOS-Sensor Light-Sensitive Surface under Gamma-Irradiation // IEEE Radiation and Its Effects on Components and Systems (RADECS). 2015. P. 1–3.
42. Agakhanyan T.M. Circuit-Design Techniques of Radiation Hardening for Monolithic Op Amps // Russian Microelectronics. 2004. Vol. 33, No. 3. P. 183–187.
43. Cherniak M., Smolin A., Ulanova A. Investigation of Nonuniform Degradation of CMOS-Sensor Light-Sensitive Surface under Gamma-Irradiation // IEEE Radiation and Its Effects on Components and Systems (RADECS). 2015. P. 1–3.
44. Artamonov A.S., Demidov A.A., Kalashnikov O.A., Nikiforov A.Y., Polevich S.A., Telets V. A. Technique and Results of ADC/DAC Radiation Hardness Simulation Tests//Third Workshop on Electronics for LHC Experiments. London, 1997. P. 410–414.
45. Kloukinas K. Development of a radiation tolerant 2.0V standard cell library using a commercial deep submicron technology for the LHC experiments // Fourth Workshop on Electronics for LHC Experiments. Rome, 1998. P. 574–580.
46. Lacoe R. Application of Hardness-By-Design Methodology to Radiation-Tolerant ASIC Technologies // IEEE Transactions on Nuclear Science. 2000. Vol. 47, No. 6. P. 2334–2341.
47. Agakhanyan T.M., Astvatsaturyan E.P., Skorobogatov P.K. Radiation effects in integrated microcircuits. Ì.: Enerergoatomizdat, 1989. 256p. (in Russian)
48. Korshunov F.P., Bogatirev Yu V., Vavilov V.A. The effect of radiation on integrated microcircuits. Minsk: Science and Technology, 1986. 254p. (in Russian)
49. Chumakov A.I., Egorov A.N., Mavritsky O.B., Yanenko A.V. Evaluation of Moderately Focused Laser Irradiation as a Method for Simulating Single-Event Effects // Russian Microelectronics. 2004. Vol. 33, No. 2. P. 106–110.
50. Barnaby H.J., Cirba C.R., Schrimpf R.D., Fleetwood D.M., Pease R.L., Shaneyfelt M.R., Turflinger T., Krieg J.F., Maher M.C. Origins of total dose response variability in linear bipolar microcircuits // IEEE Transactions on Nuclear Science. 2000. Vol. 47, No. 6. P. 2342–2349.
51. Faccio F. Total dose and SEU measurement of test structures in a deep submicron technology // Fourth Workshop on Electronics for LHC Experiments. Rome, 1998. P. 114–117.
52. Campbell M. A pixel readout chip for 10-30 Mrad in standard 0.25mm CMOS//IEEE Nature Sounds Society (NSS) Symposium. Toronto, 1998. P. 823–891.
53. Osborn J. Total Dose Hardness of Three Commercial CMOS Microelectronics Foundries // IEEE Transactions on Nuclear Science. 1998. Vol. 45, No. 3. P. 1458–1463.
54. Holmes-Siedle A., Adams L. Handbook of Radiation Effects. Oxford University Press, 1993. 218p.
55. Snoeys W., Faccio F., Burns M. Layout Techniques to Enhance the Radiation Tolerance of Standard CMOS Technologies Demonstrated on a Pixel Readout Chip // Nuclear Instructions and Methods. 2000. Vol. 439. P. 349–360.
56. Knoll G.F. Radiation Detection and Measurement. John Wiley & Sons, 2000. 194p.
57. Candelori A., Contarato D., Bacchetta N. High-Energy Ion Irradiation Effects on Thin Oxide p-Channel MOSFETs // IEEE Transactions on Nuclear Science. 2002. Vol. 49, No. 3. P. 1364–1371.
58. Melikyan V. Logic simulation of digital circuits exposed to radiation // Facta universitatis, series: Electronics and Energetics. Nis, 1999. Vol. 12, No. 1. P. 1–16.
59. Melikyan V. Sh., Muradyan V.O. Logic Simulation of radiation behavior of digital circuits // International conference "Computer science and information technologies". Yerevan, 2003. P. 368–398. (in Russian)
60. Massengill L.W., Baranski A.E., Van Nort D.O. Analysis of Single-Event Effects in Combinational Logic-Simulation of the AM2901 BitSlice Processor // IEEE Transactions on Nuclear Science. 2000. Vol. 47, No. 6. P. 1911–1917.

61. Zhu X., Massengill L.W., Cirba C.R. The Effects of Nonphysical Carrier Velocities in High-Gradient Single Event Track Simulations // IEEE Transactions on Nuclear Science. 2000. Vol. 47, No. 6. P. 1741–1747.
62. Silveira L.M., Devadas S., Reis R.A. VLSI: Systems on a Chip. Kluwer Academic Publishers, 2000. 696p.
63. J. Hurtarte. Understanding Fabless IC Technology. Newnes; 1 edition, 2007. 296p.
64. Bagad V.S. VLSI Technology and Design. Technical Publications; 1 edition, 2011. 428p.
65. M. F. Analog Design for CMOS VLSI Systems. Kluwer Academic Publishers, 2001. 374p.
66. Yeo K.S., Rofail S.S., Goh W.L. CMOS/BiCMOS VLSI: Low Voltage, Low Power. Prentice Hall, 2002. 624p.
67. Pursley D., Yeh T. High-level low-power system design optimization//IEEE VLSI Design, Automation and Test (VLSI-DAT). 2017. P. 1–4.
68. Mishra S., Singh N. System on Chip Interfaces for Low Power Design. Morgan Kaufmann; 1 edition, 2015. 406p.
69. Roy K. Low-Power CMOS VLSI Circuit Design. John Wiley & Sons Inc, 2003. 320p.
70. Flynn D., Aitken R., Gibbons A., Shi K. Low Power Methodology Manual: For System-on-Chip Design. Springer, 2011. 320p.
71. Taur Y., Ning T.H. Fundamentals of Modern VLSI Devices. New York: Cambridge University Press, 1998. 496p.
72. Jean Walrand K.B., Zobrist G. Advanced Computer Performance Modeling and Simulation. CRC Press, 1998. 356p.
73. Singh A., Singh R. Electronics Circuit SPICE Simulations with LTspice: A Schematic Based Approach. CreateSpace Independent Publishing Platform; 1 edition, 2015. 164p.
74. Basso Ch. Switch-Mode Power Supplies, SPICE Simulations and Practical Designs. McGraw-Hill Education; 2 edition, 2014. 992p.
75. Guofu N., Shiming Z. Gressler J. Modeling of single-event effects in circuits-hardened high-speed SiGe HBT logic // IEEE Transactions on Nuclear Science. 2001. Vol. 48, No. 6. P. 1849–1854.
76. Agakhanyan T.M. Integrated Microcircuits. Ì.: Energoatomizdat, 1983. 464p. (in Russian)
77. Alexenko A.G. Fundamentals of microcircuitry. Ì.: Sov. radio, 1977. 403p. (in Russian)
78. Tsividis Y. Mixed Analog-Digital VLSI Devices and Technology. Kluwer Academic Publishers, 2002. 300p.
79. Shagurin I.I. Transistor-transistor Logic Circuits. Ì.: Sov. radio, 1974. 158p. (in Russian)
80. Maniwa T. Focus Report: ASICs Today // Integrated System Design Magazine. 2000. P. 93–95, 98.
81. Razavi B. Design of analog CMOS integrated circuits. McGraw-Hill Education; 2 edition, 2016. 800p.
82. Maloberti F. Analog Design for CMOS VLSI Systems. Springer, 2010. 388p.
83. Baklanov M., Ho P., Zschech E. Advanced Interconnects for ULSI Technology. Wiley; 1 edition, 2012. 606p.
84. Walker M.G. Modeling the wiring of deep submicron IC's // IEEE Spectrum, 2000. P. 65–71.
85. Goryachev V.A., Zakharov S.M. Transient Analysis of Shielded On-Chip Interconnections // Russian Microelectronics. 2003. Vol. 32, No. 5. P. 307–314.
86. Goryachev V.A. Effect of Discontinuities on ULSI On-Chip Interconnection Characteristics // Russian Microelectronics. 2002. Vol. 31, No. 5. P. 326–334.
87. Hong X.L., Zhu Q., Jing T. Non-rectilinear on-chip interconnect-an efficient routing solution with high performance // Chinese Journal of Semiconductors. 2003. Vol. 24, No. 3. P. 225–233.
88. Xu J., Hong X., Jing T., Zhang L. ETEM: An Efficient Gate and Interconnect Timing Estimator Considering Cross-Coupling for High Performance Layout//IEEE International conference on ASIC (ASICON). Beijing, China, 2003. Vol. 1. P. 254–257.

89. Wang X., Yu W., Liu D., Wang Z. Fast extraction of 3-D interconnect resistance: numerical-analytical coupling method // IEEE International conference on ASIC (ASICON). Beijing, China, 2003. Vol. 1. P. 315–318.
90. Li T., Wang Z. 2-D interconnect inductance and resistance extraction based on the coupled circuit method // Journal of Computer-Aided Design and Computer Graphics. 2003. Vol. 15, No. 1. P. 102–106.
91. Wu B. High-bandwidth IC interconnects with silicon interposers and bridges for 3D multi-chip integration and packaging // IEEE Semiconductor Technology International Conference (CSTIC). 2017. P. 1–3.
92. Liu J., Salmela O., Sarkka J., Morris J.E., Tegehall P., Andersson C. Reliability of Microtechnology: Interconnects, Devices and Systems. Springer, 2011. 204p.
93. Kashyap C., Krauter B. A realizable driving point model for on-chip interconnect with inductance // ACM/IEEE 37th Design Automation Conference. Los Angeles, 2000. P. 190–195.
94. Kleveland Qi X., Yu B.Z. On-chip inductance modeling of VLSI interconnects // International Solid State Circuits Conference (ISSCC). San Francisco, 2000. P.172–173.
95. Massoud Y., Majors S., Bustami T., White J. Layout techniques for minimizing on-chip interconnect self-inductance // ACM/IEEE 35th Design Automation Conference. San Francisco, 1998. P. 566–571.
96. Wu B. High-bandwidth IC interconnects with silicon interposers and bridges for 3D multi-chip integration and packaging // IEEE Semiconductor Technology International Conference (CSTIC). 2017. P. 1–3.
97. Dengi E.A., Rohrer R.A. Hierarchical 2-D Field Solution for Capacitance Extraction for VLSI Interconnect Modeling // ACM/IEEE 34th Design Automation Conference. Anaheim, California, 1997. P. 127–132.
98. Delorme N., Belliville M., Chilo J. Inductance and capacitance analytic formulas for VLSI interconnects // Electronics letters. 1996. Vol. 32, No. 11. P. 996–997.
99. Melikyan V. Sh., Vatyan A.O. Interconnections model delays for the logic analysis of TTL circuits // SUAB, Vol. 1, Computer Engineering, Moscow, 1997. P. 189–198. (in Russian)
100. Melikyan V. Sh., Vatyan A.O. Interconnections model delays for the logic analysis of ECL circuits // SUAB, Vol. 2, Computer Engineering, Moscow, 1997. P. 187–194. (in Russian)
101. Melikyan V. Sh., Vatyan A.O. Interconnections model delays for the logic analysis of I^2L circuits // SUAB, Vol. 3, Computer Engineering, Moscow, 1997. P. 163–166. (in Russian)
102. Melikyan V. Sh., Vatyan A.O., Simonyan A. Sh. Delay models of digital VLSI Interconnects // RAs National Academy of Science and SEUA. Vol. 3, N 3, Yerevan, 1997. P. 201-205. (in Armenian)
103. Melikyan V. Sh., Sargsyan S.M., Petrosyan D.A. Calculation model of parasitic inductances of inner interconnects of VLSI // Simulation, optimization, control, SEUA, Yerevan, Vol. 1, No. 7, 2004. P. 59-68. (in Russian)
104. Melikyan V., Sargsyan S., Petrosyan D. A macromodel of internal interconnects of ICs // RAs National Academy of Science and SEUA. Vol. 57, No. 3, Yerevan, 2004. P. 506-516. (in Armenian)
105. Davis J.A., De V.K., Meindl J.D. A stochastic wire-length distribution for gigascale integration (GSI). – Part I: Derivation and validation // IEEE Transactions on Electron Devices. 1998. Vol. 45. P. 580–589.
106. Wei H., Wang Z. A weighted average formula for efficient inductance and resistance extraction // IEEE International conference on ASIC (ASICON). Beijing, China, 2003. Vol. 1. P. 996–999.
107. Ismail Y.I., Friedman E.G. Effects of inductance on the propagation delay and repeater insertion in VLSI circuits // IEEE Transactions on Very Large Scale Integration (VLSI) Systems. 2000. Vol. 8. P. 195–206.
108. Rao V.B. Delays analysis of the distributed RC line // ACM/IEEE 32nd Design Automation Conference. San Francisco, 1995. Vol. 12, No. 5. P. 370–375.

109. Shepard K.L., Narayanan V., Elmendorf P.C., Cheng G. Global Harmony: Coupled Noise Analysis for Full-Chip RC Interconnect Networks // ACM/IEEE International Conference on Computer Aided Design (ICCAD). San Jose, 1997. P. 139–141.
110. Zhang L., Jing T., Hong X., Xu J., Xiong J., He L. Performance Optimization Global Routing with RLC Crosstalk Constraints // IEEE International conference on ASIC (ASICON). Beijing, China, 2003. Vol. 1. P. 191–194.
111. Chen W.Y., Gupta S.K., Breuer M.A. Test Generation for Crosstalk-Induced Delay in Integrated Circuits // IEEE International Test Conference (ITC). Washington, 1999. P. 191–200.
112. Gupta A. Crosstalk noise and delay analysis for high speed on-chip global RLC VLSI interconnects with mutual inductance using 90nm process technology // IEEE Computing, Communication & Automation (ICCCA). 2015. P. 1215–1219.
113. Caddemi A., Cardillo E. A study on dynamic threshold for the crosstalk reduction in frequency-modulated radars // IEEE Computing and Electromagnetics International Workshop (CEM). 2017. P. 29–30.
114. Khatri S.P., Brayton R.K., Sangiovanni-Vincentelli A.L. Cross-Talk Noise Immune VLSI Design Using Regular Layout Fabrics. Kluwer Academic Publishers, 2001. 144p.
115. Xu J., Hong X., Jing T., Cai Y., Gu J. A Novel Timing-Driven Global Routing Algorithm Considering Coupling Effects for High Performance Circuit Design // IEEE Asia and South Pacific Design Automation Conference. Kitakyushu, Japan, 2003. Vol. E86-A, No. 12. P. 847–850.
116. Servel G., Huret F., Paleczny E. Inductance Effect in Interconnect Coupling Noise // IEEE 5th Workshop on signal propagation on Interconnects. Venice, 2001. P. 74–81.
117. Archambeault B. PCB Design for Real-World Emi Control. Kluwer Academic Publishers, 2002. 244p.
118. John H. Lau. 3D IC Integration and Packaging. McGraw-Hill Education; 1 edition, 2015. 480p.
119. Xiao H. 3D IC Devices, Technologies, and Manufacturing. The International Society for Optical Engineering, 2016. P. 220p.
120. Kuo J., Su K. CMOS VLSI Engineering: Silicon-on-Insulator (SOI). Springer, 2010. 460p.
121. Amaru L. New Data Structures and Algorithms for Logic Synthesis and Verification. Springer; 2017, 156p.
122. Qiu M., Li J. Real-Time Embedded Systems: Optimization, Synthesis, and Networking. CRC Press; 1 edition, 2011. 231p.
123. Roth Ch.H., Kinney L. Fundamentals of Logic Design. CL-Engineering; 6 edition, 2009. 784p.
124. Semiconductor Industry Association. The National Technology Roadmap for Semiconductors (SIA 2015). 62p.
125. The International Technology Roadmap for Semiconductors. 2016. 75p.
126. Semiconductor Industry Association. The National Technology Roadmap for Semiconductors (SIA 2014). 40p.
127. Restle P., Ruehli A., Walker S.G. Dealing with inductance in high-speed chip design // ACM/IEEE 36th Design Automation Conference. New Orleans, 1999. P. 904–909.
128. Hong X.L., Jing T., Xu J.Y., Bao H.Y., Gu J. CNB: A Critical-Network-Based Timing Optimization Method for Standard Cell Global Routing // Journal of Computer Science and Technology (JCST). 2003. Vol. 18, No. 6. P. 732–738.
129. Bubennikov A., Blinnik S. Design and Optimization of Super-Speed CMOS. CBiCMOS Circuits Based on TCAD and Time-Logical Simulator // Baltic Electronics Conference (BEC). Tallin, 1996. P. 256–261.
130. Melikyan V.Sh., Nazinyan S.M. Optimization algorithm of digital paths delays // Proceedings of "Computer Science and Information technologies" International Conference, Yerevan, 1997. P. 322–325. (in Armenian)

131. Melikyan V.Sh., Hovhannisyan D.D. Delay minimization algorithm of critical paths of digital circuits // RAs National Academy of Science and SEUA. Vol. 57, No. 2, Yerevan, 2004. P. 324–330. (in Russian)
132. Wunder B., Lehmann G., Muller-Glaser K.D. VAMP: a VHDL based concept for accurate modeling and post layout timing simulation of electronic systems. // ACM/IEEE 33rd Design Automation Conference. Las Vegas, 1996. P. 119–124.
133. Liou J.J., Wang L.C., Cheng K.T. On Theoretical and Practical Considerations of Path Selection for Delay Fault Testing // ACM/IEEE International Conference on Computer Aided Design (ICCAD). Freiberg, 2002. P. 94–100.
134. Chen H.C., Du D.H., Liu L.R. Critical Path Selection for Performance Optimization // IEEE Transactions on Computer Aided Design of Integrated Circuits and Systems. 1993. Vol. 12, No. 2. P. 185–195.
135. Graziano M., Delaurenti M., Masera G., Piccinini G., Zamboni M. Noise Safety Design Methodologies // IEEE International Symposium on Quality Electronic Design (ISQED). Arlington, 2000. P. 157–166.
136. Melikyan V.Sh., Hovasapyan N.O., Manukyan G.G. Definition of noise immunity of digital VLSI//Interuniversity proceedings of YPI "Technical means and mathematical provision of computing systems", Yerevan, 1988. P. 60–62. (in Russian)
137. Fu J., Hong X., Cai Y., Luo Z. Decoupling Capacitor Allocation for Power Delivery Network Noise Reduction Based on Standard Cell Layouts // IEEE International conference on ASIC (ASICON). Beijing, China, 2003. Vol. 1. P. 101–104.
138. Liou J.J., Krstic A., Jiang Y.M., Cheng K.T. Modeling, Testing and Analysis for Delay Defects and Noise Effects in Deep Submicron Devices // IEEE Transactions on Computer-Aided Design of Integrated Circuits and Systems. 2003. Vol. 22, No. 6. P. 756–769.
139. Melikyan V., Sargsyan S. A simulation method of considering parasitic effects of supply buses of ICs // Information Technologies and Management. Vol. 1, Yerevan, 2004. P. 34–48. (in Armenian)
140. Sotiriadis P., Chandrakasan A. Reducing bus delay in sub-micron technology using coding // IEEE Asia and South Pacific Design Automation Conference. Japan, Yokohama, 2001. P. 109–114.
141. Kleveland B., Qi X., Madden L. Line inductance extraction and modeling in real chip with power grid // International Electron Devices Meeting. 1999. P. 901–904.
142. Saleh R., Hussain Z., Rochel S., Overhauser D. Clock Verification in the Presence of IR-drop in the Power Distribution Network // IEEE Transaction on CAD of IC and Systems. 2000. Vol. 19, No. 6. P. 635–644.
143. Aragones X., Gonzales J., Rubio A. Analysis and Solutions for switching Noise Coupling in Mixed-Signal ICs. Kluwer Academic Publishers, 1999. 236p.
144. Charbon E., Gharpurey R., Miliozzi P., Meyer R.G., Sangiovanni-Vincentelli A.L. Substrate Noise Analysis and Optimization for IC Design. Kluwer Academic Publishers, 2001. 200p.
145. Van Heijningen M., Compiet J., Wambacq P., Donnay S., Engels M., Bolsens L. Analysis and Experimental Verification of Digital Substrate Noise Generation for Epi-Type Substrates // IEEE Journal Solid-State Circuits. 2000. Vol. 35. P. 1002–1008.
146. Panda R., Blaauw D., Chaudry R., Zolotov V., Young B., Ramaraju R. Model and Analysis for Combined Package and On-Chip Power Grid Simulation // International Symposium on Low Power Electronics and Design (ISLPED). Huntington Beach, California, 2000. P. 179–184.
147. Bondyopadhyay P.K. Moore's law governs the silicon revolution // Proceedings of the IEEE. 1998. Vol. 86, No. 1. P. 78–81.
148. Burger D., Goodman J.R. Billion-transistor architectures // Proceedings of the IEEE Computer. 1997. Vol. 30, No. 9. P. 46–48.
149. Norenkov I.P. Automated design basics. M.: MNTU after N.E. Bauman, 2002. 336p. (in Russian)
150. Zobrist G., Leonard J.V. Simulation Systems. CRC Press, 2000. 324p.
151. Hatchel G.D., Sangiovanni-Vincentelli A. A Survey of Third Generation Simulation Techniques // Proceedings of the IEEE. 1981. Vol. 69, No. 10. P. 1264–1280.

152. Barmakov Yu N., Bakharov V.A., Ilyin V.N. et al. Research results for a range of electronic circuit analysis programs // Proc. of USSR universities. Radioelectronics. 1981. N 6. P. 27–37. (in Russian)
153. Newton A.R. Computer-aided design of superintegrated integrated circuits//Works of TIIER. 1981. T.69, N10. P. 7–20.
154. Farrahi A.H., Hathaway D.J., Wang M., Sarrafzadeh M. Quality of EDA CAD Tools: Definitions, Metrics and Directions // 1st International Symposium on Quality of Electronic Design. San Jose, 2000. P. 395–403.
155. Breuer M.A. Latest achievements in design automation and analysis of digital circuits//In book: Automation in Design/under editorship of D. Kalakhan and others. M.: Mir, 1972. P. 19–47. (in Russian)
156. Batalov B.V., Egorov Yu. B., Rusakov S.G. Mathematical modeling basics of VLSI on PCs. M.: Radio i svyaz, 1982. 168p. (in Russian)
157. Jr. Roth, L. John. Digital Systems Design Using VHDL. CL. Engineering; 3 edition, 2017. 628p.
158. Clietti M. Advanced Digital Design with The Verilog Hdl. Pearson India, 2017. 992p.
159. Abraitis L.B., Sheynauskas R.I., Jilevichyus V.A. Computer-aided design. M.: Sov. Radio, 1978. 272p. (in Russian)
160. Budulin S.S., Barmaulov Yu M., Berdishyev V.A. Automated design of digital circuits. M.: Radio i svyaz, 1981. 240p. (in Russian)
161. Waterman S. Digital Logic Simulation and CPLD Programming with VHDL. Prentice Hall, 2003. 301p.
162. Nguyen Q., Van Le T. Time-parameterized Temporal Logic-based Framework for Discrete-Event Simulation // 9th International Symposium on Languages for Intensional Programming. Aachen, 1996. Vol. 11, No. 8. P. 43–51.
163. Avril H., Tropper C. Scalable Clustered Time Warp and Logic Simulation // VLSI Design. 1999. Vol. 9, No. 3. P. 36–42.
164. Kim H., Jean J. Concurrency Preserving Partitioning Algorithm for Parallel Logic Simulation // VLSI Design. 1999. Vol. 9, No. 3. P. 23–29.
165. Melikyan V.Sh. Logic simulation algorithm of digital circuits with consideration of environment temperature // Proceedings of 3rd International scientific-technical conference "Innovative Information Technologies and Systems", Penza, 1998. P. 158–160. (in Russian)
166. Melikyan V.Sh. Logic simulation of digital circuits with consideration of destabilizing factors // Proceedings of the 5th International Conference on The Experience of Designing and Application of CAD Systems in Microelectronics (CADSM'99), Lvov, 1999. P. 142–144. (in Russian)
167. Melikyan V., Poghosyan A., Durgaryan A., Petrosyan H., Simonyan M. Method of Parametrical Optimization of Multi-Core Processors // Proceedings of the 31st International Scientific-Technical Conference on "Electronics and Nanotechnologies", Kiev, Ukraine, 2011. P. 126–130. (in Russian)
168. Melikyan V.Sh., Simonyan A.Sh. Consideration of external affects in the program of logic analysis // Interuniversity proceedings of YPI "Technical means and mathematical provision of computing systems", Yerevan, 1990. P. 61–64. (in Russian)
169. Melikyan V.Sh. Principles of logic simulation of digital circuits with consideration of destabilizing factors // Proceedings of "Computer Science and Information Technologies" International Conference, Yerevan, 1999. 389–393. (in Russian)
170. Melikyan V.Sh., Balagezyan A.R. Key design tools of logic macromodels of digital circuits // Proceedings of "Computer Science and Information Technologies" International Conference, Yerevan, 1999. P. 394–398. (in Russian)
171. Heydemann M.H. A survey of MOS logic simulation tools // 9th European Solid-State Circuits Conference. Lausanne, 1983. P. 19–24.
172. Levin V.I. Dynamics of Logical devices and systems. M.: Energy, 1980. 224p. (in Russian)

173. Zolotorevich L.A. Delay-Conscious Switch-Level Modeling of MOS LSI Circuits // Russian Microelectronics. 2003. Vol. 32, No. 3. P. 182–188.
174. Melikyan V.Sh., Ovasapyan N.O., Petrukhin V.P. United system of logic simulation and layout design of VLSI // Automation design in electronics, "Technics", Vol. 40, Kiev, 1989. P. 61–64. (in Russian)
175. Melikyan V., Soghomonyan V., Mkrtchyan E. Model of digital cells' states in algorithmic calculation//Interuniversity scientific and methodical proceedings. 6.51. Yerevan, 2003. P. 44–54. (in Armenian)
176. Melikyan V., Kulakhszyan A. Logic models with consideration of leakage delays, quantum and averaged states // Information technologies and management. Vol. 3, Yerevan, 2003. P. 8–15. (in Armenian)
177. Melikyan V., Kulakhszyan A. Logic models with averaged states // RAs National Academy of Science and SEUA, Yerevan, RA, Vol. 56, No. 3, Yerevan, 2003. P. 491–499. (in Armenian)
178. Lu S.K., Chen J.L., Wu C.W., Chang W.F., Huang S.Y. Combinational circuit fault diagnosis using logic emulation // IEEE International Symposium on Circuits and Systems (ISCAS). Bangkok, 2003. Vol. 5. P. 549–552.
179. Schittenkopf Ch., Deco G., Brauer W. Finit Automata-Models for the Investigation of Dynamical Systems // Information Processing Letters. 1997. Vol. 63, No. 3. P. 137–141.
180. Snubald R., Svensson C. Accurate CMOS Models for Event driven logic simulators // European Conference on Circuit Theory and Design (ECCTD). Stutgard, 1985. P. 483–485.
181. Wang Z., Maurel P.M. LECSIM: A levelized event driven compiled logic simulator // ACM/IEEE 27th Design Automation Conference. Orlando, Florida, 1990. P. 491–496.
182. Hayes J.P. Digital Simulation with multiple logic values // IEEE Transaction on CAD. 1986. Vol. 5, No. 2. P. 274–283.
183. Flake P.L., Moorby P.R., Musgrave G. An Algebra for Logic Strength Simulation // ACM/IEEE 20th Design Automation Conference. 1983. P. 615–618.
184. Herout A., Szanto L. Logic Simulator Based on Resistor Nets // 7th European Conference on Circuit Theory and Design (ECCTD). Prague, 1985. P. 149–152.
185. Keller J., Rauber T., Rederlechner B. Scalability Analysis for Conservative Simulation of Logical Circuits // VLSI Design. 1999. Vol. 9, No. 3. P. 8–15.
186. Miszo A. Digital logic testing and simulation. New York: John Wiley and Sons, 1987. 285p.
187. Bailay M.L., Briner J.V., Chamberlain R.D. Parallel logic simulation of VLSI systems // ACM Computing Surveys. 1994. Vol. 26, No. 3. P. 255–294.
188. Steinman J. SPEEDES: A multiple synchronization environment for parallel discrete-event simulation // Journal on Computer Simulation. 1992. Vol. 2. P. 251–286.
189. Chamberlain R.D. Parallel logic simulation of VLSI systems // ACM/IEEE 32nd Design Automation Conference. San Francisco, 1995. P. 139–143.
190. Ferscha A. Parallel and distributed simulation of discrete event systems // Parallel and Distributed Computing Handbook. McGraw-Hill, 1995. P. 666–673.
191. Naroska E. Parallel VHDL simulation // ACM/IEEE Conference Design, Automation and Test in Europe (DATE). Paris, 1998. P. 159.
192. Longeann D., Richard Shi C.J. Distributed simulation of VLSI circuits via lookahead-free self-adaptive optimistic and conservative synchronization // ACM/IEEE International Conference on Computer Aided Design (ICCAD). San Jose, 1998. P. 362.
193. Walker P.A., Ghosh S. Asynchronous, distributed event driven simulation algorithm for execution of VHDL on parallel processors // ACM/IEEE 32nd Design Automation Conference. San Francisco, 1995. P. 144.
194. Noble B.L., Chamberlain R.D. Performance of Speculative Computation in Synchronous Parallel Discrete-Event Simulation on Multiuser Execution Platforms // 8th IASTED International Conference on Parallel and Distributed Computing and Systems (PDCS). Chicago, 1996. P. 489–494.

195. Noble B.L., Peterson G.D., Chamberlain R.D. Performance of Synchronous Parallel Discrete-event Simulation // 28th International Conference on System Sciences. Waileau Maui, 1995. Vol. 2. P. 185–186.
196. Noble B.L., Wade J.C., Chamberlain R.D. Performance Predictions for Speculative, Syschronous, VLSI Logic Simulation // 34th Annual Simulation Symposium. Waileau Maui, 2001. P. 56–64.
197. Chen Y., Noble B.L., Chamberlain R.D. Comparing Edge-cuts to Communications Volume in Parallel VLSI Logic Simulation // 8th IASTED International Conference on Parallel and Distributed Computing and Systems (PDCS). Chicago, 1996. P. 481–484.
198. Luksch P. Evaluation of three approaches to parallel logic simulation on a distributed memory multiprocessor // 26th Annual Simulation Symposium. Arlington, 1993. P. 2–11.
199. Lewis D. M. A hierarchical compiled code event-driven logic simulator // IEEE Transaction Computer-Aided Design. 1991. Vol. 10. P. 726–737.
200. Melikyan V.Sh. A logic simulation method for reproduction of signal bumps in interconnects // Manual of Engineering Academy of Armenia. Vol. 1, No. 3, Yerevan, 2004. P. 449–451. (in Russian)
201. Melikyan V., Shahinyan T., Melikyan H. A digital cell macromodel considering radiation affect//Manual of Engineering Academy of Armenia. Vol. 1. No. 3, Yerevan, 2004. P. 585–588. (in Armenian)
202. Vlakh Yu, Singkhal K. Machine methods of analysis and design of electronical circuits M.: Radio i svyaz, 1988. 560p. (in Russian)
203. Arkhangelskiy A.Y., Arkhangelskaya I.T., Gribkova E.M., Levshin N.G., Melikyan V.Sh., Savinova T.A., Sergienko V.Y. Design automation of electronic circuits / under the editorship of T.M Akhakhanyan. M.: MEPhI, 1985. 92p. (in Russian)
204. Ogrodzki J. Circuit Simulation Methods and Algorithms. CRC Press, 1994. 342p.
205. Arkhangelskiy A.Y., Levshin N.G., Svetsov S.V. Complex of programs of electrical analysis of electronic devices ELAIS. J.: MEPhI, 1982. 90p. (in Russian)
206. Valtonen M. APLAC-Object-Oriented Circuit Simulation and Design Tool // IEEE Microwaves and RF Conference. London, 1997. P. 245–250.
207. Abramov I.I., Goncharenko I.A., Ignatenko S.A., Korolev A.V., Novik E.G., Rogachev A.I. Nanodev: A Nanoelectronic-Device Simulation Software System // Russian Microelectronics. 2003. Vol. 32, No. 2. P. 97–104.
208. S. Li, Y. Fu. 3D TCAD Simulation for Semiconductor Processes, Devices and Optoelectronics. Springer; 2016. 308p.
209. Bubennikov A.N., Sadovnikov A.D. Physical-technological design of bipolar elements in silicon ICs. M.: Radio i svyaz, 1991. 288p. (in Russian)
210. Abramov I.I. Simulation of physical processes in silicon integral microcircuit elements. BSU, 1999. 89p. (in Russian).
211. Litovski V., Zwolinski M. VLSI Circuit Simulation and Optimization. Kluwer Academic Publishers, 1996. 368p.
212. Grivet Talocia S., Stievano I.S., Maio I.A., Canavero F.G. Combined FDTD macromodel simulation of interconnected digital devices // ACM/IEEE Conference Design, Automation and Test in Europe (DATE). Munich, 2003. P. 536–541.
213. Forzan C., Franzini B., Guardiani C. Accurate and Efficient Macromodel of Submicron Digital Standard Cells//ACM/IEEE 34th Design Automation Conference. Anaheim, California, 1997. Vol. 15, No. 3. P. 356–361.
214. Stievano S., Chen Z., Becker D., Canavero F.G., Katopis G., Maio I.A. Macromodeling of Digital I/O Ports for System EMC Assessment // ACM/IEEE Conference Design, Automation and Test in Europe (DATE). Paris, 2002. P. 1044–1049.
215. Odabasioglu A., Celik M., Pileggi L.T. PRIMA: Passive Reduced-order Interconnect Macromodeling Algorithm // ACM/IEEE 34th Design Automation Conference. Anaheim, California, 1997. P. 58–65.

216. Kong J.T., Overbauser D. Methods to improve digital MOS macromodel accuracy // IEEE Transaction on Computer-Aided Design of Integrated Circuits and Systems. 1995. Vol. 14, No. 7. P. 868–881.
217. Petkovic P., Litovski V. Macromodeling and macroanalysis of CMOS LSI electronic circuits // The first International Conference on Computer Technology, Systems and Applications. Hamburg, 1987. P. 512–513.
218. Dartu F., Tutuianu B., Pileggi L.T. RC-Interconnect Macromodels for Timing Simulation // ACM/IEEE 33rd Design Automation Conference. Las Vegas, 1996. P. 611–616.
219. Melikyan V.Sh., Arkhangelskiy A.Y. Macromodels on switched capacitors elements // Radioelectronics. Vol. 29, No. 6, 1986. P. 86–87. (in Russian)
220. Melikyan V.Sh. Macromodel of logic gate-chains//Simulation, optimization, control, State Engineering University of Armenia, Yerevan, Armenia, Vol. 2, Yerevan, 1998. P. 52–57. (in Armenian)
221. Arkhangelsky A., Svettsov S. Methodology of synthesis of electric macromodels of logic circuits of medium and large integration levels // Computer-aided design in electronics. Kiev, 1980. T.22. P. 64–70, (in Russian)
222. Flexer L.A., Tumanov V.S. Analysis of complex circuits by macrosimulation method//News of higher educational institutions of the USSR. Radio electronics. 1983. T.26, N 6. P. 81–83 (in Russian).
223. Baghov V.A. Macrosimulation of digital and pulse circuits using macro elements // News of higher educational institutions of the USSR. Radio electronics. 1980. N 6. P. 13–20. (in Russian)
224. Greenbaum J.R. Digital-IC models for computer aided design // Electronics. 1973. Vol. 46, No. 25. P. 154–175.
225. Li X., Li P., Xu Y., Pileggi L.T. Analog and RF Circuit Macromodels for System-Level Analysis // ACM/IEEE 41th Design Automation Conference. Anaheim, California 2003. P. 478–483.
226. Bogliolo A., Benini L. Robust RTL Power Macromodels // IEEE Transactions On Very Large Scale Integration (VLSI) Systems. 1998. Vol. 6, No. 4. P. 578–581.
227. Arkhangelskiy A.Y. Mixed-Mode Simulation of VLSI. Kiev: Znanie, 1985. 24p. (in Russian)
228. Melikyan V. Structural and eventual decomposition of large electronic schemes // Trans black sea region symposium on applied electromagnetism. Athens, 1996. P. MMWS 7.
229. Melikyan V., Mkrtchyan E., Mkrtchyan K., Hovhannisyan D., Soghomonyan V. Isolation methods of electronic circuits // RAs National Academy of Science and SEUA, Vol. 57, No 1, Yerevan, 2004. P. 130–137. (in Armenian)
230. De Man H.J., Newton A.R. Hybrid Simulation // IEEE International Symposium on Circuits and Systems (ISCAS). Tokyo, Japan, 1979. P. 249–259.
231. Reynaert Ph., De Man H., Arnout Y., Cornelissen J. DIANA: A Mixed-Mode Simulator with a Hardware Description Language for Hierarchical Design of VLSI // IEEE International Conference on Circuits and Computers. New York, 1980. P. 356–360.
232. Saleh R.A., Newton A.R. Mixed-mode Simulation. Kluwer Academic Publishers, 1990. 248p.
233. De Man H.J. Mixed-Mode Simulation for MOS-VLSI Why, Where and How? // IEEE International Symposium on Circuits and Systems (ISCAS). Rome, Italy, 1982. P. 699–701.
234. Maniwa R. Analog and Mixed-Signal Simulation Tools//Integrated System Design. 1996. Vol. 2, No. 3. P. 58–64.
235. Melikyan V.Sh. Logic-circuit simulation of analog-digital nodes of radio metering devices // Proceedings of 9th All-Union Scientific-Technical Conference of “Radio metering”, Kaunas, 1983. P. 177–180. (in Russian)
236. Arkhangelskiy A.Y. Lavrenov O.E., Rojukalns P.P., Melikyan V. Sh., Svettsov S.V., Fedorkov B.G. A program of mixed-mode simulation of analog-digital VLSI//Theses of reports of All-Union Conference of “Methods and microelectronic means of digital conversion and signal processing”, Riga, 1983. P. 231–236. (in Russian)

237. Arkhangelskiy A.Y., Melikyan V.Sh. Functionality models of logic cells in the program of mixed-mode logic-circuit simulation // Theses of reports of All-Union Scientific-Technical Conference of "Design Automation of Computers and Systems", Yerevan, 1983. P. 80–82. (in Russian)
238. Melikyan V.Sh., Arkhangelskiy A.Y. Mixed-mode circuit and logic simulation of analog digital circuits // Electronic simulation, Vol. 6, N5, Kiev, Ukraine, 1984. P. 35–39. (in Russian)
239. Arkhangelskiy A.Y., Melikyan V.Sh., Levshin N.G. Principles of designing a system of mixed-mode simulation of electronic circuits // Seminar materials of "Design Automation in radioelectronics and electrical engineering", Moscow, 1984. P. 91–93. (in Russian)
240. Arkhangelskiy A.Y., Melikyan V.Sh. A program of mixed-mode analysis of analog-digital circuits // Proceedings of "Electrical engineering and devices for experimental physics", Moscow, 1985. P. 134–138. (in Russian)
241. Bandarenko V.M., Aghmetov B.S., Bilenko V.I. Application of hermitian splines in the numerical realization of logic-electric macromodels // Proceedings of AN USSR. Series "A", Physical-mathematical and technical sciences. 1983. N 6. P. 49–52. (in Russian)
242. Martinyuk V.A., Fedoruk V.G. Logic-electrical analysis algorithm of VLSI // Theses of reports of All-Union Scientific-Technical Conference of. "Automation of design PC and systems". Yerevan, 1983. P. 97–98. (in Russian)
243. Gai S. MOZART: A Concurrent Multilevel Simulator // IEEE Transaction on Computer-Aided Design. 1988. Vol. 7, No. 9. P. 1005–1016.
244. Saleh R., Antao B., Singh J. Multilevel and Mixed-Domain Simulation of Analog Circuits and Systems // IEEE Transactions on Computer-Aided Design of Integrated Circuits and Systems. 1996. Vol. 15, No. 1. P. 68–81.
245. Saleh R., Jou S.J. Mixed Mode Simulation and Analog Multilevel Simulation. Kluwer Academic Publishers, 1994. 320p.
246. Chadha R., Visweswariah C., Chen C.F. Multilevel Mixed-mode A/D Simulator // IEEE Transactions on Computer Aided Design of Integrated Circuits and Systems. 1992. Vol. 11, No. 5. P. 575–586.
247. Mayaram K., Chern J.H., Yang P. Algorithms for Transient Three-Dimensional Mixed-Level and Device Simulation // IEEE Transactions on Computer Aided Design of Integrated Circuits and Systems. 1993. Vol. 12, No. 11. P. 1714–1726.
248. Cloutier J., Bourgault M., Fauvel S., Roy C., Cerney E. and Gecsei J. An Object-Oriented Mixed-Mode Hierarchical VLSI Simulator // Proceedings of Canadian Conference on VLSI CCVLSI'86. Montreal, 1986. P. 203–208.
249. Saleh R., Yang A. Modeling Mixed Systems with Spice 3 // IEEE Circuits & Devices. 1993. Vol. 1. P. 7–10.
250. Gorshkov K. The simulation technique for large-scale tree structured interconnects // IEEE Industrial Engineering, Applications and Manufacturing (ICIEAM). Chelyabinsk, Russia, 2016. P. 1–6.
251. Melikyan V.Sh. Optimization of timing parameters of digital circuits elements // Elektronika i svyaz, Vol. 4, No 2, Kiev, 1998. P. 249–253. (in Russian)
252. Melikyan V.Sh., Nazinyan S.M. Power consumption algorithm of digital ICs//Simulation, optimization, control, SEUA, Yerevan, RA, Vol. 4, Yerevan, 2001. P. 159–166. (in Armenian)
253. Liu C., Li Y., Du Y., Du L., Wang T. Hybrid thermal aware reconfigurable 3D IC with dynamic power gating architecture // IEEE Semiconductor Technology International Conference (CSTIC). Shanghai, China, 2017. P. 1–3.
254. Jing T., Hong X. A Novel And Efficient Timing-Driven Global Router For Standard Cell Layout Design Based On Critical Network Concept//IEEE International Symposium on Circuits and Systems (ISCAS). Scottsdale, Arizona, USA, 2002. P. I165–I168.
255. Liao I.M.J., Su C.F., Chang A.C.Y., Wu A.C.H. A Carry-Select-Adder Optimization Technique for High-Performance Booth-Encoded Wallace-Tree Multipliers // IEEE International Symposium on Circuits and Systems (ISCAS). Scottsdale, Arizona, 2002. P. I257–I259.

256. Li Z., Wu W., Hong X. Incremental Placement Algorithm for Wirelength and Congestion Optimization//Chinese Journal of CAD/CG. 2003. Vol. 15, No. 6. P. 651–655.
257. Yunfeng W., Jinian B., Qiang W., Heng H. Re-synthesis after Floor-planning for Timing Optimization // IEEE International conference on ASIC (ASICON). Beijing, China, 2003. Vol. 1. P. 212–215.
258. Ekpanyapong M., Balakrishnan K., Nanda V., Lim S.K. Simultaneous Delay and Power Optimization for Multi-level Partitioning and Floorplanning with Retiming // IEEE International Conference on Circuits and Systems. 2004. P. 756–760.
259. Chang S.C., Marek-Sadowska M., Cheng K.T. Perturb and Simplify: Multi-level Boolean Network Optimizer // IEEE Transactions on Computer Aided Design of Integrated Circuits and Systems. 1996. Vol. 15, No. 12. P. 1494–1504.
260. Kannan L.N., Suaris P.R., Fang H.G. A methodology and algorithms for post-placement delay optimization//ACM/IEEE 31st Design Automation Conference. San Francisco, 1994. P. 327–332.
261. Huisman L.M. Correlations between Path Delays and the Accuracy of Performance Prediction // IEEE International Test Conference. Washington, 1998. P. 801–808.
262. McGeer P.C., Saldanha A., Brayton R.K., Sangiovanni-Vincentelli A.L. Delay Models and Exact Timing Analysis. Logic Synthesis and Optimization. Kluwer Academic Publishers, 1993. 196p.
263. Lavagno L., Keutzer K., Sangiovanni-Vincentelli A.L. Synthesis of Hazard-Free Asynchronous Circuits with Bounded Wire Delays // IEEE Transactions on Computer Aided Design of Integrated Circuits and Systems. 1995. Vol. 14, No. 1. P. 61–86.
264. Saldanha A., Brayton R.K., Sangiovanni-Vincentelli A.L. Circuit Structure Relations to Redundancy and Delay // IEEE Transactions on Computer Aided Design of Integrated Circuits and Systems. 1994. Vol. 13, No. 7. P. 875–883.
265. Jiang Y.M., Krstic A., Cheng K.T., Marek-Sadowska M. Post-Layout Logic Restructuring for Performance Optimization // ACM/IEEE 34th Design Automation Conference. Anaheim, California, 1997. Vol. 11, No. 4. P. 241–246.
266. Parodi C.G., Agrawal V.D., Bushnell M.L., Wu S. A Non-Enumerative Path Delay Fault Simulator for Sequential Circuits // IEEE International Test conference. Atlantic, 1999. Vol. 15, No. 5. P. 107–110.
267. Gremoux S., Azemard N., Auvergne D. Algotithms for Path selection: a comparative study. Kluwer Academic Publisher, 1997. 356p.
268. Hsino M.S. Peak Power Estimation Using Genetic Spot Optimization for Large VLSI Circuits // IEEE Conference Design Automation and Test in Europe (DATE). Munich, 1999. P. 175–179.
269. Balkir S., Dundar G., Ogrenci A.S. Analog VLSI Design Automation. CRC Press, 2003. 275p.
270. Kankkunen A., Andersson M., Valtonen M. MOSFET Level 3 Model in APLAC. Report CT-9. Finland, 1991. 23p.
271. Bagdasarian H., Melikyan V., Nshanian M., Uzunoglu N. Computer modeling of fiber-optic communication systems // Trans black sea region symposium on applied electromagnetism. Athens, 1996. P.OPSY 9.
272. Gala K., Zolotov V., Panda R. On-chip inductance modeling and analysis // ACM/IEEE 37th Design Automation Conference. Los Angeles, 2000. P. 63–68.
273. Tsividis Y. Operation and Modeling of the MOS Transistor. New York: McGraw-Hill, 1999. 356p.
274. Ciletti M.D. Advanced Digital Design with the Verilog HDL. Prentice Hall, 2003. 982p.
275. Kleitz W. Digital Electronic with VHDL. Prentice Hall, 2004. 960p.
276. Melikyan V. The Simulation of Digital Circuits Taking into Account the Destabilizing Factors // Fourth national conference on Semiconductor Microelectronics. Tsakhkadzor, 2003. P. 240–243.
277. Melikyan V.Sh. Logic block of digital cell model for consideration of destabilizing factors // Elektronica I svyaz, N 8, Vol. 2, Kiev, 2000. P. 266–268. (in Russian)

278. Melikyan V.Sh. Simulation of logics of a digital cell for consideration of destabilizing factors' affects // Proceedings of "Computer Science and Information Technologies" International Conference, Yerevan, 2001. P. 394–398. (in Russian)
279. Melikyan V.Sh., Harutyunyan A.A., Vatyan A.O. Logic model of part of digital circuits with consideration of destabilization factors // Proceedings of the 5th International Conference on The Experience of Designing and Application of CAD Systems in Microelectronics (CADSM'99), Lvov, 1999. P. 145–146. (in Russian)
280. Melikyan V.Sh. Inertance block of models of digital element for consideration of destabilizing factors // Elektronica I svyaz, N 8, Vol. 2, Kiev, 2000. P. 269–271. (in Russian)
281. Melikyan V., Harutyunyan A. Consideration algorithm of external influence's affect on digital circuits functioning // International conference "Computer science and information technologies". Yerevan, 1997. P. 330–333.
282. Harris J. An Introduction to Fuzzy Logic Applications. Kluwer Academic Publishers, 2002. 232p.
283. Virant J. Design Considerations of Time in Fuzzy Systems. Kluwer Academic Publishers, 2000. 512p.
284. Zadeh L.A. Fuzzy Logic, Neural Networks and Soft Computing // Communications of the ACM. 1994. Vol. 37, No. 3. P. 77–84.
285. Kalashnikov O.A. Fuzzy Simulation of Total Dose Functional Failures of Digital Units // Third Workshop on Electronics for LHC Experiments. London, 1997. P. 481–486.
286. Babuska R. Accuracy and Transparency of Fuzzy Systems // Advances in Computational Intelligence and Learning: Methods and Applications. International Series in Intelligent Technologies. Kluwer Academic Publishers, 2002. P. 3–16.
287. Zadeh L.A. The Role of Fuzzy Logic in Modeling, Identification, and Control // Modeling, Identification and Control. 1994. Vol. 15, No. 3. P. 191–203.
288. Kofman A. Introduction to the theory of fuzzy sets. M.: Radio and. communication, 1982. 428p. (in Russian)
289. Ubar R. Dynamic analysis of digital circuits with multi-valued simulation // Microelectronics. 1998. Vol. 29, No. 11. P. 821–826.
290. Melikyan V.Sh., Mnatsakanyan V.A., Ziad B.Kh. Universal adaptive system of parametrical identification of models of electronic components // Proceedings of international scientific-technical conference "Problems of physical and biomedical electronics", Kiev, 1996. P. 68–72. (in Russian)
291. Melikian V., Mnatsakanian V., Ziad B., Airapetian T. VLSI adaptive simulation // 1st International conference on Application of critical Technologies for the needs of society. Yerevan, 1995. P. 135.
292. Sigorskiy V.P. Problematic adaptation of computer-aided system design //Computer-aided design in electronics. Kiev: Technika, 1982. Vol. 26, P. 3–12. (in Russian)
293. Mano M., Ciletti M. Digital Design: With an Introduction to the Verilog HDL, VHDL, and System Verilog. Pearson; 6 edition, 2017. 720p.
294. Arustamyan V.E. To the problem of transient processes in PC communication wires. Radioelectronics questions. 1976. Vol. 14. P. 26–30. (in Russian)
295. De Bor K. Practical guide to spline. M.: Radio and. communication, 1985. 304p. (in Russian)
296. Giannessi F., Pardalos P.M., Rapcsak T. Optimization Theory. CRC Press, 2003. 512p.
297. Yazgi M., Kuntman H. A new approach for parameter extraction of complex models and an application for SPICE MOSFET level-3 static model // Microelectronics Journal. 1999. Vol. 30. P. 149–155.
298. Yazgi M., Tarim T.B., Kuntman H.H. On the Extraction of Spice Level-3 MOS Transistor Parameters // IEEE 2nd International Analog VLSI Workshop. Santa Clara, 1998. P. 71–75.
299. Melikian V., Mnatsakanian V., Uzunoglu N. Optimization of SPICE system LEVEL3 MOSFET transistor models based on DC measurements // Microelectronics Journal. Great Britain: Elsevier Science, 1998. No. 29. P. 151–156.

300. Melikyan V.Sh. A universal program for identification of parameters of electronic components models // Elektronika I svyaz, Vol. 4, No 2, Kiev, 1998. P. 245–248. (in Russian)
301. Rozenbrock H.H. An automatic Method for finding the Greatest or Largest Value of a Function // Computer Journal. 1960. Vol. 4. P. 175–184.
302. Pakula L.S., French P.J., Heng Yang. Low voltage, high speed RF switch with high switching capacitance ratio // IEEE. SENSORS 2005, USA. P. 1–4.
303. Rogatkin Yu B. CMOS Switched-Capacitor Integrator // Russian Microelectronics. 2003. Vol. 32, No. 6. P. 333–338.
304. Schaumann R., Ghausi M.S., Laker K.R. Design of Analog Filters-Passive, Active RC and Switched Capacitor. Prentice-Hall, 1995. 426p.
305. Wu J., Chang J. and Chang K. 1.2V CMOS Switched-Capacitor Circuits // IEEE International Solid-State Circuits Conference. Marvell, Sunnyvale, 1996. P. 388–389.
306. Nelin B.D. Analysis of Switched-Capacitor Network Using General-Purpose Simulation Programs // IEEE Transaction on Circuits and Systems. 1983. Vol. CAS-30, No. 1. P. 43–48.
307. Mir S., Rueda A., Olbrich T., Peralias E., Huertas J.L. Automatic switch-level fault simulation and test evaluation of switched-capacitors systems // ACM/IEEE 34th Design Automation Conference. Anaheim, California, 1997. P. 281–286.
308. Li B., Gu D. SSCNAP: A Program for Symbolic Analysis of Switched Capacitor Circuits // IEEE Transactions on Computer Aided Design of Integrated Circuits and Systems. 1992. Vol. 11, No. 3. P. 334–340.
309. Farhat H.A. Digital Design and Computer Organization. CRC Press, 2001. 278p.
310. Entrena L.A., Cheng K.T. Sequential Logic Optimization by Redundancy Addition and Removal // IEEE Transactions on CAD. July 1995. Vol. 14, No. 7. P. 909–916.
311. Rajeev Murgai Layout-Driven Area-Constrained Timing Optimization by Net Buffering // Proc. DAC. June 2001. P. 82–86.
312. Lillis John, Kuan Cheng Chung, Lin Ting-Ting Y., Ching-Yen Ho. New Performance Driven Routing Techniques With Explicit Area Delay Tradeoff and Simultaneous Wire Sizing // Proc. Design Automation Conference. June 2001. P. 82–85.
313. Rezvani P., Ajami A., Pedram M., Savoj H. Leopard: A Logical Effort-Based Fanout Optimization for Area and Delay // Proc. of ICCAD. November 1999. P. 516–519.
314. Berkelaar M. Area-Power-Delay Trade-off in Logic Synthesis // Ph.D Thesis / Eindhoven University of Technology. 1992. 124p.
315. Chaudhary K., Pedram M. Computing the Area versus Delay Trade-off Curves in Technology Mapping // IEEE Transactions on Computer- Aided Design of Integrated Circuits and Systems. 1995. Vol. 14, No. 12. P. 1480–1489.
316. Jiang Y.M., Kristic A., Cheng K.T., Marek-Sadowska M. Post-Layout Logic Restructuring for Performance Optimization // Proc. Design Automation Conference. June 1997. P. 662–665.
317. Cheng Kwang-Ting. Post-Layout Logic Restructuring for Performance Optimization of Deep Submicron Design. Final Report for MICRO Projects. 1998. P. 28.
318. Chen W., Hsieh C.T., Pedram M. Simultaneous Gate Sizing and Placement // IEEE Transactions on CAD. 2001. Vol. 19. P. 206–214.
319. Melikyan V., Hovhannisyan D. Optimization Techniques for High-Performance Digital Circuits // Fourth national conference on Semiconductor Microelectronics. Tsakhcadzor, 2003. P. 260–263.
320. Melikyan V.Sh., Hovhannisyan D.D. Algorithm of digital circuits restructuring // Simulation, optimization, control, SEUA, Yerevan, RA, Vol. 6, No 1, Yerevan, 2003. P. 39–44. (in Armenian)
321. Chen W., Hsieh C.T., Pedram M. Simultaneous Gate Sizing and Fanout Optimization // IEEE Transactions on CAD. 2001. Vol. 17, No 2, P. 208–213.
322. Alpert C. J., Devgan A., Quay S. Buffer Insertion for Noise and Delay Optimization // IEEE Trans. Computer-Aided Design. Nov. 1999. Vol. 18, No. 11. P. 1633–1645.

323. Girard P., Landrault C., Pravossoudovitch S., Severac D. A gate resizing technique for high reduction in power consumption // Proceedings of the 1997 international symposium on Low power electronics and design. 1997. P. 281–286.
324. Iris Bahar R., Gary Hachtel D., Macii E., Somenzi F. A symbolic method to reduce power consumption of circuits containing false paths // ACM/IEEE International Conference on Computer Aided Design (ICCAD). 1994. P. 368–371.
325. Pedram M. Power Minimization in IC Design: Principles and Applications // ACM/IEEE 33rd Design Automation Conference. Las Vegas, 1996. P. 3–56.
326. Kundert K.S., Zinke O. The Designer's Guide to Verilog-AMS. Kluwer Academic Publishers, 2004. 270p.
327. Melikyan V.Sh. Description of discrete systems on ELAIS-L language // Theses of reports of 4th All-Union Seminar "Simulation of discrete control and calculating systems", Sverdlovsk, 1984. P. 64–66. (in Russian)
328. Janpoladyan B.L., Melikyan V.Sh. Dialog Teaching System // Proceedings of "Information Transferring and Processing"/YrPI. 1984. P. 39–42. (in Russian)
329. Bhatnagar H. Advanced ASIC Chip Synthesis: Using Synopsys Design Compiler, Physical Compiler and Primetime Design. Kluwer Academic Publishers, 2001. 360p.

Index

V. Melikyan, *Simulation and Optimization of Digital Circuits*,
https://doi.org/10.1007/978-3-319-71637-4

T

MIX
Papier aus verantwortungsvollen Quellen
Paper from responsible sources
FSC® C105338

If you have any concerns about our products,
you can contact us on
ProductSafety@springernature.com

In case Publisher is established outside the EU,
the EU authorized representative is:
Springer Nature Customer Service Center GmbH
Europaplatz 3, 69115 Heidelberg, Germany

Printed by Libri Plureos GmbH
in Hamburg, Germany